STOCHASTIC MODELING AND THE THEORY OF QUEUES

PRENTICE HALL INTERNATIONAL SERIES
IN INDUSTRIAL AND SYSTEMS ENGINEERING

W. J. Fabrycky and J. H. Mize, Editors

ALEXANDER *The Practice and Management of Industrial Ergonomics*
AMOS AND SARCHET *Management for Engineers*
BANKS AND CARSON *Discrete-Event System Simulation*
BEIGHTLER, PHILLIPS, AND WILDE *Foundations of Optimization,* 2/E
BANKS AND FABRYCKY *Procurement and Inventory Systems Analysis*
BLANCHARD *Logistics Engineering and Management,* 3/E
BLANCHARD AND FABRYCKY *Systems Engineering and Analysis*
BROWN *Systems Analysis and Design for Safety*
BUSSEY *The Economic Analysis of Industrial Projects*
CHANG AND WYSK *An Introduction to Automated Process Planning Systems*
ELSAYED AND BOUCHER *Analysis and Control of Production Sytems*
FABRYCKY, GHARE, AND TORGERSEN *Applied Operations Research and Management Science*
FRANCES AND WHITE *Facility Layout and Location: An Analytical Approach*
GOTTFRIED AND WEISMAN *Introduction to Optimization Theory*
HAMMER *Occupational Safety Management and Engineering,* 4/E
HAMMER *Product Safety Management and Engineering*
HUTCHINSON *An Integrated Approach to Logistics Management*
IGNIZIO *Linear Programming in Single- and Multiple-Objective Systems*
MIZE, WHITE, AND BROOKS *Operations Planning and Control*
MUNDEL *Improving Productivity and Effectiveness*
MUNDEL *Motion and Time Study: Improving Productivity,* 6/E
OSTWALD *Cost Estimating,* 2/E
PHILLIPS AND GARCIA-DIAZ *Fundamentals of Network Analysis*
SANDQUIST *Introduction to System Science*
SMALLEY *Hospital Management Engineering*
TAHA *Simulation Modeling and SIMNET*
THUESEN AND FABRYCKY *Engineering Economy,* 7/E
TURNER, MIZE, AND CASE *Introduction to Industrial and Systems Engineering,* 2/E
WHITEHOUSE *Systems Analysis and Design Using Network Techniques*
WOLFF *Stochastic Modeling and the Theory of Queues*

STOCHASTIC MODELING AND THE THEORY OF QUEUES

Ronald W. Wolff

University of California, Berkeley

Prentice Hall, Englewood Cliffs, New Jersey 07632

Library of Congress Cataloging-in-Publication Data
WOLFF, RONALD W.,
Stochastic modeling and the theory of queues / Ronald W. Wolff.
p. cm.
Bibliography: p.
Includes index.
ISBN 0-13-846692-0
1. Stochastic processes—Mathematical models. 2. Queueing theory.
I. Title.
QA274.W65 1988 88-19624
519.2—dc19 CIP

Editorial/production supervision and
interior design: Patrice Fraccio
Manufacturing buyer: Mary Noonan

A Division of Simon & Schuster
Englewood Cliffs, New Jersey 07632

Printed in the United States of America
10 9 8 7 6 5 4 3 2 1

ISBN 0-13-846692-0

PRENTICE-HALL INTERNATIONAL (UK) LIMITED, *London*
PRENTICE-HALL OF AUSTRALIA PTY. LIMITED, *Sydney*
PRENTICE-HALL CANADA INC., *Toronto*
PRENTICE-HALL HISPANOAMERICANA, S.A., *Mexico*
PRENTICE-HALL OF INDIA PRIVATE LIMITED, *New Delhi*
PRENTICE-HALL OF JAPAN, INC., *Tokyo*
SIMON & SCHUSTER ASIA PTE. LTD., *Singapore*
EDITORA PRENTICE-HALL DO BRASIL, LTDA., *Rio de Janeiro*

CONTENTS

PREFACE

Many years ago, I started to write a first-year graduate-level book on stochastic processes and the theory of queues, for two main reasons:

(1) Expositions of renewal theory began with "hard stuff," e.g., Blackwell's Theorem. Time-average behavior (the "easy stuff"), where most of the applications are, was hardly mentioned. Even when the easy stuff was mentioned, e.g., in a problem at the end of a chapter, it was not pointed out that the easy stuff does not depend on the hard stuff.

(2) Expositions of queueing theory were in even worse shape. The subject seemed to be a catalog of models that were solved by unrelated and often highly technical mathematical methods. Messy mathematical expressions were derigueur. Unifying ideas were missing.

My early treatment of the easy stuff, which was based on Tauberian theorems, wasn't so easy, and while I had most of the right unifying ideas for queueing theory, my understanding of them wasn't what it is today. Partially for these reasons, this first effort fizzled.

Some years later, Stuart Dreyfus and I decided to write an introductory operations research text, where he would handle optimization, and I would handle stochastic models. The treatment of stochastic models in most introductory texts is little more than formula plugging. I was confident that unifying ideas can be explained at the undergraduate level.

For this purpose, the time-average approach was simplified. A basic idea was to envisage processes evolving over time, either by drawing pictures, or observing real processes that we were modeling. A time average then became an average over sample paths of the process. Because all the models we would encounter were regenerative, and had renewal processes embedded in them, we needed only the strong law of large numbers to derive time-average properties. With this approach, the inspection paradox and related effects became easy to understand. Tauberian theorems were banished!

This fit beautifully with fundamental queueing results such as "$L = \lambda w$," a relation between continuous- and discrete-time averages. The concept of work was introduced, and what is now called the PASTA property of the Poisson process was explained and applied. Elementary queueing models were compared in important and understandable ways. At that time, I wrote most of what now are the more elementary parts of Chapter 5.

A publisher who was very interested in our project sent us several reviews. The reviewers all held the view that we were writing two incompatible books. The problem seemed to be that I was after deep conceptual understanding, through a unified approach, while Dreyfus, feeling that there is no unified approach to optimization in general, was writing more of a guided tour of diverse optimization methods. Although the publisher was still enthusiastic, we agreed with the reviewers, and abandoned the joint project.

Having overcome problems that stalled my first effort, and with nearly 100 pages of new notes, I returned to my original objective. This is the result.

In the intervening years, many other books have come out, and of course we know more now. In fact, several of the topics covered in latter chapters have exploded into subfields. This growth has made the need for a unified approach even more important today than it was when my first effort began.

While written as a graduate text, this book contains a substantial amount of material not found in other general-purpose queueing books. In addition, sections at the end of seven chapters discuss important points in detail, and relate them to the literature. These features should make this book a useful general reference both for research and applications.

Most chapters have extensive sets of problems. Some of them extend the theory, while others supply it, sometimes in artful ways. Many involve problem formulation, i.e., the ability to translate into appropriate mathematics a problem stated in phsyical terms. This is the reason for the word *modeling* rather than either *models* or *processes* in the title.

While this has turned out to be a larger book than planned, it is my hope that the unified approach and the stress on intuitive ideas have not been diluted by its size, and that they will improve our understanding of stochastic phenomena.

We strive for rigor as well as intuitive understanding, but we also have placed a limit on assumed mathematical background that sometimes makes this difficult to achieve, e.g., in the definition of what are called stopping times. These instances occur rarely, and the benefit of this limit is greater accessibility. Furthermore, it is my firm belief that a thorough understanding of the stochastic phenomena in this book does not require a higher level of mathematical abstraction.

It is a pleasure to acknowledge the corrections and suggestions received from many friends, colleagues, students, and former students. Betsy Greenberg, George Shanthikumar, Karl Sigman, and Ward Whitt looked over the entire manuscript, and Gordon Newell recently used portions of the manuscript in a course. Their suggestions have led to numerous improvements. I also thank Philip Bitar and Yat-Wah Wan for numerous corrections and other suggestions.

Preparation of this manuscript would have been impossible without the generous technical support provided by the Department of Industrial Engineering and Operations Research at Berkeley. Finally, I thank Genji Yamazaki and the Tokyo Metropolitan Institute of Technology for their support during the editorial phase of the manuscript.

Ronald W. Wolff

ON USING THIS BOOK

As a text, this book can be used in several ways. There is a natural division: Chapters 1 thorugh 4 for a renewal-theory (regenerative process) based course on stochastic processes, and the rest of the book for a course on queueing theory. In the latter case, there is more material than can be covered in one course. Many of the details of Chapter 11 may be omitted, as may whole topics such as Chapter 7, but it is my preference to at least explain what every major topic is about. For example, Chapter 7 helps put Chapter 6 in perspective.

At the cost of some redundancy, the introduction to queueing theory in Chapter 5 is written so that fundamental ideas can be understood without detailed knowledge of earlier chapters. Subsequent chapters rely more heavily on the earlier ones. Chapters 6 through 8 can be read independently of each other, and, aside from some knowledge of renewal theory, Chapter 9 is nearly self-contained.

The minimum prerequisites for this book are a year of calculus and a calculus-based course in probability theory. For Markov chains, some knowledge of matrix algebra is also helpful.

From time to time, we use mathematical concepts and results that go somewhat beyond these prerequisites. These include different definitions of an integral, and the validity of interchanging operations such as integral and sum. These issues arise primarily in proofs, rather than in examples or problems, and are discussed in the Appendix.

Shortcomings in probability background are more difficult to pin down. Inadequate practice at problem solving generally, and in conditional arguments in particular, is common. Thus one often hears "I thought I understood the lecture, but I can't do the homework." Certain elementary but useful matters of technique, e.g., indicator functions, may be unfamiliar. In stochastic processes, it is often very useful to fix a point in the sample space, so that a sequence of random variables becomes a sequence of real numbers. This may also seem new, but it is important for a proper understanding of the strong law of large numbers.

Chapter 1 is intended to help remedy for these shortcomings. It is intended as a review. Topics of particular importance for the book that may also be unfamiliar are treated in some detail toward the end of this chapter. For first-year graduate students, a heterogenous group, I find that at least a week of review of this material is worthwhile.

We occasionally footnote sections, portions of sections, or even portions of proofs with an asterisk (*). The reason, often given in the text, is one or more of these: The mathematical level is higher, this is a special topic that may be omitted, or what follows is intricate and does not contribute to understanding the results.

INTRODUCTION

In this book, we strive for a balance between the theory of stochastic processes and applications of the theory to model building and the analysis of specific models. The term *stochastic modeling* is intended to convey the latter emphasis.

Much of the theory is based on the notion of *starting over* of what we call *regeneration cycles*. This is the main concept underlying renewal theory; it provides the basis for our analysis of regenerative processes in general, and discrete- and continuous-time Markov chains in particular. Nearly all the queueing models in this book are regenerative, and this approach provides a theoretical foundation for these models as well.

Because this concept is so basic, it is explicitly introduced and used early in Chapter 2, much earlier than other books of comparable level.

Renewal theory is primarily concerned with *asymptotic* or *limiting* behavior. We place great emphasis on *time averages*, which have the technical name of *Cesàro limits*. For example, we ask "what fraction of time is a process in state B," rather than the corresponding limiting probability of the process being in state B as time approaches infinity. We call limits of the latter type *pointwise*.

There are two reasons for this emphasis: First, a fraction of time is generally of greater practical significance than the corresponding "limit at infinity." It has a more meaningful connection with performance measures for the models we will build. Secondly, the theory underlying the existence and representation of time averages is more elementary, and the averages hold under more general conditions than in the case for the corresponding pointwise limits.

Related to time averages is the notion of observing an ongoing process "at random," where more formally, we consider stationary versions of processes.

The time-average point of view helps us to think in a clear and consistent manner, and to avoid mistakes. For example, the *inspection paradox* and related batch effects, which at first seem counterintuitive, are introduced early in Chapter 2 and carefully explained from the time-average point of view. The objective is to improve intuition so that when these effects occur in more complex models later (or, for that matter, in everyday life!), intuition leads in the right direction.

Realizations (sample paths) of elementary processes are shown in several figures. It is important to be able to think in these terms and to visualize processes evolving over time. This, after all, is what we would see observing the real thing—realized values of quantities that we model as random variables.

Theorems about the existence of pointwise limits, notably Blackwell's Theorem and the Key Renewal Theorem, are important for a full theoretical treatment of renewal theory. Sometimes they are relevant in applications. This topic is also treated in Chapter 2, but separately from and after time averages. Except for periodic chains in Chapter 3, this more advanced material is not needed in the rest of this book, and may be omitted. (Graduate students should be aware of these results and understand through elementary examples that pointwise limits may not exist in the lattice case.)

Another reason for the clear separation of time averages and pointwise limits is this: By mixing the two, or what is worse, hardly mentioning time averages, the reader is likely to come away with the mistaken impression that distributions must be *nonlattice* or *spread out* (see Chapter 2) for certain results to hold. This impression is reinforced by far too many papers in the literature that make these assumptions when they are not necessary. Thirty lashes!

Our point of view is the same for discrete-time processes, notably discrete-time Markov chains (Chapter 3). The distinction between periodic and aperiodic states is the analogue of lattice versus nonlattice distributions. A time average is now an average over discrete time points, usually the integers, and in Markov chains, a fraction of time becomes a *fraction of transitions*. By formulating in terms of time averages, an important theorem about the existence and uniqueness of a stationary probability vector is proven without mention of the distinction between periodic and aperiodic states.

Theoretically more difficult results about pointwise limits for discrete-time chains are covered toward the end of Chapter 3, and may be omitted.

Continuous-time Markov chains are treated in Chapter 4, along with (briefly) semi-Markov processes. New theoretical difficulties can occur for these chains, notably the possibility of an infinite number of transitions in finite time, called an *explosion*. This possibility will not occur (with positive probability) for applications of these chains in the rest of the book. Consequently, our approach is to introduce sufficient conditions that both eliminate this possibility and cover our subsequent needs. Both here and in Chapter 3, we briefly treat what are called *stationary measures*, because they are useful in Chapter 6.

Many examples in Chapters 2 through 4 and problems at the end of each chapter emphasize modeling for a wide range of applications, including queues, inventories, transportation, replacement and maintenance policies, and reliability. The organization of these chapters, however, is based on theoretical considerations: classes of stochastic processes and their properties.

Queueing theory, which is the subject of the rest of the book, is about modeling and analysis of congestion and delays at service facilities. Nearly all of the models we discuss are stochastic processes that fall under one or more headings in earlier chapters. However, queueing theory cannot be systematically treated by a series of examples in chapters on stochastic processes. Nor should it be viewed as a "handbook" of models that, as we proceed from chapter to chapter, are treated by what may appear to be unrelated mathematical methods.

We present a unified treatment of queueing theory, based on the time-average point of view introduced in Chapter 2.

Chapter 5 is an overview, a queueing short course. Measures of performance such as L and w are defined as (continuous or discrete) time averages. Important concepts and results are introduced and applied as early as possible, including "$L = \lambda w$," the concept of *work-in-system* and its relation to other quantities, and the result *Poisson arrivals see time averages (PASTA)*. These results are treated in a more theoretical manner toward the end of the chapter. Queues that can be modeled as continuous-time Markov chains are included for two purposes: to show how to model complicated situations, and to facilitate the comparision of single and multichannel queues, a theme we return to several times in later chapers.

The remaining chapters are more specialized, and each has a substantial literature. We will not attempt to summarize them here; the introduction and the last section of each chapter should meet this need.

Except in Chapter 5, where we briefly consider rush-hour behavior, our queueing models are regenerative processes. See Sections 2-21 and 11-6 for discussion of recent results on the regenerative nature of queues once thought not to have this property. For these results, it is necessary to generalize how we define a regenerative process; see Section 2-21. While beyond our scope, properties of these more general regenerative processes and their application to queueing networks is currently an active research area.

1

PROBABILITY THEORY

This book is about building and analyzing stochastic models; thus, some knowledge of probability theory is necessary. This chapter provides what is needed in a way that is self-contained but compact.

You should already be familiar with probability theory at the level and content of a typical upper division calculus-based course. While much of this chapter will be review for most readers, some of it will be new.

For self-study, browsing through the entire chapter is recommended. On this initial reading, more time may be spent on the less familiar topics. These topics may then be studied in more detail later, as the need arises.

1-1 INTERPRETATIONS OF PROBABILITY

Probability may be used to describe everyday events, e.g., in the statement "the probability of rain tomorrow is .3." More formally and abstractly, we may decide to conduct an experiment where, under specified conditions, event A may or may not occur. We may wish to assign a probability to the event that A occurs, $P(A)$, where $P(A)$ is a real number in the range $0 \leq P(A) \leq 1$.

Results in the physical sciences are usually of the form that under specified conditions event A is certain (or impossible) to occur. For example, if we toss a coin, it will fall to the ground. If A is certain, then the event "not A" (called the *complement* of A and denoted by A^c) is impossible, and the converse. If A is certain, we will require that $P(A) = 1$, and if A is impossible, that $P(A) = 0$. In general, the larger $P(A)$, the "more likely" (in some sense) is A.

On the other hand, suppose A is the event that the coin lands on its head. If we knew the spin and other properties of a specific toss, it may be possible to predict whether A occurs with certainty. However, this situation is usually modeled probabilistically by assigning probability $P(A)$ to the event, where usually $0 < P(A) < 1$.

Events that may or may not occur are called *possible*, i.e., they are neither certain nor impossible. Probability theory deals with collections of possible events (together with one certain and one impossible event, see Section 1-2). It may seem odd at first that if A is possible, we choose *not* to rule out either 0 or 1 as values we may assign to $P(A)$.

For example, if X denotes the numerical value of some measured quantity in an experiment, e.g., the weight gain of a rat, the possible values of X may vary over a continuum. (We will call quantities such as X *random variables*.) For any possible value x, we may assign to the event $\{X = x\}$ a probability $P(X = x) = 0$. More fundamental reasons for permitting possible events to have probabilities of either 0 or 1 occur in connection with what are called *repeated trials*.

In applications, we often assign numerical values to event probabilities. Before doing this, it is appropriate to ask, "What do these numbers mean?" In the remainder of this section, we discuss several so-called definitions of probability.

For our coin-tossing example, there are only two possible outcomes: heads and tails. We might argue that by symmetry the outcomes are "equally likely," and conclude that $P(A) = 1/2$. This way of determining $P(A)$ is based on the classical definition of probability. Consider an experiment (such as tossing a coin or rolling a die) in which there are K equally likely possible outcomes. Let event A correspond to a subset of k of these outcomes; i.e., A occurs if and only if the actual outcome belongs to a particular set of k outcomes. The *classical definition* of the probability of A is

$$P(A) = \frac{k}{K}. \tag{1}$$

For example, if the six numbers on a die are the equally likely outcomes of a single roll, and A is the event that the outcome is either a 5 or a 6, we have $K = 6$, $k = 2$, and $P(A) = 2/6 = 1/3$.

To utilize this definition, we must identify the set of equally likely outcomes. A logical requirement is that the experiment must result in the occurrence of one and only one outcome. However, this requirement can often be met in many ways. For example, suppose a coin is tossed twice. While there is no logical inconsistency in choosing the set of outcomes to be {exactly r heads: $r = 0, 1, 2$}, we will soon see that "normally" they are not equally likely.

Given agreement on the set of equally likely outcomes, $P(A)$ can be found by enumeration of k and K. This often is a problem in combinatorial analysis. For example, suppose that 5 cards are selected "at random" from a ordinary deck of 52 cards. Let A be the event that among the selected cards there are exactly two aces. By *at random* we mean that all of the $K = \binom{52}{5}$ ways of selecting the 5 cards are equally likely.† The number of outcomes k that result in the event A is the product of $\binom{4}{2}$ and $\binom{48}{3}$. We now find $P(A)$ from (1).

The classical definition does not provide any way of utilizing empirical or experimental evidence in the selection of the set of equally likely outcomes. How do we decide that one set is appropriate while another set is not? Apparently, other considerations must be at work. One consideration is based on an alternative definition of probability:

Consider an experiment that can be repeated indefinitely under some idealized set of conditions. Each performance of the experiment will be called a *trial*. On each trial, the event A may or may not occur. Let m denote the number of occurrences of A in M trials. Then the *statistical probability* (or *relative frequency*) of A is defined to be

$$P(A) = \lim_{M\to\infty} \frac{m}{M}. \tag{2}$$

In the case of coin tossing, (2) provides us with an empirical check on probabilities based on the classical definition. If A is the event heads and m is the number of heads in M trials, we can empirically verify that m/M stabilizes around some value, p, a constant, for large M. For coins that are not bent and not too worn, $p = 1/2$ (or at least is very close to this value). Coins for which $p = 1/2$ are called *fair*.

Suppose that the experiment consists of tossing a coin n times, and A is the event that heads come up r times, $0 \le r \le n$. Let the probability of a head on one toss be p. It can be empirically verified that the statistical probability of A is

$$P(A) = \binom{n}{r} p^r(1 - p)^{n-r}. \tag{3}$$

[Support for (3) is not only empirical. See section 1-6.]

From (3), we reject that when a coin is tossed twice, the outcomes 0, 1, and 2 heads are equally likely. For a fair coin ($p = \frac{1}{2}$), these probabilities are $\frac{1}{4}$, $\frac{1}{2}$, and $\frac{1}{4}$, respectively. From (3) we see that regardless of the value of p, these outcomes

† For integers $0 \le r \le n$, denote the number of combinations of n things taken r at a time by

$$\binom{n}{r} = \frac{n!}{r!\,(n - r)!},$$

where $n! = (n)(n - 1) \cdots (1)$ and, by definition, $0! = 1$. For example,

$$5! = 120 \quad\text{and}\quad \binom{5}{3} = \binom{5}{2} = 10.$$

are *not* equally likely. In the case of the fair coin the *ordered* outcomes HH, HT, TH, and TT may be regarded as equally likely.

Another drawback of the classical definition is that it overly restricts applications of probability theory. First, the number of outcomes must be finite. Even when this is so, the enumeration of the numerator and denominator of (1) is sometimes highly artificial. In fact, even in coin tossing, there is nothing to prevent the limit in (2) from being irrational. (Irrational numbers can be obtained as limits of rational numbers.) In this case, $P(A)$ cannot be represented as the ratio of two integers.

A closer look at (2) and the discussion preceding it will show that there are difficulties with the statistical definition as well. First, although the set of conditions for each trial is asserted to be the same, the outcomes that occur can be different. This is ascribed to "randomness." Although not explicitly stated, it is also assumed that successive trials are independent in some sense. However, a precise definition of independent trials involves certain properties of the probabilities of events. We should also ask what is meant by the limit in (2) since m is not a particular function of M. It turns out that the ordinary concept of limit used in analysis is *not* what is meant. In fact, a precise definition of the sense in which a limit is approached can be made only in terms of probability.

Thus (2) is an unsatisfactory definition because it is circular. However, the fact that coin tossing and other experiments exhibit predictable empirical behavior cannot be ignored in any theory that purports to be applicable to these phenomena. Under conditions that appear to be reasonable for tossing coins, we will find that probability theory predicts known empirical results. For example, a precisely stated form of (2) will be a theorem (either the strong or weak law of large numbers). See Section 1-15.

There are many situations where neither of the above definitions is appropriate for the theory of probability. For example, in a presidential election, let A be the event that one major candidate wins. While there are only two major candidates and we are virtually certain that one of them will win, this is hardly sufficient reason to set $P(A) = 1/2$. On the other hand, we have a one-of-a-kind situation; it doesn't make sense to consider $P(A)$ as a relative frequency. Even if we insist on the latter interpretation, it is impossible to verify by (2) any particular numerical value we may assign to $P(A)$.

Examples abound. A newly developed drug is thought to be "more likely" to be effective for the treatment of cancer than an old one. A marketing vice-president is "80 percent sure" an advertising campaign will be successful. In statistics, such statements, e.g., a new drug is more effective than an old one, are called *hypotheses*.

To test a particular hypothesis H, a sample is taken (in the above terminology, a sequence of experiments is performed) and, based on the outcome of the sample, H is either accepted or rejected. Much of the statistics literature on hypothesis tests is concerned with the selection of test procedures where whether H is accepted depends *only* on the sample and *not* on prior knowledge about whether this hypothesis or some other hypothesis is in fact true. (This is not strictly true. For example, in selecting a test procedure, one considers alternatives to the hypothesis being tested. The selection of the set of alternatives is based in part on prior knowledge.)

Often, a procedure is justified by desirable properties of "sufficiently large" samples. In these cases, the information obtained from the sample overwhelms prior knowledge. (We are excluding "perfect knowledge," e.g., knowledge that H is true. If we know that a coin is fair, no amount of tossing can change this. If a long sequence of heads should occur, all we can conclude is that an extremely unlikely event has occurred.)

When samples are not sufficiently large, we are confronted with the difficult problem of how to make judgments about the real world that utilize in some way both our prior knowledge and knowledge obtained by sampling. The problem is compounded by vagueness in the meaning of imperfect prior knowledge. (There is no clear distinction between "sufficiently large" and "not large enough." Hence, this is a real problem for any finite sample.)

One way of representing prior knowledge about whether some hypothesis H is true is to assign to it what is called subjective probability $P(H)$, a concept that we now define.

Subjective probability is a measure of the "degree of belief" of a person (you, say) that a hypothesis is true or that an event will occur (or occurred in the past). Several ways of measuring subjective probability have been proposed. For example, let your subjective probability of hypothesis H be $P(H)$, where $P(H)$ is a number determined by you such that you are indifferent to two situations:

1. You get a prize if H is true, and no prize otherwise.
2. An urn contains a proportion of $P(H)$ black balls. You draw a ball "at random." You get the same prize if the drawn ball is black, and no prize otherwise.

Philosophical objections to subjective probability are of two kinds. One is that since H is either true or not true, it must have a probability of either 1 or 0. The other is that $P(H)$ is a measure of one person's beliefs about the real world and not a statement about the real world itself. It is asserted that this makes subjective probability unsuitable for investigating real-world phenomena. Proponents of subjective probability assert the opposite, i.e., since a decision maker's choice depends on what he believes to be true, not on what is true, subjective probability is a useful concept in modeling the decision-making process. Of course, different people may have quite different subjective probabilities about the same hypothesis or of a one-of-a-kind event. If everyone agreed, only inveterate gamblers would bet on the outcome of an election.

It is not our purpose to enter into this philosophical debate. These definitions serve as interpretations that assist us in applying probability theory to a wide range of problems. Fortunately, there is general agreement about how probabilities are to be manipulated, regardless of which interpretation is used.

1-2 THE PROBABILITY SPACE AND THE ALGEBRA OF EVENTS

An arbitrary outcome of a random experiment will be denoted by ω, a *point* (or *element*) of a set of possible outcomes, Ω, called the *sample* (or *probability*) *space*. *Events* are subsets of Ω and will be denoted by capital Roman letters from the early portion of the alphabet (A, B, etc.).

We will use standard set notation; $\omega \in A$ means "ω is an *element* of (or *belongs* to) A," and $\Rightarrow$ means "implies." In addition:

The *complement* of A is A^c $= \{\omega: \omega \notin A\}$.

The *union* of A and B is $A \cup B$ $= \{\omega: \omega \in A \text{ or } \omega \in B \text{ or both}\}$.

The *intersection* of A and B is $A \cap B = AB = \{\omega: \omega \in A \text{ and } \omega \in B\}$.

By $A \subset B$ (A is a *subset* of B, or B *contains* A), we mean that $\omega \in A \Rightarrow \omega \in B$. Sets A and B are *equal* ($A = B$) if both $A \subset B$ and $B \subset A$. We will say that event A occurs if and only if the outcome of the experiment is $\omega \in A$. Similarly, AB occurs if both A and B occur.

In any application, we will want to assign probabilities to some collection of events, $\mathcal{A}$. How rich a collection should we choose? For any collection of events meaningful to us, we can generate other events through the set operations of complement, union, and intersection. Thus, for any event A, $A \cup A^c = \Omega$, the entire sample space, where Ω is called the *certain* event. Similarly, the null set $\varnothing = \Omega^c$, where $\varnothing$ is called the *impossible* event. Thus, Ω and $\varnothing$ are always events, and $\mathcal{A}$ is closed under complementation, union, and intersection.

If Ω is a finite set of points, we can let each single element subset $\{\omega\}$ be an event. In this case, $\mathcal{A}$ is the set of all subsets of Ω. For example, consider a single roll of a die with $\Omega = \{1, 2, 3, 4, 5, 6\}$. In this case, the number of events (the number of elements of $\mathcal{A}$) is $2^6 = 64$. (See Problem 1-1.)

When Ω is infinite, the set of all subsets turns out to be "too large" a collection for technical reason beyond our scope. Instead, what is typically done is that some basic events of interest to us are defined first. For example, if Ω is the interval [0,1], we can let the basic events be the interval subsets of [0,1], e.g., (a, b) for $0 \le a < b \le 1$. For infinite Ω, we require that $\mathcal{A}$ be closed under *countable* unions and intersections, e.g.,

$$A_i \in \mathcal{A},\ i = 1, 2, \ldots \Rightarrow \bigcup_{i=1}^{\infty} A_i \in \mathcal{A}. \tag{4}$$

Any set of basic events can always be enlarged to some collection $\mathcal{A}$ in this manner.

Countable unions and intersections occur naturally. For example, consider a sequence of coin tosses where A_i is the event "heads on toss i," $i = 1, 2, \ldots$. (A_i^c denotes "tails.") Suppose we stop tossing when the first head occurs. The event "the number of tosses is finite" is

$$\bigcup_{n=0}^{\infty} A_1^c \cdots A_n^c A_{n+1}. \tag{5}$$

However, there is no guarantee that this event will occur, i.e., it is *not* the certain event. Similarly, an infinite sequence of tails is not impossible, although it *may* have zero probability.

There is an elementary algebra for the manipulation of events (point sets). The basic laws are:

$$AA = A \qquad A \cup A = A$$

$$AB = BA \qquad A \cup B = B \cup A$$

$$A(BC) = (AB)C \qquad A \cup (B \cup C) = (A \cup B) \cup C$$

$$A(B \cup C) = (AB) \cup (AC) \qquad A \cup (BC) = (A \cup B) \cap (A \cup C)$$

$$A\varnothing = \varnothing \qquad A\Omega = A$$

$$A \cup \varnothing = A \qquad A \cup \Omega = \Omega$$

$$AA^c = \varnothing \qquad A \cup A^c = \Omega$$

$$(AB)^c = A^c \cup B^c \qquad (A \cup B)^c = A^c B^c$$

$$(A^c)^c = A$$

Proofs of these laws are elementary. They can also be verified intuitively by using Venn's diagram. (See Problem 1-2.)

1-3 PROBABILITY MEASURE

Probability measure is simply a function P that assigns a nonnegative number (a *probability*) to each event in the class $\mathcal{A}$. Without explicitly attaching meaning to probability, we must impose conditions on P that are consistent with the interpretations in Section 1-1. These conditions are stated in the form of axioms.

When the sample space Ω is finite, a real-valued function P defined on $\mathcal{A}$ is a *probability measure* if it satisfies the following axioms for all events:

1. $P(A) \geq 0$
2. $P(\Omega) = 1$
3. If $AB = \varnothing$, $P(A \cup B) = P(A) + P(B)$

Whenever $AB = \varnothing$, events A and B are said to be *mutually exclusive*.

In any application, there may be considerable latitude in the choice of the *triple* $(\Omega, \mathcal{A}, P)$; a particular choice may be either a good or a poor approximation of reality. However, once chosen, $(\Omega, \mathcal{A}, P)$ must be regarded as fixed. Most of the so-called paradoxes in probability theory (e.g., exhibiting a probability exceeding 1) occur because probabilities determined by different triples are used in the same computation. (For example, see Problems 1-6 and 1-7.)

Many results are immediate consequences of the axioms and the algebra of events. For example,

$$P(A^c) = 1 - P(A) \tag{6}$$

follows from $A \cup A^c = \Omega$, $AA^c = \varnothing$, and axioms 2 and 3. For $A = \Omega$, (6) becomes

$$P(\varnothing) = 0. \tag{7}$$

From (6) and axiom 1,

$$P(A) \leq 1. \tag{8}$$

The following are also easy:

$$P(A \cup B) = P(A) + P(B) - P(AB), \tag{9}$$

$$A \subset B \Rightarrow P(A) \leq P(B). \tag{10}$$

If Ω is infinite, it is convenient to strengthen axiom 3 to axiom 3′: If A_1, $A_2, \ldots$ are *pairwise mutually exclusive* (i.e., $i \neq j \Rightarrow A_iA_j = \varnothing$), then

$$P\left(\bigcup_{i=1}^{\infty} A_i\right) = \sum_{i=1}^{\infty} P(A_i),$$

where axiom 3′ is called *countable additivity*.

Countable additivity may be shown to be equivalent to what is called the *monotone property*: Let $\{A_n\}$ be a monotone sequence of events (either $A_1 \subset A_2$, $\cdots$ with $A = \bigcup_{n=1}^{\infty} A_n$, or $A_1 \supset A_2, \cdots$ with $A = \bigcap_{n=1}^{\infty} A_n$). In either case,

$$\lim_{n\to\infty} P(A_n) = P(A). \tag{11}$$

To illustrate countable additivity and the monotone property, consider an infinite sequence of coin tosses. Let A_n be the event that the cumulative number of heads exceeds the cumulative number of tails *for the first time* on toss n, $n = 1, 2, \ldots$. (Clearly, n must be an odd integer, but that does not concern us here.) The events A_n, $n = 1, 2, \ldots$ are mutually exclusive. Let B be the event that the cumulative number of heads *ever* exceeds the cumulative number of tails, i.e., $B = \bigcup_{n=1}^{\infty} A_n$, and, by countable additivity,

$$P(B) = \sum_{n=1}^{\infty} P(A_n).$$

To continue this example, now let B_n be the event that the cumulative number of heads exceeds the cumulative number of tails *on or before* the nth toss, $n = 1, 2, \ldots$. Clearly, $\{B_n\}$ is monotone increasing with $B_n = \bigcup_{j=1}^{n} A_j$, $n = 1, 2, \ldots$, where (the same) $B = \bigcup_{n=1}^{\infty} B_n$. From the monotone property,

$$P(B) = \lim_{n\to\infty} P(B_n) = \lim_{n\to\infty} \sum_{j=1}^{n} P(A_j).$$

Clearly, the two expressions for $P(B)$ are the same.

1-4 CONDITIONAL PROBABILITY; BAYES' THEOREM

Suppose a company has N employees of which N_A are women, N_B are married, and N_{AB} are married women. An employee's file is selected "at random" (what is meant by this?). Let A, B, and AB be the events that the file is that of a woman, a married person, and a married woman, respectively. We have

$$P(A) = \frac{N_A}{N}, \quad P(B) = \frac{N_B}{N}, \quad \text{and} \quad P(AB) = \frac{N_{AB}}{N}.$$

On the other hand, suppose that the files of women are separate from the files of men, and we are given that the selected file is drawn from the files of women. This is equivalent to saying that A is certain to occur. The *conditional* probability of B given A, $P(B \mid A)$, is the probability that B occurs given that A occurs. Obviously,

$$P(B \mid A) = \frac{N_{AB}}{N_A} = \frac{N_{AB}/N}{N_A/N} = \frac{P(AB)}{P(A)}. \tag{12}$$

One is also led to (12) from the relative frequency viewpoint. For arbitrary events A and B, let $P(A)$, $P(B)$, and $P(AB)$ be the proportion of trials that result in A, B, and AB, respectively. By $P(B \mid A)$ we mean the proportion of those trials on which A occurs that B also occurs. Note that A occurs and B also occurs if AB occurs. Again,

$$P(B \mid A) = \frac{P(AB)}{P(A)}. \tag{13}$$

With these interpretations in mind, we adopt (13) as the formal *definition* of the conditional probability of B given A, $P(B \mid A)$, provided that $P(A) > 0$. If we allow B to vary over $\mathcal{A}$ and keep A fixed, the function defined by (13) is a probability measure on $\mathcal{A}$.

Conditional probability is a powerful tool in probability theory. It is often the case that the probability of an event is found most easily by expressing it in terms of conditional probabilities. This can be done by observing that because $AB \cap A^cB = \varnothing$, $P(B) = P(AB) + P(A^cB)$. From (13), this becomes

$$P(B) = P(B \mid A)P(A) + P(B \mid A^c)P(A^c). \tag{14}$$

As an example of (14), consider two urns, one containing one white and one black ball, and the other one white and two black balls. An urn is selected at random, and a ball is then selected at random from that urn. Let A be the event that the first urn is selected and B be the event that a white ball is selected. Applying (14) gives us

$$P(B) = \left(\frac{1}{2}\right)\left(\frac{1}{2}\right) + \left(\frac{1}{3}\right)\left(\frac{1}{2}\right) = \frac{5}{12}.$$

In the example above, suppose we know the color of the ball that was selected (white, say), and we wish to find the "inverse" probability that it came from urn 1, i.e., we want $P(A \mid B)$. But

$$P(A \mid B) = \frac{P(AB)}{P(B)} = \frac{P(B \mid A)P(A)}{P(B)},$$

or

$$P(A \mid B) = \frac{(1/2)\,(1/2)}{5/12} = \frac{3}{5}.$$

This elementary exercise in conditional probability is an example of *Bayes' Theorem*: For event B and sequence of events $\{A_i\}$,

$$P(A_i \mid B) = \frac{P(B \mid A_i)P(A_i)}{P(B)}. \tag{15}$$

Note that the denominator of (15) does not depend on i. A set of events $\{A_i\}$ is *exhaustive* if $\bigcup_i A_i = \Omega$. When $\{A_i\}$ is both exhaustive and pairwise mutually exclusive, we will say that $\{A_i\}$ is a *partition* of Ω. (For example, $\{A, A^c\}$ is a partition.) If $\{A_i\}$ is exhaustive,

$$B = \bigcup_i A_iB. \tag{16}$$

If $\{A_i\}$ is a partition, we can utilize (13) and (16) to rewrite (15):

$$P(A_i \mid B) = \frac{P(B \mid A_i)P(A_i)}{\sum_i P(B \mid A_i)P(A_i)}. \tag{17}$$

Bayes' Theorem, (15) or (17), is merely a simple application of the definition of conditional probability. It is a correct way to manipulate probabilities, regardless of their meaning. Consider the following example.

Suppose Joe's subjective probability that he has cancer (event C) is .6 prior to seeing a doctor. He goes to Jane, a doctor for whom the (statistical) probability of *correct* diagnosis is estimated to be .8 for patients who have cancer and .9 for patients who do not. Given Jane's diagnosis that cancer is not present (event D^c), Joe's subjective probability (called a *posterior* probability) that he has cancer now is

$$P(C \mid D^c) = \frac{P(CD^c)}{P(D^c)} = \frac{P(D^c \mid C)P(C)}{P(D^c \mid C)P(C) + P(D^c \mid C^c)P(C^c)}$$

$$= \frac{(.2)(.6)}{(.2)(.6) + (.9)(.4)} = .25.$$

An important concept in probability theory is that of *independence*. Event B is *independent* of event A if

$$P(B \mid A) = P(B). \tag{18}$$

To write $P(B \mid A)$, we must have $P(A) > 0$. If $P(B) > 0$, we can use (13) to rewrite (18) as

$$P(A \mid B) = P(A), \tag{19}$$

i.e., if B is independent of A and $P(B) > 0$, then A is independent of B. That is, in spite of the asymmetric appearance of (18), independence is a *symmetric pairwise relation* between events, and we will speak of events A and B which satisfy (18) or (19) as being *independent* events. A symmetric way of writing this is that if A and B are independent events,

$$P(AB) = P(A)P(B). \tag{20}$$

Note that (20) holds trivially if either $P(A) = 0$ or $P(B) = 0$. It is easily shown that if the pair (A, B) are independent, then so are the pairs (A, B^c), (A^c, B), and (A^c, B^c).

As an example of independent events, suppose a fair coin is tossed twice, where A denotes heads on the first toss and B denotes heads on the second. Ordinarily, we assume that A and B are independent. Now let C be the event that the number of trials on which heads occur is even (i.e., 0 or 2). If the coin is fair, $P(C \mid A) = P(C \mid B) = P(C) = 1/2$. Thus, A, B, and C are pairwise independent. However, $P(C \mid AB) = 1 \neq P(C)$.

We have shown that it is possible for the pairwise relation of independence to hold over a class of events and yet not hold between events obtained by performing set operations within this class. This possibility is eliminated if the events are mutually independent.

Events in an arbitrary class of events $\mathscr{C}$ are called *mutually independent* if for *every* finite subclass $\{A_1, A_2, \ldots, A_n\} \subset \mathscr{C}$,

$$P(A_1A_2 \cdots A_n) = P(A_1)P(A_2) \cdots P(A_n). \tag{21}$$

The intuitive meaning of independence is often overstated. See Problem 1-15.

1-5 RANDOM VARIABLES

If you went on an automobile trip and filled up your gas tank every 200 miles, you would not expect the number of gallons to be the same each time. Why not? Different driving conditions is one reason. Even if you know how to take this factor into account, could you predict the number of gallons to within .1? Probably not. The important concept of random variable will enable us to apply probability theory to this kind of situation.

As already defined, probability measure is a *set* function, i.e., its domain is the class of subsets $\mathcal{A}$ of the sample space Ω. A random variable is a real-valued *point* function of ω, i.e., a function $X(\omega)$ is a real number for every $\omega \in \Omega$. The simplest example of a random variable is an *indicator* function. An indicator I_A of event A is the function

$$\begin{aligned} I_A(\omega) &= 1 \qquad \text{if } \omega \in A, \\ &= 0 \qquad \text{if } \omega \notin A. \end{aligned} \tag{22}$$

The indicator I_A is a random variable that equals 1 if A occurs and equals 0 if A^c occurs. Thus, $P(I_A = 1) = P(A)$ and $P(I_A = 0) = 1 - P(A)$.

A general definition of random variable is given in terms of an arbitrary but fixed triple $(\Omega, \mathcal{A}, P)$: A *random variable* X is a real-valued function defined on Ω such that for every real x the set

$$\{\omega \colon X(\omega) \le x\} \in \mathcal{A}. \tag{23}$$

Thus, for a random variable to be well defined, we must be able to assign probabilities to sets defined in terms of the magnitude of the variable.

Viewing (23) as a function of x, we may define F as the (cumulative) *distribution function* of X where

$$F(x) = P(\{\omega \colon X(\omega) \le x\}). \tag{24}$$

It is convenient to abbreviate expressions such as (23) and (24) by deleting explicit mention of dependence on ω. For example, (24) becomes

$$F(x) = P(X \le x). \tag{25}$$

We will use the shorthand notation $X \sim F$ where $\sim$ means *distributed as*. Since (25) is a probability, it is clear that for every x, $0 \le F(x) \le 1$. From axiom 3, it is obvious that F is a monotone nondecreasing function of x (denoted by $F \uparrow x$). By $F(\infty)$ and $F(-\infty)$, we mean $\lim_{x \to \infty} F(x)$ and $\lim_{x \to -\infty} F(x)$, respectively. If $P(|X(\omega)| < \infty) = 1$, it follows that $F(\infty) = 1$ and $F(-\infty) = 0$. A random variable (or its distribution) with these properties is called *proper* or *honest*.

Instead of working with F, it is often convenient to work with $P(X > x) = 1 - F(x) \equiv F^c(x)$, called the *tail distribution*, or sometimes just the *tail* of X.

In applications, it is sometimes convenient to permit $X(\omega)$ to take on values $\pm\infty$ for some ω. If $P(|X(\omega)| < \infty) = F(+\infty) - F(-\infty) < 1$, X and its distribution

F are called *improper* or *defective*. Usually, the *defect* $[1 - P(|X| < \infty)]$ is assigned to either $+\infty$ or $-\infty$. For example, if the defect is assigned to $+\infty$, $F(-\infty) = 0$, $F(+\infty) < 1$, and $P(X = \infty) = 1 - F(+\infty)$. In Chapter 2, examples of this kind occur where X represents the epoch (point on the time axis) at which an event occurs, but it is not certain that the event will ever occur. In some cases, e.g., for recurrence times in Chapter 3, we will have $P(X(\omega) = \infty) > 0$.

Unless the contrary is stated, all random variables are assumed to be proper.

It is common practice to define a random variable merely by specifying its distribution function. Similarly, in Section 1-8, we will define several random variables simultaneously by specifying their joint distribution.

1-6 DISCRETE DISTRIBUTIONS

The indicator function (22) is a random variable that can take on only two values, 0 or 1. More generally, we say X is a *discrete* random variable if for some finite or at most countable set of values $x_1, x_2, \ldots, P(X = x_i) > 0$, $i = 1, 2, \ldots,$ and $P(X \in \{x_1, x_2, \ldots\}) = 1$. For any possible value x, we will often denote $P(X = x)$ by p_x or $p(x)$. The distribution function (25) of a discrete random variable has a discontinuity with jump p_x at any point x where $p_x > 0$. $F(x)$ is flat between successive discontinuities. When the x_i are nonnegative integers, we have the important special case of *integral-valued* random variables.

For example, consider independent coin tosses (*trials*) where p is the probability of a head on any toss, $0 < p < 1$. Suppose we want the distribution of X, the number of heads that occur in n tosses. By *independent trials*, we mean that the events $A_1, \ldots, A_n$ are mutually independent, where A_i is the event: "heads on trial i." Hence, the event $A_1 \cdots A_x A_{x+1}^c \cdots A_n^c$ has probability

$$p^x q^{n-x}, \tag{26}$$

where $q = 1 - p$. Other orderings of the n head-tail outcomes such that exactly x heads occur also have probability (26). Since these orderings are mutually exclusive events, $P(X = x)$ is the product of the number of such orderings and (26):

$$P(X = x) = \binom{n}{x} p^x q^{n-x}, \qquad x = 0, 1, \ldots, n. \tag{27}$$

Clearly, X must be proper, i.e., $\sum_{x=0}^{n} P(X = x) = 1$. This is easy to verify because the right-hand side of (27) are the terms in the expansion of $(p + q)^n$, where $\binom{n}{x}$ are the binomial coefficients. For this reason (27) is called the *binomial distribution*.

Now consider an infinite sequence of coin tosses where X is the number of tosses prior to the occurrence of the first head. It is easily shown that

$$P(X = x) = q^x p, \qquad x = 0, 1, \ldots, \tag{28}$$

where (28) is called the *geometric* distribution. Notice that $\{X < \infty\}$ is the event defined in (5) and is not the certain event Ω. Clearly, then, the event $\{X = \infty\}$ is possible. Nevertheless, by summing (28), we easily see that X is proper, and, in particular, $P(X = \infty) = 0$.

The *Poisson* distribution,

$$p_x = \frac{\lambda^x e^{-\lambda}}{x!}, \qquad x = 0, 1, \ldots, \tag{29}$$

where parameter $\lambda > 0$, is important in many applications, e.g., in Chapter 2. We denote that X has distribution (29) by $X \sim P(\lambda)$.

1-7 CONTINUOUS DISTRIBUTIONS

If X has no values x with the property $P(X = x) > 0$, X is said to be *continuous* and F is a continuous function. In this case, it is usually assumed that there exists a *density function* $f(x) \geq 0$ such that $\int_{-\infty}^{t} f(x)\,dx = F(t)$ for every real t. This allows us to represent probability as an area under a curve. A density function may be found from a distribution by differentiation, $f(x) = F'(x)$, or from the corresponding tail,

$$f(x) = -\frac{d}{dx} F^c(x).$$

An important density function in operations research and statistics is the *normal distribution*:

$$f(x) = \frac{1}{\sigma\sqrt{2\pi}} e^{-(x-\mu)^2/2\sigma^2}, \qquad -\infty < x < \infty, \tag{30}$$

where μ and σ are real parameters with $\sigma > 0$. We denote that X has distribution (30) by $X \sim N(\mu, \sigma^2)$. A major reason for the prominence of the normal distribution is the *Central Limit Theorem*. See Section 1-16.

Next, we define the *uniform* distribution:

$$\begin{aligned} f(x) &= \frac{1}{b-a}, && a \leq x \leq b, \\ &= 0, && \text{otherwise}, \end{aligned} \tag{31}$$

where $b > a$. Our shorthand terminology for (31) is that X is uniformly distributed on $[a, b]$. The uniform distribution on $[0, 1]$ has important applications in simulation, as well as other uses.

An important distribution in stochastic modeling, particularly in this book, is the *exponential*:

$$f(x) = \mu e^{-\mu x}, \qquad x \geq 0, \tag{32}$$
$$= 0, \qquad x < 0,$$

where $\mu > 0$. We denote that X has density (32) by $X \sim \exp(\mu)$.

The Memoryless Property of the Exponential Distribution

To motivate the discussion, let $X \geq 0$ be the life of a component and, for any $t \geq 0$, let $(X - t \mid X > t)$ be the remaining life of the component, given it has survived age t. The tail distribution of remaining life is easily seen to be

$$P(X - t > x \mid X > t) = \frac{F^c(t + x)}{F^c(t)}, \qquad x \geq 0. \tag{33}$$

For components that "wear out," we might expect the remaining life to be shorter (in some sense) than the original life. On the other hand, some electronic components, after a "burn-in period," may tend to live longer than new ones.

By the memoryless property, we mean that for any age t, the remaining life distribution is the *same* as the original life distribution. In this sense, it "forgets" its age. Mathematically, $X \sim F$ is said to have the *memoryless property* if

$$P(X - t > x \mid X > t) = F^c(x) \qquad \text{for all } t \geq 0 \quad \text{and} \quad x \geq 0,$$

which, from (33), is equivalent to

$$F^c(t + x) = F^c(t)F^c(x) \qquad \text{for all } t \geq 0 \quad \text{and} \quad x \geq 0. \tag{34}$$

If $X \sim \exp(\mu)$, it has tail distribution

$$P(X > x) = \int_x^\infty \mu e^{-\mu t}\, dt = e^{-\mu x}, \qquad x \geq 0. \tag{35}$$

From (34) and (35), we easily see that the exponential distribution has the memoryless property. In fact, it may be shown (see Problem 1-29) that the exponential is the *only* distribution with this property. In effect, (34) is a *defining property* of the exponential.

If X is integral-valued and the memoryless property holds for all integer-valued ages, it is easily shown that X has a *geometric* distribution. The particular form of this result depends on how the memoryless property is defined (in terms of whether $X > t$ or $X \geq t$). If we use (34), $X - 1$ will have a geometric distribution. That is, X is the trial (toss) on which the first success (head) occurs, where X has tail distribution of form

$$P(X > x) = q^x, \qquad x = 0, 1, \ldots .$$

The distinction between discrete and continuous distributions is convenient for defining the well-known ones. Conceptually, this distinction should be suppressed because in applications, mixtures of discrete and continuous distributions occur. For example, in Chapter 5 the delay in queue is defined to be the time between the arrival of a customer who requires service and the commencement of service, a random variable. Delay is zero if a server is free on arrival (an event of positive probability). Otherwise, delay is positive and (usually) continuous.

When we speak of a . . . distribution, e.g., the exponential, we mean whatever version (density, distribution function, tail) that is convenient for the purpose at hand.

1-8 JOINT DISTRIBUTIONS

If $X_1, X_2, \ldots, X_n$ are random variables, the set $\{\omega: X_i(\omega) \leq x_i, i = 1, 2, \ldots, n\}$ is the intersection of events of the form in (23) and is therefore an event. The *joint distribution function* of the collection $X_1, X_2, \ldots, X_n$ is defined for all real $x_1, x_2, \ldots, x_n$ as follows:

$$F(x_1, x_2, \ldots, x_n) = P(X_i \leq x_i, i = 1, 2, \ldots, n). \tag{36}$$

Obviously, we must have $F \uparrow x_i$ for every i, $F(-\infty, -\infty, \ldots, -\infty) = 0$, and $F(\infty, \infty, \ldots, \infty) = 1$. If $X_i \sim F_i$, we can find F_i (in this context called the *marginal* distribution of X_i) from

$$P(X_i \leq x_i) = F_i(x_i) = F(\infty, \ldots, \infty, x_i, \infty, \ldots, \infty). \tag{37}$$

Random variables $X_1, X_2, \ldots, X_n$ are *mutually independent* if for all real $x_1, x_2, \ldots, x_n$,

$$F(x_1, x_2, \ldots, x_n) = \prod_{i=1}^{n} F_i(x_i). \tag{38}$$

This is a natural extension of the definition of independent events. Random variables and events that are not independent are called *dependent*.

As an example, consider n independent coin tosses where $I_i = 1$ if heads occurs on trial i, $I_i = 0$ otherwise, and $p = P(I_i = 1)$, $i = 1, 2, \ldots, n$. The random variables $I_1, I_2, \ldots, I_n$ are mutually independent. The number of heads in n trials may be represented as the sum

$$X = \sum_{i=1}^{n} I_i, \tag{39}$$

where we have already shown that X has the binomial distribution (27).

If X and Y are discrete random variables, we can represent their joint distribution by specifying probabilities

$$P(X = x, Y = y) = p(x, y). \tag{40}$$

To obtain $P(X = x)$ we must sum over y. Similarly, if X and Y are continuous, we usually can represent a distribution in terms of a *joint density function* $f(x, y)$, where

$$P(a < X < b, c < Y < d) = \int_a^b \int_c^d f(x, y)\, dy\, dx.$$

To obtain the density function of X we integrate $f(x, y)$ with respect to y. If X and Y are independent, (40) factors: For all (x, y), $p(x, y) = P(X = x)P(Y = y)$. In the continuous case, the joint density function can be written as a product of the marginal densities. Extensions to higher dimensions are straightforward.

As an example, let X and Y be independent and uniform on (0, 1). That is,

$$\begin{aligned} f(x, y) &= 1, \qquad x, y \in (0, 1), \\ &= 0, \qquad \text{otherwise.} \end{aligned}$$

The probability $P(X > Y) = \int_0^1 \int_0^x 1\, dy\, dx = 1/2$. (One might have guessed this by symmetry.) Similarly, we can find $P(X + Y \leq 1/2) = 1/8$ by integrating over the appropriate region.

The "$\leq$" convention in (36) is exactly that. We could just as easily work with $<$, $\geq$, and $>$. In the last case, we have joint *tail* distributions. We now illustrate the connection between joint distribution functions and joint tail distributions for two dimensions: Let $P(X \leq x, Y \leq y) = F(x, y)$, which also determines the marginal distributions $P(X \leq x) = F(x, \infty)$, and $P(Y \leq y) = F(\infty, y)$. The joint tail distribution is

$$\begin{aligned} P(X > x, Y > y) &= 1 - P(X \leq x \text{ or } Y \leq y) \\ &= 1 \quad [P(X \leq x) + P(Y < y) - P(X \leq x, Y \leq y)] \\ &= 1 - F(x, \infty) - F(\infty, y) + F(x, y). \end{aligned} \tag{41}$$

If X and Y have a joint distribution function $F(x, y)$,

$$f(x, y) = \frac{\partial^2}{\partial x\, \partial y} F(x, y). \tag{42}$$

We see from (41) that a joint density function may be found just as easily from a joint tail distribution,

$$f(x, y) = \frac{\partial^2}{\partial x\, \partial y} P(X > x, Y > y), \tag{43}$$

with *no sign change*. [Recall that if X has density $f(x)$, $f(x) = -\dfrac{d}{dx} P(X > x)$.] In general, a sign change is necessary only when the number of random variables is odd.

Later in this chapter we will see that for some purposes, working with tail distributions is particularly convenient.

1-9 MATHEMATICAL EXPECTATION; MAGNITUDE AND DISPERSION OF RANDOM VARIABLES

If $a_1, a_2, \ldots, a_n$ are real numbers, their *arithmetic mean* is

$$\overline{a} = \sum_{i=1}^{n} \frac{a_i}{n}. \tag{44}$$

If a_i is the annual income of wage earner i in a city with n wage earners, then $\overline{a}$ is the average income of the wage earners in that city. Similarly, let $a_i = 1$ if voter i votes for candidate R and $a_i = 0$ otherwise; $\overline{a}$ is the proportion of the voters who vote for candidate R. In both cases, (44) is a measure of location of a distribution (of income or of voter preference).

If a wage earner in the above example were selected at random, i.e., the probability of selecting a particular wage earner is $1/n$, and X is the income of the selected wage earner, the expected value of X is

$$E(X) = \overline{a}. \tag{45}$$

This is a special case of a discrete random variable. If X is an arbitrary discrete random variable with $P(X = x) = p_x$, the *expected value* (or *mean*) of X is

$$E(X) = \sum_x x p_x. \tag{46}$$

If X is continuous with density f, the *expected value* of X is

$$E(X) = \int_{-\infty}^{\infty} x f(x)\, dx. \tag{47}$$

If $X \sim F$ is a nonnegative random variable, $P(X < 0) = 0$, but otherwise arbitrary (discrete, continuous, or a mixture of both), we show in Section 1-14 that $E(X)$ may be found by integrating the *tail* distribution,

$$E(X) = \int_{0}^{\infty} F^c(x)\, dx. \tag{48}$$

All of the definitions of $E(X)$ above represent expected value as some kind of integral. In all cases, the intent is to average the possible values of X, *weighted by their probability of occurrence*. For real number c and random variables X, Y, and $u(X)$, where $u(X)$ denotes a random variable that is a function of X, a few of the basic rules are

$$E(cX) = cE(X), \tag{49}$$

$$E(X + Y) = E(X) + E(Y), \tag{50}$$

$$E\{u(X)\} = \int_{-\infty}^{\infty} u(x) f(x)\, dx, \tag{51}$$

where we have written (51) for the case where X has a density function. The question of how general $u(x)$ may be in order to ensure that $u(X)$ is a random variable is discussed briefly in Section 1-11.

A few words are in order about other aspects of the generality of what we have done. First notice that (46) and (47) may diverge. We want to ensure that these expressions are well defined. To do this, let

$$X = X^+ - X^-, \tag{52}$$

where $X^+ \equiv \max(X, 0)$ and $X^- \equiv -\min(X, 0) = (-X)^+$ are called the *positive* and *negative parts* of X, respectively. We can apply (49) and (50) to (52), writing

$$E(X) = E(X^+) - E(X^-), \tag{53}$$

which is *well defined* ($\neq \infty - \infty$), but possibly infinite, if either $E(X^+)$ or $E(X^-)$ is finite. Since X^+ and X^- are nonnegative, their expectations can be found by applying (48) to their distributions, where (48) is always well defined, but possibly infinite.

Applying (48) to X^+ and X^-, (53) becomes

$$E(X) = \int_0^\infty F^c(x)\,dx - \int_{-\infty}^0 F(x)\,dx, \tag{54}$$

where (54) is valid for an arbitrary distribution F, whenever (53) is well defined. [To obtain (54), we need to observe that $\int_{-\infty}^0 P(X \le x)\,dx = \int_{-\infty}^0 P(X < x)\,dx$, i.e., the *area* under the function is the same.] Similarly, (46) and (47) are well defined whenever (53) and (54) are.

Sometimes it is convenient to combine discrete and continuous cases, e.g., in place of either (46) or (47), we write

$$E(X) = \int_{-\infty}^\infty x\,dF(x).$$

For the time being, we may regard this as a notational convention. For further discussion of "dF" notation, see Chapter 2 and the Appendix.

Also notice that (46) and (47) define $E(X)$ in a straightforward manner in terms of its own distribution. Inspection of most advanced books on probability, however, will reveal that $E(X)$ is defined as an integral of X with respect to the probability measure P over the sample space, i.e., an expression like this:

$$E(X) = \int_\Omega X(\omega)\,dP(\omega). \tag{55}$$

For a discrete sample space, i.e., Ω is a countable set of points, (55) is a sum that is easily seen to be equivalent to (46). [Compare (45) and (46) when some wage earners have the same income.]

For more general sample spaces, the definition of (55) becomes quite technical, and we omit details. It is then a *theorem* that $E(X)$ defined by (55) may be computed from its distribution, e.g., (46) or (47).

Some mention of this fact is necessary because we are using essentially the same theorem when we write (51): Let Ω be the real line on which X takes its values and let $P\{I\} = \int_I f(x)\,dx$ for any interval I. With these definitions, (51) is a version of (55). Thus, for any random variable Z expressible as $Z = u(X)$, we may find $E(Z)$ either from (51) or from the distribution of Z.

Seen in this light, (49) is a special case of (51), and (50) generalizes (51) for a particular function of two variables, $u(x, y) = x + y$. The obvious generalization of (51) for functions of a finite number of random variables is also valid. The practical importance of (51) is that the distribution of $u(X)$ may be unknown to us (although formally well defined) or, even if known, more difficult to work with directly than (51). Linear functions are particularly convenient; for

$$Z = \sum_{i=1}^{n} c_i X_i, \tag{56}$$

$$E(Z) = \sum_{i=1}^{n} c_i E(X_i). \tag{57}$$

As an example, let X be the number of heads in n independent coin tosses. We know that X has the binomial distribution (27), and $E(X)$ can be found directly from (46) to be

$$E(X) = \sum_{x=0}^{n} x \binom{n}{x} p^x q^{n-x} = np, \tag{58}$$

where obtaining the right-hand equality requires a little algebra. It is easier to obtain $E(X)$ directly from representation (39), where $E(I_i) = 1p + 0q = p$. By applying (57), we obtain the same result.

$E(X)$ is sometimes called the first moment of X. The rth moment of $X \sim F$ is defined to be

$$E(X^r) = \int_{-\infty}^{\infty} x^r f(x)\,dx \qquad \left(\text{or } \sum_x x^r p_x\right), \tag{59}$$

depending on whether X is continuous or discrete. This may also be viewed as an application of (51), where $E(X^r)$ is the first moment of the random variable X^r, expressed as a function of X.

The *variance* of X, $V(X)$ [or $\sigma^2(X)$], is defined [when $E(X)$ is finite] as the expectation

$$V(X) = E\{[X - E(X)]^2\}. \tag{60}$$

Notice that the definition (60) implies that $V(X) \geq 0$. The *standard deviation* of X is $\sigma(X) = \sqrt{V(X)}$. Variance is a measure of the spread or dispersion of a random variable (distribution) about its mean. By expanding the square in (60) and applying (57), we obtain a convenient equation for finding variance:†

$$V(X) = E(X^2) - E^2(X). \tag{61}$$

We summarize in Table 1-1 a few results for well-known distributions that occur frequently in later chapters. Each of these results may be obtained directly by either summation or integration.

TABLE 1-1

Name	Equation number	Mean	Variance
Binomial	(27)	np	npq
Geometric	(28)	q/p	q/p^2
Poisson	(29)	λ	λ
Normal	(30)	μ	σ^2
Uniform	(31)	$(b + a)/2$	$(b - a)^2/12$
Exponential	(32)	$1/\mu$	$1/\mu^2$

The *covariance* of a pair of random variables X and Y (with finite means) is defined as

$$\text{Cov}(X, Y) = E\{[X - E(X)][Y - E(Y)]\}. \tag{62}$$

By expanding the product term above, (62) can be written as

$$\text{Cov}(X, Y) = E(XY) - E(X)E(Y). \tag{63}$$

We say X and Y are *positively correlated*, *uncorrelated*, or *negatively correlated*, depending on whether (62) is positive, zero, or negative. Covariance is a measure of the degree of "association" of two random variables, e.g., positive correlation usually means large values (and small values) tend to occur together. (*Association* has a more technical meaning in the reliability theory literature.)

Stochastic Ordering

While the expected value (mean) of a random variable is a measure of its magnitude, and variance is a measure of its dispersion, these are crude measures. For example, let X and Y be the useful lifetimes of competing electronic components. Knowing $E(X) > E(Y)$ by no means guarantees that X will "outlive" Y.

To guarantee $X \geq Y$, we mean that

$$X(\omega) \geq Y(\omega) \qquad \text{for all } \omega \in \Omega. \tag{64}$$

Almost as strong as (64) is the requirement that the inequality in (64) hold on a set $\{\omega\}$ where $P(\{\omega\}) = 1$, i.e.,

$$P(X \geq Y) = 1. \tag{65}$$

† By $E^2(X)$, we mean $[E(X)]^2$.

Now (64) implies that

$$\{\omega: Y > t\} \subset \{\omega: X > t\} \qquad \text{for all } t, \tag{66}$$

which in turn implies that

$$P(Y > t) \leq P(X > t) \qquad \text{for all } t. \tag{67}$$

When (67) is true, we say that Y is *stochastically smaller* than X, i.e., (67) is the definition of this property. We denote (67) by

$$Y \overset{\text{st}}{\leq} X. \tag{68}$$

[Actually, (65) implies (67).] Note that (68) includes as a special case $X \overset{\text{st}}{=} Y$, i.e., X and Y have the *same* distribution.

We use the same notation for the corresponding distribution. If $X \sim F$ and $Y \sim G$, where $Y \overset{\text{st}}{\leq} X$, we write

$$G \overset{\text{st}}{\leq} F. \tag{69}$$

Note that (69) means $G(t) \geq F(t)$ for all t!

If $Y \overset{\text{st}}{\leq} X$, it is easily shown [from (48) if X and Y are nonnegative or from (54) if they are not] that $E(Y) \leq E(X)$, if finite. The converse can easily be false. Thus, (68) is a stronger statement about the relative magnitude of X and Y.

However, (68) by no means implies (65). For example, if X and Y have the same distribution and $P(X = Y) < 1$, then the events $\{X > Y\}$ and $\{Y > X\}$ must both have positive probability.

Even when the inequality (67) is strict for some t, (68) does not imply (65). For example, let X be uniformly distributed on (1,3) and Y be uniformly distributed on (0,2), where X and Y are independent. It is easily seen (draw a picture!) that while (67) is strict for $t \in (0,3)$, $P(Y > X) = 1/8$.

Sometimes equipment specifications are set in terms of tail probabilities, e.g., $P(X > t) \geq c$ for some c and t. In these situations, knowledge of stochastic orderings is important.

Some elementary properties of stochastic ordering are:

$$Z \overset{\text{st}}{\leq} Y \quad \text{and} \quad Y \overset{\text{st}}{\leq} X \Rightarrow Z \overset{\text{st}}{\leq} X, \tag{70}$$

and if X and U are independent, and Y and V are independent,

$$V \overset{\text{st}}{\leq} U \quad \text{and} \quad Y \overset{\text{st}}{\leq} X \Rightarrow V + Y \overset{\text{st}}{\leq} U + X. \tag{71}$$

Stochastic ordering is an important concept for the purpose of comparing stochastic processes. In Chapter 5 and later, we will compare queues under different modes of operation and show that certain performance measures are stochastically ordered.

1.10 SUMS OF RANDOM VARIABLES; CORRELATION, INDEPENDENCE, AND CONVOLUTION

Correlation is related to the variance of a sum:

$$\begin{aligned} V(X + Y) &= E\{(X + Y)^2\} - E^2(X + Y) \\ &= (E(X^2) - E^2(X)) + (E(Y^2) \\ &\quad - E^2(Y)) + (2E(XY) - 2E(X)E(Y)) \\ &= V(X) + V(Y) + 2\,\text{Cov}\,(X, Y). \end{aligned} \tag{72}$$

If X and Y are independent with marginal densities $f(x)$ and $g(y)$,

$$\begin{aligned} E(XY) &= \int_{-\infty}^{\infty}\int_{-\infty}^{\infty} xyf(x, y)\,dy\,dx \\ &= \left[\int_{-\infty}^{\infty} xf(x)\,dx\right]\left[\int_{-\infty}^{\infty} yg(y)\,dy\right] = E(X)E(Y). \end{aligned} \tag{73}$$

In fact, (73) does not depend on X and Y being continuous, but is true in general. Hence, if X and Y are independent, (63) becomes

$$\text{Cov}\,(X, Y) = E(X)E(Y) - E(X)E(Y) = 0. \tag{74}$$

That is, if X and Y are independent, they are uncorrelated.

The converse, i.e., $\text{Cov}\,(X, Y) = 0 \Rightarrow X$ and Y are independent, is *not* necessarily true. For example, let X be the number of aces and Y be the number of spades in a bridge hand. It can be shown (see Problem 1-45) that X and Y are uncorrelated. Obviously, e.g., given $Y = 13$, $P(X = 1) = 1$, they are dependent.

If X and Y are uncorrelated, (72) becomes

$$V(X + Y) = V(X) + V(Y). \tag{75}$$

It is easily shown that if $U = aX + b$,

$$V(U) = a^2V(X). \tag{76}$$

By combining and extending these results, it is easy to show that for *n pairwise uncorrelated* random variables, $X_1, \ldots, X_n$,

$$V\left(\sum_{i=1}^{n} c_iX_i\right) = \sum_{i=1}^{n} c_i^2V(X_i). \tag{77}$$

When correlations are present, the variance of a sum is given in Problem 1-42. Applications of (77) in this book usually will be to independent variables; however, the validity of this result under the weaker condition of uncorrelated variables has many applications in areas outside our scope.

As an example of how to use these results, let X have binominal distribution

(27) and representation (39). The I_i are mutually independent with $V(I_i) = E(I_i^2) - E^2(I_i) = p - p^2 = pq$. From (77),

$$V(X) = \sum_{i=1}^{n} V(I_i) = npq,$$

in agreement with Table 1-1.

The distribution of a *sum* of independent random variables is called the *convolution* of their distributions.

First consider integral-valued random variables. Let X and Y be independent with distributions $P(X = n) = a_n$, $P(Y = n) = b_n$, $n = 0, 1, \ldots$. Denote the distribution of $Z = X + Y$ by $P(Z = n) = c_n$, $n = 0, 1, \ldots$. We want to find $\{c_n\}$ in terms of $\{a_n\}$ and $\{b_n\}$. Since

$$\{Z = n\} = \bigcup_{j=0}^{n} \{X = j, Y = n - j\}, \tag{78}$$

a union of mutually exclusive events, and X and Y are independent,

$$c_n = \sum_{j=0}^{n} P(X = j, Y = n - j) = \sum_{j=0}^{n} a_j b_{n-j}, \qquad n = 0, 1, \ldots, j, \tag{79}$$

where $\{c_n\}$ is called the *convolution* of $\{a_n\}$ and $\{b_n\}$.

As an example, let X and Y be independent Poisson random variables; $X \sim P(\mu)$, $Y \sim P(\lambda)$. We want the distribution of $X + Y$. From (29) and (79),

$$\begin{aligned} c_n &= \sum_{j=0}^{n} \frac{e^{-\mu}\mu^j}{j!} \frac{e^{-\lambda}\lambda^{(n-j)}}{(n-j)!} \\ &= \frac{e^{-(\lambda+\mu)}(\lambda + \mu)^n}{n!} \sum_{j=0}^{n} \binom{n}{j} \left(\frac{\mu}{\lambda + \mu}\right)^j \left(\frac{\lambda}{\lambda + \mu}\right)^{n-j} \qquad (80) \\ &= \frac{e^{-(\lambda+\mu)}(\lambda + \mu)^n}{n!}, \qquad n = 0, 1, \ldots, \text{ i.e.,} \qquad (81) \end{aligned}$$

$$X + Y \sim P(\lambda + \mu). \tag{82}$$

In words, the *sum* of independent Poisson random variables is Poisson with parameter (mean) that is the sum of the original parameters. [We obtained (80) by pulling the exponential factors outside the sum and putting inside (with reciprocals outside) $n!$ and $(\lambda + \mu)^{-n}$.]

Now consider the continuous case where X and Y are independent with density functions $f(x)$ and $g(y)$, and distribution functions $F(x)$ and $G(y)$, respectively, $x,y \in (-\infty,\infty)$. From independence, the joint density is $f(x)g(y)$. We first find H, the distribution function of $Z = X + Y$, where the joint density of X and Y is integrated over the region $x + y \le t$,

$$H(t) = P(Z \le t) = \int_{-\infty}^{\infty} \int_{-\infty}^{t-y} f(x)g(y)\, dx\, dy = \int_{-\infty}^{\infty} F(t - y)g(y)\, dy. \tag{83}$$

The corresponding density function h is the derivative of (83),

$$h(t) = H'(t) = \int_{-\infty}^{\infty} \frac{\partial F(t - y)}{\partial t} g(y)\, dy = \int_{-\infty}^{\infty} f(t - y)g(y)\, dy, \tag{84}$$

where the density h is called the *convolution* of densities f and g. For nonnegative random variables, the range of integration in (83) and (84) reduces to $[0, t]$ because f and g are zero for negative arguments, e.g.,

$$h(t) = \int_0^t f(t - y)g(y)\, dy. \tag{85}$$

Compare with the integral-valued case, (79).

As an example, let X and Y have the same exponential distribution, exp (λ). The density of $Z = X + Y$ is

$$h(t) = \int_0^t \lambda e^{-\lambda(t-y)} \lambda e^{-\lambda y}\, dy \tag{86}$$

$$= \lambda^2 \int_0^t e^{-\lambda t}\, dy = \lambda^2 t e^{-\lambda t}, \qquad t \ge 0. \tag{87}$$

Whether discrete, continuous, or mixed, the distribution function H of $Z = X + Y$, where X and Y are independent, is called the *convolution* of the corresponding distribution functions F and G. In effect, F and G *determine* H, no matter how it is actually computed. This is denoted by

$$H = F * G. \tag{88}$$

Since $X + Y = Y + X$,

$$F * G = G * F. \tag{89}$$

Similarly, for arbitrary distributions F, G, and K,

$$F * (G * K) = (F * G) * K. \tag{90}$$

For completeness, we present an example of *dependent* random variables for which the distribution of their sum is a convolution of their distributions. Let $P(X = i) = P(Y = j) = 1/3$, $i, j \in \{1, 2, 3\}$, where the joint probabilities are given in the following table:

		Y values		
		1	2	3
	1	1/9	2/9	0
X values	2	0	1/9	2/9
	3	2/9	0	1/9

It is easy to verify that the distribution of $X + Y$ is correctly given by the right-hand expression in (79), even though X and Y obviously are dependent. (In practice, examples like this are rare.)

In summary, for $X \sim F$, $Y \sim G$,

$$X, Y \text{ independent} \Rightarrow X + Y \sim F * G \Rightarrow X, Y \text{ uncorrelated}, \tag{91}$$

but the converse implications can each be false. (The second $\Rightarrow$ requires finite means.)

1-11 FUNCTIONS OF RANDOM VARIABLES; TRANSFORMATIONS

The sum of random variables is itself a random variable. That is, we can define a function Z on Ω by

$$Z(\omega) = X(\omega) + Y(\omega). \tag{92}$$

It is easy to see that Z is a random variable because its distribution is determined by (92) and the joint distribution of X and Y.

Similarly, the functions $X^2 + Y^3$, $\max(X, Y)$, and e^{-X} are random variables. A function $u(X)$ is a random variable provided that we are able to assign probabilities to events of form $P\{u(X) \leq u\}$. The class of functions that may be chosen is very broad. For example, if Z is an arbitrary continuous function of n random variables, it is a random variable. This does *not* imply that Z is a continuous random variable, e.g., $Z = X + Y$ is discrete when both X and Y are.

We would like to develop some facility for finding the distribution of a function of one or more continuous random variables.

Consider the one-dimensional case where X has density f. We seek the density function $g(u)$ of $U = u(X)$, where $u(x)$ is differentiable and strictly monotone. If $u \uparrow x$, $P(U \leq u) = P(X \leq u^{-1}(u))$. Differentiating yields

$$g(u) = \frac{d}{du} P(U \leq u) = f(x) \frac{dx}{du} = f(x) \left[\frac{du}{dx}\right]^{-1},$$

where du/dx is positive. If $u \downarrow x$, $P(U \leq u) = P(X \geq u^{-1}(u))$, and

$$g(u) = -f(x) \left(\frac{du}{dx}\right)^{-1},$$

where du/dx is negative. In either case,

$$g(u) = f(x) \left|\frac{du}{dx}\right|^{-1}. \tag{93}$$

To illustrate (93), let $X \sim N(0,1)$ and $U = aX + b$, $a \neq 0$. We have

$$g(u) = e^{-x^2/2} \mid a \mid^{-1}/\sqrt{2\pi}$$
$$= e^{-(u-b)^2/2a^2}/\sqrt{2\pi} \mid a \mid, \tag{94}$$

i.e., $U \sim N(b, a^2)$.

We now state without proof the generalization† of (93) to two dimensions: Let X and Y have joint density $f(x, y)$, and $u(x, y)$, $v(x, y)$ be a pair of differentiable functions that define a one-to-one transformation from the (x, y) plane to the (u, v) plane. The joint density $g(u, v)$ of $U = u(X, Y)$ and $V = v(X, Y)$ is

$$g(u, v) = f(x, y) \mid J \mid, \tag{95}$$

where J, the *Jacobian* of the transformation, is the determinant

$$J = \begin{vmatrix} \dfrac{\partial x}{\partial u} & \dfrac{\partial x}{\partial v} \\ \dfrac{\partial y}{\partial u} & \dfrac{\partial y}{\partial v} \end{vmatrix} = \begin{vmatrix} \dfrac{\partial u}{\partial x} & \dfrac{\partial u}{\partial y} \\ \dfrac{\partial v}{\partial x} & \dfrac{\partial v}{\partial y} \end{vmatrix}^{-1}.$$

To illustrate (95), let X and Y be independent and identically distributed as $N(0, \sigma^2)$, $U = X + Y$, and $V = X - Y$. We have

$$J = \begin{vmatrix} 1 & 1 \\ 1 & -1 \end{vmatrix}^{-1} = -\frac{1}{2},$$

and

$$g(u, v) = \exp\,[-x^2/2\sigma^2 - y^2/2\sigma^2] \mid -1/2 \mid /2\pi\sigma^2$$
$$= \exp\,[-u^2/4\sigma^2 - v^2/4\sigma^2]/4\pi\sigma^2, \tag{96}$$

where the substitution $x^2 + y^2 = (u^2 + v^2)/2$ follows from $u = x + y$ and $v = x - y$. Notice that (96) factors into the product of two identical normal density functions. This shows that U and V are independent and identically distributed as $N(0, 2\sigma^2)$. In general, if X and Y were independent, normal, but not identically distributed, $g(u, v)$ would be what is known as a *bivariate* normal. The marginal distributions of U and V would still be normal.

When the transformation is many-to-one, the problem is more difficult. To employ (95), the (x, y) plane must be partitioned into regions such that the transformation is one-to-one from each region to the (u, v) plane. It is usually easier to find the distribution function by a direct argument and then the density by differentiation, as we did for (84).

† This technique is not used in subsequent chapters.

For example, let X be uniform on $(-1, 1)$ and $u = x^2$. For $0 < u < 1$,

$$P(U \leq u) = P(|X| \leq \sqrt{u}) = \int_{-\sqrt{u}}^{\sqrt{u}} dx/2 = \sqrt{u}.$$

Differentiating, we have $g(u) = 1/(2\sqrt{u})$, $0 < u < 1$.

1-12 GENERATING FUNCTIONS AND OTHER INTEGRAL TRANSFORMS

Expected values of certain functions of random variables are valuable analytic tools in probability theory. For integral-valued random variables, the most useful function is the (probability) *generating function*,

$$G(z) = E\{z^X\}, \tag{97}$$

where z is real. Of course, (97) may be well defined for an arbitrary random variable. However, if X is integral valued with $P(X = i) = p_i$, $i = 0, 1, \ldots,$ (97) becomes a power series:

$$G(z) = \sum_{i=0}^{\infty} p_i z^i. \tag{98}$$

As a function of z, (98) is well defined (the series converges) provided that $|z| \leq 1$. In fact, we may differentiate (98) with respect to z term by term for $|z| < 1$ to obtain

$$G'(z) = \sum_{i=1}^{\infty} i p_i z^{i-1}. \tag{99}$$

Taking the limit of (99) as $z \to 1$ (from the left), we have

$$\lim_{z \to 1} G'(z) = \sum_{i=1}^{\infty} i p_i = E(X). \tag{100}$$

Differentiating again gives us

$$\lim_{z \to 1} G''(z) = \sum_{i=2}^{\infty} (i)(i - 1)p_i = E\{(X)(X - 1)\}. \tag{101}$$

The expectation (101) is called the second *factorial* moment of X. Higher derivatives can be used to find higher factorial moments.

As an example, for $X \sim P(\lambda)$,

$$\begin{aligned} G(z) &= \sum_{i=0}^{\infty} \frac{\lambda^i e^{-\lambda}}{i!} z^i = e^{-\lambda} \sum_{i=0}^{\infty} \frac{(\lambda z)^i}{i!} \\ &= e^{-\lambda} e^{\lambda z} = e^{\lambda(z-1)}. \end{aligned} \tag{102}$$

By differentiating, we easily obtain $E(X) = \lambda$, $E\{(X)(X - 1)\} = \lambda^2$, and

$$V(X) = E\{(X)(X - 1)\} + E(X) - E^2(X) = \lambda^2 + \lambda - \lambda^2 = \lambda,$$

in agreement with Table 1-1.

Let X and Y be independent integral-valued random variables and $U = X + Y$. To find the generating function of U, first observe that z^U can be written as a product

$$z^U = z^{X+Y} = z^X \cdot z^Y.$$

Now z^X and z^Y are independent because X and Y are (why?). Hence, from (73),

$$E\{z^U\} = E\{z^X \cdot z^Y\} = E\{z^X\}E\{z^Y\}, \tag{103}$$

i.e., the generating function of a *sum* of independent random variables is the *product* of their generating functions.

If X has binomial distribution (27), with representation (39), we can apply (103) to obtain

$$E\{z^X\} = [E\{z^{I_i}\}]^n = (q + pz)^n. \tag{104}$$

The mean and variance of X can now be obtained from (104).

An important property of generating functions is this: *Distinct distributions have distinct generating functions.* Thus, the generating function of a random variable determines its distribution.

For example, let $X \sim P(\lambda)$ and $Y \sim P(\mu)$ be independent. The generating function of $U = X + Y$ is

$$E\{z^U\} = e^{\lambda(z-1)}e^{\mu(z-1)} = e^{(\lambda+\mu)(z-1)}. \tag{105}$$

Thus, we again have (82), $X + Y \sim P(\lambda + \mu)$, i.e., the *sum* of independent Poisson random variables is Poisson. Compare this derivation with the direct derivation, (80)–(82).

To obtain $U \sim P(\lambda + \mu)$ from (105) we recalled that a parametric family of distributions (Poisson with parameter λ) has a parametric family of generating functions (102) where the *parameter* appears in a particular place (as a factor in the exponent) of the generating function. Another example of this reasoning, slightly more complicated because there are two parameters, is the derivation of (112) below.

When a nonnegative random variable X has distribution function F with density function f, another useful function is the *Laplace transform* of f, $\tilde{F}$:†

$$\tilde{F}(s) = E\{e^{-sX}\} = \int_0^\infty e^{-sx}f(x)\,dx, \tag{106}$$

where $s \geq 0$. (In some applications, s is allowed to be complex with restrictions

† The form $E(e^{-sX})$ is valid when F does not have a density function. This accounts for the notation $\tilde{F}$ rather than $\tilde{f}$. See the Appendix for a formal definition.

on the magnitude of the real part.) By differentiating $\tilde{F}$ with respect to s and letting $s \to 0$ (from the right), we can find moments of X, e.g.,

$$\lim_{s\to 0} \tilde{F}'(s) = \lim_{s\to 0} \int_0^\infty -xe^{-sx} f(x)\, dx = -\int_0^\infty xf(x)\, dx = -E(X), \quad (107)$$

$$\lim_{s\to 0} \tilde{F}''(s) = \lim_{s\to 0} \int_0^\infty +x^2e^{-sx} f(x)\, dx = \int_0^\infty x^2f(x)\, dx = +E(X^2). \quad (108)$$

Higher derivatives will yield higher moments, but with alternating signs.

For example, let X have exponential density $\mu e^{-\mu x}$, $x \geq 0$:

$$\tilde{F}(s) = \int_0^\infty e^{-sx}\mu e^{-\mu x}\, dx = \frac{\mu}{\mu + s}. \quad (109)$$

By applying (107) and (108) it is easily shown that $E(X) = 1/\mu$ and $V(X) = 1/\mu^2$, in agreement with Table 1-1.

As with generating functions, the transform of the *sum* of independent random variables is the *product of* their transforms, and *distinct distributions* have *distinct transforms*. Characteristic functions and moment generating functions, defined below, have the same properties.

In the statistics literature, one may find the *characteristic function* of X,

$$\phi(t) = E\{e^{itX}\}, \quad (110)$$

where t is real and $i = \sqrt{-1}$. The advantage of the characteristic function over (97) and (106) is by restricting *it* to the imaginary axis, (110) is always finite. In fact, $|\phi(t)| \leq 1$.

It can be shown that the characteristic function of $X \sim N(\mu, \sigma^2)$ is

$$\phi(t) = \exp(i\mu t - \sigma^2t^2/2). \quad (111)$$

If $X \sim N(\mu_1, \sigma_1^2)$ and $Y \sim N(\mu_2, \sigma_2^2)$ are independent,

$$E\{e^{it(X+Y)}\} = E\{e^{itX}\}\, E\{e^{itY}\} = \exp[i(\mu_1 + \mu_2)t - (\sigma_1^2 + \sigma_2^2)t^2/2],$$

which implies that

$$X + Y \sim N(\mu_1 + \mu_2, \sigma_1^2 + \sigma_2^2). \quad (112)$$

An alternative way to derive (112) is by direct use of the convolution equation (84). As with (105), the derivation based on transforms involves considerably less algebra.

Other integral transforms occur from time to time in the literature on probability and statistics. A popular choice in statistics texts is the *moment generating function* of X,

$$E\{e^{\theta X}\}, \quad (113)$$

where θ is usually assumed to be real.

Obviously, (113) can be obtained from (106) by setting $s = -\theta$. In fact, when viewed as expected values, we see that these integral transforms are all minor variations of the same thing. (Changing variables does affect derivatives, and because of this, formulas for moments are changed somewhat.) The particular choice of transform in applications depends on convenience and fashion. In subsequent chapters, we will use only two types: generating functions and Laplace transforms.

Formal techniques exist for "inverting" transforms, i.e., determining the distribution function corresponding to a particular transform. These techniques will not be discussed here. However, you should learn to recognize the transforms of commonly occurring distributions. They obviously came in handy for deriving (112) from (111), and (82) from (105).

We now state without proof a useful result for determining when a sequence of random variables *converges in distribution* (see Section 1-16). This result is called the

Continuity Theorem. Let $\{X_n\}$ be a sequence of nonnegative random variables where $X_n \sim F_n$ and $\tilde{F}_n(s) = E\{e^{sX_n}\}$. The sequence $\{X_n\}$ ($\{F_n\}$) converges in distribution to (possibly defective) X (F) if and only if for all $s > 0$,

$$\lim_{n \to \infty} \tilde{F}_n(s) = \tilde{F}(s).$$

where $\tilde{F}(s) = E\{e^{-sX}\}$. Furthermore, F is proper if and only if $\lim_{s \to 0} \tilde{F}(s) = 1$, i.e., $\tilde{F}$ is *continuous* at the origin.

Remark. Versions of the continuity theorem are valid for other transformations. For generating functions, $G_n(z) = E\{z^{X_n}\}$, which just amounts to the change of variable $z = e^{-s}$, where the range of z is [0, 1). For random variables unrestricted in sign, we switch to the characteristic function (110), where t is real.

As an application of the generating function version of the continuity theorem, let X_n have binomial distribution (27) with generating function G_n given by (104). As we vary n, suppose that $p = \lambda/n$, i.e., $E(X_n) = np = \lambda > 0$, a constant. The limit of

$$G_n(z) = [1 + \lambda(z - 1)/n]^n$$

is easily shown (take logs) to be $e^{\lambda(z-1)}$, the generating function of the Poisson distribution, $P(\lambda)$. Hence $\{X_n\}$ converges to a Poisson random variable with distribution $P(\lambda)$.

1-13 CONDITIONAL DISTRIBUTIONS AND CONDITIONAL EXPECTATION

If X and Y are discrete random variables with joint probability

$$P(X = x, Y = y) = p(x, y), \tag{114}$$

the conditional probability $P(X = x \mid Y = y) = p(x \mid y)$ is

$$p(x \mid y) = \frac{p(x, y)}{P(Y = y)} = \frac{p(x, y)}{\sum_x p(x, y)}. \tag{115}$$

For fixed y, (115) defines a legitimate distribution, the *conditional distribution* of X given $Y = y$.

The expected value of this distribution, the *conditional expectation* of X given $Y = y$, is

$$E(X \mid Y = y) = \sum_x xp(x \mid y). \tag{116}$$

Note that (116) is a function of y, $u(y)$, say, and may be used to define a random variable $u(Y) = E\{X \mid Y\}$. The expected value of $u(Y)$ is

$$\begin{aligned} E\{E(X \mid Y)\} &= \sum_y E(X \mid Y = y)P(Y = y) \\ &= \sum_y \sum_x xp(x, y) = \sum_x \sum_y xp(x, y) = \sum_x xP(X = x) \\ &= E(X), \end{aligned} \tag{117}$$

where interchanging the order of summation is permissible provided that $E(|X|) < \infty$, or, more generally, if either X^+ or X^- has finite expectation, i.e., whenever $E(X)$ is well defined. (See Section 1-9.)

For example, suppose two balls are drawn simultaneously from an urn containing four balls numbered 1, 2, 3, 4. Let X be the number on one of the drawn balls and Y be the number on the other. We find $E(Z)$ where $Z = XY$ first by

$$E(Z) = E(XY) = E\{E(XY \mid Y)\} = E\{YE(X \mid Y)\}.$$

Note that $E(X) = (1 + 2 + 3 + 4)/4 = 5/2$. However, once Y is selected, X must be one of the three remaining numbers. Hence,

$$\begin{aligned} E(X \mid Y) &= (1 + 2 + 3 + 4 - Y)/3 = (10 - Y)/3, \\ E(Z) &= E\{Y(10 - Y)/3\} = [10E(Y) - E(Y^2)]/3 \\ &= [25 - (1 + 4 + 9 + 16)/4]/3 = 35/6. \end{aligned}$$

This problem can also be solved by enumeration: There are $\binom{4}{2} = 6$ possible samples: (4, 3), (4, 2), (4, 1), (3, 2), (3, 1), (2, 1). The product, Z, takes on the values 12, 8, 4, 6, 3, and 2 each with probability 1/6, and

$$E(Z) = (12 + 8 + 4 + 6 + 3 + 2)/6 = 35/6.$$

Here, enumeration was probably easier. However, if the urn contained 20 balls, the method using conditional expectation would save considerable labor.

The conditional distribution function of an arbitrary random variable given an event of positive probability is also defined directly in terms of conditional probability. If $X \sim F$, the conditional distribution of X given A is

$$F(x \mid A) = \frac{P(X \leq x, A)}{P(A)}. \tag{118}$$

Conditional density functions are defined in a way analogous to (115). Let X and Y have joint density $f(x, y)$ with marginal densities $f(x)$ and $g(y)$, respectively. The conditional density of X given $Y = y$ is

$$f(x \mid y) = \frac{f(x, y)}{g(y)}. \tag{119}$$

We can motivate (119) from (118) by letting A be the event that Y belongs to an open interval about y. By taking an appropriate limit, (119) can be obtained. $E(X \mid Y) = u(Y)$, say, is defined by the function

$$u(y) = E(X \mid Y = y) = \int_{-\infty}^{\infty} xf(x \mid y)\, dx. \tag{120}$$

It is easy to verify again that

$$E\{E(X \mid Y)\} = E(X). \tag{121}$$

If X and Y are independent, $p(x \mid y) = p(x)$, $f(x \mid y) = f(x)$, and $E(X \mid Y) = E(X)$.

If X and Y are completely general random variables defined on an abstract space, Ω, such that $E(X)$ is well defined, there will always exist a random variable $E(X \mid Y)$ analogous to (120) such that (121) holds. In the general formulation, which is beyond our scope, $E(X \mid Y)$ is not uniquely defined; however, different versions are equal on a set $\{\omega\}$ with probability 1 (abbreviated w.p.1).

For practical purposes, conditional expectations possess all the properties of expectations except that relations hold w.p.1, e.g., $E\{aX + bZ \mid Y\} = aE(X \mid Y) + bE(Z \mid Y)$ w.p.1.

There is a distinction between the formal theory of a subject (theorems on existence or the validity of certain operations) and a working knowledge of a subject to obtain particular results. No better example exists than probability in general and using conditional expectation in particular. The validity of (121), now assured in general, is of little help in choosing a suitable Y in order to obtain $E(X)$ efficiently.

As a discrete example, consider a sequence of coin tosses with probability p of heads, $q = 1 - p$, and let X be the number of tails prior to the first head. Thus, X has a geometric distribution (28) with mean and variance in Table 1-1. We will find $E(X)$ from the following easily verified property of the geometric distribution: For any $j \geq 1$,

$$P(X = i + j \mid X \geq j) = P(X = i), \qquad i = 0, 1, \ldots, \tag{122}$$

i.e., the distribution of the number of *additional* tails prior to the first head, $X - j$, given tails occur on the first j tosses, has the same geometric distribution (28).

[This is the integral-valued version of the memoryless property of the exponential, (34).] It follows that

$$E(X \mid X \geq j) = E(X - j \mid X \geq j) + j = E(X) + j, \tag{123}$$

$$V(X \mid X \geq j) = V(X - j \mid X \geq j) = V(X). \tag{124}$$

Now condition on the outcome of the first toss, $Y \in \{A_1, A_1^c\}$, where $Y = A_1$ means $X = 0$, and $Y = A_1^c$ means $X \geq 1$.

$$\begin{aligned} E(X) &= E\{E(X \mid Y)\} \\ &= E(X \mid A_1)P(A_1) + E(X \mid A_1^c)P(A_1^c) \\ &= 0 \cdot p + [E(X) + 1]q. \end{aligned} \tag{125}$$

Solving for $E(X)$,

$$E(X) = q/p.$$

Note that we have not bothered to assign numerical values to Y. The calculation depends *only* on values and conditional values of X.

We now present a useful result for finding variance called the *conditional variance relation*:

$$V(X) = E\{V(X \mid Y)\} + V\{E(X \mid Y)\}, \tag{126}$$

where $V(X \mid Y) = E(X^2 \mid Y) - E^2(X \mid Y)$. This follows from

$$\begin{aligned} V(X) &= E(X^2) - E^2(X) \\ &= E\{E(X^2 \mid Y)\} - E^2\{E(X \mid Y)\} \\ &= E\{V(X \mid Y)\} + E\{E^2(X \mid Y)\} - E^2\{E(X \mid Y)\}. \end{aligned}$$

By combining the last two terms on the right, we have (126).

To use (126), we work "inside out" by first evaluating $V(X \mid Y)$ and $E(X \mid Y)$ as functions of Y. For example, to find the variance of a geometric distribution by conditioning on the first toss, we first evaluate $V(X \mid Y)$: $V(X \mid A_1) = 0$ because $(X \mid A_1) = 0$, and, from (124), $V(X \mid A_1^c) = V(X)$. From (125), we already know that $E(X \mid Y)$ is zero with probability p and $E(X) + 1 = 1/p$ with probability q. Hence,

$$V\{E(X \mid Y)\} = E\{E^2(X \mid Y)\} - E^2\{E(X \mid Y)\} = (1/p^2)q - (q/p)^2 = q/p.$$

From (126), we now have

$$\begin{aligned} V(X) &= 0 \cdot p + V(X) \cdot q + q/p, \\ &= q/p^2, \end{aligned}$$

in agreement with Table 1-1.

For a more conventional application of (121) and (126), let Y be the (random) number of customers who enter a store in a day, where D_i is the demand of customer i, $i = 1, \ldots, Y$. The total demand for a day is

$$X = \sum_{i=1}^{Y} D_i. \tag{127}$$

Assume that the D_i are independent and identically distributed, each with mean $E(D)$ and variance $V(D)$, independent of Y. We will find the mean and variance of X in terms of the first two moments of Y and the D_i:

From (127), $E(X \mid Y) = YE(D)$, and, from (121),

$$E(X) = E\{YE(D)\} = E(Y)E(D). \tag{128}$$

From (126) and (127),

$$\begin{aligned} V(X) &= E\{V(X \mid Y)\} + V\{E(X \mid Y)\} \\ &= E\{YV(D)\} + V\{YE(D)\} \\ &= E(Y)V(D) + E^2(D)V(Y). \end{aligned} \tag{129}$$

In (129), the algebra was streamlined because the conditional mean and variance of X are simple functions of Y. By conditioning appropriately, we will often be able to take advantage of special structure and information about conditional distributions in order to simplify manipulations.

1-14 APPLICATIONS OF INDICATOR FUNCTIONS

An *indicator* of set A, I_A, is a random variable defined in (22):

$$\begin{aligned} I_A &= 1 \quad \text{if } A \text{ occurs,} \\ &= 0 \quad \text{otherwise.} \end{aligned}$$

It is important to observe that

$$E(I_A) = P(A), \tag{130}$$

i.e., any probability may be interpreted as the expected value of a random variable. Similarly, any conditional probability is a conditional expectation. Consequently, the compact notation of the preceding section, e.g., (117), may be used to find probabilities.

For example, to derive (83) for $Z = X + Y$, where X and Y are independent, $X \sim F$, and Y has density g, let $A = \{Z \le t\}$. We have

$$E(I_A \mid Y = y) = F(t - y) \quad \text{and} \quad P(Z \le t) = E(I_A) = \int_{-\infty}^{\infty} F(t - y)g(y)\, dy.$$

An elementary application of indicator functions is (39), where I_i is the indicator of the event "heads on trial i." This representation is the easiest way to find the mean, variance, and generating function of a binomial distribution.

We now use indicators to obtain convenient expressions for moments of nonnegative random variables, e.g., (48): Let X be a nonnegative random variable with distribution $F(x) = P(X \leq x)$ for all x, and define

$$I(x) = 1 \quad \text{if } X > x, \qquad (131)$$
$$= 0 \quad \text{otherwise.}$$

For any point ω in the sample space, $X(\omega)$ is just a number; we plot $I(x) \equiv I(x, \omega)$ as a function of x for fixed ω in Figure 1-1. From Figure 1-1,

$$\int_0^\infty I(x)\, dx = X, \qquad (132)$$

i.e., for any point in the sample space, the integral of $I(x)$ is the area of a rectangle of length X and height 1. We now find $E(X)$ by taking the expected value of the left-hand side of (132):†

$$E(X) = E\left\{\int_0^\infty I(x)\, dx\right\} = \int_0^\infty E\{I(x)\}\, dx$$
$$= \int_0^\infty P(X > x)\, dx = \int_0^\infty F^c(x)\, dx,$$

which is equation (48). In words, the expected value of a nonnegative random variable is the integral of the *tail* of its distribution.

As an example of using (48), we find the mean of an exponential distribution. If X has distribution (32),

$$P(X > x) = \int_x^\infty \mu e^{-\mu t}\, dt = e^{-\mu x},$$

$$E(X) = \int_0^\infty e^{-\mu x}\, dx = 1/\mu,$$

in agreement with Table 1-1.

Equation (48) is valid for discrete as well as continuous variables. If X is integral-valued, (48) can be written

$$E(X) = \sum_{j=0}^{\infty} P(X > j). \qquad (133)$$

It is easy to extend (48). For example, let X and Y be nonnegative with an arbitrary joint distribution. Define

$$I(x, y) = 1 \quad \text{if } X > x, \quad Y > y, \qquad (134)$$
$$= 0 \quad \text{otherwise.}$$

† Expected values are integrals. Thus, $E \int = \int E$ amounts to reversing the order of integration. This is always valid when the function, $I(x, \omega)$ here, is nonnegative. For further discussion, see the Appendix.

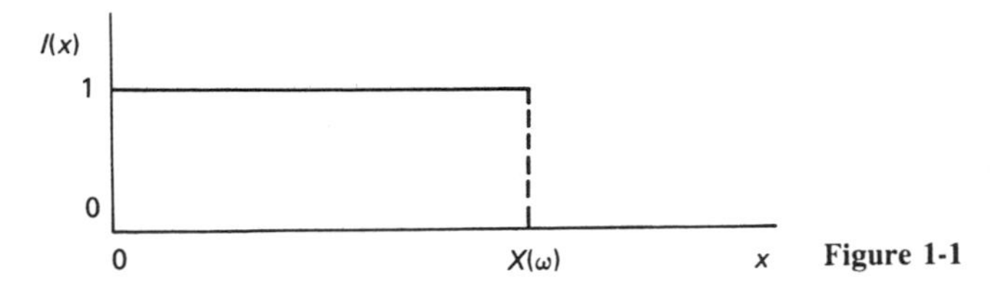

Figure 1-1

We now have

$$\int_0^\infty \int_0^\infty I(x, y)\, dx\, dy = XY, \tag{135}$$

and

$$E(XY) = E\iint = \iint E = \int_0^\infty \int_0^\infty P(X > x, Y > y)\, dx\, dy. \tag{136}$$

Equation (48) may also be extended to find higher moments of X. For $r > 0$, apply (48) to X^r:

$$\begin{aligned} E(X^r) &= \int_0^\infty P(X^r > x)\, dx \\ &= \int_0^\infty P(X > x^{1/r})\, dx. \end{aligned} \tag{137}$$

Now change variables: $x = u^r$, $dx = ru^{r-1}\, du$, and

$$\begin{aligned} E(X^r) &= \int_0^\infty ru^{r-1}P(X > u)\, du \\ &= r\int_0^\infty u^{r-1}F^c(u)\, du, \end{aligned} \tag{138}$$

where F is the distribution of X.

For $r = 2$, (138) is applied to obtain equation (49) in Chapter 2. Other applications of indicator functions will be found in the next two chapters. In particular, see the examples in Section 2-3.

To provide an intuitive explanation of (48), consider the discrete case as shown in Figure 1-2, where $P(X = x_j) = p_j \geq 0, j = 0, 1, \ldots, 0 = x_0 < x_1 < \cdots$, and $\sum_j p_j = 1$.

Now $E(X) = \sum_j x_j p_j$ is simply the sum of the horizontal rectangular areas shown by the dashed lines in Figure 1-2, which is the area under $F^c(x)$, $x \geq 0$. In terms of vertical rectangles, we also have

$$E(X) = \sum_{j=0}^{\infty} (x_{j+1} - x_j)F^c(x_j),$$

which is a generalization of (133).

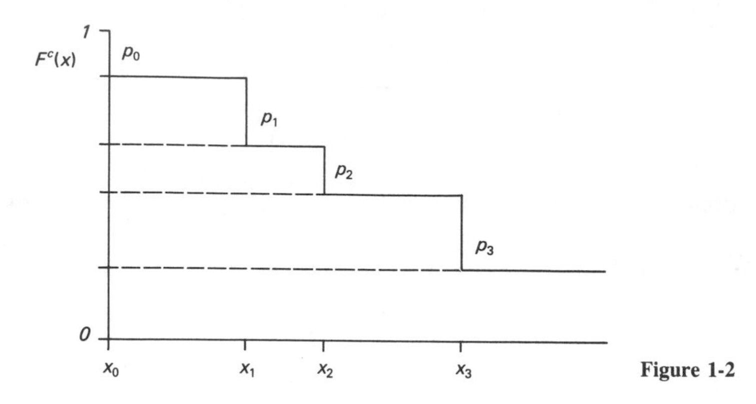

Figure 1-2

Remarks on Notation. Some authors employ the notation at the beginning of this section rigidly, i.e., $I_A(\omega) = 1$ for all $\omega \in A$, $I_A(\omega) = 0$ otherwise. When the defining property of A is complex, it is clumsy to carry along this information as a subscript, as is often done. Although more ad hoc, our notation is easier on the eyes.

1-15 LAWS OF LARGE NUMBERS

Games of chance, e.g., when a coin is tossed repeatedly, may be regarded as playful examples of scientific experimentation. Suppose X_n is the outcome of the nth trial in an experiment where a sequence of "independent trials" is performed under "identical conditions." This language means that we are assuming that $\{X_n, n = 1, 2, \ldots\}$ is a sequence of independent, identically distributed (i.i.d.) random variables.

Consider the following: X_n is the weight gain of rat n under an experimental diet, X_n is the sales gain in test market n under an experimental marketing program, or X_n is the outcome of toss n of a coin. For the last example, let $X_n = 1$ if "heads" occurs on toss n, $X_n = 0$ otherwise; i.e., X_n is an indicator variable. Similar to heads and tails, we will sometimes speak of "success" or "failure" of each of a sequence of trials. We assume below that $\mu \equiv E(X_n)$ is finite.

In this discussion, we put no limit on the number of trials. In fact, we will be concerned about what can be learned from experiments as the number of trials approaches infinity. In particular, we will describe the behavior of the sequence of sample means $\{\overline{X}_n, n = 1, 2, \ldots\}$ as $n \to \infty$, where

$$\overline{X}_n = \sum_{i=1}^{n} X_i/n, \qquad n = 1, 2, \ldots. \tag{139}$$

Consider the coin-tossing experiment and suppose the coin is fair, i.e., $\mu = P(X_n = 1) = 1/2$. One possible sequence is the infinite sequence of heads,

$\{X_n = 1, n = 1, 2, \ldots\}$. In this case, $\{\overline{X}_n = 1, n = 1, 2, \ldots\}$, and

$$\lim_{n \to \infty} \overline{X}_n = 1. \tag{140}$$

Similarly, $\lim_{n \to \infty} \overline{X}_n = 0$ is possible. However, because trials are independent, the probability of n successive heads is 2^{-n}, and the probability of an infinite sequence of heads is

$$\lim_{n \to \infty} 2^{-n} = 0. \tag{141}$$

Other infinite sequences of values for $\{\overline{X}_n\}$ can have limit 1/2, a different limit, or no limit at all. However, experience with coin-tossing experiments leads us to expect that for large n, $\overline{X}_n$ is "likely" to be "close" to μ.

Laws of large numbers are precise statements about when and in what sense $\overline{X}_n$ approaches μ. The first (weak) law, defined later as equation (144), is an easy consequence of an elementary inequality that we now derive.

For an arbitrary random variable Y such that $E(Y^2) < \infty$,

$$|Y| \geq \min(|Y|, t) \qquad \text{for any } t > 0.$$

Hence,

$$E(Y^2) \geq E\{Y^2 \mid |Y| \leq t\}P(|Y| \leq t) + t^2P(|Y| > t) \geq t^2P(|Y| > t),$$

$$P(|Y| > t) \leq E(Y^2)/t^2, \qquad t > 0, \tag{142}$$

where (142) is known as *Chebyshev's* inequality.

Now apply (142) to $Y = \overline{X}_n - \mu$, where $E(Y^2) = V(\overline{X}_n)$, and from (77), $V(\overline{X}_n) = \sigma^2/n$, where $\sigma^2 = V(X_n)$. For $n = 1, 2, \ldots$ and any $t > 0$,

$$P(|\overline{X}_n - \mu| > t) \leq \sigma^2/nt^2. \tag{143}$$

Assume that $\sigma^2 < \infty$. As $n \to \infty$ in (143), we have the

Weak Law of Large Numbers. For a sequence $X_1, X_2, \ldots$ of i.i.d. random variables with finite mean μ and variance σ^2,

$$\lim_{n \to \infty} P(|\overline{X}_n - \mu| > t) = 0 \qquad \text{for any } t > 0. \tag{144}$$

Since t may be arbitrarily small, $\overline{X}_n$ becomes arbitrarily close to μ as $n \to \infty$, in the sense of (144). We say that $\overline{X}_n$ converges *in probability* or *in (probability) measure* to μ and denote this by

$$\overline{X}_n \xrightarrow{P} \mu \qquad \text{as} \qquad n \to \infty, \tag{145}$$

where (145) is simply a shorthand way of expressing (144).

An alternative way of thinking about convergence is to go back to limits such as (140). For any point ω in the sample space, $X_n(\omega)$ and $\overline{X}_n(\omega)$ are simply sequences of numbers. We wish to consider the set $\{\omega\}$ such that

$$\lim_{n\to\infty} \overline{X}_n(\omega) = \mu. \tag{146}$$

From (140), we see that (146) is false for some ω. We will show that the set $\{\omega\}$ where (146) holds has probability 1. This is called the

Strong Law of Large Numbers. For a sequence $X_1, X_2, \ldots$ of i.i.d. random variables, with finite mean μ and variance σ^2,

$$P(\{\omega\}: \lim_{n\to\infty} \overline{X}_n(\omega) = \mu) = 1. \tag{147}$$

We say that $\overline{X}_n$ converges *with probability one (w.p.1)* to μ, and denote this by

$$\overline{X}_n \to \mu \quad \text{w.p.1} \qquad \text{as} \quad n \to \infty. \tag{148}$$

Almost sure and *almost everywhere* are equivalent terms for w.p.1.

Clearly, the strong law implies the weak and, more generally, convergence w.p.1 implies convergence in measure. We use "in measure" rather than "in probability" because the latter sounds too similar to w.p.1.

While the weak law says only that $\overline{X}_n$ is "likely to be close" to μ for any *particular* sufficiently large n, the strong law says that $\overline{X}_n$ is "likely to be close" for *all* sufficiently large n *simultaneously*. In applications, the distinction can be vitally important. For example, as shown in Section 2-2, n is replaced by a random function of a parameter where the function goes to infinity w.p.1 as the parameter $\to \infty$. The sample-path approach used there, where deterministic functions are studied at fixed points ω in the sample space, is very useful, both conceptually and mathematically.

To prove the strong law, we need to recall the precise meaning of limit. In particular, (146) holds at some point ω if for any $t > 0$, there is an integer $n(\omega, t)$ such that

$$|\overline{X}_n - \mu| \le t \qquad \text{for all } n > n(\omega, t). \tag{149}$$

For any $t > 0$ and $n = 1, 2, \ldots$, call the event $\{|\overline{X}_n - \mu| > t\}$ an *exception,* and define the indicator variables

$$\begin{aligned} I_n(t) &= 1 \qquad \text{if } |\overline{X}_n - \mu| > t, \\ &= 0 \qquad \text{otherwise.} \end{aligned} \tag{150}$$

The sum

$$N(t) = \sum_{n=1}^{\infty} I_n(t), \tag{151}$$

the total *number* of exceptions, is a (possibly defective) random variable. At some point ω, we see from (149) that $N(t, \omega) < \infty$ for arbitrarily small t if and only if

(146) holds. Hence, (147) is equivalent to the statement that for every $t > 0$, $P(\{\omega\}: N(t, \omega) < \infty) = 1$, i.e., that $N(t)$ is a *proper* random variable for every $t > 0$.

A sufficient condition for $N(t)$ to be proper is $E\{N(t)\} < \infty$. Hence, for any $t > 0$, if we can show

$$E(N(t)) = \sum_{n=1}^{\infty} P\{|\overline{X}_n - \mu| > t\} < \infty, \tag{152}$$

(147) will follow. Notice the connection between (152) and (144)! Unfortunately, the upper bound (143) is not tight enough to show (152) because $\sum_{n=1}^{\infty} 1/n = \infty$. However, applying the same argument to the *subsequence* $\overline{X}_{n^2}$, $n = 1, 2, \ldots$, where n^2 replaces n in (143) and $\sum_{n=1}^{\infty} 1/n^2 < \infty$, we have

$$\overline{X}_{n^2} \to \mu \quad \text{w.p.1} \qquad \text{as} \quad n \to \infty. \tag{153}$$

To complete the proof,* we need to show that for $n^2 < k < (n+1)^2$, $\overline{X}_k$ and $\overline{X}_{n^2}$ are close enough to behave the same as $k \to \infty$. Let

$$M_n = \max_{n^2 \le k \le (n+1)^2} |k\overline{X}_k - n^2\overline{X}_{n^2}|. \tag{154}$$

From

$$M_n \le \max\left(\sum_{k=n^2+1}^{(n+1)^2-1} X_k^+, \sum_{k=n^2+1}^{(n+1)^2-1} X_k^-\right), \tag{155}$$

it is easy to find an upper bound of the form

$$E(M_n^2) \le 4n^2B, \tag{156}$$

where B is a finite constant. From (142),

$$P(M_n > n^2t) \le 4B/n^2t^2, \tag{157}$$

for any $t > 0$, which implies

$$M_n/n^2 \to 0 \quad \text{w.p.1} \qquad \text{as} \quad n^2 \to \infty. \tag{158}$$

Now let $k \to \infty$ in the following expressions,

$$(\overline{X}_{n^2} - M_n/n^2)(n^2/k) \le \overline{X}_k \le (\overline{X}_{n^2} + M_n/n)(n^2/k), \tag{159}$$

where $n^2 \to \infty$ in accordance with $n^2 \le k < (n+1)^2$, and $n^2/k \to 1$. From (153) and (158), we have completed the proof of the strong law.

Important Remark. Both the weak and strong laws hold under much weaker

* Because the remainder of the proof is technical in nature and uses ideas not needed elsewhere in this book, reading this paragraph may not be necessary.

conditions than we have stated. Inspection of our proofs will show we need only the X_n to be *uncorrelated*. More important for the needs of this book is that the strong law holds even when $\sigma^2 = \infty$, as does the converse; i.e., for finite constant μ,

$$\overline{X}_n \to \mu \quad \text{w.p.1} \qquad \text{implies} \quad E(X_n) = \mu.$$

Obviously, the method of proof would have to be modified to cover this case. More general versions of these laws exist, e.g., where the X_n are not required to be identically distributed. Chapters 4 and 5 of Chung [1974] are good references on convergence concepts and general versions of these laws.

1-16 MODES OF CONVERGENCE; THE CENTRAL LIMIT THEOREM

The strong and weak laws of large numbers are examples of two modes of convergence that we now define in terms of a sequence of random variables $\{Y_n, n = 1, 2, \ldots\}$, where F_n denotes the distribution function of Y_n, $n = 1, 2, \ldots$.

There are various ways in which a sequence $\{Y_n\}$ may converge to a random variable Y. If

$$P\{\{\omega\}: \lim_{n\to\infty} Y_n(\omega) = Y(\omega)\} = 1, \tag{160}$$

we say that $\{Y_n\}$ *converges w.p.1* to Y and denote this by

$$Y_n \to Y \quad \text{w.p.1} \qquad \text{as} \quad n \to \infty. \tag{161}$$

If for any positive $\varepsilon > 0$,

$$\lim_{n\to\infty} P\{| Y_n - Y | > \varepsilon\} = 0, \tag{162}$$

we say that $\{Y_n\}$ *converges in measure* to Y and denote this by

$$Y_n \xrightarrow{P} Y \qquad \text{as} \quad n \to \infty. \tag{163}$$

For the strong and weak laws, $Y_n = \overline{X}_n$ and Y is the constant μ. In the study of stochastic processes, there are many situations where convergence to a nonconstant random variable occurs. In Chapter 3, for example, *absorption probabilities* are found for Markov chains. If Y_n denotes the "state" of the chain after n transitions and absorption into some absorbing state is certain,

$$Y_n \to Y \quad \text{w.p.1} \qquad \text{as} \quad n \to \infty,$$

where the probability that Y is equal to an absorbing state is the absorption probability for that state.

Another mode of convergence is convergence in distribution. If

$$\lim_{n\to\infty} F_n(t) = F(t) \tag{164}$$

for all points t where F is continuous, and F is a (possibly defective) distribution function, we say that $\{Y_n\}$ *converges in distribution* to Y, denoted either by

$$Y_n \xrightarrow{D} Y \quad \text{or} \quad F_n \xrightarrow{D} F \qquad \text{as} \quad n \to \infty, \tag{165}$$

where Y simply denotes a random variable with distribution F. The convergence is called *proper* if F (and Y) are proper. Sometimes we mix this notation and write

$$Y_n \xrightarrow{D} F \qquad \text{as} \quad n \to \infty. \tag{166}$$

Notice that convergence in distribution as defined in (164) is a conventional mathematical limit of a sequence of deterministic functions F_n. Thus the *distributions* of the Y_n get "close" to F and to each other as n increases. This does *not* imply that the Y_n get close to Y or to each other in any sense. For example, a sequence of i.i.d. random variables with common distribution F obviously converges in distribution (to F). The Y in (165) is an artifact; it does not represent any particular random variable defined on the sample space of $\{Y_n\}$. It is introduced simply for notational convenience. On the other hand, the Y in (161) and (163) does denote a particular random variable on the same sample space of $\{Y_n\}$.†

An important example of convergence in distribution is known as the

Central Limit Theorem. For a sequence of i.i.d. random variables $\{X_n\}$ with finite mean μ and finite variance σ^2,

$$\sqrt{n}(\overline{X}_n - \mu)/\sigma \xrightarrow{D} N(0, 1), \tag{167}$$

where $\overline{X}_n$ is the sample mean (139).

The Central Limit Theorem is very important in statistics where the sample mean is frequently used for estimation and testing purposes. This theorem accounts for the "central role" the normal distribution plays in much of the statistics literature. Generalized versions of the Central Limit Theorem are useful for estimating various quantities from a simulation of a stochastic process.

As with the laws of large numbers, we have stated only the most elementary version of the Central Limit Theorem. For a good general exposition on this topic, see Chapter 7 of Chung [1974].

Using transform methods, it is not difficult to prove the Central Limit Theorem. However, we omit a proof here because it is essentially analytic (rather than probabilistic) in nature and the technique used is not needed for the rest of this book.

We mention one final mode of convergence. If

$$\lim_{n\to\infty} E\{| Y_n - Y |^2\} = 0, \tag{168}$$

† Technically, this is not quite true; Y denotes any member of a family of random variables that are equal w.p.1.

we say that Y_n converges to Y in *quadratic mean* and denote this by

$$Y_n \xrightarrow{\text{q.m.}} Y \quad \text{as} \quad n \to \infty. \tag{169}$$

If the square in (168) is replaced by p, $0 < p < \infty$, we have what is called L^p *convergence*.

Just as the strong law of large numbers implies the weak law, the modes of convergence we have defined are logically related. The relationship is

$$\begin{array}{c} \text{w.p.1} \Rightarrow P \Rightarrow D. \\ \Uparrow \\ \text{q.m.} \end{array}$$

Our use of Chebyshev's inequality to prove the weak law in Section 1-15 is an application of q.m. $\Rightarrow P$. In the special case where Y is a constant, convergence modes P and D are equivalent.

PROBLEMS

1-1. *Inclusion, exclusion.* Let $\mathcal{A}$ be the set of all subsets of a finite sample space consisting of n points. Any subset can be represented by an n-digit binary number, where a 1 in position j means that the jth point is included in the subset and a zero in position j means that the jth point is excluded from the subset. Use this representation to argue that the number of subsets (number of elements of $\mathcal{A}$) is 2^n.

1-2. *Venn's diagram* represents sets of points as areas. Intersections of sets are overlapping areas. For example, we can represent sets A, B, and AB as follows:

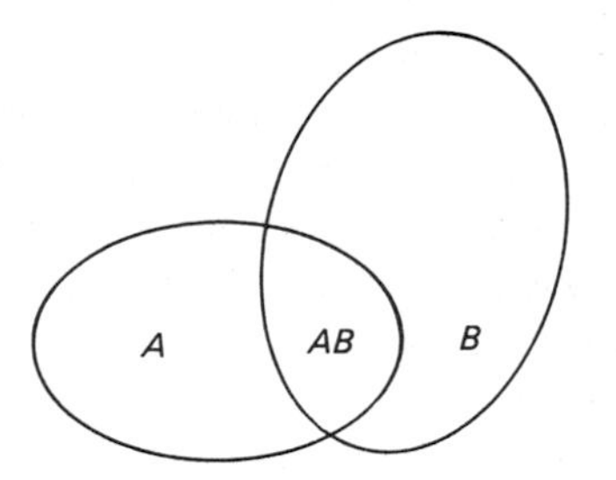

We see from Venn's diagram that the sum $P(A) + P(B)$ counts the contribution of AB twice. Use this idea to derive (9).

1-3. *Continuation.* Draw Venn's diagram for three events, A, B, C, and their intersections. Use it to determine an expression for $P(A \cup B \cup C)$.

1-4. Let A, B, and C be three arbitrary events. Find expressions for each of the following events:

(a) Only A occurs.

(b) None of the events (A, B, C) occurs.

(c) At least one of the events occurs.
(d) At least two of the events occur.
(e) Exactly one of the events occurs.

1-5. A (hypothetical) survey of newspaper reading in the Bay Area involved three papers: the *Chronicle* (C), the *Examiner* (E), and the *Tribune* (T). Results were as follows:

20 percent read C (and possibly E, T, or both)
16 percent read E
14 percent read T
8 percent read both C and E
5 percent read both C and T
4 percent read both E and T
2 percent read all three

(a) What percentage read none of the three papers?
(b) Of those who read at least one of the papers, what percentage read the *Chronicle*?
(c) Of those who read at least two of the papers, what percentage read the *Tribune*?

1-6. An urn contains n balls, each of which is either black or not black. The number of black balls in the urn is one of the integers $0, 1, \ldots, n$. Suppose that we consider these $(n + 1)$ possibilities to be equally likely. A ball is selected at random from the urn. Show that the probability the ball is black is $P(B) = 1/2$. (Use conditional probability. Verify for $n = 3$ first. Can you generalize?)

1-7. For the urn in Problem 1-6, it is also true that each ball is either white or not white. Following the same line of reasoning, we may conclude that $P(W) = 1/2$. Similarly, for yellow balls, $P(Y) = 1/2$. However, these events are mutually exclusive. Hence,

$$P(B \cup W \cup Y) = P(B) + P(W) + P(Y) = 3/2$$

Comment.

1-8. A coin is flipped and a die is rolled. How many combined outcomes are there?

1-9. How many committees of size 4 may be chosen from among 11 individuals? What is your answer if one of the members is to be made chairman?

1-10. Consider the set of three-digit numbers (to the base 10), e.g., 678 and 010.
(a) How many numbers are there with no repetition of digits?
(b) How many numbers are there with 1 repetition of digits?
(c) A three-digit number is selected at random. What is the probability it has exactly 1 repetition of digits?

1-11. A four-digit number is selected at random. What is the probability it has exactly two pair of digits? (For example, 0110.)

1-12. A hand of five cards is drawn at random from an ordinary deck. What is the probability the hand contains exactly two pairs?

1-13. If n balls are placed at random into n cells, find the probability that *exactly* one cell remains empty. *Answer:* $\binom{n}{2} n!\, n^{-n}$.

1-14. Let A and B be independent events. Show that A and B^c are independent. Show that A^c and B^c are also independent.

1-15. From (18), it is tempting to say the following: B is independent of A if the occurrence of A "tells us nothing" about whether B occurs. Why is this misleading?

1-16. At a recent party, a bearded psychologist sat next to you and related his latest research finding. It has been conclusively demonstrated, he said, that when people call the outcome of a coin toss, they are more likely to say "heads" than "tails." He concluded that it is better to be the one to flip the coin than to call it. Without questioning his finding on human behavior, comment on his conclusion. What two assumptions have you made about the behavior of the coin?

1-17. Two cards are drawn from a deck of cards in order. Find the probability that the second card is an ace by conditioning on what the first card is.

1-18. For any finite collection of events $A_1, \ldots, A_n$ show that

$$P\left(\bigcup_{i=1}^{n} A_i\right) \leq \sum_{i=1}^{n} P(A_i). \tag{P.1}$$

This result is valid when $\{A_i\}$ is a countable collection; in this form, it is known as *Boole's inequality*.

1-19. A brewery has two bottling lines. Line A bottles at a rate of 600 per hour, of which 3 percent have short fill. Line B bottles at a rate of 400 per hour, of which 2 percent have short fill. The lines operate for the same number of hours per day. A bottle selected "at random" from the brewery's production is found to have short fill. What is the probability that this bottle came from line A?

1-20. *Genetics*. Labrador retrievers may be any one of three colors, black (B), yellow (Y), or chocolate (C). The color of a dog is determined by an (unordered) pair of genes: BB, BY, etc. Suppose that black is dominant to yellow, and yellow is dominant to chocolate. That is, a dog with any of the gene pairs BB, BY, or BC will be black; a dog with either of the pairs YY or YC will be yellow. Only dogs with the pair CC will be chocolate. Each parent contributes one gene to a puppy, which is equally likely to be either of the genes of that parent. The contribution of one parent is independent of the other.

When necessary to answer the questions below, assume the following model: All unordered gene pairs are equally likely for each parent and independent between parents.

(a) Suppose one parent has pair BC and the other has pair YC. What is the probability that a puppy will be black, yellow, or chocolate?

(b) What is your answer to (a) if we know only the color of the parents to be black and yellow?

(c) Given that a yellow mother has a yellow puppy, what is the probability that the father (unseen but assumed to be a Labrador) is black?

1-21. *Continuation*. In (a) above, the probability of a black, yellow, or chocolate puppy was determined. For the questions below, let these probabilities be p_1, p_2, and p_3, respectively. In the same litter, assume that the color of each puppy is independent of the colors of the others.

(a) What is the probability that a litter of size n will have exactly x black puppies, $x = 0, 1, \ldots, n$?

(b) Given a litter of size n has exactly x black puppies, what is the probability that it has exactly y yellow puppies, $y = 0, 1, \ldots, n - x$?

(c) Combine (b) and (c) to show that in a litter of size n, the probability of x black,

y yellow, and z chocolate puppies is

$$\binom{n}{x,\ y,\ z} p_1^x p_2^y p_3^z, \tag{P.2}$$

where x, y, and z are nonnegative integers with $x + y + z = n$, and

$$\binom{n}{x,\ y,\ z} = \frac{n!}{x!\,y!\,z!}.$$

(d) Obtain the result (P.2) by a direct argument similar to the way the binomial distribution (27) was obtained. This is the *trinomial* distribution for three categories (colors) of objects (puppies). For the binomial, there are two categories, e.g., black or not black in (a). For k categories, we would have the *multinomial* distribution.

1-22. *Continuation.* In Problem 1-20(c), we considered the color of only one puppy in a litter. Suppose a mother known to be "pure" yellow (YY) has a litter of n puppies, all yellow.

(a) Find the probability that the father is black.

(b) Show that the probability in (a) is a decreasing function of n.

1-23. *Continuation.* Suppose that we don't know the color of either parent.

(a) Show that the probability that a puppy is black, yellow, or chocolate is 5/9, 1/3, and 1/9, respectively.

(b) Given that the first puppy in a litter is chocolate, what is the probability that the second puppy is also chocolate?

(c) In general, if we do not know the gene pairs of the parents, the colors of the puppies in the same litter are *not* independent. Explain why this is true.

1-24. *Continuation.* Notice that the probabilities found in problem 23(a) for the color of a puppy are not the same as the corresponding probabilities for the parents, which are 1/2, 1/3, and 1/6, respectively. The problem is this: The assumption that unordered gene pairs are equally likely is inconsistent with the assumption that the gene pairs of the parents are independent.

Show that if the *ordered* gene pairs are equally likely for each parent (where the first gene comes from the male and the second from the female grandparent), they are also equally likely for the puppies. (In practice, the independence assumption is false because breeders favor pure color lines, e.g., BB with BB.)

(The probability that a puppy is each color here is the same as in Problem 1-23(a). This is true because the "gene pool" has the same proportion (1/3) of each color. Otherwise, these probabilities would be different.)

1-25. Suppose a coin is tossed twice where the probability of a head is p on the first toss and q on the second. Assume that outcomes on different tosses are independent. Let X be the number of heads that occur on the two tosses.

(a) Find the probability of each of the events $\{X = 0\}$, $\{X = 1\}$, and $\{X = 2\}$.

(b) Show that $P(X = 0) = P(X = 2)$ if and only if $p + q = 1$.

(c) Use (b) to show that if the tosses are independent, the three events in (a) can *never* be equally likely.

1-26. Derive the *geometric distribution* (28).

1-27. *Continuation.* Consider a sequence of independent experiments (trials) where the probability of success on each experiment is p, $0 < p < 1$, and $q = 1 - p$. We terminate the experiment with the kth success. Let X be the number of failures prior to the kth success. Show that

$$P(X = x) = \binom{k + x - 1}{x} p^k q^x, \qquad x = 0, 1, \ldots. \tag{P.3}$$

(P.3) is called the *negative binomial* distribution. [*Hint:* Refer to how the binomial distribution (27) was derived.]

1-28. In *sampling inspection,* a lot of size N items is accepted or rejected based on the composition of a sample of size n items drawn from the lot, where all $\binom{N}{n}$ possible samples are equally likely. Suppose the lot contains D defective items, and let X be the number of defective items in the sample, where X is a random variable. Show that

$$P(X = x) = \binom{D}{x}\binom{N - D}{n - x}\Big/\binom{N}{n}, \qquad x = 0, 1, \ldots, n, \tag{P.4}$$

where (P.4) is called the *hypergeometric distribution.* [We define a combinatorial expression $\binom{n}{x}$ to equal zero if $x > n$.]

1-29. Let $g(t)$, $t > 0$, be a nonnegative monotone decreasing function with the properties $0 < g(1) < 1$ and $g(t + x) = g(t)g(x)$ for all $t, x > 0$. Show that for some $\lambda \in (0, \infty)$, $g(t)$ is of form

$$g(t) = e^{-\lambda t}, \qquad t > 0.$$

(*Hint:* First show $g(t) = [g(1)]^t$ for any *rational* $t > 0$.)

1-30. Let X and Y be independent where $X \sim \exp(\lambda)$ and $Y \sim \exp(\mu)$. Find $P(U > u)$, where $U = \min(X, Y)$. Use this result to show that $U \sim \exp(\lambda + \mu)$. [*Example:* Let X be the life of a component in the absence of a catastrophe (e.g., shock caused by an earthquake), where, on the occurrence of a catastrophe, the component fails. Let Y be the time until the occurrence of the (first) catastrophe.]

1-31. *Continuation.* Let $Z \sim \exp(\gamma)$ be independent of X and Y, and define $V = \min(Y, Z)$.

(a) Find $P(U > u, V > v)$ for $u, v > 0$. Consider the cases $u > v$ and $u < v$ separately.

(b) Differentiate (a) with respect to u and v, obtaining a "joint density" function, $g(u, v)$ say.

(c) Show that $\int_0^\infty \int_0^\infty g(u, v)\, du\, dv = (\gamma + \lambda)/(\gamma + \lambda + \mu)$. Why isn't this integral equal to 1?

(d) Use the example in Problem 1-30 to give an interpretation to U and V here. Find $P(U = V)$. Does this help explain (c)? *Remark.* In the literature, the joint distribution (or tail) found in (a) is one version of what is called the *bivariate exponential* distribution.

1-32. Find the mean of the exponential distribution directly from (32).

1-33. Use (48) to derive (54).

1-34. Derive (61).

1-35. Suppose X has a uniform distribution on $(0, b)$. Use (61) to find $V(X)$. Compare with Table 1-1.

1-36. Derive (63).

1-37. Suppose one card is drawn from an ordinary deck of 52 cards. Let $X = 1$ if the card is an ace, $X = 0$ otherwise. Let $Y = 1$ if the card is a spade, $Y = 0$ otherwise. Use (63) to show that X and Y are uncorrelated.

1-38. Derive (76).

1-39. An objection to covariance as a measure of "association" of two random variables is that it depends on scale parameters, e.g., whether measurements are made in feet or meters. The *correlation coefficient* of random variables X and Y is defined as

$$\rho_{xy} = \frac{\text{Cov}(X, Y)}{\sigma(X)\sigma(Y)}. \tag{P.5}$$

Show that for $U = aX + b$ and $V = cY + d$,

$$\rho_{uv} = \rho_{xy}, \tag{P.6}$$

provided that $a > 0$ and $c > 0$. What happens if one or both of $\{a, c\}$ are negative?

1-40. *Continuation of Problem 1-27.* Let X_1 be the number of failures prior to the first success and X_i be the number of failures between success $i - 1$ and i, $i = 2, 3, \ldots, k$.

(a) How are the X_i distributed? Are they independent?

(b) From the definition of the X_i,

$$X = \sum_{i=1}^{k} X_i, \tag{P.7}$$

where X has the negative binomial distribution (P.3). Use (P.7) to show that

$$E(X) = kq/p, \tag{P.8}$$

$$V(X) = kq/p^2. \tag{P.9}$$

1-41. Two dice are rolled. Find the probability that the sum of the outcomes is X, $X = 2, 3, \ldots, 12$, as an application of (79).

1-42. Show that

$$V\left(\sum_{i=1}^{n} X_i\right) = \sum_{i=1}^{n} V(X_i) + 2 \sum_{1 \le i < j \le n} \text{Cov}(X_i, X_j). \tag{P.10}$$

1-43. Suppose that the first N integers are listed in random order. If integer i is ith on the list, e.g., 5 is 5th on the list, we say i is matched. Let $X_i = 1$ if the ith integer is matched, $X_i = 0$ otherwise, and let

$$X = \sum_{i=1}^{N} X_i$$

be the total number of matches on the list.

(a) For any integer i, find $E(X_i)$, $E(X_i^2)$, and

$$V(X_i) = (N - 1)/N^2.$$

(b) For $i \neq j$, find $E(X_iX_j)$ and

$$\text{Cov}\,(X_i, X_j) = [N^2(N - 1)]^{-1}.$$

(c) From (a), (b), and (P.10), show that

$$E(X) = V(X) = 1.$$

1-44. *Continuation of Problem 1-28.* Let $X_i = 1$ if the ith item in the sample is defective, $X_i = 0$ otherwise. Now

$$X = \sum_{i=1}^{n} X_i,$$

the number of defective items in the sample, has distribution (P.4).

(a) Find the mean and variance of X_i and, for $i \neq j$, Cov (X_i, X_j).

(b) Use (a) and (P.10) to show that

$$E(X) = np, \tag{P.11}$$

$$V(X) = npq\left(\frac{N - n}{N - 1}\right), \tag{P.12}$$

where $p = D/N$ and $q = 1 - p$. Compare with the mean and variance of the binomial distribution with the same n and p.

1-45. Let X be the number of aces and Y the number of spades in a bridge hand. Show that X and Y are uncorrelated. (*Hint:* See Problem 37.) Why are X and Y *not* independent?

1-46. Let X be uniformly distributed on (0, 1) and $Y = X^2$.

(a) From (51), show that

$$E(Y) = 1/3.$$

(b) Show that $P(Y > y) = 1 - \sqrt{y}$, $0 < y < 1$.

(c) Find $E(Y)$ from (b) and (48).

(d) From (b), show that the density function of Y, $g(y)$, is

$$g(y) = 1/2\sqrt{y}, \qquad 0 < y < 1.$$

(e) Find the density function of Y from (93).

(f) Find $E(Y)$ from $g(y)$ and (47).

1-47. Find the mean and variance of the binomial distribution from its generating function (104).

1-48. For the geometric distribution q^xp, $x = 0, 1, \ldots,$

(a) show that the generating function is

$$G(z) = \sum_{x=0}^{\infty} q^xpz^x = p/(1 - qz). \tag{P.13}$$

(b) Use (P.13) to find the mean and variance of the geometric distribution.

1-49. *Animal trapping.* An expedition is sent to the Himalayas with the objective of catching a pair of wild yaks for breeding. Assume yaks are loners and roam about the Himalayas at random. The probability p, $0 < p < 1$, that a given trapped yak is male is independent of prior outcomes. Let N be the number of yaks that must be caught until a pair is obtained.

(a) By conditioning on the outcome of the first catch, show that

$$E(N) = 1 + p/q + q/p,$$

where $q = 1 - p$.

(b) Find the variance $V(N)$.

1-50. Derive the conditional covariance relation

$$\text{Cov}(X, Y) = E\{\text{Cov}[(X \mid Z), (Y \mid Z)]\} + \text{Cov}\{E(X \mid Z), E(Y \mid Z)\} \quad \text{(P.14)}$$

1-51. *Continuation*. For $X = Y$, show that (P.14) reduces to (126).

1-52. Use $X = Y$ in (136) to obtain (138) for the special case $r = 2$.

1-53. *Continuation of Problem 1-31*. Use (136) to obtain the covariance of U and V.

2

RENEWAL THEORY AND REGENERATIVE PROCESSES

In subsequent chapters, we will study probability models where collections of random variables interact in various ways. For example, queueing theory deals with the interaction of an arrival process of customers with customer service requirements and characteristics of a service facility.

A *stochastic process* is simply a collection of random variables. The arrival process of customers at a queue may be modeled as a stochastic process, and the service requirements may be modeled as another stochastic process. The interaction of these processes generates other stochastic processes, e.g., a sequence of customer delays.

A *stochastic model* defines the interacting processes and the nature of their interaction, which, in principle, determines the characteristics of generated processes. The analysis of generated processes is the subject of the latter half of this book and of many of the examples from this point on.

Stochastic processes are the building blocks of stochastic models. Renewal processes and Markov processes, the subject matter of Chapters 2–4, are important as building blocks and in their own right.

In this chapter, we develop renewal theory and properties of regenerative processes, with applications introduced as examples and as problems at the end. Since a renewal process is a particular kind of stochastic process, we begin with a general discussion of stochastic processes.

2-1 DEFINITION OF A STOCHASTIC PROCESS; REALIZATIONS

We call a collection of random variables a *stochastic process* when the collection is infinite. The collection will be indexed in some way, e.g., by calling $\{X(t): t \in T\}$ a stochastic process, we mean that $X(t)$ is a random variable for each t belonging to some *index set* T. [More generally, $X(t)$ may be a vector of random variables.]

Usually, the index set is either an interval e.g., $T = [0, \infty)$, or a countable set such as the set of nonnegative integers, $T = \{0, 1, \ldots\}$, and t will denote *time*. With these alternatives for T, we have either a *continuous-time* or *discrete-time* process, respectively. Of course, we can index in other ways, e.g., t could denote distance from some reference point, or even be a vector with (say) time and distance components.

Whether continuous or discrete, we will refer to a particular point on the time axis as *epoch* t, e.g., 10:30 A.M., in contrast to a time interval, e.g., an hour. In the discrete case, we often will replace t by n and, particularly in this case, place the index as a subscript, i.e., X_n rather than $X(n)$.

Possible values of $X(t)$ will be called *states*, and by the phrase "the process $\{X(t)\}$ is in state x at epoch t" we mean that the event $\{X(t) = x\}$ has occurred.

Stochastic processes made an (unheralded) appearance toward the end of Chapter 1. In Section 1-15, $\{X_n; n = 1, 2, \ldots\}$ is a sequence of i.i.d. random variables, denoting the outcomes on an infinite sequence of trials. In our new terminology, this is an example of a discrete-time process. In terms of $\{X_n\}$, we generated another stochastic process, $\{\overline{X}_n\}$. In Section 1-14, $\{I(x): x \in [0, \infty)\}$ defined by (131) is a continuous-time process.

We say a stochastic process $\{X(t)\}$ is *well defined* if the joint distribution of $X(t_1), \ldots, X(t_k)$ is determined for every finite set of epochs, $t_1, \ldots, t_k$. For example, the joint distribution of any finite collection from the $\{X_n\}$ above is, by independence, the product of the marginals. This determines joint distributions for $\{\overline{X}_n\}$. Thus "determined" means that joint distributions are well defined, not necessarily that they are written down somewhere.

Any collection of random variables (stochastic process) is defined on some sample space Ω. It might appear that this leads to awkward complications in modeling. For example, in Section 1-2, we considered rolling a die once and chose $\Omega = \{1, \ldots, 6\}$. If we were to roll a die twice, we would have to redefine Ω, e.g., we could choose a set of 36 points corresponding to all possible (ordered) outcomes. Any outcome on the first roll would now correspond to a *set* of 6 points.

Does this mean that every time we change a model by changing assumptions about distributions, joint distributions, or even about what random variables to include, that we have to go back and formally redefine Ω, together with the class of events $\mathcal{A}$ and probability measure P? Fortunately, the answer is no!

What is actually done is just the opposite. If a stochastic process is *well defined* in the sense just specified, it can be shown that there always exists (infinitely many) $(\Omega, \mathcal{A}, P)$ that could be chosen. (When stochastic processes interact, the joint collection of all random variables in all of these processes is the stochastic process that must be well defined.) Thus, it is not necessary to specify Ω at all. Rather, it is an undefined set that remains in the background.

At this point, it is tempting to ask, "if Ω doesn't really matter, why bother to mention it?" Aside from the fact that in certain technical discussions Ω does matter, there is a very practical reason for having (even an undefined) Ω at our disposal: When it is convenient to do so, we can fix a sample point $\omega \in \Omega$. Examples: The statement of the strong law of large numbers is in terms of the behavior of $\overline{X}_n(\omega)$ for fixed ω. Similarly, in Section 1-14, we integrated $\int_0^\infty I(x, \omega)\, dx$ at an arbitrary fixed ω.

For any fixed ω, every random variable is a real number and every event either occurs or does not occur. When we wish to emphasize the dependence of a stochastic process on ω as well as epoch t, it is denoted by $\{X(t, \omega)\}$. For fixed ω, a stochastic process is a function of time, and is called a *realization* or *sample path* of the process. Sometimes, we think of a realization as a function that *evolves* over time. When we collect data on an ongoing stochastic process (or simulate it) we are observing a portion of some realization. (Since we observe only over a finite interval, there will be a set of ω that could give rise to the same set of observed values.)

For example, if we conduct an experiment involving n independent trials, we observe values of $X_1, \ldots, X_n$ from a (potentially) infinite sequence of i.i.d. random variables. From the observed values, we may wish to infer something about the magnitude of their mean. This is a problem of statistical inference.

Modeling a stochastic phenomenon as either a discrete or continuous time process is sometimes a matter of choice. For example, we may model the Dow Jones stock market average as a discrete-time stochastic process $\{X(t): t = 0, 1, \ldots\}$, where $X(t)$ is the average at the end of day t, and $t = 0$ denotes today. On the other hand, if we wish to consider monitoring the stock market continuously, the corresponding model would be a continuous-time process.

Another distinction we wish to make is whether the *random variables* $X(t)$ are discrete or continuous. We call $\{X(t)\}$ a *discrete-state* process if for some countable set $\mathcal{S}$, $P(X(t) \in \mathcal{S}) = 1$ for all t. The usual alternative is to permit $X(t)$ to vary over a continuum of possible values, in which case $\{X(t)\}$ is called a *continuous-state* process.

An important special case of discrete-state processes is when $\mathcal{S} = \{0, 1, \ldots\}$, i.e., the random variables are nonnegative integers. This often occurs when we are counting the number of some kind of "event" over time. Conceptually, this is the starting point for what we will call *renewal processes*.

The term "event" in renewal theory has a meaning different from that in probability theory. In renewal theory, an *event* is an abstract term for something that occurs from time to time at (usually random) points on the (usually continuous) time axis, where we will assign probabilities to the *number* of events that occur in particular intervals, rather than to the events themselves.

To provide some physical intuition, it is convenient, at least initially, to think of events as arrivals of (say) customers, orders, telephone calls, or vehicles.

2-2 RENEWAL PROCESSES

Suppose we model the arrival of transit vehicles, buses say, where we observe the arrival of buses at a bus stop beginning at epoch 0, some arbitrary point on the time axis.

Let X_1 be the arrival epoch of the first bus and, for $n \geq 2$, let X_n be the inter-arrival time between buses $n - 1$ and n. Let $Z_0 = 0$ and

$$Z_n = \sum_{j=1}^{n} X_j, \qquad n = 1, 2, \ldots, \tag{1}$$

be the arrival epoch of the nth bus. Since X_1 is not an interarrival time between buses, it may have a distribution different from $X_j, j \geq 2$.

This description provides a conceptual framework for an introduction to renewal theory where, in general, events called *renewals* occur over time, and the times between successive events are *independent random variables*. In the above notation, let $X_1, X_2, \ldots$ be independent, nonnegative random variables where $P(X_1 \leq t) = A(t)$ and $P(X_j \leq t) = F(t), j = 2, 3, \ldots$.

Thus, except possibly for X_1, the random variables have the same distribution. As defined in (1), Z_n is called the *renewal epoch* of the nth renewal, $n = 1, 2, \ldots$. Assume $F(0) < 1$, i.e., $P(X_j > 0) > 0$ for $j > 1$. This implies that the mean (expected value) of F is strictly positive. Let μ be the reciprocal of this mean: For $j > 1$,

$$E(X_j) = 1/\mu > 0. \tag{2}$$

To fix ideas, we will refer to the bus example frequently. However, it is important to realize that the abstract formulation is really what counts. Renewal theory has a broad range of applications where it would be artificial at best to call renewals arrivals. (See Examples 2-1 and 2-2.) In applications, one tries to find epochs at which some more complex stochastic process "starts over" or "renews" itself. These processes are called *regenerative*.

Renewal theory is primarily concerned with limiting behavior (as $t \to \infty$). We will emphasize results that can be interpreted as *time averages*. To obtain these results, we need to represent certain quantities of interest and their distributions.

For notational convenience, assume A and F are differentiable with densities $a(t)$ and $f(t)$, respectively. The discrete case, particularly when the X_j are integral-valued, is also important. Fortunately, the limits obtained here and in the next section are valid for this case as well; only the representation of certain distributions would be different.

To find an expression for the distribution of Z_n, recall that the distribution of Z_2 is a convolution,

$$P(Z_2 \le t) = P(X_1 + X_2 \le t) = \int_0^t A(t - u)f(u)\, du. \tag{3}$$

The corresponding density function, the derivative of (3), is

$$\int_0^t a(t - u)f(u)\, du. \tag{4}$$

The distribution (3) is called the *convolution* of A and F and is denoted by $A * F(t)$. Similarly, (4) is called the *convolution* of the densities a and f. Notice that the mathematical form of a convolution depends on whether we are dealing with distribution functions or density functions. Clearly, (3) is equivalent to

$$P(Z_2 \le t) = \int_0^t F(t - u)a(u)\, du. \tag{5}$$

For further discussion of convolutions, see Section 1-10.

Denote $F^{(2)}(t) = F * F(t)$ and, in general, $F^{(n)}(t) = F^{(n-1)} * F(t)$, which is called the *n-fold* convolution of F. The corresponding density function is denoted by $f^{(n)}(t)$. Note that Z_n is the sum of n independent random variables that can be added in any order we choose. By writing $Z_n = X_1 + (X_2 + \cdots + X_n)$ the distribution of Z_n has the representation

$$P(Z_n \le t) = A * F^{(n-1)}(t), \qquad n = 1, 2, \ldots, \tag{6}$$

where we define $A * F^{(0)} = A$.

Permitting X_1 to have a distribution different from the others will be convenient later. In the bus example, the process may start at some point (epoch 0) that is between two buses.

Now let $M(t)$ be the number of renewals that have occurred by epoch $t \ge 0$, i.e.,

$$M(t) = \max\{n \ge 0: Z_n \le t\}, \tag{7}$$

where from $Z_0 = 0$, $M(t) = 0$ if $X_1 > t$. For each t, $M(t)$ is an integral-valued random variable. Since t is continuous, $\{M(t): t \ge 0\}$ is a continuous-time stochastic process, called a *renewal process*. It is interesting to observe that the renewal process is generated (determined) by $\{X_n: n = 1, 2, \ldots\}$, a discrete-time stochastic process where the random variables are continuous.

Since the event $\{M(t) \ge j\}$ occurs if and only if $\{Z_j \le t\}$, i.e., they are the same event and have the same probability,

$$P\{M(t) \ge j\} = P\{Z_j \le t\} = A * F^{(j-1)}(t). \tag{8}$$

Also, if we let $I_j = 1$ if $Z_j \le t$, $I_j = 0$ otherwise, $M(t)$ has the representation

$$M(t) = \sum_{j=1}^{\infty} I_j, \tag{9}$$

and expected value

$$m(t) = E\{M(t)\} = E\left\{\sum_{j=1}^{\infty} I_j\right\}$$

$$= \sum_{j=1}^{\infty} E\{I_j\}$$

$$= \sum_{j=1}^{\infty} A * F^{(j-1)}(t), \tag{10}$$

where $m(t)$ is called a *renewal function*.

It will turn out that other important quantities can be represented in terms of $m(t)$ or $M(t)$. For example:

$$Z_{M(t)} = \sum_{j=1}^{M(t)} X_j = \text{last renewal epoch prior to } t \text{, and}$$

$$Z_{M(t)+1} = \sum_{j=1}^{M(t)+1} X_j = \text{first renewal epoch after } t.$$

If the renewal process is a model of an arrival process of buses, a customer arriving at a bus stop at t is primarily interested in the time until the next bus arrives, and may also be interested in the elapsed time since the last bus arrived (see Figure 2-1). These quantities, called the *excess* $Y(t)$ and the *age* $U(t)$, respectively, are

$$Y(t) = Z_{M(t)+1} - t, \qquad \text{and} \tag{11}$$

$$U(t) = t - Z_{M(t)}, \qquad t \ge 0. \tag{12}$$

Finally, the *spread* is

$$X(t) = Y(t) + U(t) = Z_{M(t)+1} - Z_{M(t)}, \qquad t \ge 0. \tag{13}$$

Viewing (11), (12), and (13) as functions of time, e.g., $\{Y(t): t \ge 0\}$, we have defined three *continuous-time* stochastic processes.

We now investigate the behavior of $M(t)/t$ for large t. For this purpose, assume that $\mu > 0$, i.e., that F has finite mean. (Of course, we would usually expect this to be true. In certain applications, however, we may wish to permit $\mu = 0$.)

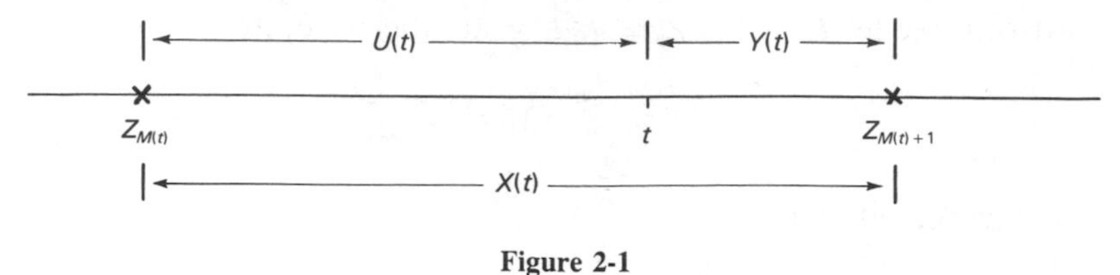

Figure 2-1

As $t \to \infty$, it is easily shown that $M(t) \to \infty$ with probability 1 (w.p.1). Hence, as $t \to \infty$, we apply the strong law of large numbers (see Section 1-15) to show that

$$Z_{M(t)}/M(t) \to 1/\mu \quad \text{w.p.1.} \tag{14}$$

Notice that (14) is essentially the strong law with n replaced by $M(t)$, a random variable. To prove (14), consider the set of ω for which both of the following hold:

$$\lim_{n\to\infty} Z_n(\omega)/n = 1/\mu, \tag{15}$$

$$\lim_{t\to\infty} M(t, \omega) = \infty. \tag{16}$$

Since each of these limits holds on a set of ω with probability 1, the intersection of these sets also has probability 1. For any ω where (15) and (16) hold, (14) follows immediately. Hence, (14) holds w.p.1.

Thus, when using the strong law, we can obtain results by fixing a point in the sample space and treating expressions such as (14) as ordinary limits. The weak law of large numbers provides much less information. From the weak law, all we can say is that for any *particular* sufficiently large n, $P(|\, Z_n/n - 1/\mu \,| > \varepsilon)$ can be made arbitrarily small. Note that in (14), n is replaced by a random variable $M(t)$ that "tends" to get large as $t \to \infty$. By itself, the weak law cannot tell us how the left-hand side of (14) behaves as $t \to \infty$.

Similarly, as $t \to \infty$,

$$Z_{M(t)}/Z_{M(t)+1} \to 1 \quad \text{w.p.1.} \tag{17}$$

From (17) and Figure 2-1,

$$Z_{M(t)}/Z_{M(t)+1} < Z_{M(t)}/t \le 1, \quad \text{and} \tag{18}$$

$$\lim_{t\to\infty} Z_{M(t)}/t = 1 \quad \text{w.p.1.} \tag{19}$$

Now write

$$\frac{M(t)}{t} = \frac{Z_{M(t)}}{t} \cdot \frac{M(t)}{Z_{M(t)}}, \tag{20}$$

and apply (14) and (19) to (20). We have the random-variable version of

Theorem 1: The Elementary Renewal Theorem

(Random-variable version) $$\lim_{t\to\infty} \frac{M(t)}{t} = \mu \quad \text{w.p.1, and} \tag{21}$$

(Expected-value version) $$\lim_{t\to\infty} \frac{m(t)}{t} = \mu, \tag{22}$$

where these limits are $\mu = 0$ when the mean of F, $1/\mu = \infty$.

While we have shown (21) only for $\mu > 0$, the extension to the case $\mu = 0$ by a truncation method is easy; see Problem 2-10. For the expected-value version (22), we simply replace $M(t)$ by its expected value, the renewal function $m(t)$, where (22) is a limit in the ordinary sense. However, (22) does not follow immediately from (21). A proof of (22) is presented in Section 2-14.

Because of (21) and (22), μ can be thought of as the number of renewals per unit time (in the long run), and is called the *renewal rate*. Thus, the renewal rate is simply the reciprocal of the mean time between renewals.

We wish to emphasize the following: The limits (21) and (22) do not depend on the distribution of X_1 [we could even have $E(X_1) = \infty$], and as should be evident in the proof of (21), they do not depend on whether F has a density function. Thus, F could be discrete or a mixture of continuous and discrete.

2-3 REWARD AND COST MODELS; CUMULATIVE PROCESSES

Many applications of renewal theory involve costs or rewards (negative costs). For example, consider a taxi where X_j represents the time between the collection of fares $(j - 1)$ and j, and X_j includes the duration of the trip resulting in fare j. We would expect the size of fare j (a reward) to depend on X_j.

Motivated by the above, let $\{X_j\}$ be defined as in Section 2-2 and let R_j be the *reward* associated with X_j, $j = 1, 2, \ldots$. For simplicity, assume $A = F$, i.e., that X_1 has the same distribution as the others. Assume $\{R_j\}$ is a sequence of i.i.d. random variables where, for each j, R_j may depend on X_j but is independent of $\{X_i : i \neq j\}$. The R_j are not sign constrained, i.e., they can be positive or negative. Finally, assume $E\{|R_j|\} < \infty$ and $E\{X_j\} < \infty$. *Costs* are simply negative rewards.

In the taxi example, the fare is normally collected at the end. With this in mind, we may define the *cumulative reward* $C(t)$ on the interval $[0, t]$ to be

$$C(t) = \sum_{j=1}^{M(t)} R_j, \tag{23a}$$

where the stochastic process $\{C(t)\}$ is called a *cumulative process*. In some applications, e.g., fares collected in advance, it is more natural to define $C(t)$ as

$$C(t) = \sum_{j=1}^{M(t)+1} R_j. \tag{23b}$$

In either case, the *expected* (cumulative) *reward* on $[0, t]$ is denoted by

$$c(t) = E\{C(t)\}. \tag{24}$$

The reward per unit time (*reward rate*) is defined to be the limit of $C(t)/t$ as $t \to \infty$. In the taxi example, this would be the income from fares per hour. How would we expect this limit to be related to properties of R_j and X_j? Two candidates for this limit are $E\{R_j/X_j\}$ and $E\{R_j\}/E\{X_j\}$. These expressions are usually different, even when R_j and X_j are independent.

The desired limit is easily obtained by writing (23a) as

$$\frac{C(t)}{t} = \frac{\sum_{j=1}^{M(t)} R_j}{M(t)} \cdot \frac{M(t)}{t}. \tag{25}$$

Applying the strong law of large numbers to the first factor on the right-hand side of (25) [as was done to obtain (14)], and (21) to the second, we obtain the limit (26) below. Using (23b), the limit is the same, and we have the random-variable version of

Theorem 2: The Renewal-Reward Theorem. If $E\{|R_j|\} < \infty$ and $\mu > 0$,

(Random-variable version) $\lim_{t\to\infty} C(t)/t = \mu E(R_j)$

$$= E(R_j)/E(X_j) \quad \text{w.p.1, and} \tag{26}$$

(Expected-value version) $\lim_{t\to\infty} c(t)/t = E(R_j)/E(X_j) \quad \text{for } j \geq 2,$ (27)

where (27) is a limit in the ordinary sense.

These results do not depend on whether rewards are collected at the beginning of intervals, at the end, or they accumulate in the more complicated way described under *partial rewards* later in this section. The expected-value version (27) is proven in Section 2-15. (The first cycle can be different, i.e., X_1 and R_1 can have different distributions.)

We will call renewal intervals *cycles*. With this terminology, the *reward* (or *cost*) *rate* (reward or expected reward per unit time) is *the expected cumulative reward (cost) per cycle divided by the expected cycle length.*

Example 2-1: Age Replacement Policies

A component may fail while in service, at a cost c_f. Alternatively, we may choose to replace a working component that has been in service a length of time T. The advantage of doing this is that the cost of replacing a working component is only c_r, where $c_r < c_f$. Let *policy T* be a policy that replaces a component on failure or on reaching age T, whichever occurs first. Consider the cost rate of policy T, $k(T)$, applied to a sequence of components that are put in service one at a time, where the next component is put in service immediately after replacing the previous one. Let $\{V_j\}$ be a sequence of i.i.d. nonnegative random variables, where V_j is the in-service life of component j. (Components fail only while in service.) Under policy T, the time spent in service of component j is

$$X_j = \min(V_j, T), \tag{28}$$

where $\{X_j\}$ is a sequence of i.i.d. random variables. The cost of the jth replacement, R_j, depends on V_j as follows:

$$\begin{aligned} R_j &= c_r \qquad \text{if } V_j > T \\ &= c_f \qquad \text{if } V_j \leq T. \end{aligned} \tag{29}$$

From (26), or in terms of expected values, (27),

$$k(T) = E(R_j)/E(X_j), \tag{30}$$

where

$$E(R_j) = c_f P(V_j \leq T) + c_r P(V_j > T). \tag{31}$$

Since X_j is nonnegative, we can find $E(X_j)$ from its tail distribution,

$$E(X_j) = \int_0^\infty P(X_j > x)\, dx = \int_0^T P(X_j > x)\, dx, \tag{32}$$

where, from (28), the range of integration may be reduced because $P(X_j > x) = 0$ for $x > T$. Also observe that for $t < T$, $X_j > t$ if and only if $V_j > t$. Therefore,

$$E(X_j) = \int_0^T P(V_j > x)\, dx, \quad \text{and} \tag{33}$$

$$k(T) = \frac{c_f P(V_j \leq T) + c_r P(V_j > T)}{\int_0^T P(V_j > x)\, dx}. \tag{34}$$

Depending on the distribution of V_j, the optimal age replacement policy would be the value of T that minimizes (34). For numerical examples, see Problems 2-3 and 2-4.

Partial Rewards and Reward Rates

In some applications, rewards may not occur at one end of an interval, but may accumulate during the interval. The most common case will be where the reward associated with X_j accumulates at some constant rate r during X_j or some portion of X_j.

The portion of the reward associated with $X(t)$ acquired between $Z_{M(t)}$ and t is called a *partial reward*. Now R_j is not sign constrained, and partial rewards need not accumulate in a monotone fashion. Thus there is not necessarily a simple relationship between a partial reward and $R(t) \equiv R_{M(t)+1}$, the reward associated with $X(t)$.

We will make assumptions about partial rewards such that (26) and (27) will still be valid and the applications we have in mind will be covered. Assume that each R_j is the result of positive and negative reward streams that accumulate in a monotone fashion during X_j. Let $P_j \geq 0$ and $N_j \geq 0$ be the magnitudes of the total accumulation of these streams during X_j, respectively, where $\{P_j\}$ and $\{N_j\}$ are each sequences of i.i.d. random variables, $E(P_j) < \infty$ and $E(N_j) < \infty$, and

$$R_j = P_j - N_j. \tag{35}$$

For example, we can think of a reward as the difference between revenue and cost streams that accumulate continuously over time. The "smallest" P_j and N_j that satisfy the requirements above are $P_j = \max(R_j, 0)$ and $N_j = -\min(R_j, 0)$.

We assume that the partial reward accumulated between $Z_{M(t)}$ and t is bounded above by $P(t)$ and bounded below by $-N(t)$, where these quantities are the positive and negative rewards associated with $X(t)$; e.g., given $M(t) = n$, $P(t) = P_{n+1}$. Thus

$$\sum_{j=1}^{M(t)} R_j - N(t) \leq C(t) \leq \sum_{j=1}^{M(t)} R_j + P(t), \tag{36}$$

no matter how rewards accumulate during an interval. Under these assumptions, (26) follows immediately, and (27) is still true.

The Renewal-Reward Theorem applies to any quantity that accumulates over time, in accordance with our assumptions; it need not have a monetary interpretation. The next two examples present important applications of this result, where the R_j are now *lengths of time*.

Example 2-2: The Alternating Renewal Process

Suppose a machine breaks down and is repaired. Let $\{X_{uj}\}$ be the sequence of times during which the machine is "up" prior to breakdown, and $\{X_{dj}\}$ be the sequence of times during the machine is "down" prior to completion of repair. Assume these sequences are independent of each other, and that the random variables in each sequence are i.i.d. Define $\{X_j\}$ by $X_j = X_{uj} + X_{dj}$, $j = 1, 2, \ldots$. Thus if an up period begins at epoch 0, the first breakdown occurs at epoch X_{u1}, the first repair at epoch X_1, the second breakdown at epoch $X_1 + X_{u2}$, and so on. Call X_j a repair *cycle*, and consider the corresponding renewal process generated by $\{X_j\}$, where $M(t)$ is the number of repairs completed by epoch t.

For every $t \geq 0$, let

$$\begin{aligned} I(t) &= 1 \quad \text{if the machine is up at epoch } t, \\ &= 0 \quad \text{otherwise, where} \end{aligned}$$

$$C(t) = \int_0^t I(u)\, du \tag{37}$$

is the cumulative amount of up time during $[0, t]$. The corresponding rewards per cycle are

$$R_j = \int_{Z_{j-1}}^{Z_j} I(u)\, du = X_{uj}, \qquad j = 1, 2, \ldots. \tag{38}$$

Note that there is a partial reward which is the up time between $Z_{M(t)}$ and t. The reward rate is $r = 1$ when the machine is up, and $r = 0$ otherwise. From (26), the *fraction of time* that the machine is up is

$$\lim_{t\to\infty} C(t)/t = E(X_{uj})/E(X_j). \tag{39}$$

Remark. The stochastic process $\{I(t): t \geq 0\}$ alternates between two "states" and, under our assumptions, is called an *alternating renewal process.*

Further Remark. We could introduce a monetary reward in Example 2-2 by assuming that reward accumulates at rate \$1.00/unit time when the machine is up. Doing this is artificial, however, and diverts attention from the important interpretation of (39) as a fraction of time.

Example 2-3: The Limiting Distribution of Excess

Recall the earlier notation where the *excess* $Y(t)$ is the time until the first renewal after epoch t. For any fixed $y \geq 0$ and *every* $t \geq 0$, now let

$$\begin{aligned} I(t) &= 1 \quad \text{if } Y(t) \leq y \\ &= 0 \quad \text{otherwise,} \end{aligned}$$

where a realization of $\{I(t)\}$ is shown in Figure 2-2.

Define $C(t)$ as in Example 2-2.

$$C(t) = \int_0^t I(u)\, du, \tag{40}$$

where now from Figure 2-2 we have

$$R_j = \begin{cases} X_j & \text{if } X_j \leq y, \\ y & \text{if } X_j > y. \end{cases} \tag{41}$$

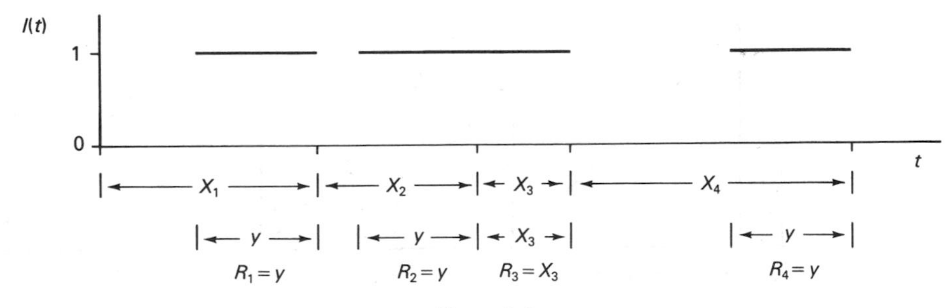

Figure 2-2

Combining the two expressions in (41), we have

$$R_j = \min (X_j, y), \tag{42}$$

an expression of the same form as (28). From (26), we have

$$\lim_{t\to\infty} C(t)/t = E(R_j)/E(X_j) = \mu \int_0^y P(X_j > x)\, dx. \tag{43}$$

We now interpret this result in terms of the expected reward,

$$c(t) = E\{C(t)\} = E\left\{\int_0^t I(u)\, du\right\} = \int_0^t E\{I(u)\}\, du = \int_0^t P[Y(u) \le y]\, du. \tag{44}$$

From the expected-value version (27),

$$\lim_{t\to\infty} c(t)/t = \lim_{t\to\infty} \int_0^t P[Y(u) \le y]\, du/t = \mu \int_0^y P(X_j > x)\, dx, \qquad y \ge 0. \tag{45}$$

i.e., from the Renewal-Reward Theorem, this limit must be the same as (43).

While the limits are the same, their interpretations are different. Notice that $c(t)/t$ in (45) is a distribution function (with argument y) for every t, and the limit is also a distribution function. [What is the limit of the right side of (45) as $y \to \infty$?] Thus while (43) is a proportion of time, (45) tells us that *$c(t)/t$ converges in distribution* to the same limit. For this reason, we call the result [either (43) or (45)] a *limiting distribution*. This interpretation of (45) is discussed in more detail in the next section.

In addition to the limiting distribution of excess found in Example 3, which is a time average of an indicator process $\{I(t)\}$, we may also wish to find the time average of the process $\{Y(t)\}$ itself, where a realization of $\{Y(t)\}$ is shown in Figure 2-3.

For $\{Y(t)\}$, let $C(t) = \int_0^t Y(u)du$ and R_j be the integral of $\{Y(t)\}$ over the jth cycle, which is the area of a triangle, $R_j = X_j^2/2$. We have

$$\lim_{t\to\infty} \int_0^t Y(u)\, du/t = \mu E(X_j^2)/2$$

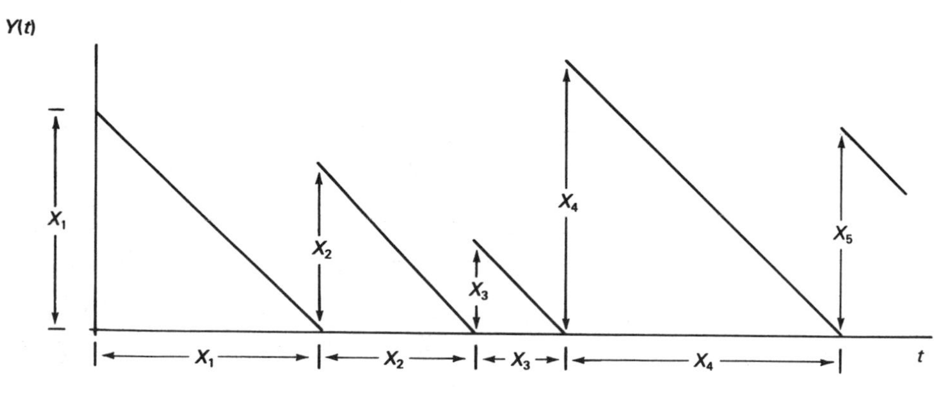

Figure 2-3

We will show as (49) that this limit is also the mean of limiting distribution (43). This equality is an illustration of the important Theorem 11 in Section 2-12.

2-4 OBSERVING A PROCESS AT RANDOM; THE EQUILIBRIUM DISTRIBUTION AND THE INSPECTION PARADOX

In the bus example, suppose a customer arrives at a bus stop "at random" in accordance with a uniform distribution on $(0, t)$. Given that the customer arrives at epoch u, this customer's waiting time for the next bus is the excess random variable, which has some distribution

$$P[Y(u) \leq y], \qquad y \geq 0.$$

The *unconditional* waiting time distribution for the next bus is the integral of the conditional distribution above with respect to the density function of the customer's arrival epoch, $1/t$.

$$\int_0^t P[Y(u) \leq y]\, du/t, \qquad y \geq 0.$$

which is $c(t)/t$ in (45).

As $t \to \infty$, this waiting time distribution converges to the limit in (45). Thus, if customers do not take into account the bus schedule, if any, limit (45) is (at least approximately) their waiting time distribution.

Alternatively, we can think of a "random observer" collecting data on some process, e.g., $\{Y(t)\}$. The observer, who can arrive anywhere in $(0, t)$, has no effect on the process being observed. Given the observer arrives by epoch t, the observer's arrival epoch is uniformly distributed on $(0, t)$. The distribution of what the observer "sees" is found by letting $t \to \infty$.

The Equilibrium Distribution

The limit (45) is a distribution defined in terms of an integral of the tail of the distribution F. Because this relationship is important and will occur in other contexts, we introduce some special notation and terminology.

For any nonnegative random variable $X \sim F$, with mean $0 < 1/\mu < \infty$, we define a random variable $X_e \sim F_e$, where F_e is the limit in (45), and is called the *equilibrium distribution* of F:

$$P(X_e \leq x) \equiv F_e(x) = \mu \int_0^x F^c(u)\, du, \qquad x \geq 0. \tag{46}$$

This terminology arises from the construction of stationary (equilibrium) versions of processes; see Section 2-16. The corresponding density function is

$$f_e(x) = F_e'(x) = \mu F^c(x) = \mu[1 - F(x)], \qquad x \geq 0. \tag{47}$$

The transform of F_e may also be found in terms of the transform of F:

$$\tilde{F}_e(s) \equiv E(e^{-sX_e}) = \int_0^\infty e^{-sx} f_e(x)\,dx = \mu \int_0^\infty e^{-sx}[1 - F(x)]\,dx = \mu[1 - \tilde{F}(s)]/s. \tag{48}$$

Remark. When F has density f, the last equality in (48) follows from

$$\tilde{F}(s) = E(e^{-sX}) = \int_0^\infty e^{-sx} f(x)\,dx = s \int_0^\infty e^{-sx} F(x)\,dx,$$

where we have integrated by parts. However, (48) is valid even when F does not have a density; see (21) in the Appendix.

The mean and higher moments of X_e may be found from (47) by applying equation (138) in Chapter 1, e.g.,

$$E(X_e) = \int_0^\infty x f_e(x)\,dx = \mu \int_0^\infty x F^c(x)\,dx = \mu E(X^2)/2. \tag{49}$$

Thus the mean waiting time (for the next bus) depends not only on the frequency of service μ, but also on the regularity of service, as measured by $E(X^2) = V(X) + 1/\mu^2$. That is, for fixed μ, the mean waiting time increases with the *variance* of the inter-arrival distribution F.

We now illustrate these results for two important special cases; the reader should verify details:

If X is a *constant,* $P(X = c) = 1$, F is a unit step function at c, and, from (46), X_e has a *uniform distribution* on $(0, c)$. Thus, if buses arrive every $c = 20$ minutes, the mean waiting time for a bus is $E(X_e) = 10$ minutes.

If X is *exponential,* $X \sim F = \exp(\mu)$, we have from (46) that X_e is distributed as *the same exponential.* This is another consequence of the memoryless property of the exponential, and will be discussed in more detail later in connection with the Poisson process. Thus if buses arrive with frequency 3 per hour, the mean waiting time is 20 minutes, not 10. It can be shown that X and X_e have the same distribution *only if* X is exponential; see Problem 2-15.

Example 2-4: The Limiting Distribution of Spread

We apply the methods of Section 2-3, where $\{I(t)\colon t \geq 0\}$ is now defined as follows for any fixed $x \geq 0$:

$$\begin{aligned} I(t) &= 1 \quad \text{if } X(t) \leq x, \\ &= 0 \quad \text{otherwise, where} \end{aligned}$$

$$R_j = \int_{Z_{j-1}}^{Z_j} I(u)\,du = \begin{cases} X_j & \text{if } X_j \leq x, \\ 0 & \text{otherwise,} \end{cases}$$

and it follows that

$$E(R_j) = \int_0^\infty P(R_j > u)\,du = \int_0^x P(R_j > u)\,du = \int_0^x P[X_j \in (u, x]]\,du$$
$$= \int_0^x [F(x) - F(u)]\,du = xF(x) - \int_0^x F(u)\,du.$$

Applying (27) gives

$$\lim_{t\to\infty} \int_0^t P[X(u) \le x]\,du/t = \mu E(R_j) = \mu[xF(x) - \int_0^x F(u)\,du] \equiv F_s(x), \qquad x \ge 0, \tag{50}$$

where the limit in (50) is called the *limiting distribution of spread*. As with the equilibrium distribution, we introduce special notation F_s to denote this limit as a function of F. When F has density f, F_s has the density

$$F_s'(x) \equiv f_s(x) = \mu x f(x), \qquad x \ge 0. \tag{51}$$

The factor x in (51) is called *length biasing*. The intuitive idea is that a customer (or a random observer) is more likely to land in a long interval than a short one (there is more to land in). This is reflected by biasing (weighting) the density function at each x by interval length x.

While the particular form of (50) and (51) is the result of taking a limit (i.e., averaging over all time), the tendency to land in or "select" long intervals is also valid at arbitrary finite epochs. This tendency is called the *inspection paradox*. In fact, if $A = F$, i.e., X_1 has the same distribution as the others, it can be shown that $X(t)$ is *stochastically larger* than X_j for *every* t. (See Problem 2-22.)

For another view of length biasing, consider a realization of the spread process $\{X(t)\}$ as shown in Figure 2-4. Notice that the numerical values that $\{X(t)\}$ passes through persist for lengths of time that are proportional (equal) to these values. The process $\{X(t)\}$ is quite different from $\{X_j\}$.

The phenomena discussed in this section have important practical consequences, and the equilibrium distribution in particular plays a prominent role in queueing theory.

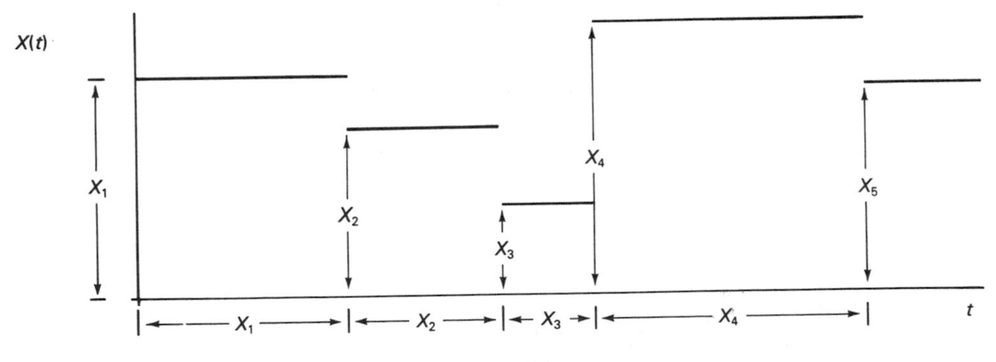

Figure 2-4

It is also important to develop the ability to visualize how processes evolve over time, as shown in Figures 2-2 through 2-4.

2-5 THE DISCRETE CASE—BATCH EFFECTS

The distributions (45) and (50) were obtained by averaging over all time, where time was treated as continuous. Sometimes we are interested in averages over a discrete set of points, usually the integers. This occurs in the context of batch arrivals in queueing theory, and we will have occasion to call upon results from this section in subsequent chapters. Other applications occur, e.g., in survey sampling, where a lack of awareness of the effects discussed here can lead to interesting mistakes; see Problem 2-33.

What we will call *batch effects* is a discrete version of length biasing, and this section can be viewed as an application of earlier results. Instead, we will develop what we need from scratch.

Let's consider a specific example. Suppose ν_i is the number of children in family i, $i = 1, 2, \ldots$, where $\{\nu_i\}$ is a sequence of i.i.d. random variables with distribution

$$P(\nu_i = j) = \alpha_j, \qquad j = 0, 1, \ldots,$$

with mean $E(\nu_i) \equiv E(\nu)$. In general, we will call ν_i the *size* of *batch i*.

What fraction of children

(a) are members of batches of size k, $k = 1, 2, \ldots$, and
(b) are jth born, $j = 1, 2, \ldots$?

To answer question (b), let $\nu_i(j)$ be the number of children in family i who are jth born. Obviously, for $i \geq 1$,

$$\begin{aligned} \nu_i(j) &= 1 \qquad \text{if } \nu_i \geq j, \\ &= 0 \qquad \text{otherwise.} \end{aligned} \tag{52}$$

For the first n families, the fraction of children who are jth born is

$$\frac{\sum_{i=1}^{n} \nu_i(j)}{\sum_{i=1}^{n} \nu_i}, \qquad j = 1, 2, \ldots, \tag{53}$$

where the numerator and denominator are sums of i.i.d. random variables. Now divide the numerator and denominator of (53) by n and apply the strong law of large numbers as $n \to \infty$:

$$\lim_{n\to\infty}\frac{\sum_{i=1}^{n} \nu_i(j)/n}{\sum_{i=1}^{n} \nu_i/n} = \frac{P(\nu \geq j)}{E(\nu)} = \frac{\sum_{i\geq j} \alpha_i}{E(\nu)} \equiv P(J = j), \tag{54}$$

where J denotes the order of birth of a randomly selected child.

The mean of (54) is

$$E(J) = \sum_{j=1}^{\infty} jP(J = j) = \frac{\sum_{j=1}^{\infty}\sum_{i\geq j} j\alpha_i}{E(\nu)} = \frac{\sum_{i=1}^{\infty}\sum_{j=1}^{i} j\alpha_i}{E(\nu)}$$

$$= \frac{\sum_{i=1}^{\infty} (i)(i+1)\alpha_i/2}{E(\nu)} = \frac{E\{(\nu)(\nu+1)\}}{2E(\nu)}. \tag{55}$$

Results (54) and (55) are different from (45) and (49) because we are averaging over a discrete set of points, *not* because the ν_i are discrete. In fact, Example 2-3 is valid for any distribution F of a nonnegative random variable with finite mean.

It is straightforward to answer (a) using the same approach. The fraction of children in families of size k is

$$P(K = k) \equiv \frac{k\alpha_k}{E(\nu)}, \qquad k = 1, 2, \ldots, \tag{56}$$

where K is the number of children in the family of a randomly selected child. Notice the *batch size* biasing in (56).†

Of course, the context of children in families has no bearing on these results.

2-6 THE POISSON PROCESS

We generated a renewal process $\{M(t)\}$ from a sequence of independent inter-event times $\{X_j\}$, where $M(t)$ is the number of some kind of events (renewals) in interval $[0, t]$.

A Poisson process is a stochastic process $\{\Lambda(t): t \geq 0\}$ with properties specified in this section, where $\Lambda(t)$ may also be interpreted as the number of events in $[0, t]$. From these properties, we will show that the inter-event times of a Poisson process are i.i.d. exponential. Hence, a Poisson process is "merely" a special case of a renewal process.

Nevertheless, the Poisson process is extremely important in model building. For example, it is used as a model of arriving customers at a queue, breakdowns

† If the X_j in Example 2-4 are integral-valued, (50) is equivalent to (56).

of components in a complex system, the occurrence of catastrophes such as fires or earthquakes, and the flow of traffic on a highway.

The Poisson process deserves and will get careful treatment. We will continue to think in terms of arrivals and will speak of $\{\Lambda(t)\}$ as an arrival process.

If $\{\Lambda(t): t \geq 0\}$ is an arrival process where $\Lambda(t)$ is the number of arrivals in $[0, t]$, then $\Lambda(t)$ must be integral-valued for every t. Furthermore, if we were to observe some realization of an arrival process, i.e., $\Lambda(t, \omega)$ as a function of t at some point ω in the sample space, it would be a nondecreasing step function.

The number of arrivals in the interval $(t, t + h]$, $\Lambda(t + h) - \Lambda(t)$, is called the *increment* (or change) in the process $\{\Lambda(t)\}$ between t and $t + h$. An arrival process is said to have *independent increments* if the number of arrivals in disjoint intervals are independent random variables. In particular, if $\{\Lambda(t)\}$ has this property, $\Lambda(t)$ and $[\Lambda(t + h) - \Lambda(t)]$ are independent random variables.

A *Poisson process* is an arrival process $\{\Lambda(t)\}$ with two properties:

1. $\{\Lambda(t)\}$ has independent increments, and for some finite $\lambda > 0$,
2. $P\{\Lambda(t) = j\} = e^{-\lambda t}(\lambda t)^j/j!, \quad j = 0, 1, \ldots .$ (57)

Thus, for each t, $\Lambda(t)$ has a Poisson distribution with mean λt, the product of the interval length t and *arrival rate* λ.†

From properties 1 and 2 it is easy to find the generating function and from it, the distribution of $\Lambda(t + h) - \Lambda(t)$:

$$\begin{aligned} E\{z^{\Lambda(t+h)}\} &= E\{z^{\Lambda(t)}z^{(\Lambda(t+h)-\Lambda(t))}\} \\ &= E\{z^{\Lambda(t)}\}E\{z^{(\Lambda(t+h)-\Lambda(t))}\}, \end{aligned} \tag{58}$$

from independent increments. From (57), the generating functions of $\Lambda(t)$ and $\Lambda(t + h)$ are $e^{\lambda t(z-1)}$ and $e^{\lambda(t+h)(z-1)}$, respectively. In solving (58), we have

$$E\{z^{[\Lambda(t+h)-\Lambda(t)]}\} = e^{\lambda h(z-1)}, \tag{59}$$

which implies that

$$\Lambda(t + h) - \Lambda(t) \sim P(\lambda h). \tag{60}$$

Notice that the distribution of the increment, the number of arrivals in $(t, t + h]$, is *independent* of the epoch t where the increment is located. This property is called

Property 3: *stationary increments*.

It is easy to show that the probability of two or more arrivals in an interval of length h gets small in the following sense as $h \to 0$: For every $t \geq 0$,

$$\lim_{h\to 0} P\{\Lambda(t + h) - \Lambda(t) \geq 2\}/h = \lim_{h\to 0} (1 - e^{-\lambda h} - \lambda h e^{-\lambda h})/h = 0. \tag{61}$$

† It may be a bit premature to call λ an arrival rate. However, from properties 1 and 2 and the stationary increments property which they imply, it is easily shown that $\lim_{t\to\infty} \Lambda(t)/t = \lambda$ w.p.1.

Equation (61) is called

Property 4: *orderliness*.

For every $t \geq 0$, it is also easily shown that

$$\text{Property 5: } \lim_{h \to 0} P\{\Lambda(t + h) - \Lambda(t) = 1\}/h = \lambda.$$

Property 4 means that arrivals occur "one at a time," and Property 5 means that the probability of one arrival in a short interval is (approximately) proportional to interval length. These properties, combined with stationary and independent increments, are the intuitive reasons often given for expecting an arrival process to be Poisson. An example would be the arrival of telephone calls during a "busy hour," where many potential callers behave independently, and each caller is unlikely to make a call during that hour.

In fact, it can be shown that Properties 1, 3, 4, and 5 together imply Property 2. Furthermore, from Theorem 8 in Section 2-9 it easily follows that Properties 3 and 4 imply Property 5. It follows that an *equivalent definition* of a Poisson process is an arrival process with Properties 1, 3, and 4.

Corresponding to $\{\Lambda(t)\}$, which counts the number of arrivals in intervals, there are arrival epochs, $0 \leq t_1 \leq t_2 \cdots$, and *inter-arrival times*, $\tau_1 = t_1$, $\tau_j = t_j - t_{j-1}$, $j = 2, 3, \ldots$. The event $\{\tau_1 > t\}$ occurs if and only if $\{\Lambda(t) = 0\}$, i.e., they are the same event. Therefore,

$$P(\tau_1 > t) = P\{\Lambda(t) = 0\} = e^{-\lambda t}, \qquad t \geq 0, \tag{62}$$

which is the *tail* of an exponential. That is, τ_1 has exponential density

$$\lambda e^{-\lambda t}, \qquad t \geq 0. \tag{63}$$

The distribution of τ_2 is found by conditioning on τ_1:

$$P(\tau_2 > t \mid \tau_1 = x) = P\{\Lambda(t + x) - \Lambda(x) = 0\} = e^{-\lambda t}, \tag{64}$$

where we used independent increments. (Formally, we used more, i.e., that $\Lambda(t + x) - \Lambda(x)$ is independent of the arrival epochs within $[0, x]$. While this is true, it has not been shown. To overcome this objection, we can derive (64) by conditioning on $\tau_1 \in [x, x + h]$, and then letting $h \to 0$.)

Observe that conditional distribution (64) is independent of the value of τ_1. Thus τ_2 is independent of τ_1 and has the same distribution. Continuing in this manner, we have this result:

$$\{\tau_j\} \text{ is a sequence of i.i.d. random variables with exponential density function (63).} \tag{65}$$

It turns out that a third *equivalent definition* of a Poisson process is an arrival process $\{\Lambda(t)\}$ generated by a sequence $\{\tau_j\}$ of inter-arrival times that have property (65). That is, $\{\Lambda(t)\}$ will have Properties 1 and 2.

Thus a Poisson process is a special case of a renewal process, where $\{\tau_j\}$ has the role of $\{X_j\}$, and $\{\Lambda(t)\}$ has the role of $\{M(t)\}$.

Given $\Lambda(t) = n$, the distribution of arrival epochs and inter-arrival times during $[0, t]$ will change. First consider $P[\Lambda(u) = x \mid \Lambda(t) = n]$ for $u \in [0, t]$ and $x \in \{0, 1, \ldots, n\}$. The corresponding joint event can be written $\{\Lambda(u) = x, \Lambda(t) = n\} = \{\Lambda(u) = x, \Lambda(t) - \Lambda(u) = n - x\}$ and, from independent increments,

$$P\{\Lambda(u) = x, \Lambda(t) - \Lambda(u) = n - x\} = P(\Lambda(u) = x)P\{\Lambda(t) - \Lambda(u) = n - x\}.$$

Hence,

$$P(\Lambda(u) = x \mid \Lambda(t) = n) = \frac{P(\Lambda(u) = x)P\{\Lambda(t) - \Lambda(u) = n - x\}}{P(\Lambda(t) = n)}. \tag{66}$$

The probabilities on the right-hand side of (66) are known and the expression reduces to

$$P(\Lambda(u) = x \mid \Lambda(t) = n) = \binom{n}{x} (u/t)^x(1 - u/t)^{n-x}, \tag{67}$$

a *binomial distribution* for the probability of x "successes" in n trials, where u/t is the success probability. Note that (67) does not involve λ, and that u/t is the ratio of interval lengths.

By partitioning the interval $[0, t]$ into k regions, the joint distribution of the number of arrivals in each region is easily shown to be a multinomial distribution, where the probability p_i of an arrival falling into region i is the length of the region divided by t.

Now consider the actual arrival epochs during $[0, t]$, given $\Lambda(t) = n$. From (67), and for any $u \in [0, t]$,

$$P(t_1 \le u \mid \Lambda(t) = 1) = P(\Lambda(u) = 1 \mid \Lambda(t) = 1) = u/t, \tag{68}$$

i.e., given exactly one arrival during $[0, t]$, the arrival epoch is *uniformly distributed* on that interval.

The extension of (68) to n arrival epochs, given $\Lambda(t) = n$, is important in applications. It is simple to state in terms of *unordered* arrival epochs: Given $0 \le t_1 \le t_2 \cdots \le t_n$, there are $n!$ ways of "relabeling" these n random variables. Let these ways be equally likely and $t_1', t_2', \ldots, t_n'$ be the corresponding relabeled unordered variables. In particular, $P(t_j' = t_i) = 1/n$ for $i = 1, \ldots, n$ and $j = 1, \ldots, n$. For example, when $n = 3$, $P(t_1' = t_3, t_2' = t_1, t_3' = t_2) = 1/6$.

To obtain the desired extension, we begin with the *ordered* epochs. Let $I_1, \ldots, I_n$ be n disjoint subintervals of $[0, t]$ such that for $i < j$, I_i is to the left of I_j. Let $R = [0, t] - \bigcup_{i=1}^{n} I_i$ be the region consisting of the points of $[0, t]$ that are not in any of the subintervals. Let l_i be the length of I_i, $i = 1, \ldots, n$ and $l_{n+1} = t - \sum_{i=1}^{n} l_i$ be the length of R.

The event

$$\{t_i \in I_i,\ i = 1, \ldots, n \mid \Lambda(t) = n\}$$

is equivalent to the event

{exactly one arrival occurs in each subinterval I_i, $i = 1, \ldots, n$, and no arrivals occur in $R \mid \Lambda(t) = n$}.

Hence, by the methods used to obtain (67), we can write

$$P(t_i \in I_i,\ i = 1, \ldots, n \mid \Lambda(t) = n) = n!(l_{n+1}/t)^0 \prod_{i=1}^{n} (l_i/t)^1$$

$$= n! \prod_{i=1}^{n} (l_i/t), \tag{69}$$

where (69) is a particular term in the multinomial distribution referred to above with $k = n + 1$ regions.

Considering $(t_1, \ldots, t_n)$ to be a vector in n-dimensional space, (69) states that $(t_1, \ldots, t_n)$ falls in a set of volume $\prod_{i=1}^{n} l_i$ with probability *proportional* to that volume. Hence, the joint density function of $(t_1, \ldots, t_n)$ is the proportionality constant,

$$n!/t^n, \qquad 0 \le t_1 \le \cdots \le t_n \le t. \tag{70}$$

Notice that the region for which (70) holds is one of $n!$ equivalent subregions of the n-dimensional positive "quadrant." [Since for any i, $P(t_i = t_{i+1}) = 0$, we can ignore "double counting" on boundaries.] Hence, by randomly relabeling the $\{t_i\}$ as described previously, we spread out the density function over the entire quadrant, and we have proven

Theorem 3. Given $\Lambda(t) = n$, the joint density function of the n *unordered* arrival epochs in $[0, t]$, $t'_1, \ldots, t'_n$, is

$$1/t^n, \qquad 0 \le t'_i \le t, \qquad \text{for } i = 1, \ldots, n, \tag{71}$$

i.e., the unordered epochs are i.i.d. *uniformly distributed* on $[0, t]$.

Remark. In statistics, one sometimes takes a sample of n i.i.d. random variables $X_1, \ldots, X_n$ from a density f. The joint density at $X_i = x_i$, $i = 1, \ldots, n$ is $\prod_{i=1}^{n} f(x_i)$. Suppose we order the sample, from smallest to largest. The joint density for the corresponding n *order statistics* is $n! \prod_{i=1}^{n} f(x_i)$ over the restricted region $x_1 \le \cdots \le x_n$. In the proof of Theorem 3, we have reversed this process.

Equation (67) and the corresponding multinomial result are examples of what we will call a *random partition*: Let Y have a Poisson distribution with mean $E(Y) = \lambda$. Given $Y = n$, the n "objects" are partitioned into k mutually exclusive and exhaustive categories, where p_i is the probability that an object falls in category i, $i = 1, 2, \ldots, k$. Let Y_i be the number of objects in category i, $i = 1, \ldots, k$, where $Y = \sum_{i=1}^{k} Y_i$. Given $Y = n$, the joint distribution of $(Y_1, \ldots, Y_k)$ is multinomial. (We assume that the p_i do not depend on the value of Y.)

For simplicity, we will discuss in detail a random partition only in the binomial case: $k = 2$, $p_1 = p$, and $p_2 = 1 - p = q$. Hence,

$$P(Y_1 = i \mid Y = n) = \binom{n}{i} p^i q^{n-i}, \qquad i = 1, \ldots, n. \tag{72}$$

We now find the joint distribution of Y_1 and Y_2 for any $i, j \geq 0$.

$$\begin{aligned} P(Y_1 = i, Y_2 = j) &= P(Y_1 = i, Y_2 = j \mid Y = i + j)P(Y = i + j) \\ &= P(Y_1 = i \mid Y = i + j)P(Y = i + j) \\ &= \binom{i+j}{i} p^i q^j e^{-\lambda}\lambda^{(i+j)}/(i + j)! \\ &= (e^{-\lambda p}(\lambda p)^i/i!)(e^{-\lambda q}(\lambda q)^j/j!) \\ &= P(Y_1 = i)P(Y_2 = j), \end{aligned} \tag{73}$$

i.e., Y_1 and Y_2 are independent and Poisson! The argument extends in a straightforward manner and we have this important result:

Theorem 4. Let $Y \sim P(\lambda)$ be randomly partitioned into k categories where p_i is the probability of category i and Y_i is the number in category i, $\sum p_i = 1$ and $\sum Y_i = Y$. The random variables $Y_1, \ldots, Y_k$ are mutually independent with $Y_i \sim P(\lambda p_i)$, $i = 1, \ldots, k$.

Example 2-5: Partitioning a Poisson Process into Independent Poisson Processes

As an application of Theorem 4, we partition a Poisson process $\{\Lambda(t)\}$ at rate λ into independent Poisson processes. For example, suppose we partition an arrival stream of cars by make of car (Chevrolet, Ford, etc.). Let p equal the probability that a car is a Ford, independent of the makes of other cars. Let $\{\Lambda_1(t)\}$ be the arrival stream of Fords and $\{\Lambda_2(t)\}$ be the arrival stream of other cars, where for every t, $\Lambda(t) = \Lambda_1(t) + \Lambda_2(t)$. For any fixed t, $\Lambda_1(t) \sim P(p\lambda t)$ and $\Lambda_2(t) \sim P((1 - p)\lambda t)$, where these random variables are independent. [Partition $\Lambda(t)$ into two categories, the number of Fords by epoch t, and the number of other cars by epoch t.]

We have not yet shown that $\{\Lambda_1(t)\}$ and $\{\Lambda_2(t)\}$ are either independent processes or Poisson processes. For the latter we need to show that each process has independent increments. For the former, we need to show that the increments of one are independent of the increments of the other. To do both, we simply create more categories by partitioning $[0, t]$ into k_1 subintervals for $\{\Lambda_1(t)\}$ and k_2 different subintervals for $\{\Lambda_2(t)\}$, where k_1, k_2, and the particular subintervals are arbitrary. Now partition $\Lambda(t)$ (by make and subinterval) into $k_1 + k_2$ categories. From Theorem 4, both results are immediate.† Thus, $\{\Lambda_1(t)\}$ and $\{\Lambda_2(t)\}$ are independent Poisson processes at rates $p\lambda$ and $(1 - p)\lambda$, respectively.

It is easily shown that the sum of independent Poisson random variables is Poisson. Similarly, if $\{\Lambda_i(t)\}$ are independent Poisson processes, the composite process $\{\Lambda(t)\}$, defined by $\Lambda(t) = \sum \Lambda_i(t)$ for each t, is a Poisson process. The previous example shows that these results are reversible. A more complicated application of Theorems 3 and 4 is Example 2-6, which is also of independent interest.

Example 2-6: The Infinite Channel Queue with Poisson Arrivals; The M/G/∞ Queue

We present this example in a particular context so that it can be readily understood without embarking on a general discussion of queues. Let calls at a telephone switchboard be Poisson at rate λ. Let phone conversations, which are assumed to be independent of each other and of the arrival stream, have distribution function G with mean $1/\mu$, $0 < \mu < \infty$. Assume that the number of phone lines is so large that calls never interfere with each other. Thus, a call that arrives at epoch t_i would begin its conversation immediately on arrival. If the duration of the call's conversation is S_i, the call would be completed at epoch $t_i + S_i$. The shorthand notation for this model is the M/G/∞ queue.

Let $\Lambda(t)$ be the number of calls that arrive during $[0, t]$, and $N(t)$ be the number of calls in progress at epoch t. Given $\Lambda(t) = n$, consider the arrival of any one of the *unordered* calls, which, from (71) has a uniform distribution on $[0, t]$. Given the call arrives at x, the probability that the corresponding conversation is in progress at t is $G^c(t - x)$. By integrating with respect to the arrival distribution, we obtain p_t, the probability that the call is in progress at t,

$$p_t = \int_0^t G^c(t - x)\, dx/t = \int_0^t G^c(u)\, du/t, \tag{74}$$

independent of other calls, where $u = t - x$ and $du = -dx$.

Partition calls that arrive during $[0, t]$, $\Lambda(t)$, into those that are in progress at t, $N(t)$, and those that have departed by t, $\Lambda(t) - N(t)$. Clearly, this is a random partition. It follows immediately that $N(t)$ and $\Lambda(t) - N(t)$ are independent random variables with

$$N(t) \sim P(\lambda t p_t) = P(\lambda \int_0^t G^c(u)\, du), \tag{75}$$

$$\Lambda(t) - N(t) \sim P(\lambda t(1 - p_t)). \tag{76}$$

What happens to the distribution of $N(t)$ as $t \to \infty$?

† We use Theorem 3 to verify that we have a random partition. Given $\Lambda(t) = n$, an unordered arriving car has probability pl/t of being a Ford that arrives in a subinterval of length l, independent of other cars.

As illustrated in Example 2-7, a Poisson process is often a component in some more complex stochastic model.

Example 2-7

Let the arrival process of customers at a train station be Poisson at rate λ, and the arrival process of trains be a renewal process with inter-arrival times X_j, where $P(X_j \le t) = F(t)$ and $E(X_j) = 1/\mu$. We wish to find the fraction of customers for whom the waiting time for the next train does not exceed y (is $\le y$).

Let K_j be the number of customers who arrive during X_j and $K_j(y)$ be the number of them who wait $\le y$. Clearly, given X_j, K_j has a Poisson distribution with mean λX_j. Hence,

$$E(K_j) = E\{E(K_j \mid X_j)\} = E\{\lambda X_j\} = \lambda E(X_j). \tag{77}$$

$K_j(y)$ is the number of arrivals in the right-hand portion of X_j, an interval of length y if $X_j > y$. If $X_j \le y$, $K_j(y) = K_j$. Therefore,

$$\begin{aligned} E\{K_j(y) \mid X_j\} &= \lambda X_j && \text{if } X_j \le y \\ &= \lambda y && \text{if } X_j > y \text{, i.e.,} \end{aligned}$$

$$E\{K_j(y)\} = \lambda E\{\min(X_j, y)\} = \lambda \int_0^y F^c(x)\,dx. \tag{78}$$

By the fraction of customers who wait $\le y$, we mean the limit

$$\begin{aligned} \lim_{n\to\infty} \frac{\sum_{j=1}^{n} K_j(y)}{\sum_{j=1}^{n} K_j} &= \frac{E\{K_j(y)\}}{E(K_j)} = \mu \int_0^y F^c(x)\,dx \\ &= F_e(y) \quad \text{w.p.1,} \end{aligned} \tag{79}$$

the *equilibrium* distribution of F. Compare this result with that in Example 2-3. Also see Section 2-4. Note that to obtain (79) we again used the strong law of large numbers.

For Theorem 4 to apply, the quantity being partitioned must be Poisson and the partition must be random. In Example 2-8, the partition is *not* random.

Example 2-8

Cars arrive at the Oakland Coliseum to attend a rock concert. (Cars love rock!) A traffic cop at the entrance directs even-numbered cars (numbered in their order of arrival) north toward parking lot 1 and odd-numbered cars south toward parking lot 2. Let the arrival process of cars $\{\Lambda(t)\}$ be Poisson at rate λ, and denote the arrival processes of cars heading north and south by $\{\Lambda_n(t)\}$ and $\{\Lambda_s(t)\}$, respectively.

Clearly, this partition is not random. [Given $\Lambda(t) = 2$, what is $\Lambda_n(t)$?] None of the conclusions of Theorem 4 hold. How are the interarrival times of $\{\Lambda_n(t)\}$ distributed?

2-7 STOPPING TIMES FOR A POISSON PROCESS

As Examples 2-6 and 2-7 illustrate, a Poisson process is a component of many stochastic models. In applications, the independent increments property of the Poisson process is often of vital importance for identifying renewal cycles in these models. This permits application of the Renewal-Reward Theorem in order to obtain specific results.

From the independent-increments property, we have that for any $t \geq 0$, $\{\Lambda(t + s) - \Lambda(t)\}$ and the excess $Y(t)$ are independent of $\{\Lambda(h); h \leq t\}$, where $Y(t) \sim \exp(\lambda)$. It is easily shown that if t is replaced by a random variable $\mathcal{T}$, where $\mathcal{T}$ is independent of $\{\Lambda(t)\}$, the conclusions would be the same, where $\mathcal{T}$ is a special case of what we will call a *stopping time*. (In Example 2-7, X_1, $X_1 + X_2, \ldots$, are stopping times.) Before introducing the general concept, we will present an example in considerable detail.

Example 2-9

A single-server *loss system* operates as follows: There is a single server who can serve only one customer at a time. Arriving customers who find the server free are served. Arrivals who find the server busy (serving some other customer) are not served; they are *lost*. Let S_j be the service time of the jth customer served, $j = 1, 2, \ldots$, where the S_j are nonnegative i.i.d. random variables with service time distribution $G(t) = P(S_j \leq t)$ and mean $E(S_j) = 1/\mu$, where $0 < \mu < \infty$. (We will call μ the service *rate*.) Assume the arrival process of customers is Poisson at rate λ.

First, we will find the *fraction of time the server is busy*, f_b. If the server is initially idle (empty), the time until the first arrival is $\tau \sim \exp(\lambda)$. During the interval $(\tau, \tau + S_1)$, the server is busy and arrivals are lost. At $\tau + S_1$, the server becomes idle again. Let I be the length of time the server remains idle. How is I distributed?

Clearly, $I = Y(\tau + S_1)$ is a "randomized" excess of a Poisson process. However, this is a slightly different problem from how $\mathcal{T}$ was defined just prior to this example because $\tau + S_1$ depends on the Poisson process. Nevertheless, because we know that the inter-arrival times of a Poisson process are i.i.d. exponential, the Poisson process in effect "starts over" at τ. Since S_1 is independent of the Poisson process, our problem reduces to the previous case, and we conclude that $I \sim \exp(\lambda)$ and is independent of τ and S_1. In fact, if we let I_j be the idle period preceding completion of service S_j, the sequence $\{I_j, S_j\}$ defines an *alternating renewal process* (Example 2-2), with cycles $X_j = I_j + S_j, j = 1, 2, \ldots (\tau = I_1)$, where we have two states: busy and idle. Hence,

$$f_b = \frac{E(S_j)}{E(I_j) + E(S_j)} = \frac{1/\mu}{1/\lambda + 1/\mu} = \frac{\lambda}{\mu + \lambda}. \tag{80}$$

We now find the *fraction of arrivals who find the server busy*, a_b, which is also the *fraction of arrivals lost*. We will find a_b in two ways. First, it will be found from the rate (number per unit time) of arrivals who find the server busy, n_b. To find n_b, let K_j be the number of arrivals lost during the jth cycle, which is also the number of Poisson arrivals during S_j, where $E(K_j) = \lambda/\mu$ (see Example 2-7). Hence, by the Renewal-Reward Theorem,

$$n_b = E(K_j)/E(X_j) = \frac{\lambda/\mu}{1/\lambda + 1/\mu}.$$

Now a_b is simply n_b divided by the *arrival rate* λ,

$$a_b = n_b/\lambda = \frac{\lambda}{\mu + \lambda}. \tag{81}$$

Formally, let $N_b(t)$ be the number of arrivals lost by epoch t, where

$$a_b = \lim_{t\to\infty} N_b(t)/\Lambda(t) = \lim_{t\to\infty} \frac{N_b(t)/t}{\Lambda(t)/t} = n_b/\lambda.$$

The second way to find a_b is by defining cycles directly in terms of customers. The "length" of cycle j is the number of customers who arrive during X_j, $K_j + 1$. Again by the Renewal-Reward Theorem,

$$a_b = \frac{E(K_j)}{E(K_j) + 1} = \frac{\lambda/\mu}{\lambda/\mu + 1} = \frac{\lambda}{\lambda + \mu}.$$

Notice that $a_b = f_b$, i.e., the fraction of *arrivals* who find the server busy is equal to the fraction of *time* it is busy. This is an elementary example of what is called *Poisson Arrivals See Time Averages* (*PASTA*). This is a shorthand way of saying that when arrivals are Poisson, the fraction of arrivals who find a process (an alternating renewal process here) in some state (busy or idle here) is equal to the fraction of time the process is in that state. Even for this example, PASTA is by no means a trivial result because, contrary to Example 2-7, the arrival process affects when the server is busy. In fact, if the arrival process in this example were *not* Poisson, then typically we would have $a_b \neq f_b$.

The PASTA property is particularly important in the analysis of queueing models and is discussed in detail in Chapter 5.

Also notice that (80) and (81) depend on the mean service time $1/\mu$, but are otherwise independent of the service distribution G. This is a special case of what is called *Erlang's Loss Formula*, which is treated in detail in Chapters 5 and 6.

To continue the main theme of this section, $\tau + S_1$ in the preceding example is another example of a stopping time, which now depends on the Poisson process and a random variable S_1 that is independent of $\{\Lambda(t)\}$.

More abstractly, let $\{S(r): r \in T\}$ be a stochastic process independent of $\{\Lambda(t)\}$, e.g., $\{S(r)\}$ can be a sequence of service times.

Definition. $\mathcal{T}$ is called a *stopping time* for a Poisson process $\{\Lambda(t)\}$ if $\mathcal{T}$ is a random variable such that for every $t \geq 0$, the occurrence of the event $\{\mathcal{T} \leq t\}$ depends only on $\{\Lambda(s): 0 \leq s \leq t\}$ and $\{S(r)\}$. In particular, it is independent of the *future* increments of the Poisson process, $\{\Lambda(s) - \Lambda(t): s > t\}$.

With this definition of a stopping time, we state without proof

Theorem 5. If $\mathcal{T}$ is a stopping time for a Poisson process $\{\Lambda(t)\}$ at rate λ, $\{\Lambda(t + \mathcal{T}) - \Lambda(\mathcal{T}): t > 0\}$ is a Poisson process at rate λ that is independent of $\mathcal{T}$, $\{\Lambda(t): 0 \leq t \leq \mathcal{T}\}$, and $\{S(r)\}$. In particular, the excess $Y_{\mathcal{T}} \sim \exp(\lambda)$ and is independent of the same collection of random variables.

Thus, at stopping times, the Poisson process in effect "starts over," independent of the past. Another example of a stopping time is a busy period for the $M/G/\infty$ treated in Example 2-6. This is discussed in detail in the next section.

Defining Stopping Times

A stopping time for a Poisson process is usually defined more narrowly than we have done, without the introduction of $\{S(r)\}$. Thus, whether $\{\mathcal{T} \leq t\}$ occurs or not depends *only* on $\{\Lambda(s): 0 \leq s \leq t\}$. However, as illustrated in Example 2-9, this is inadequate for applications where a Poisson process is a component of a stochastic model. Actually, stopping times may be defined without introducing $\{S(r)\}$. The critical property is that for any fixed t, $\{\mathcal{T} \leq t\}$ and $\{\Lambda(s) - \Lambda(t): s \geq t\}$ are independent.

The notion of a stopping time has other important applications and, in spite of various definitions, the essential meaning is always the same. A random variable $\mathcal{T}$ is a stopping time for a stochastic process only if events of form $\{\mathcal{T} \leq t\}$ and $\{\mathcal{T} > t\}$ depend on the process only up to epoch t. If the process has certain independence properties, the definition of $\mathcal{T}$ may make use of them.

In Section 2-11, a stopping time is used to define regenerative processes. In Section 2-13, it is essential for the validity of Wald's equation. In both cases, independence plays a key role in defining a stopping time. This notion is also needed for continuous-time Markov chains (Chapter 4).

2-8 THE CONNECTION BETWEEN TIME AVERAGES AND POINTWISE LIMITS

In Example 2-6, we found that the number of calls in progress at t, $N(t)$, is Poisson with parameter $\lambda \int_0^t G^c(u)\,du$. As $t \to \infty$, the parameter converges to $\lambda \int_0^\infty G^c(u)\,du = \lambda/\mu$, where $1/\mu$ is the mean service time (conversation length). This means that $N(t)$ *converges in distribution*, i.e.,

$$\lim_{t\to\infty} P(N(t) = j) = e^{-\lambda/\mu}(\lambda/\mu)^j/j!, \qquad j = 0, 1, \ldots. \tag{82}$$

This is a different kind of limit from

$$\lim_{t\to\infty} \int_0^t P(N(u) = j)\,du/t,$$

which we called a *time average* in Section 2-4. How are these limits related?

We are discussing two kinds of limits of a function, $g(t)$ say,

$$\lim_{t \to \infty} g(t), \tag{83}$$

and

$$\lim_{t \to \infty} \int_0^t g(u)\, du/t. \tag{84}$$

A limit of form (83) is called a *pointwise limit* of g, i.e., $g(t)$ will be arbitrarily close to that limit for all sufficiently large "points" t. A limit of form (84) is called a *time average* or *Cesàro limit* of g. To distinguish the two types of limits, we denote (83) by *lim g* and (84) by *C.L.g* (for Cesàro limit).

Now (84) does not imply that $g(t)$ will be close to *C.L.g* for any t, e.g., let

$$\begin{aligned} g(t) &= 1 \quad \text{for } t \in (n-1, n], \quad n = 1, 3, \ldots, \\ &= -1 \quad \text{for } t \in (n-1, n], \quad n = 2, 4, \ldots, \end{aligned}$$

for which it is easily shown that $C.L.g = 0$, but *lim g* does not exist.

Of course, (84) is also a pointwise limit, of the function $h(t) = \int_0^t g(u)\, du/t$, rather than g.

An elementary but important relationship between the two limits is given by

Theorem 6. If *lim g* exists and $\int_0^t g(u)\, du$ is finite for every finite t,

$$C.L.g = lim\ g, \tag{85}$$

i.e., the limits (83) and (84) are the same.

Proof. Let $lim\ g = a$. For any $\varepsilon > 0$, there is a T such that $|g(t) - a| < \varepsilon$ for all $t > T$. For any $t > T$, write

$$\int_0^t g(u)\, du = \int_0^T g(u)\, du + \int_T^t g(u)\, du, \quad \text{from which we obtain}$$

$$(t - T)(a - \varepsilon) \leq \int_0^t g(u)\, du - \int_0^T g(u)\, du \leq (t - T)(a + \varepsilon). \tag{86}$$

Now divide (86) by t and let $t \to \infty$. Since ε is arbitrarily small, we have (85).

Example 2-10: Continuation of Example 2-6, the $M/G/\infty$ Queue

As mentioned at the beginning of this section, we have pointwise limits (82) for $P(N(t) = j)$. We say "the system" is *busy* at epoch t if $N(t) \geq 1$ and *idle* if $N(t) = 0$. Hence, the pointwise limit for the probability that the system is idle is

$$\lim_{t \to \infty} P(N(t) = 0) = e^{-\lambda/\mu}. \tag{87}$$

From Theorem 6, the corresponding time average is

$$\lim_{t\to\infty} \int_0^t P(N(u) = 0)\, du/t = e^{-\lambda/\mu}. \tag{88}$$

Clearly, the duration of a busy period depends on the arrival process and service times during the period, and *not* on the arrival process after the busy period ends. The service time sequence is completely independent of the Poisson process. Hence, the epochs at which busy periods end are stopping times for the Poisson process. Let B_j be the duration of the jth busy period, I_j be the duration of the jth subsequent idle period, and $B_j + I_j$ be the duration of the jth cycle. We again have the structure of an alternating renewal process with two states, busy and idle, where the limit (88) is the limit we get from the expected-value version of the Renewal-Reward Theorem. Hence,

$$\frac{E(I_j)}{E(B_j) + E(I_j)} = \frac{1/\lambda}{E(B_j) + 1/\lambda} = e^{-\lambda/\mu}, \quad \text{or} \tag{89}$$

$$E(B_j) = (e^{\lambda/\mu} - 1)/\lambda. \tag{90}$$

From the random-variable version of the Renewal-Reward Theorem, (89) is also the *fraction of time* the system is idle.

The busy period distribution for an $M/G/\infty$ queue is very complicated. Fortunately, we are able to obtain the mean without it. Note that $E(B)$ depends only on the mean service time.

Remarks. There are really two stopping times associated with a cycle, one for the Poisson process and the other for a customer number in the sequence of service times. Furthermore, it should be clear that we could start a cycle with either a busy period or an idle period. A more serious matter is this: How do we know that busy periods (or cycles) end? We will discuss this question further in Section 2-10.

2-9 POINT PROCESSES; GENERALIZATIONS OF POISSON PROCESSES

At the beginning of Section 2-6, we mentioned properties any arrival process should have: $\Lambda(t)$ is integral-valued for each t, and realizations of the process, $\Lambda(t, \omega)$, are monotone nondecreasing functions of t. The first property follows because we are counting events. The second simply means that as t increases, the cumulative count can only get larger. Stochastic processes with these properties are called *counting* or *point* processes. Clearly, a renewal process $\{M(t)\}$ is a point process.

We now introduce some generalizations of the Poisson process that are important in applications where the assumption of Poisson arrivals is too restrictive. Clearly, a renewal process generalizes a Poisson process by permitting the inter-arrival times to have an arbitrary distribution. Other generalizations are obtained by keeping some but not all of the properties that define a Poisson process. With one or two exceptions, these generalizations are still point processes.

A point process that has stationary and independent increments, but not orderliness, is called a *batch* (or *bulk*) Poisson process. For example, arriving customers at a warehouse may be a Poisson process, $\{\Lambda(t)\}$, say, at rate λ. Customer j has demand of υ_j items, where the υ_j are i.i.d. random variables, independent of $\{\Lambda(t)\}$. The total demand in $[0, t]$ is

$$\upsilon(t) = \sum_{j=1}^{\Lambda(t)} \upsilon_j, \tag{91}$$

where $\{\upsilon(t): t \geq 0\}$ is a batch Poisson process.

Properties of $\upsilon(t)$ are easy to obtain. For example,

$$\begin{aligned} E\{\upsilon(t)\} &= E\{E(\upsilon(t) \mid \Lambda(t))\} \\ &= E\{\Lambda(t)E(\upsilon_j)\} \\ &= \lambda t E(\upsilon_j), \quad \text{and} \end{aligned} \tag{92}$$

$$\begin{aligned} V\{\upsilon(t)\} &= E\{V(\upsilon(t) \mid \Lambda(t))\} + V\{E\{\upsilon(t) \mid \Lambda(t)\}\} \\ &= E\{\Lambda(t)V(\upsilon_j)\} + V\{\Lambda(t)E(\upsilon_j)\} \\ &= \lambda t V(\upsilon_j) + \lambda t E^2(\upsilon_j). \end{aligned} \tag{93}$$

More generally, demand need not be a number; it could be an amount, i.e., υ_j could be continuous. When this is permitted $\{\upsilon(t)\}$ is called a *compound* Poisson process. It will no longer be a point process, but it is what is called a *jump* process. That is, realizations of the process are constant between jumps. The times between jumps are still exponential. Furthermore, (92) and (93) are still valid.

We can give up stationary increments and even orderliness and have what is called a *nonstationary* Poisson process, a counting process $\{\Lambda(t)\}$ which has *independent increments*, where for every $t \geq 0$,

$$\Lambda(t) \sim P(\gamma(t)), \tag{94}$$

where the only restriction on $\gamma(t)$ is that it be nonnegative and a nondecreasing function of t that is continuous from the right. [If $\gamma(t)$ has a jump of size c at t_0, say, the number of arrivals *at* t_0 is Poisson with parameter c.] If $\gamma(t)$ is differentiable, $\gamma'(t) = \lambda(t)$, say, where $\lambda(t)$ can be thought of as the "arrival rate" at t. If $\lambda(t) = \lambda$, a constant for all $t \geq 0$, we are back to the usual (stationary) Poisson process.

Nonstationary Poisson processes can be used to model arrival processes where the arrival rate varies rapidly over time, e.g., rush hour traffic. Any nonstationary Poisson process may be generated as a function of a (stationary) Poisson process by transforming the time axis as follows: Let $\{\beta(x): x \geq 0\}$ be a Poisson process at rate $\lambda = 1$. For every $t \geq 0$, define $\Lambda(t) = \beta(\gamma(t))$. It is easy to see that $\{\Lambda(t)\}$ is the nonstationary stochastic process (94).

Theorem 3 and the relation between $\{\beta(x)\}$ and $\{\Lambda(t)\}$ may be used to prove

Theorem 7. Given $\Lambda(t) = n$, where $\{\Lambda(t)\}$ is the nonstationary Poisson process (94), the n *unordered* arrival epochs in $[0, t]$, $t'_1, \ldots, t'_n$, are i.i.d. with distribution

$$P(t'_i \leq u) = \gamma(u)/\gamma(t) \qquad \text{for } 0 \leq u \leq t, \quad i = 1, \ldots, n. \tag{95}$$

Poisson processes can also be defined in terms of more than one "time" variable. For example, in dispatching fire trucks, we would need to know where as well as when fires occur. This can be modeled as a *spatial* Poisson process, where the number of fires in any region (e.g., city block) over any time interval has a Poisson distribution. Here, a time point would specify location as well as time. In this context, independent increments means that the number of fires that occur on disjoint sets of points are independent random variables.

The assumption of independent increments is stronger than it might at first appear. For example, among renewal processes, it can be shown that only the Poisson process (or a special case of the batch Poisson process) has independent increments. Renewal processes generally do not have stationary increments either. However, this is a less fundamental property in that a renewal process will have stationary increments if the initial distribution, A, is chosen to be $A = F_e$. In effect, this means that the initial epoch, $t = 0$, may be chosen at random. This is sometimes called the *stationary version* of a renewal process. Properties of this process are discussed further in Section 2-16 under the name of the *equilibrium* renewal process.

Stationary Point Processes*

As a final generalization of a Poisson process, we retain only one property: stationary increments. A point process $\Lambda(t)$ is called *stationary* if it has stationary increments, i.e., if the number of arrivals in $(h, t + h]$, $\Lambda(t + h) - \Lambda(h)$, has a distribution independent of h for all $h \geq 0$.† In this section, $\Lambda(t)$ is the number of arrivals in $(0, t]$.‡

In the remainder of this section, we explore how far the assumption of stationary increments will take us. To do this, let $P_n(t) = P(\Lambda(t) = n)$, $n = 0, 1,$

* The remainder of this section is technical and may be omitted.

† Usually, stationary increments is the term for the stronger property that the *joint* distribution of the number of arrivals in any finite collection of intervals does not change when the location of the entire collection is shifted. This stronger property is not needed in our proofs. When independent increments is also assumed, joint distributions are determined by their marginals, and the two definitions are equivalent.

‡ This is a technical point. If $\Lambda(t)$ is to include arrivals *at t*, an increment necessarily refers to an interval that is open on the left and closed on the right. With this convention, $\Lambda(t)$ is also an increment. For renewal processes, we used a different convention: $M(t)$ is the number of renewals in $[0, t]$. This allows inclusion of renewals at epoch $t = 0$.

. . . , which, by stationary increments, is also the distribution $\Lambda(t + h) - \Lambda(h)$. An important result for stationary point processes is

Theorem 8. For any stationary point process, excluding the case $P_0(t) = 1$ for any $t > 0$, the following limit, denoted by λ, exists:

$$\lim_{t \to 0} \frac{1 - P_0(t)}{t} = \lambda > 0, \tag{96}$$

where $\lambda = \infty$ is possible.

Proof. Let $g(t) = 1 - P_0(t)$ and note that $g(t)$ is positive and nondecreasing on, say, $0 < t \le a$. Furthermore, $g(x + y) \le g(x) + g(y)$, since the left side represents the probability of the union of two events, where the sum of their probabilities is on the right. Hence,

$$g(t) \le kg(t/k) \quad \text{for any integer } k \text{ and } 0 < t \le a, \quad \text{and at} \quad t = a,$$

$$\frac{g(a/k)}{a/k} \ge \frac{g(a)}{a}, \qquad k = 1, 2, \ldots .$$

Now let

$$\lambda = \limsup_{t \to 0} g(t)/t \ge g(a)/a > 0,$$

where $\lambda = \infty$ is possible. First assume $\lambda < \infty$. In this case, there exist values of the function $g(t)/t$ arbitrarily close to λ. In particular, for some $c > 0$, $g(c)/c > \lambda - \varepsilon$, where ε is arbitrarily small. For any $t \in (0, c)$ we can determine an integer $k > 1$ such that

$$c/k \le t \le c/(k - 1).$$

From monotonicity of $g(t)$, we obtain for all t in this interval:

$$g(t)/t \ge \frac{g(c/k)}{c/(k-1)} = \frac{k-1}{k}\frac{kg(c/k)}{c} \ge \frac{k-1}{k}\frac{g(c)}{c}, \quad \text{and hence}$$

$$g(t)/t \ge \left(1 - \frac{1}{k}\right)(\lambda - \varepsilon).$$

As $k \to \infty$, $t \to 0$, and we have

$$\lim_{t \to 0} g(t)/t = \lambda.$$

Now assume $\lambda = \infty$. In this case, for some, $c > 0$, $g(c)/c > A$, and $g(t)/t \ge (1 - 1/k)A$ for arbitrarily large A. This implies $\lim_{t \to 0} g(t)/t = \infty$, which completes the proof.

We exclude the case $P_0(t) = 1$ above because if this were true for some $t > 0$, then $P_0(kt) = 1$ for any integer k, and $P_0(t) = 1$ for *all* $t > 0$, i.e., we would not have any arrivals!

For a Poisson process, the λ in (96) is the arrival rate. Is this still the case? To explore this question, let $a(t)$ be the expected number of arrivals in $(h, t + h]$ for $t > 0$, where $a(t) > 0$, and $a(t) = \infty$ is possible. Since $\Lambda(t + h) = \Lambda(h) + (\Lambda(t + h) - \Lambda(h))$, $a(t)$ has the property

$$a(t + h) = a(t) + a(h) \qquad \text{for all } t, h \geq 0. \tag{97}$$

Clearly, if $a(t) = \infty$ for some $t > 0$, $a(t) = \infty$ for *all* $t > 0$. Otherwise, $a(t)$ is finite, positive, and nondecreasing.

Now assume that $a(t) < \infty$. For any rational t, $t = m/n$, where m and n are positive integers, an interval $(0, nt]$ may be partitioned either into n intervals of length t or m intervals of length 1. From (97),

$$na(t) = ma(1),$$

$$a(t) = a(1)t = \theta t \qquad \text{for rational } t > 0, \tag{98}$$

where we have defined $\theta = a(1) > 0$. From the monotonicity of $a(t)$, it is easy to extend (98) to all real t, and we have

Theorem 9. For a stationary point process, excluding the case $P_0(t) = 1$ for any $t > 0$, the expected number of arrivals in $(0, t]$ is of form

$$E\{\Lambda(t)\} = \theta t, \qquad t > 0, \tag{99}$$

where $\theta > 0$, and $\theta = \infty$ is possible.

Remark. The proof that an *additive* function (97) must have linear form (99) is essentially the same as the proof that a function f with the *product* property: $f(t + h) = f(t)f(h)$ for all $t, h > 0$ must be exponential, because the log of a product function is an additive function. We used monotonicity in the proof. Actually, it is necessary to assume only that $a(t)$ [or $f(t)$] is bounded on finite intervals.

We will call θ the *arrival rate*. To relate θ and λ, we write

$$\theta t = \sum_{n=0}^{\infty} nP_n(t) \geq \sum_{n=1}^{\infty} P_n(t) + \sum_{n=2}^{\infty} P_n(t) = 1 - P_0(t) + P(\Lambda(t) \geq 2).$$

From the above and (96), $\theta \geq \lambda$ and, when $\lambda < \infty$, $\theta = \lambda$ *only if*

$$\lim_{t \to 0} P(\Lambda(t) \geq 2)/t = 0,$$

i.e., only if the orderliness property holds. [Owing to stationary increments, we need (61) only for $t = 0$.]

A stationary point process for which (61) holds is called *regular* or *orderly*.

We will show the converse of the above result: For a regular stationary point process with $\lambda < \infty$, $\theta = \lambda$.

To do this, partition the interval (0, 1] into n subintervals of equal length. For any $n \geq 1$, let A_{ni} be the number of arrivals in subinterval i, $i = 1, \ldots, n$. Let E be the event that two or more arrivals occur simultaneously somewhere in $(0, t]$. Now E occurs only if $A_{ni} \geq 2$ for some i. Thus, for any $n \geq 1$,

$$P(E) \leq P\left(\bigcup_{i=1}^{n} \{A_{ni} \geq 2\}\right) \leq nP(\Lambda(1/n) \geq 2).$$

From orderliness, the right side above $\to 0$ as $n \to \infty$, and we have

$$P(E) = 0, \tag{100}$$

i.e., with probability 1, the arrivals in $(0, t]$ occur at distinct epochs.

For any $n \geq 1$, now let $I_{ni} = 1$ if $A_{ni} \geq 1$, and $I_{ni} = 0$ otherwise, $i = 1, \ldots, n$, and $S_n = \sum_{i=1}^{n} I_{ni}$. For any sample point ω for which the arrivals in $(0, t]$ are distinct, the difference

$$\Lambda(1, \omega) - S_n(\omega) \geq 0,$$

where the difference is *strict* only if $A_{ni} \geq 2$ for some i. Let $\delta(\omega)$ be the minimum distance between successive arrivals in $(0, t]$. For all $n > 1/\delta$, $A_{ni} \leq 1$ and $S_n = \Lambda(1)$. Hence, we have shown

$$\lim_{n\to\infty} S_n = \Lambda(1) \quad \text{w.p.1.} \tag{101}$$

Convergence w.p.1 implies convergence in distribution. What about moments? The S_n are nonnegative and, from Fatou's Theorem (see the Appendix),

$$E(\Lambda(1)) = \theta \leq \lim_{n\to\infty} E(S_n) = \lim_{n\to\infty} nE(I_{ni}) = \lim_{n\to\infty} nP(\Lambda(1/n) > 0) = \lambda.$$

Since we already know $\theta \geq \lambda$, we must have $\theta = \lambda$. By combining these results, we have

Theorem 10. For a stationary point process, $\theta \geq \lambda$; $\lambda = \infty$ implies $\theta = \infty$. For $\lambda < \infty$, $\theta = \lambda$ if and only if the process is regular.

Remark. The "if" part of Theorem 10 is known as Korolyook's Theorem.

If we assume independent increments as well as stationary increments, $P_0(t + h) = P_0(t)P_0(h)$, from which it is easily shown that $\lambda = \infty$ if and only if $P_0(t) = 0$ for all $t > 0$. This in turn implies that, with probability 1, $\Lambda(t) = \infty$ (not a very interesting case). Hence, assuming $\sum_{n=0}^{\infty} P_n(t) = 1$ *implies* $\lambda < \infty$.

2-10 DEFECTIVE RENEWAL PROCESSES

When we defined stopping times for a Poisson process in Section 2-7, we implicitly assumed that they *stopped*, i.e., sooner or later, a stopping time occurs. This means that $P(\mathcal{T} < \infty) = 1$. Thus, we are sure to start another renewal cycle.

In some circumstances, e.g., in Example 2-9, this is obviously true. In the remark following Example 2-10, however, we expressed some concern that stopping times (in this case, epochs where busy periods end) might not occur.

To deal with this possibility, we introduce defective renewal processes. The assumptions and notation of Section 2-2 are unchanged except that now we permit $X_1, X_2, \ldots$ to be defective random variables.

Since each X_j is nonnegative, the set corresponding to "X_j occurs" is $E_j = \{\omega: 0 \leq X_j < \infty\}$, where the E_j are independent events. How do we define $X_j(\omega)$ for any $\omega \in E_j^c$? The most convenient thing to do is to *define* $X_j(\omega) = \infty$ for all $\omega \in E_j^c$. Thus, if the "next event" occurs at ∞, it will never be observed in any finite interval. We retain the notation $X_j \sim F$ for $j \geq 2$, and $X_1 \sim A$, where we permit $A \neq F$. With this notation and our definition, $\lim_{t\to\infty} A(t) = A(\infty)$ is $P(E_1)$, the probability that X_1 occurs (is finite), and $\lim_{t\to\infty} F(t) = F(\infty)$ is $P(E_j)$, the probability that X_j occurs, $j \geq 2$.

The distribution of the sum Z_n is still the convolution $A * F^{(n-1)}(t)$, $n = 1, 2, \ldots$. Notice that the sum $Z_n = X_1 + \cdots + X_n$ is finite if and only if *all* the terms in the sum are finite, i.e., *all* the events $E_1, \ldots, E_n$ occur. By this argument, or by dealing directly with the convolution integral, we have that for $n \geq 1$,

$$P(Z_n < \infty) = \lim_{t\to\infty} A * F^{(n-1)}(t) = \prod_{j=1}^{n} P(E_j) = A(\infty)[F(\infty)]^{(n-1)}. \tag{102}$$

A renewal process is called *defective* (the terms *transient* and *terminating* are also used) if $F(\infty) < 1$, i.e., if F is *defective*, whether or not A is defective. The renewal rate $\mu = 0$ in this case (the integral of the tail of F is $1/\mu = \infty$). A renewal process is called *proper* if $F(\infty) = A(\infty) = 1$, i.e., if *both* F and A are proper.

This terminology leaves out what may appear to be an oddball case: $A(\infty) < 1$ and $F(\infty) = 1$. However, this case occurs in a natural way in connection with Markov processes.

The argument by which we obtained the representation for the renewal function in equation (10) is still valid. Hence, for a *defective* renewal process, we obtain, by interchanging limit and sum, the following:

$$\lim_{t\to\infty} m(t) = \lim_{t\to\infty} \sum_{n=1}^{\infty} A * F^{(n-1)}(t) = A(\infty) \sum_{n=1}^{\infty} [F(\infty)]^{(n-1)} = \frac{A(\infty)}{1 - F(\infty)} < \infty. \tag{103}$$

For a proper renewal process,

$$\lim_{t\to\infty} m(t) = \infty. \tag{104}$$

The actual number of renewals by epoch t, $M(t)$, is a nondecreasing function of t on every realization.

For a *defective* renewal process,

$$\lim_{t\to\infty} M(t) = M(\infty) < \infty \quad \text{w.p.1}, \tag{105}$$

i.e., it has a finite limit with probability 1, where $M(\infty)$ is the number of finite X_j prior to the first one that is infinite. Hence, $M(\infty)$ has distribution

$$\begin{aligned} P(M(\infty) = n) &= P(E_1, \ldots, E_n, E_{n+1}^c) \\ &= P(E_{n+1}^c) \prod_{j=1}^{n} P(E_j) & n = 0, 1, \ldots, \\ &= 1 - A(\infty) & \text{for } n = 0, \\ &= A(\infty)[F(\infty)]^{(n-1)} [1 - F(\infty)] & \text{for } n = 1, 2, \ldots. \end{aligned} \tag{106}$$

Notice that $M(\infty)$ has a *geometric* distribution when $A = F$.

For a *proper* renewal process, $P(M(\infty) = n) = 0$, $n = 0, 1, \ldots$, and

$$\lim_{t\to\infty} M(t) = \infty \quad \text{w.p.1}. \tag{107}$$

For the case $A(\infty) < 1$ and $F(\infty) = 1$, the form of the limit (103) is still valid, and we have $m(\infty) = \infty$ unless, trivially, $A(\infty) = 0$, in which case no renewals occur, and $m(\infty) = 0$. Similarly, $M(\infty) = \infty$ if the first renewal occurs; otherwise, $M(\infty) = 0$.

Now return to Example 2-10. Can the (busy, idle) alternating renewal process be defective? Since idle periods are proper, this can happen only if busy periods are defective, in which case, sooner or later, an infinite busy period will occur. This implies that the limiting probability that the system is idle, (87), must be zero. However, this limit is strictly positive, and the renewal process must therefore be proper.

We are not quite done. How do we know the mean cycle length is finite (a condition for applying the Renewal-Reward Theorem)? Clearly, an infinite mean cycle length must correspond to an infinite mean busy period, $E(B_j) = \infty$. We know from (90) that this is false, but this reasoning won't do. Formally, we need to extend the Renewal-Reward Theorem to cover the case $E\{|R_j|\} < \infty$ and $E(X_j) = \infty$. We show in Section 2-12 that for this case, the limit (26) exists and is equal to zero. If $C(t)/t$ is a bounded function of t, (26) implies (27) (see the Appendix). Fortunately, fractions of time are bounded between 0 and 1, and hence the limit (27) is also zero. Thus the strictly positive limit (87) implies $E(X_j) < \infty$.

2-11 REGENERATIVE PROCESSES

In Section 2-2, we started with a renewal process, and then used it to define new processes. In Examples 2-9 and 2-10, we started with processes that have more structure. In each case, we found i.i.d. cycle lengths X_j, and renewal epochs X_1, $X_1 + X_2, \ldots$ such that the original process started over at these epochs. We then used the Renewal-Reward Theorem to analyze the original process. This illustrates how renewal theory is applied.

Stopping times for a Poisson process were used to identify cycles; we now generalize this concept. A stochastic process $\{B(t)\colon t \geq 0\}$ *regenerates itself* or *starts over* at some (usually random) epoch X_1 if, from X_1 on, the process is independent of its past, and has the same stochastic structure that it had originally. Formally, we introduce the following definition, where X_1 is a *stopping time*.

Definition. $\{B(t)\colon t \geq 0\}$ is called a *regenerative process* if there exists a nonnegative random variable X_1 such that for every epoch $s \geq 0$,

(i) $\{B(t + X_1)\colon t \geq 0\}$ is independent of $\{B(t)\colon t \leq s;\, X_1 > s\}$, and
(ii) $\{B(t + X_1)\colon t \geq 0\}$ is stochastically equivalent to $\{B(t)\colon t \geq 0\}$.

Stochastically equivalent means that the two stochastic processes in (ii) have the same joint distributions. Because of property (i), we call X_1 a *stopping time* for $\{B(t)\}$, and note that (i) implies that $\{B(t + X_1)\}$ is independent of X_1. While $\{B(t)\colon t \leq s\}$ and $\{X_1 > s\}$ will typically be dependent, we are *not* assuming that knowledge of $\{B(t)\colon t \leq s\}$ *determines* whether or not the event $\{X_1 > s\}$ occurs. We assume that X_1 occurs, i.e., $P(0 \leq X_1 < \infty) = 1$, and that $P(X_1 = 0) < 1$.

We call X_1 a *regenerative point (epoch)*. The existence of X_1 implies the existence of subsequent regeneration points, $X_1 + X_2, \ldots$, such that $\{B(t)\}$ starts over (satisfies the previous definition) at each of these epochs, where the X_j are i.i.d. Thus, $\{X_j\}$ generates a renewal process said to be *embedded* in $\{B(t)\}$ (at the points $X_1, X_1 + X_2, \ldots$). Define $\{Z_j\}$ as in Section 2-2. The X_j are called *cycle lengths*, and $\{B(t)\colon Z_{j-1} \leq t < Z_j\}, j \geq 1$, are called cycles.

Notice that $A = F$ here, i.e., X_1 is distributed like $X_j, j \geq 2$, and the cycles are stochastically equivalent. When $A = F$, we say the renewal process is *ordinary*.

It will sometimes be convenient to allow the first cycle to be different: $\{B(t)\colon t \geq 0\}$ is called a *generalized* (or *delayed*) regenerative process if there exists random variable $X_1 \geq 0$ such that $\{B(t)\}$ satisfies (i) in the above definition, and from X_1 on, $\{B(t + X_1)\colon t \geq 0\}$ is regenerative. Thus X_j and the corresponding cycles are each i.i.d. for $j \geq 2$, but X_1 and the first cycle may be different. In this case, we typically have $A \neq F$, and the corresponding renewal process is called *general* or *delayed*.

Generalized regenerative processes occur naturally. Suppose $\{B(t)\}$ is regenerative, and *you* (an observer or customer) arrive at some fixed or random epoch. Set your arrival epoch equal to zero. The process $\{B(t)\}$ *from then on* is a generalized regenerative process.

An equivalent way to define a regenerative process is as follows: With each X_j, we associate a stochastic process $\{\beta_j(t): 0 \leq t < X_j\}$ called a *cycle* or *tour*, $j = 1, 2, \ldots$, where cycles are mutually independent and stochastically equivalent. [Formally, we may define $\beta_j(t) = 0$ for any $t \geq X_j$.] We now define $\{B(t)\}$ by

$$B(t) = \beta_{M(t)+1}[U(t)], \qquad t \geq 0, \tag{108}$$

where $U(t)$, the *age* at epoch t, is the attained duration of the cycle in progress at epoch t. To define a generalized regenerative process, we simply permit $\{\beta_1(t)\}$ and X_1 to have stochastic properties different from subsequent cycles.

Although the concept of regenerative processes is a bit complicated to define in a general way, we have already been dealing with elementary examples of them:

The alternating renewal process $\{I(t)\}$ (Example 2-2) is regenerative where $I(t) = 1$ if the machine is up at epoch t, $I(t) = 0$ otherwise. Suppose an up period begins at epoch 0. We have regeneration points $X_1, X_1 + X_2, \ldots$, where $X_j = X_{uj} + X_{dj}, j \geq 1$, and $P(I(t) = 1) = P(I(t + X_1) = 1)$. If a down period begins at epoch 0, we could let $X_1 = X_{d1}$ and $X_j = X_{uj} + X_{dj}, j \geq 2$, and $\{I(t)\}$ would be a generalized regenerative process. We now have $P(I(t + X_1 + X_2) = 1) = P(I(t + X_1) = 1) \neq P(I(t) = 1)$. In this case, we can easily convert the generalized process into an ordinary one simply by defining the cycles differently: $X_j = X_{dj} + X_{u,j+1}, j \geq 1$. On the other hand, if we begin observing an alternating renewal process while either an up or a down interval is in progress, it will be a generalized regenerative process from then on.

The stochastic process of excess $\{Y(t)\}$ is regenerative for an ordinary renewal process and generalized regenerative otherwise. In either case, to obtain $\{Y(t)\}$ from (108), let

$$\beta_j(t) = X_j - t \qquad \text{for } 0 \leq t < X_j \quad \text{and} \quad j = 1, 2, \ldots.$$

Similarly, the stochastic process of spread $X(t)$ is (generalized) regenerative and may be obtained from (108) by letting

$$\beta_j(t) = X_j \qquad \text{for } 0 \leq t < X_j \quad \text{and} \quad j = 1, 2, \ldots.$$

For many purposes the distinction between regenerative and generalized regenerative is not important. To simplify terminology and reduce tongue twisting, we will often use *regenerative* to cover both cases.

It is often easy to see that a process is regenerative, but tedious or at best uninformative to formally define $\{\beta_j(t)\}$ and use (108). For example, the process $\{N(t): t \geq 0\}$ in Examples 2-6 and 2-10 is regenerative, where $N(t)$ is the number of busy servers (conversations in progress) at epoch t for the $M/G/\infty$ queue. For this conclusion, recall that the epochs at which busy periods end are stopping times for the Poisson process. Furthermore, because both the Poisson process and the service time sequence start over there, these epochs are also regeneration points for $\{N(t)\}$.

Connection with Reward and Cumulative Processes

For much of the rest of this book, we will be dealing with regenerative processes. This is true of what we will call recurrent Markov processes and queueing models that are Markovian. The Renewal-Reward Theorem is a basic tool for the analysis of these processes and models. This is true because for any regenerative process, we may define a reward structure as follows:

$$R_j = \int_0^{X_j} \beta_j(t)\, dt = \int_{Z_{j-1}}^{Z_j} B(t)\, dt, \qquad j = 1, 2, \ldots, \tag{109}$$

where the R_j are independent and, except possibly for R_1, identically distributed. The corresponding cumulative process $\{C(t)\}$ is defined by

$$C(t) = \int_0^t B(u)\, du, \qquad t \geq 0. \tag{110}$$

When the mean cycle length is finite, expressions for the time average $\overline{B}$ of $\{B(t)\}$, defined as

$$\overline{B} = \lim_{t\to\infty} \int_0^t B(u)\, du/t \tag{111}$$

may be found from (110) and the Renewal-Reward Theorem.

Sometimes other time averages are of interest, e.g.,

$$\lim_{t\to\infty} \int_0^t f(B(u))\, du/t, \tag{112}$$

where f is a real-valued function. Since $\{f(B(t)\}$ is regenerative whenever $\{B(t)\}$ is, these limits may be found in the same manner.

A special case of (112) is of particular importance: For $f(x) = 1$ if $x > b$, $f(x) = 0$; otherwise, we have

$$\begin{aligned} f(B(t)) \equiv I_b(t) &= 1 \qquad \text{if } B(t) > b, \\ &= 0 \qquad \text{otherwise.} \end{aligned} \tag{113}$$

The time average

$$\overline{I}_b = \lim_{t\to\infty} \int_0^t I_b(u)\, du/t \tag{114}$$

is the *fraction of time* that $\{B(t)\}$ takes on values greater than b.

The expected-value version of (114) has the same limit, i.e., for any finite b,

$$\bar{I}_b = \lim_{t \to \infty} \int_0^t P(B(u) > b) \, du/t, \tag{115}$$

where, as a function of b, $\bar{I}_b$ is the *limiting* (tail) *distribution* of $B(t)$, in a time average sense. It is convenient to let B_∞ denote a random variable with limiting distribution (115), i.e., for any finite b,

$$P(B_\infty > b) = \bar{I}_b. \tag{116}$$

There is an important relationship between distribution (116) and time average $\bar{B}$ that we discuss in the next section (Theorem 11).

We close this section with the obvious but important observation that to apply renewal-reward theory, we must recognize when a process is regenerative and select appropriate cycles. As the next example illustrates, sometime a cycle can be selected in more than one way.

Example 2-11

A machine breaks down and is repaired. The times until breakdown, $V_1, V_2, \ldots$ are i.i.d. $\sim$ exp (μ). Repair times are instantaneous. However, the machine cannot be repaired until the breakdown has been detected at an inspection epoch. After each repair, inspections occur at epochs $T, 2T, \ldots$, where T is a constant. Suppose we want to know the fraction of time the machine is working (up), f_u.

To answer this question, we must identify regeneration points. One choice is the epochs at which failures are detected (random multiples of T). Because of the memoryless property of the exponential, however, a second choice is *every* multiple of T, i.e., the cycle length is T. We will make the *second* choice below.

At the beginning of each cycle, the time until breakdown of the machine, whether just repaired or not, is V, say, where $V \sim$ exp (μ). The up time during that cycle is min (V, T). Hence,

$$f_u = E\{\min(V, T)\}/T = \int_0^T e^{-\mu u} \, du/T. \tag{117}$$

A worthwhile exercise is to show that we obtain the same expression for f_u with the original choice of regeneration points.

2-12 THE RELATION BETWEEN A TIME AVERAGE AND THE MEAN OF A LIMITING DISTRIBUTION

Let $\{B(t)\}$ be a regenerative process, where we may be interested in both the time average value of the process (111), and the limiting distribution, where the tail distribution version of this limit is defined by (115) and (116). These limits are related by

Theorem 11. For a regenerative process with $E(X_1) < \infty$ and $E\{\int_0^{X_1} | B(t) | \, dt\} < \infty$,

$$\bar{B} = E(B_\infty), \tag{118}$$

i.e., the time average is the mean of the limiting distribution.

Proof. First assume $B(t) \geq 0$. From (113),

$$\int_0^\infty I_b(t)\, db = B(t) \qquad \text{for any } t > 0.$$

From the theory of renewal-rewards,

$$\lim_{t\to\infty} \int_0^t B(u)\, du/t = \frac{E\left\{\int_0^{X_1} B(t)\, dt\right\}}{E(X_1)} = \overline{B}, \tag{119}$$

and

$$\lim_{t\to\infty} \int_0^t I_b(u)\, du/t = \frac{E\left\{\int_0^{X_1} I_b(t)\, dt\right\}}{E(X_1)} = P(B_\infty > b), \tag{120}$$

where the limits in (119) and (120) are w.p.1, and we get the same limit in (120) when $I_b(u)$ on the left is replaced by $E(I_b(u)) = P(B(u) > b)$. Now the following identity holds for all realizations of $\{B(t)\}$:

$$\int_0^{X_1} B(t)\, dt = \int_0^{X_1} \int_0^\infty I_b(t)\, db\, dt = \int_0^\infty \int_0^{X_1} I_b(t)\, dt\, db.$$

Taking expectations,

$$E\left\{\int_0^{X_1} B(t)\, dt\right\} = \int_0^\infty E\left\{\int_0^{X_1} I_b(t)\, dt\right\} db. \tag{121}$$

Substituting (121) into (119), we have from (120) that

$$\overline{B} = \int_0^\infty P(B_\infty > b)\, db = E(B_\infty).$$

The interchanges of operations above are valid for nonnegative $\{B(t)\}$. The stated condition on $\{B(t)\}$ is needed only to complete the proof: Write

$$B(t) = [B(t)]^+ - [B(t)]^-, \qquad t \geq 0,$$

and treat the positive and negative parts separately.

Theorem 11 holds for *generalized* regenerative processes, provided that

$$\int_0^{X_1} |\, B(t)\,|\, dt < \infty \qquad \text{w.p.1},$$

and subsequent cycles satisfy the conditions of the Theorem, i.e.,

$$E(X_2) < \infty \quad \text{and} \quad E\left\{\int_{Z_1}^{Z_2} |\, B(t)\,|\, dt\right\} < \infty.$$

Example 2-12

For the excess process $\{Y(t)\}$, let $C(t) = \int_0^t Y(u)\, du$.

$$\overline{Y} = \lim_{t\to\infty} C(t)/t = E\left\{\int_0^{X_1} Y(t)\, dt\right\} \Big/ E(X_1) = \mu E\left\{\int_0^{X_1} (X_1 - t)\, dt\right\} = \mu E(X_1^2)/2,$$

the mean of equilibrium distribution F_e. We previously showed that $Y_\infty \sim F_e$.

The Case of Infinite Mean Cycle Lengths—Null States

The Renewal-Reward Theorem is easily extended to the case where X_j is proper, but $E(X_j) = \infty$, provided that the partial rewards P_j and N_j, defined in Section 2-3, have finite expectations. In this case,

$$\lim_{t\to\infty} C(t)/t = E(R_j)/E(X_j) = 0.$$

For regenerative process $\{B(t)\}$ with cycle lengths X_j, let $J_b(t) = 1$ if $|B(t)| \leq b$, $J_b(t) = 0$ otherwise, for all $b \geq 0$ and $t \geq 0$. Let R_j be the integral of $J_b(t)$ over the jth cycle, and suppose that for every finite b, $E\{R_j\} < \infty$. It follows that

$$P(|B_\infty| \leq b) = E(R_j)/E(X_j) = 0 \qquad \text{for every } b \geq 0. \tag{122}$$

i.e., B_∞ is defective with no mass on $(-\infty, \infty)$! These are properties of *null-recurrent* Markov chains and semi-Markov processes. If $\{B(t)\}$ is nonnegative, $P(B_\infty = \infty) = 1$, and it can be shown that

$$\overline{B} = E(B_\infty) = \infty. \tag{123}$$

The expected-value version of (111) is

$$\lim_{t\to\infty} \int_0^t E\{B(u)\}\, du/t = \overline{B}. \tag{124}$$

Now (115) and (124) are time averages of two functions. Suppose the corresponding pointwise limits exist, i.e.,

$$\lim_{t\to\infty} E\{B(t)\} = \overline{B}, \quad \text{and} \tag{125}$$

$$\lim_{t\to\infty} P(B(t) \leq b) = P(B_\infty \leq b). \tag{126}$$

In this light, we see that (111) and (115) are disguised versions of two types of convergence, convergence of moments (the first moment) and convergence in distribution, respectively. By Theorem 11, the limiting first moment of a regenerative process is the first moment of the corresponding limiting distribution. This is extended to higher moments in Theorem 13.

In the absence of regeneration points, this conclusion can be false. Consider an elementary example of a discrete time process: Let $\{Y_n\}$ be a sequence of random variables, where

$$P(Y_n = n) = 1/n \quad \text{and} \quad P(Y_n = 0) = 1 - 1/n \qquad \text{for } n = 1, 2, \ldots.$$

Now $\lim_{n\to\infty} P(Y_n = 0) = 1$, which means that the limiting distribution is the unit step at the origin. In random variable notation, $\{Y_n\}$ converges in distribution to the constant $Y_\infty = 0$. However, $E(Y_n) = 1$ for all n, and

$$\lim_{n\to\infty} E(Y_n) = 1 \neq E(Y_\infty) = 0.$$

In classical probability theory, one can show that $\lim_{n\to\infty} E(Y_n) = E(Y_\infty)$ whenever the corresponding distributions are *uniformly integrable*. See the Appendix.

Theorem 11 is important in applications. In queueing theory, for example, we define a stochastic process $\{N(t): t \geq 0\}$, where $N(t)$ is the number of customers in system (e.g., the number of busy servers in Examples 2-6 and 2-10). One measure of performance for a queue is

$$L = \lim_{t\to\infty} \int_0^t N(u)\, du/t, \tag{127}$$

called the *average number of customers in system*. However, in many situations, L is not found directly from the definition (127) but rather from a limiting distribution.

Example 2-13

The stochastic process $\{N(t)\}$ introduced in Example 2-6 converges in distribution (in a pointwise sense!) to (82), a Poisson distribution with mean λ/μ. Hence, for L as defined in (127), we have by Theorem 11 that

$$L = \lambda/\mu.$$

For nonnegative regenerative processes, i.e., $B(t) \geq 0$ for all $t \geq 0$. Theorem 11 is easily extended, but we omit proof:

Theorem 12. For a nonnegative regenerative process $\{B(t)\}$, where $E(X_1) < \infty$,

$$\overline{B} = E(B_\infty), \tag{128}$$

whether or not (128) is finite.

Example 2-14

Nearly all of the queueing models we will discuss are regenerative, and Theorem 12 applies directly; only the notation is different. Thus, suppose $\{N(t)\}$ is regenerative where for any $t \geq 0$, $N(t)$ is the number of customers in system at some queue, and p_n is the fraction of time there are n customers in system, $n = 0, 1, \ldots$. For L as defined in (127), we have

$$L = \sum_{n=0}^{\infty} np_n, \tag{129}$$

whether or not (129) is finite.

Remark. Theorem 12 is equivalent to a generalization of the Renewal-Reward Theorem for the case where $\{C(t)\}$ is monotone (nondecreasing or nonincreasing).

As mentioned earlier, functions of regenerative processes are regenerative. That is, if $\{B(t)\}$ is regenerative and f is a real-valued function, the process $\{H(t)\}$, defined by $H(t) = f(B(t))$ for all $t \geq 0$, is regenerative and has the same regeneration points. We can apply Theorem 11 to find $\overline{H}$ in terms of H_∞. It is easily shown that the distribution of H_∞ is determined by the following functional relationship between the random variables:

$$H_\infty = f(B_\infty). \tag{130}$$

Owing to (130), we can bypass finding the distribution of H_∞:

Theorem 13. Let $\{H(t)\}$ be a regenerative process defined by $H(t) = f(B(t))$ for all $t \geq 0$, where f is a real-valued function and $\{B(t)\}$ is regenerative. If $\{H(t)\}$ satisfies the conditions of Theorem 11,

$$\overline{H} = E\{f(B_\infty)\} = \int_{-\infty}^{\infty} f(t)\, dG(t), \tag{131}$$

where $G(t) = P(B_\infty \leq t)$ for all real t.

Remark. We do not require $|\,\overline{B}\,| < \infty$ here.

Example 2-15: Continuation of Example 2-14

Suppose we want to find the time average of $\{K(t)\}$ where $K(t) = N^2(t)$, $t \geq 0$. From Theorem 13,

$$\overline{K} = \sum_{n=0}^{\infty} n^2 p_n.$$

Pointwise Limits

The results in this section relate the time average of a process to a limiting distribution that is also defined as a time average.

For the $M/G/\infty$ queue, pointwise limits $\lim_{t\to\infty} P(N(t) = n)$ exist simply because we are able to carry out explicitly the limiting operations. This is true for some other elementary models as well. However, you may have noticed an absence of general statements about conditions for the existence of these limits. This topic is treated later in this chapter.

Fortunately, for most applications, time averages are of primary importance. Nevertheless, pointwise limits such as (125) and (126) are helpful for understanding some of our results.

The State Space and Sample Paths of a Regenerative Process

In elementary applications, $B(t)$ and $f(B(t))$ are random variables for every t. Occasionally in later chapters, they will be vectors of random variables. Our results apply to each component of these vectors. The only change required would be in notation.

Recall that a *sample path* is the function $\{B(t, \omega): t \geq 0\}$ for fixed point ω in the sample space. While we have often integrated sample paths of processes without comment, it is necessary to impose some restriction on them in order to ensure that these integrals are well defined. There are also other technical reasons for restricting sample paths. For our needs, it will be sufficient to assume that the sample paths of regenerative processes are continuous from the right and have left-hand limits. This is the space $\mathscr{D}$ defined in Section 2-19. Virtually all processes that arise in queueing theory have this property. [Sometimes it is convenient to assume left-continuity and right-hand limits instead, as we will do for the process $U(t)$ in Section 5-16.]

2-13 WALD'S EQUATION

We now present a result known as Wald's equation that has applications in renewal theory and in many other areas. The original context for Wald's equation was sequential experimentation.

Let $X_1, X_2, \ldots$ be an infinite sequence of independent random variables, unrestricted in sign, where X_n is the outcome of the nth experiment, should that experiment be performed. Let N be the number of experiments that are performed. After each experiment, the experimenter must decide whether or not to perform the next experiment. That decision may depend on the outcomes of the performed experiments, but cannot depend on the outcomes of the experiments that have *not* been performed. Thus, we assume that for every $n > 0$, the event $\{N \geq n\}$ is *independent* of the random variables $X_n, X_{n+1}, \ldots$, where $\{N \geq n\}$ means that the experimenter has decided to perform experiment n. If N has this property and is also proper [finite with probability one, i.e., $\sum_{n=0}^{\infty} P(N = n) = 1$], it is what we have called a *stopping time*.

Wc now prove

Theorem 14: Wald's Equation. Let N be a stopping time for an independent sequence of random variables $X_1, X_2, \ldots$. For $E(N) < \infty$, $E\{|X_n|\}$ a bounded function of n, and $E(X_n) = E(X)$ for all n,

$$E(S_N) = E(X)E(N), \tag{132}$$

where $S_N = \sum_{n=1}^{N} X_n$.

Proof. Let

$$\begin{aligned} I_n &= 1 \qquad \text{if } N \geq n, \\ &= 0, \qquad \text{otherwise, and write} \end{aligned}$$

$$S_N = \sum_{n=1}^{\infty} X_n I_n.$$

We now have

$$E(S_N) = E\left\{\sum_{n=1}^{\infty} X_n I_n\right\} = \sum_{n=1}^{\infty} E(X_n I_n). \tag{133}$$

Since X_n and I_n are independent for each n (N is a stopping time), (133) becomes

$$E(S_N) = \sum_{n=1}^{\infty} E(X_n)E(I_n) = E(X) \sum_{n=1}^{\infty} P(N \geq n) = E(X)E(N).$$

Note. The interchange $E \sum = \sum E$ in (133) is valid because the stated conditions ensure absolute convergence. See the Appendix.

Remark. We do not require that the X_n be identically distributed, but only that they have the same mean. If they are identically distributed with finite mean (the usual case in applications), the condition on $E\{|X_n|\}$ is automatically satisfied. Also notice that we permit $P(N = 0) > 0$. That is, the experimenter could decide not to perform any experiments.

Further Remark. If N is independent of the entire sequence $\{X_n\}$, Wald's equation can be obtained by conditioning on N, i.e., $E(S_N \mid N) = NE(X)$. However, this intermediate step can be incorrect. See the next example.

Example 2-16

Consider a sequence of independent coin tosses where $X_n = 1$ if heads occurs on toss n, $X_n = 0$ otherwise, with $E(X_n) = p$, $0 < p < 1$. If N is the toss on which the first head occurs, N will be a stopping time, $E(N) = 1/p < \infty$, and Wald's equation becomes

$$E(S_N) = p/p = 1. \tag{134}$$

Notice that $S_N = 1$ for any value of N because $X_N = 1$, but for any $n < N$, $X_n = 0$. In particular,

$$E(S_N \mid N = 1) = 1 \neq 1 \cdot p,$$

i.e., a conditional derivation of (134) would not be valid.

Now let $K = N - 1$ be the number of tails *prior* to the first head, and $S_K = \sum_{n=1}^{K} X_n$. Since $X_n = 0$ for every term in the sum,

$$E(S_K) = 0 \neq pE(K) > 0.$$

Thus, Wald's equation does not apply to S_K. Apparently, K is not a stopping time. Why not?

Example 2-17

Consider a sequence of independent coin tosses where $X_n = 1$ if heads occurs on toss n, but now $X_n = -1$ otherwise, with $P(X_n = 1) = p$, $P(X_n = -1) = q = 1 - p$, and $E(X) = p - q$. Thus, for any $n = 1, 2, \ldots,$

$$S_n = \sum_{j=1}^{n} X_j = \begin{array}{l}\text{number of heads minus the}\\ \text{number of tails on the first } n \text{ tosses.}\end{array}$$

Now let

$$N = \min\{n\colon S_n = 1, n = 1, 2, \ldots\},$$

i.e., N is the first toss on which the (cumulative) number of heads exceeds the number of tails.

Since for any $n \geq 1$, $\{N \geq n\}$ is independent of $X_n, X_{n+1}, \ldots$, N is a stopping time *if it is proper*. Let $\overline{X}_n = S_n/n$, $n = 1, 2, \ldots$. From (either the weak or strong) law of large numbers, $\overline{X}_n$ converges to $p - q$. When $p > q$, $n\overline{X}_n = S_n \to \infty$, which implies that N is proper. [In the next chapter, we consider this example again and show the following: N is proper if and only if $p \geq q$, and $E(N) < \infty$ if and only if $p > q$.]

Now apply Wald's equation to $E(S_N)$ in order to find $E(N)$ when $p > q$:

$$E(S_N) = 1 = (p - q)E(N), \text{ or}$$

$$E(N) = 1/(p - q), \qquad p > q. \tag{135}$$

For $p = q$, suppose we know that N is still a stopping time, but we do not know whether $E(N)$ is finite or infinite. If $E(N) < \infty$, Wald's equation gives

$$E(S_N) = 1 = 0 \cdot E(N) = 0,$$

a contradiction. The only alternative is $E(N) = \infty$, i.e., (135) is valid for $p = q$ as well. For $p < q$, (135) is nonsense, but in this case, N is defective and therefore not a stopping time.

2-14 PROOF OF THE EXPECTED-VALUE VERSION OF THE ELEMENTARY RENEWAL THEOREM

We now apply Wald's equation to obtain the expected-value version of Theorem 1, the Elementary Renewal Theorem. For a general renewal process with $E(X_j) = 1/\mu > 0, j \geq 2$,

$$\lim_{t\to\infty} m(t)/t = \mu, \tag{136}$$

where $\mu = 0$ if $E(X_j) = \infty, j \geq 2$.

Proof. First assume $E(X_j) < \infty$ for all j and write

$$Z_{M(t)+1} = X_1 + \sum_{j=2}^{M(t)+1} X_j, \tag{137}$$

i.e., we are separating out X_1, which is certain to be included in the sum. Now apply Wald's equation to the second term on the right of (137), where $N = [M(t) + 1] - 1$ is a stopping time for the i.i.d. sequence $\{X'_k: k \geq 1\}$, where for each k, $X'_k = X_{j+1}$:

$$E(Z_{M(t)+1}) = E(X_1) + m(t)/\mu. \tag{138}$$

The expected value (138) can also be represented in terms of the excess $Y(t)$:

$$E(Z_{M(t)+1}) = t + E\{Y(t)\}. \tag{139}$$

Equating (138) and (139), we have

$$m(t)/t = \mu - \mu E(X_1)/t + \mu E\{Y(t)\}/t. \tag{140}$$

Since $E\{Y(t)\} > 0$, we will have $m(t)/t > \mu - \varepsilon$ for any $\varepsilon > 0$ and all sufficiently large t, i.e.,

$$\liminf_{t\to\infty} m(t)/t \geq \mu.\dagger \tag{141}$$

To obtain the corresponding lim sup, we use a device called *truncation*: Consider the renewal process $\{M'(t)\}$ generated by $\{X'_j\}$ where

$$\begin{aligned} X'_j &= X_j \qquad \text{if } X_j \leq c, \\ &= c \qquad \text{if } X_j > c. \end{aligned} \tag{142}$$

Since $Z'_n \leq Z_n$ for all n, $M'(t) \geq M(t)$ for all t, and

$$m(t) \leq m'(t) \qquad \text{for all } t \geq 0. \tag{143}$$

Furthermore, $E\{Y'(t)\} \leq c$. Applying (140) to the truncated process,

$$\lim_{t\to\infty} m'(t)/t = \mu', \tag{144}$$

† *Lim inf* is defined in the Appendix.

where $\mu' = E(X_j')$, $j \geq 2$. Hence, from (143) and (144), $m(t)/t < \mu' + \varepsilon$ for all sufficiently large t, i.e.,

$$\limsup_{t\to\infty} m(t)/t \leq \mu'. \tag{145}$$

Now (145) holds for any choice of c in (142). As $c \to \infty$ in (145), $\mu' \to \mu$, and we have

$$\limsup_{t\to\infty} m(t)/t \leq \mu, \tag{146}$$

which, together with (141), implies (136).

For $\mu = 0$, (145) still holds, $\mu' \to 0$ as $c \to \infty$, and (136) is still true. If $\mu > 0$, but $E(X_1) = \infty$, (146) still holds. Furthermore, $E\{M(t) \mid X_1 \leq T\}/t \to \mu$ as $t \to \infty$, for any finite T. Since $m(t) \geq E\{M(t) \mid X_1 \leq T\}P(X_1 \leq T)$, lim inf $m(t) \geq \mu$, and we again have (136). This completes the proof.

Remark. When $E(X_j) < \infty$, (136) and (140) imply

$$\lim_{t\to\infty} E\{Y(t)\}/t = 0, \tag{147}$$

which is by no means obvious because $E\{Y(t)\}$ depends (in a rather complicated way) on t.

Further Remark. An obvious consequence of (136) is $m(t) = E\{M(t)\} < \infty$ for all t. A variation of (142) can be used to show that $M(t)$ *has finite moments of all orders*: Let $X_j' = c$ if $X_j > c$, $X_j' = 0$ if $X_j \leq c$. See Problem 2-11 for details.

Analogous to (136), the reward rate does not depend on the first cycle. Nor does it depend on how rewards accumulate during a cycle, provided that the accumulation satisfies the conditions set down under the heading of *partial rewards*. Instead of a general proof, however, we will be content for now to show this result for an ordinary renewal process with i.i.d. rewards, where rewards are *collected at the beginning*. (The proof is completed in the next section.) In this case, we have (23b),

$$C(t) = \sum_{j=1}^{M(t)+1} R_j.$$

Now $M(t) + 1$ is a stopping time for $\{R_j\}$ and, from Wald's equation,

$$c(t) = [m(t) + 1]E(R_j), \tag{148}$$

and

$$\lim_{t\to\infty} c(t)/t = E(R_j) \lim_{t\to\infty} [m(t) + 1]/t = \mu E(R_j), \tag{149}$$

provided that $E(|R_j|) < \infty$. Since (148) is exact, (149) is valid for $\mu = 0$ as well as $\mu > 0$.

2-15 THE RENEWAL EQUATION FOR ORDINARY RENEWAL PROCESSES; TRANSIENT BEHAVIOR

Until now, we have been concerned primarily with time averages, e.g., the fraction of time an alternating renewal process spends in a particular state, or either the reward or expected reward per unit time. We were able to obtain results without having to represent transient behavior explicitly, e.g. $P(I(t) = 1)$ as a function of t.

Sometimes transient behavior is important at some particular t or for all "sufficiently large" t. In this section, we will present a method for finding transient expressions for quantities of interest defined on regenerative processes. The corresponding renewal process is ordinary, i.e., the initial distribution $A = F$. In the next section we will extend this method to generalized regenerative processes (and general renewal processes).

For the remainder of this chapter, we will be dealing with convolutions of distributions, of other functions with distributions, and of other functions with renewal functions.

In order to understand what we are doing, we will define convolutions and other integrals in a more general way than in Section 2-2, where we assumed that A and F have density functions.

Convolutions and other related integrals will be written in terms of what are called *Stieltjes* integrals. Thus, by $A * F$, we mean that for all $t \geq 0$,

$$A * F(t) = \int_0^t A(t - u)\, dF(u). \tag{150}$$

This "dF" notation generalizes the discrete and density function cases.

For applications, one can read $dF(u) = f(u)\, du$ in the density function case, and interpret (150) as a sum in the discrete case. Thus, if $X \sim F$ is discrete with $P(X = u) = p_u$, say, (150) becomes

$$\sum_{\{u: u \in [0,t]\}} A(t - u) p_u.$$

To understand the proofs, however, one should understand how a Stieltjes integral is defined. For this purpose, see the Appendix.

We now present two examples of the kinds of problems the methods of this section will solve.

Example 2-18

Consider an alternating renewal process where an up interval begins at epoch 0, and for all $t \geq 0$, we let $I(t) = 1$ if the process is up at epoch t, $I(t) = 0$ otherwise. We want to find the probability that the machine is up at epoch t, $P_u(t) = P(I(t) = 1)$. To do this, let X_{u1} be the duration of the first up period, X_{d1} be the duration of the first down period, and $X_1 = X_{u1} + X_{d1}$ be the duration of the first cycle, where $X_1 \sim F$ and $X_{u1} \sim F_u$.

If $X_{u1} > t$, $I(t) = 1$. If $X_{u1} \le t$ and $X_1 > t$, $I(t) = 0$. On the other hand, if $X_1 = x \in [0, t]$, the process $\{I(t)\}$ *regenerates* or *renews* itself (or *starts over*) at x. That is, $P(I(t) = 1 \mid X_1 \in [0, t]) = P(I(t - X_1) = 1 \mid X_1 \in [0, t])$. Hence, for any $t \ge 0$,

$$\begin{aligned} P(I(t) &= 1 \mid X_{u1} > t, X_1 > t) = 1, \\ P(I(t) &= 1 \mid X_{u1} \le t, X_1 > t) = 0, \\ P(I(t) &= 1 \mid X_1 = x \in [0, t]) = P_u(t - x), \end{aligned} \tag{151}$$

where the concept of starting over that we used to obtain (151) is called the *renewal argument*. By unconditioning, we have

$$P_u(t) = F_u^c(t) + \int_0^t P_u(t - x)\, dF(x), \qquad t \ge 0, \tag{152}$$

where (152) is called a *renewal equation* for $P_u(t)$.

Example 2-19

We now find the tail distribution of excess, $P(Y(t) > y) = K(t)$, say, for an ordinary renewal process. Again, we condition on X_1: For any $t \ge 0$,

$$\begin{aligned} P(Y(t) > y \mid X_1 > t + y) &= 1, \\ P(Y(t) > y \mid X_1 \in (t, t + y]) &= 0, \\ P(Y(t) > y \mid X_1 = x \in [0, t]) &= K(t - x). \end{aligned}$$

By unconditioning, we have

$$P(Y(t) > y) = K(t) = F^c(t + y) + \int_0^t K(t - x)\, dF(x), \qquad t \ge 0. \tag{153}$$

Let's see what (152) and (153) have in common. In both cases, an unknown function of time, $H(t)$ say, is represented as the sum of a known function and the convolution of the unknown function and a distribution function. That is, they both have form

$$H(t) = Q(t) + \int_0^t H(t - x)\, dF(x), \qquad t \ge 0, \tag{154}$$

where (154) is called a *renewal equation* for $H(t)$.

[Notice that in Example 2-19 we introduced what might appear to be unconventional notation: $K(t) = P(Y(t) > y)$. This was done to help us recognize the convolution form in (154) and its solution in (155).]

The solution to equations of form (154) is given by

Theorem 15. The only solution to (154) that is bounded on finite intervals is

$$H(t) = Q(t) + \int_0^t Q(t - x)\, dm(x), \qquad t \geq 0, \tag{155}$$

where m is the *ordinary* renewal function, $m(t) = \sum_{j=1}^{\infty} F^{(j)}(t)$.

Proof. We use shorthand convolution notation, e.g., $H * F$ denotes the second term on the right in (154). Showing (155) satisfies (154) is equivalent to showing that $H * F = Q * m$. To show this, convolve (155) with F:

$$H * F = Q * F + Q * m * F. \tag{156}$$

The convolution operation is associative, i.e., $(Q * m) * F = Q * (m * F)$. But $m * F = \sum_{j=2}^{\infty} F^{(j)} = m - F$. Hence, (156) becomes $H * F = Q * F + Q * (m - F) = Q * m$, and we have shown that (155) is a solution to (154). To show that the solution is unique, let H_1 and H_2 be distinct bounded solutions to (154). The difference, $D(t) = H_1(t) - H_2(t)$, satisfies

$$D = D * F, \tag{157}$$

where D is bounded, i.e., $| D(x) | < k$, say, for $x \in [0, t]$, where k is a finite number. Now convolve (157) with F:

$$D * F = D * F * F,$$

i.e., $D = D * F^{(2)}$. This can be repeated indefinitely, and we have

$$D = D * F^{(n)}, \qquad n = 1, 2, \ldots. \tag{158}$$

From (158) and that $| a + b | \leq | a | + | b |$,

$$| D(t) | \leq \int_0^t | D(t - x) |\, dF^{(n)}(x) \leq kP(Z_n \leq t). \tag{159}$$

Now let $n \to \infty$ in (159). Since $E(X) > 0$, $P(Z_n \leq t) \to 0$ and we have

$$| D(t) | = 0,$$

i.e., $D(t)$ is identically equal to zero for any $t \geq 0$, a contradiction. Hence the solution to (154) is unique. This completes the proof.

Continuation of Examples 2-18 and 2-19

By applying Theorem 15, the solution to (152) is

$$P_u(t) = F_u^c(t) + \int_0^t F_u^c(t - x)\, dm(x), \qquad t \geq 0, \tag{160}$$

and the solution to (153) is

$$P(Y(t) > y) = K(t) = F^c(t + y) + \int_0^t F^c(t + y - x)\, dm(x), \qquad t \geq 0. \tag{161}$$

Defective Renewal Processes

Theorem 15 is valid when F is a defective distribution, i.e., $F(\infty) < 1$, and we are dealing with a defective renewal process (Section 2-10). As Example 2-20 shows, defective processes can arise in unexpected ways.

Example 2-20: Waiting for a Gap in Traffic

The arrival process of vehicles by the side of a road is an ordinary renewal process. You are waiting to cross the road until a gap (time between successive vehicles) greater than T occurs, the time required for you to cross safely. Let W be your waiting time until you begin to cross the road. Assume that $F(T) < 1$; otherwise, you would never be able to cross.

To find the distribution of W, $P(W \leq t) = V(t)$, say, condition on X_1. For any $t \geq 0$,

$$P(W \leq t \mid X_1 > T) = 1, \text{ i.e., } W = 0 \text{ in this case, and}$$

$$P(W \leq t \mid X_1 = x, x \in [0, T]) = V(t - x),$$

where $V(t - x) = 0$ if $x > t$. By unconditioning, we have

$$V(t) = 1 - F(T) + \int_0^T V(t - x)\, dF(x). \tag{162}$$

Now (162) is not in the standard form of a renewal equation, because the second term on the right is not a convolution. However, we can write (162) in standard form with the following device: Let

$$\begin{aligned} G(t) &= F(t) \qquad \text{for } t \leq T, \\ &= F(T) \qquad \text{for } t > T. \end{aligned} \tag{163}$$

In terms of (163), (162) can be written

$$V(t) = 1 - F(T) + \int_0^t V(t - x)\, dG(x), \qquad t \geq 0, \tag{164}$$

where (164) is a *renewal equation*! (To show (164), check out the cases $T > t$ and $T < t$ separately. For $T > t$, the range of integration can be reduced because $V(t - x) = 0$ for $x \in (t, T)$. For $T < t$, the range of integration can be extended because G has zero probability mass on $(T, t]$.)

By applying Theorem 15 to (164),

$$\begin{aligned} P(W \leq t) = V(t) &= 1 - F(T) + \int_0^t [1 - F(T)]\, dm(x) \\ &= 1 - F(T) + [1 - F(T)]m(t), \qquad t \geq 0, \end{aligned} \tag{165}$$

where now $m(t) = \sum_{j=1}^{\infty} G^{(j)}(t)$ is a defective renewal function. [To check that W is a proper random variable, let $t \to \infty$ in (165). We have $m(\infty) = G(\infty)/[1 - G(\infty)]$, where $G(\infty) = F(T)$. Hence $\lim_{t \to \infty} P(W \leq t) = 1$.]

For an ordinary renewal process, where $R(t)$ is the reward associated with spread $X(t)$, it is easy to see that $\{R(t): t \geq 0\}$ is a regenerative process! In the next example, we obtain a useful expression for $E\{R(t)\}$, $t \geq 0$.

Example 2-21

For an ordinary renewal process, let $H_r(t) = E\{R(t)\}$. By conditioning on X_1, we have for any $t \geq 0$,

$$E(R(t) \mid X_1 > t) = E(R_1 \mid X_1 > t),$$

$$E(R(t) \mid X_1 = x \in [0, t]) = H_r(t - x).$$

By unconditioning, we have

$$H_r(t) = E(R_1 \mid X_1 > t)P(X_1 > t) + \int_0^t H_r(t - x)\, dF(x), \qquad t \geq 0. \tag{166}$$

By applying Theorem 15 to (166), we have

$$E(R(t)) = H_r(t) = Q(t) + \int_0^t Q(t - x)\, dm(x), \qquad t \geq 0, \tag{167}$$

where

$$Q(t) = E(R_1 \mid X_1 > t)P(X_1 > t).$$

Completion of Proof of the Renewal-Reward Theorem

We now use (167) to complete the proof of the following: If $E(|R_j|) < \infty$ and $\mu > 0$,

$$\lim_{t \to \infty} c(t)/t = \mu E(R_j). \tag{168}$$

We showed (168) earlier as (149) under the assumption that rewards are collected at the beginning. If they are collected at the end,

$$C(t) = \sum_{j=1}^{M(t)} R_j,$$

and from (148),

$$c(t) = [m(t) + 1]E(R_j) - E(R(t)). \tag{169}$$

Hence, (168) will hold if and only if

$$\lim_{t \to \infty} H_r(t)/t = 0. \tag{170}$$

To show (170), first observe that $E(|R_j|) < \infty$ implies that for the $Q(t)$ in (167), $\lim_{t \to \infty} Q(t) = 0$, and for any t, $|Q(t)| < k = E\{|R_j|\} < \infty$. Now choose T so large that $|Q(t)| < \varepsilon$ for all $t > T$, and write

$$\left| \int_0^t Q(t-x)\,dm(x) \right| \le \left| \int_0^{nT} \cdots \right| + \left| \int_{nT}^t \cdots \right|$$

$$\text{for } t \in (nT, (n+1)T], \qquad n = 1, 2, \ldots,$$

$$\le \varepsilon m(nT) + k[m(t) - m(nT)]. \tag{171}$$

Since ε can be made arbitrarily small, and $[m(t) - m(nT)]/t \to 0$ as $t \to \infty$, (171) implies (170), and (168) holds when rewards are collected at the end. If rewards during a cycle accumulate in an arbitrary way, but in accordance with our assumptions under *partial rewards*, (170) holds for the positive and negative reward terms in equation (36). Hence (168) holds in general.

A slightly more involved example of the methods of this section is the following.

Example 2-22: Higher Moments of the Renewal Process

For an ordinary renewal process, let $K_2(t) = E\{[M(t)]^2\}$, $t \ge 0$. By conditioning on X_1, for any $t \ge 0$,

$$E\{[M(t)]^2 \mid X_1 > t\} = 0,$$

$$E\{[M(t)]^2 \mid X_1 = x \in [0, t]\} = E\{[1 + M(t-x)]^2\}$$

$$= 1 + 2m(t-x) + K_2(t-x).$$

By unconditioning, we have

$$K_2(t) = F(t) + 2\int_0^t m(t-x)\,dF(x) + \int_0^t K_2(t-x)\,dF(x), \qquad t \ge 0. \tag{172}$$

We previously observed that $m * F = m - F$, and (172) becomes

$$K_2(t) = 2m(t) - F(t) + \int_0^t K_2(t-x)\,dF(x). \tag{173}$$

By applying Theorem 15 to (173), we obtain

$$K_2(t) = 2m(t) - F(t) + 2\int_0^t m(t-x)\,dm(x) - \int_0^t F(t-x)\,dm(x). \tag{174}$$

Since $F * m = m * F = m - F$, (174) becomes

$$E\{[M(t)]^2\} = m(t) + 2\int_0^t m(t-x)\,dm(x), \qquad t \ge 0. \tag{175}$$

Higher moments can be found by the same method. In fact, it can be shown that the nth *factorial* moment of $M(t)$ is related to the n-fold convolution of the renewal function as follows:

$$E\{M(t)(M(t)-1)\cdots(M(t)-n+1)\} = (n!)m^{(n)}(t), \qquad t \ge 0. \tag{176}$$

To summarize, we have illustrated a method for finding explicit expressions for $H(t) = E\{S(t)\}$, say, where $\{S(t)\}$ is a stochastic process defined on a regenerative process. This was done by conditioning on X_1, and sometimes other random variables as well (see Example 2-18). Because the regenerative process starts over at X_1, the result of this is a renewal equation (154) for $H(t)$, which, from Theorem 15, we then solve.

2-16 THE RENEWAL EQUATION FOR GENERAL RENEWAL PROCESSES; STATIONARY REGENERATIVE PROCESSES

The methods and results of the preceding section can readily be extended to the transient behavior of generalized regenerative processes. Results for the general case are found in terms of results for the regenerative case and its *ordinary* renewal process. The reason for this is that once X_1 has occurred, the generalized process is regenerative *from then on.*

To keep the general and ordinary cases distinct, in this section we subscript functions restricted to the ordinary case, e.g., $m_o(t)$ is now the renewal function for an ordinary renewal process, $A = F$.

With this notation, Theorem 15 in the preceding section becomes: the unique bounded solution to

$$H_o = Q_o + H_o * F$$

is

$$H_o = Q_o + Q_o * m_o, \tag{177}$$

where $m_o = \sum_{j=1}^{\infty} F^{(j)}$.

When we solve for the corresponding quantity for a general renewal process, we will obtain an expression of form

$$H = Q + H_o * A, \tag{178}$$

where $X_1 \sim A$. The solution to (178) is given by

Theorem 16. The solution to (178) is

$$H = Q + Q_o * m, \qquad t \geq 0, \tag{179}$$

where m is the *general* renewal function, $m = \sum_{j=0}^{\infty} A * F^{(j)}$.

Proof. Simply plug (177) into (178):

$$H = Q + Q_o * A + Q_o * m_o * A,$$

where $Q_o * m_o * A = Q_o * (m - A) = Q_o * m - Q_o * A$, and we have (179).

Remark. Notice that the solution to a renewal equation, in both the ordinary and general cases, includes a term that is the convolution of some function with a renewal function. As we shall see in Section 2-19, this form is particularly important for the investigation of pointwise limits.

Example 2-23

We now find the tail distribution of excess, $P(Y(t) > y)$, $t \geq 0$, for a general renewal process. Following Example 2-19,

$$P(Y(t) > y \mid X_1 > t + y) = 1$$

$$P(Y(t) > y \mid X_1 \in (t, t + y]) = 0$$

$$P(Y(t) > y \mid X_1 = x \in [0, t]) = K_o(t - x),$$

when $K_o(t)$ now denotes the right side of (161). By unconditioning, where now $X_1 \sim A$, we have

$$P(Y(t) > y) = K(t) = A^c(t + y) + \int_0^t K_o(t - x)\, dA(x). \tag{180}$$

From Theorem 16, the solution to (180) is

$$P(Y(t) > y) = A^c(t + y) + \int_0^t F^c(t + y - x)\, dm(x), \tag{181}$$

where $m = \sum_{j=0}^{\infty} A * F^{(j)}$. For $A = F$, note that (181) becomes (161). (Notice that the conditional expectations are the same as in the ordinary case. While this is often true, one should "think through" each step to be sure.)

The Equilibrium Renewal Process

We now consider a special case of a general renewal process where for $X_j \sim F$, $j \geq 2$, and $\mu > 0$, we choose initial distribution $A = F_e$, the equilibrium distribution. For this special case we will subscript certain quantities, e.g., $\{M_e(t)\}$ is an *equilibrium* renewal process and $m_e(t)$, $t \geq 0$ is an *equilibrium* renewal function.

This terminology is used because, as we shall see, $\{M_e(t)\}$ has stationary increments, i.e., it is a *stationary point process*. Conceptually, choosing $A = F_e$ amounts to observing an (ordinary or general) renewal process at random with the excess at epoch 0, $Y_0 \sim F_e$. The terms "stationary" and "in equilibrium" are used interchangeably. This is the reason we called F_e an equilibrium distribution.

We begin by finding the transform of a *general* renewal function, $\tilde{m}(s)$, where we interchange integral and sum below and take transforms term by term: For $s \geq 0$,

$$\tilde{m}(s) = \int_0^{\infty} e^{-st}\, dm(t) = \sum_{n=1}^{\infty} \int_0^{\infty} e^{-st}\, dA * F^{(n-1)}(t) = \sum_{n=0}^{\infty} \tilde{A}(s)[\tilde{F}(s)]^n$$

$$= \frac{\tilde{A}(s)}{1 - \tilde{F}(s)}. \tag{182}$$

For $A = F_e$, substitute $\tilde{F}_e(s) = \mu[1 - \tilde{F}(s)]/s$ [found as (48)], and (182) becomes

$$\tilde{m}_e(s) = \mu/s.$$

Since distinct functions have distinct transforms, we can verify that

$$m_e(t) = \mu t, \qquad t \geq 0, \tag{183}$$

by finding its transform

$$\int_0^\infty e^{-st}\, dm_e(t) = \mu \int_0^\infty e^{-st}\, dt = \mu/s.$$

We now evaluate (181) for an equilibrium renewal process: For any $t \geq 0$,

$$\begin{aligned} P(Y(t) > y) &= F_e^c(t + y) + \mu \int_0^t F^c(t + y - x)\, dx \\ &= F_e^c(t + y) + \mu \int_y^{t+y} F^c(u)\, du \\ &= F_e^c(t + y) + F_e^c(y) - F_e^c(t + y) \\ &= F_e^c(y), \end{aligned}$$

i.e., if $X_1 \sim F_e$, $Y(t) \sim F_e$ for all t. Since subsequent inter-arrival times are i.i.d. $\sim F$, this implies that for any $t \geq 0$, $M_e(t + x) - M_e(t)$ has the same distribution as $M_e(x)$, i.e., $\{M_e(t)\}$ has stationary increments.

We can reverse the argument leading to (183) to show that a renewal function is of form $m(t) = \mu t$ if *and only if* $A = F_e$. For an ordinary renewal process, the renewal function is of form $m_o(t) = \mu t$ if *and only if* $A = F_e = F$, which implies (why?) that the ordinary renewal process is Poisson!

Stationary Regenerative Processes

Now denote a regenerative process by $\{B_o(t)\}$ and a corresponding *generalized* regenerative process by $\{B(t)\}$, i.e., these processes differ only in the probabilistic structure of the first cycle.

From the renewal argument, it is easily shown that for an arbitrary set of states E and every epoch $t \geq 0$.

$$\begin{aligned} P(B_o(t) \in E) &= Q_o(t) + Q_o * m_o(t), \text{ and} \\ P(B(t) \in E) &= Q(t) + \mathrm{Q}_o * \mathrm{m}(t), \end{aligned} \tag{184}$$

where

$$Q_o(t) = P(B_o(t) \in E,\, X_1 > t), \text{ and} \tag{185}$$

$$Q(t) = P(B(t) \in E,\, X_1 > t), \tag{186}$$

where $X_1 \sim F$ in (185) and $X_1 \sim A$ in (186).

For any regenerative process with *finite mean cycle length*, we now construct what we will call a *stationary version* of that process, $\{B_e(t)\}$, by defining the first cycle in a special way:

Assume all cycle durations are $\sim F$, but that measured from epoch 0, the first cycle is assumed to have started at epoch $-U$, where $U \sim F_e$ is the age (attained duration) of that cycle at epoch 0. Let X_r be the *remaining duration* of the cycle in progress at epoch 0, and, for every $u > 0$, define

$$P(B_e(t) \in E, X_r > t \mid U = u) \equiv P(B_o(t + u) \in E, X_1 > t + u \mid X_1 > u)$$
$$= P(B_o(t + u) \in E, X_1 > t + u)/F^c(u).$$

Integrating with respect to the density of U, $f_e(u) = \mu F^c(u)$, we have

$$\begin{aligned} Q_e(t) &\equiv P(B_e(t) \in E, X_r > t) \\ &= \mu \int_0^\infty P(B_o(t + u) \in E, X_1 > t + u)\, du \\ &= \mu \int_t^\infty P(B_o(u) \in E, X_1 > u)\, du, \end{aligned} \tag{187}$$

where X_r has the role of the first cycle length for $\{B_e(t)\}$.

If we let E be the entire state space of the regenerative process, (187) simplifies, and it is easily shown that $X_r \sim F_e$. Therefore, process $\{B_e(t)\}$ has corresponding renewal function $m_e(t) = \mu t$. Specializing (184) and (186) (substituting Q_e for Q and m_e for m), we have that for every $t \geq 0$.

$$\begin{aligned} P(B_e(t) \in E) &= \mu \int_t^\infty P(B_o(u) \subset E, X_1 > u)\, du + \mu \int_0^t P(B_o(u) \in E, X_1 > u)\, du \\ &= \mu \int_0^\infty P(B_o(u) \in E, X_1 > u)\, du, \end{aligned}$$

independent of t, i.e., $\{B_e(t)\}$ is a *stationary* stochastic process.

Clearly, the pointwise limit of $P(B_e(t) \in E)$ as $t \to \infty$ exists, and from Section 2-8, this limit must be the corresponding time average, $P\{B_\infty \in E\}$, i.e., $B_e(t) \sim B_\infty$. To verify, define *two* indicator functions for every $t \geq 0$:

$$I(t) = 1 \quad \text{if } B_o(t) \in E, \qquad J(t) = 1 \quad \text{if } B_o(t) \in E \text{ and } X_1 > t,$$
$$= 0 \quad \text{otherwise.} \qquad\qquad = 0 \quad \text{otherwise.}$$

Clearly, $\int_0^{X_1} I(t)\, dt = \int_0^{X_1} J(t)\, dt = \int_0^\infty J(t)\, dt$. By working with the right-hand integral and interchanging $E \int = \int E$, we have

$$E\left\{\int_0^{X_1} I(t)\, dt\right\} = \int_0^\infty P(B_o(t) \in E, X_1 > t)\, dt. \tag{188}$$

By combining (188) with time average results for regenerative processes, we have

$$P(B_e(t) \in E) = P(B_\infty \in E) = \mu \int_0^\infty P(B_o(u) \in E, X_1 > u)\, du \quad \text{for all } t \geq 0. \tag{189}$$

Because the distribution of B_∞ does not depend on properties of the first cycle, there can't be *two* stationary distributions, i.e., (189) must be unique.

Essentially the same argument may be used to show that $\{B_e(t)\}$ is *strictly stationary*, which means that joint distributions are stationary, e.g., for every $t \geq 0$ and $h > 0$, the joint distribution of $B_e(t)$ and $B_e(t + h)$ can be constructed and shown to be independent of t. The renewal argument still applies to the process $\{B(t), B(t + h): t \geq 0\}$ because $\{B(X_1 + t), B(X_1 + t + h): t \geq 0\}$ is independent of X_1, and is stochastically equivalent to the ordinary version. Note that there is now some dependence between cycles, e.g., for $X_1 - t < h$, $\{B(t), B(t + h)\}$ and $\{B(X_1), B(X_1 + h)\}$ are dependent because $B(t + h)$ falls in the next cycle.

Now return to a generalized regenerative process with arbitrary first cycle. For every $t \geq 0$ and arbitrary x and y, define the indicator $I(t) = 1$ if $B(t) \leq x$ *and* $B(t + h) \leq y$, $I(t) = 0$ otherwise, and let R_j be the integral of $\{I(t)\}$ over the jth cycle, $j \geq 1$. As noted, the R_j are no longer independent because when t is within h of the end of the cycle in which it falls, $t + h$ is in the next (or a latter) cycle. Nevertheless, it is easy to see that for $j \geq 2$, $\{R_j\}$ is a strictly stationary sequence.

* It can be shown that $\{R_j\}$ satisfies a *mixing* condition, which implies that it is *ergodic*. From the *Ergodic Theorem*,

$$\lim_{n\to\infty} \sum_{j=1}^{n} R_j/n$$

converges to a constant w.p.1. See pages 12 and 13 of Billingsley [1965] for definitions and a statement of this theorem.

Thus the limiting joint distribution of $B(t)$ and $B(t + h)$ exists (in a time average sense), where the random-variable version of this limit is a constant w.p.1, and this limit is independent of properties of the first cycle. This limit uniquely determines the stationary joint distribution of $B_e(t)$ and $B_e(t + h)$. In turn, these stationary joint distributions, at $t, t + h_1, \ldots, t + h_k$ say, uniquely determine the stochastic process $\{B_e(t)\}$. We have proven

Theorem 17. For every regenerative process $\{B_o(t)\}$ with finite mean cycle length, there is a *unique* stochastic process $\{B_e(t)\}$ called the *stationary version* of $\{B_o(t)\}$, where $\{B_e(t)\}$ is a strictly stationary stochastic process with $B_e(t) \sim B_\infty$.

Important Remarks. In applications, e.g., to Markov chains, the stationarity of $\{B_e(t)\}$ can often be used to *find* this stationary distribution, or at least its mean,

* Uses concepts and results not needed elsewhere in this book.

$E\{B_\infty\}$, which of course is $\overline{B}$. We call this approach a *stationary analysis*. Important applications of this approach will be given in subsequent chapters, e.g., Theorem 5 in Chapter 3. When mean cycle lengths are *infinite*, and the conditions for (122) hold, a stationary version $\{B_e(t)\}$ *will not exist*.

Example 2-24

We now do a stationary analysis. Consider the stationary version of the excess process, $\{Y_e(t)\}$, when the inter-renewal times $X_j = c$, a constant, $j \geq 2$. We will use stationarity to show that $Y_e(t) \sim F_e$, a uniform distribution in this case, which of course we already know. Assume that $Y_e(0)$ is continuous (continuity can be argued separately). For arbitrary $(a, b) \subset (0, c)$ and $t \in (0, c)$,

$$\begin{aligned} P\{Y_e(0) \in (a, b)\} &= P\{Y_e(t) \in (a - t, b - t)\} && \text{for } t < a, \\ &= P\{Y_e(t) \in (a + c - t, b + c - t)\} && \text{for } t > b, \\ &= P\{Y_e(t) \in (0, b - t) \cup (a + c - t, c)\} && \text{for } t \in (a, b). \end{aligned}$$

From $Y_e(t) \sim Y_e(0)$ and the above, the probability that $Y_e(0)$ falls in an interval of length l within $(0, c)$ is *independent of the location of the interval*. It follows that this probability is proportional to l (break the interval into subintervals), which implies that $Y_e(0) \sim$ uniform on $(0, c)$.

Remark. Result (189) and Theorem 17 hold for more general state spaces, e.g., $B_o(t)$ can be a vector. The only change is the definition of E.

2-17 A DIRECT METHOD FOR FINDING TRANSIENT EXPRESSIONS

In the last two sections, we developed a method for finding transient expressions based on the renewal argument. These expressions can also be found in a more direct way that in some cases is easier. We now present this method for the case of a *general* renewal process.

For a generalized regenerative process $\{B(t)\}$, suppose we wish to find $k(t) = E\{f(B(t)\}$ for some function f. This can be done formally, as follows:

$$\begin{aligned} k(t) &= \sum_{j=0}^{\infty} E\{f(B(t)) \mid M(t) = j\}P(M(t) = j) \\ &= \sum_{j=0}^{\infty} E\{f(B(t))I_j\}, \end{aligned} \tag{190}$$

where

$$\begin{aligned} I_j &= 1 && \text{if } M(t) = j, \\ &= 0, && \text{otherwise.} \end{aligned}$$

These equations are no more than standard conditional expectation expressions. Implementing them as a "method" depends on handling (190) conveniently. When $k(t)$ is a probability (the expected value of an indicator), the terms in (190) are joint probabilities.

Example 2-25

For $k(t) = P(Y(t) > y)$, (190) becomes

$$\begin{aligned} P(Y(t) > y) &= \sum_{j=0}^{\infty} P(Y(t) > y,\ M(t) = j) \\ &= P(X_1 > t + y) + \sum_{j=1}^{\infty} P(Y(t) > y,\ M(t) = j) \\ &= A^c(t + y) + \sum_{j=1}^{\infty} P(Z_j \leq t,\ X_{j+1} > t + y - Z_j) \\ &= A^c(t + y) + \sum_{j=1}^{\infty} \int_0^t F^c(t + y - u)\, dA * F^{(j-1)}(u). \end{aligned}$$

Now interchange $\sum \int = \int \sum$, and then $\sum d = d \sum$, where

$$d \sum_{j=1}^{\infty} A * F^{(j-1)}(u) = dm(u),$$

and we have (181).

While Example 2-25 is typical, sometimes there are easier ways to proceed than (190).

Example 2-26: The $GI/G/\infty$ Queue

This system operates like the $M/G/\infty$ queue first introduced in Example 2-6. The only difference is that now the arrival process is a general renewal process. The service times are still i.i.d. $\sim G$.

As before, we let $N(t)$ be the number of customers in system (busy servers) at epoch t, $t \geq 0$, but now we define

$$\begin{aligned} I_j &= 1 \quad \text{if } Z_j \leq t < Z_j + S_j, \\ &= 0 \quad \text{otherwise,} \end{aligned}$$

i.e., $I_j = 1$ if the jth arrival is in the system at epoch t. We have

$$N(t) = \sum_{j=1}^{\infty} I_j.$$

It is now easy to find an expression for $E(N(t))$.

$$E(N(t)) = \sum_{j=1}^{\infty} E(I_j) = \sum_{j=1}^{\infty} \int_0^t G^c(t-u)\, dA * F^{(j-1)}(u)$$
$$= \int_0^t G^c(t-u)\, dm(u). \tag{191}$$

2-18 LATTICE DISTRIBUTIONS; EXAMPLES WHERE POINTWISE LIMITS DO NOT EXIST

We now return to the question of the existence and form of pointwise limits. Specific results are presented in the next section. In this section, we present examples where these limits do not exist.

Returning now to the excess distribution, consider a deterministic ordinary renewal process generated by $X_j = 1, j = 1, 2, \ldots$. For $y \in (0, 1)$,

$$\begin{aligned} P(Y(t) \le y) &= 1 \qquad \text{for } t \in [n - y, n), \quad n = 1, 2, \ldots, \\ &= 0 \qquad \text{otherwise.} \end{aligned}$$

Clearly, $P(Y(t) \le y)$ is a periodic function of t; *lim* $P(Y(t) \le y)$ does not exist. However, by direct argument, it is easy to show that

$$C.L.\ P(Y(t) \le y) = y, \qquad 0 < y < 1.$$

This may also be obtained by specializing earlier results. We know $C.L.\ P(Y(t) \le y) = F_e(y)$. If F is a unit step function at c, say, F_e is a uniform distribution on $(0, c)$.

For the same special case, $m(t) - t$ is also a periodic function

$$m(t) - t = [t] - t, \qquad t \ge 0,$$

where, for any t, $[t]$ is defined as the greatest integer $\le t$.

More generally, if the X_j are integral-valued, so are their partial sums Z_n, $n = 1, 2, \ldots$. Hence, renewals can occur only on the integers. Clearly, $P(Y(t) \le y) = 0$ and $m(t + y) - m(t) = 0$ if $(t, t + y]$ does not contain an integer. Since $C.L.\ P(Y(t) \le y) = F_e(y) > 0$ for y near 0, the corresponding pointwise limit does not exist (otherwise, by Theorem 6, it would be $F_e(y)$).

We have identified circumstances under which pointwise limits will fail to exist. A minor generalization of these ideas characterizes this situation.

Definition. A proper random variable X, and its distribution F, are called *lattice* if for some $d > 0$,

$$\sum_{n=-\infty}^{\infty} P(X = nd) = 1. \tag{192}$$

Otherwise, $X(F)$ are called *nonlattice*. The largest d with property (192) is called the *span* of X, and we say X is *d-lattice* when we wish to denote the span as well. If X is defective, it is lattice if (192) holds when we replace 1 on the right by $P(|X| < \infty) > 0$.

Remark. Feller ([1971], p. 138) uses the term *arithmetic* in place of lattice. Unfortunately, he also uses the term lattice, but in a slightly different way.

Integer-valued random variables are lattice with span $d = 1$ or a multiple thereof. The common discrete distributions (those with names) have span $d = 1$. Other examples are

(a) If $P(X = 4) = P(X = 6) = 1/2$, X is 2-lattice.
(b) If $P(X = 4\pi) = P(X = 6\pi) = 1/2$, X is 2π-lattice.
(c) If $P(X = \sqrt{2}) = P(X = \sqrt{8}) = 1/2$, X is $\sqrt{2}$-lattice.
(d) If $P(X = \sqrt{2}) = P(X = \sqrt{3}) = 1/2$, X is nonlattice.
(e) If $X = 1/(Y + 1)$, where $Y \sim P(\lambda)$, X is nonlattice.

2-19 BLACKWELL'S THEOREM, THE KEY RENEWAL THEOREM, AND POINTWISE LIMITS

It is remarkable that lattice versus nonlattice turns out to be the crucial distinction for the existence of pointwise limits of many functions that arise in renewal theory. These results are summarized by the next two theorems. For regenerative processes, an additional complication occurs. It is treated in Theorem 20.

Because proofs are long and technical, they are omitted. In Section 2-21, we give a brief historical account of the development of renewal theory, together with references where proofs of Theorems 18 through 20 may be found.

For proper distributions A and F, we now state

Theorem 18: Blackwell's Theorem. For a *general* renewal process with arbitrary A and nonlattice F, and any $h > 0$,

$$\lim_{t \to \infty} [m(t + h) - m(t)] = \mu h. \tag{193}$$

For an *ordinary* renewal process ($A = F$) with d-lattice F,

$$\lim_{t \to \infty} [m(t + h) - m(t)] = \mu h \qquad \text{for } h = d, 2d, \ldots. \tag{194}$$

For $E(X_j) = \infty$, $j \geq 2$, the limits are valid with $\mu = 0$. For $E(X_j) < \infty$, $j \geq 2$, the limit (194) will not exist if h is not a multiple of the span.

In the nonlattice case, Blackwell's Theorem tells us that the renewal function behaves like a linear function as $t \to \infty$, independent of A. In the lattice case, initial conditions are important. Hence, we have stated (194) only for $A = F$. If we choose $A = F_e$, $m_e(t) = \mu t$, and (194) would hold for any $h > 0$, whether or not F is lattice!

Convolutions of form $Q * m$, where m is a renewal function, arise from solving renewal equations. Blackwell's Theorem suggests that the convolutions may also have pointwise limits. For nonlattice F, this turns out to be true provided that Q is "reasonable." In fact, if we approximate $dm(u)$ by $\mu\, du$ and change variables ($x = t - u$), we can approximate the convolution

$$\int_0^t Q(t - u)\, dm(u) \approx \mu \int_0^t Q(x)\, dx,$$

which in turn suggests the limit $\lim Q * m(t) = \mu \int_0^\infty Q(x)\, dx$.

The class of functions Q for which our intuitive argument turns out to predict the correct limit is called *Directly Riemann Integrable (DRI)*. The idea is this: Recall how a Riemann integral is defined on any finite interval $[a, b]$ in terms of upper and lower sums. Riemann integrals over infinite intervals are called *improper* and are defined as limits of Riemann integrals over finite intervals:

$$\int_0^\infty Q(t)\, dt = \lim_{b \to \infty} \int_0^b Q(t)\, dt.$$

Instead of this, suppose we partition $[0, \infty)$ into intervals of length h and define upper and lower sums on the entire interval for Q. If these sums are finite and converge to the same limit, l say, as $h \to 0$, Q is said to be *DRI* with integral l.

A function that is DRI is necessarily RI (Riemann integrable in the improper sense), but not conversely. As a practical matter, the distinction will rarely be important to us because we usually will be dealing with monotone functions, for which DRI and RI are the same.

We are now ready to state

Theorem 19: The Key Renewal Theorem (KRT). For a general renewal process where F is nonlattice and Q is DRI,

$$\lim_{t \to \infty} \int_0^t Q(t - u)\, dm(u) = \mu \int_0^\infty Q(t)\, dt, \tag{195}$$

where this limit is zero if $E(X_j) = \infty, j \geq 2$.

Remark. We omit a statement of the KRT that corresponds to (194) for an ordinary renewal process when F is lattice.

Blackwell's Theorem and the KRT may be shown to be equivalent. Showing $KRT \Rightarrow B$ is easy. Showing $B \Rightarrow KRT$ is a bit messy because of the definition of DRI. While Blackwell's Theorem is easier to understand, the *KRT* is usually easier to use in applications, because of the convolution form of transient expressions.

Example 2-27

For an ordinary renewal process we found the distribution of excess,

$$P(Y(t) > y) = F^c(t + y) + \int_0^t F^c(t + y - x)\, dm(x).$$

For F nonlattice, we apply the KRT:

$$\lim_{t\to\infty} P(Y(t) > y) = 0 + \mu \int_0^\infty F^c(t + y)\, dt = F_e^c(y),$$

the tail of the equilibrium distribution. For a general renewal process, we found

$$P(Y(t) > y) = A^c(t + y) + \int_0^t F^c(t + y - x)\, dm(x),$$

where m is now a general renewal function. Nevertheless, as $t \to \infty$, the limit is the same.

In the nonlattice case, pointwise limits corresponding to the time average results obtained earlier will in most cases exist and be independent of initial conditions (and, of course, be equal to the corresponding time averages).

A more refined example, where initial conditions do matter, is the following.

Example 2-28

Let $m_o(t)$ denote an ordinary renewal function and $m_e(t) = \mu t$ be the corresponding equilibrium renewal function, where $A = F_e$. Recalling that both renewal functions are sums of convolutions, we can write

$$\begin{aligned} m_o(t) - \mu t &= m_o(t) - m_e(t) \\ &= F(t) - F_e(t) + \int_0^t [F(t - u) - F_e(t - u)]\, dm_o(u). \end{aligned}$$

We would like to apply the KRT to the convolution above, but F and F_e are not integrable $[\int_0^\infty F(t)\, dt = \infty]$. However, $F - F_e = F_e^c - F^c$. If F is nonlattice and $E(X^2) < \infty$,

$$\begin{aligned} \lim_{t\to\infty} [m_o(t) - \mu t] &= 0 + \lim_{t\to\infty} \int_0^t [F_e^c(t - u) - F^c(t - u)]\, dm_o(u) \\ &= \mu\left[\int_0^\infty F_e^c(t)\, dt - \int_0^\infty F^c(t)\, dt\right] = \mu^2 E(X^2)/2 - 1. \qquad (196) \end{aligned}$$

Whether or not F is lattice, it can be shown (see Problem 2-89) that

$$\lim_{t\to\infty} \int_0^t [m_o(u) - \mu u]\, du/t = \mu^2 E(X^2)/2 - 1. \qquad (197)$$

Check this out for the renewal function $m(t) - t = [t] - t$ in Section 2-18. Clearly, the pointwise limit may be different if $A \neq F$, e.g., $A = F_e$.

Existence of Pointwise Limits for Regenerative Processes

For regenerative process $\{B(t)\}$, let

$$P_b(t) = P(B(t) \leq b), \tag{198}$$

for any $t \geq 0$ and any real b. We now explore whether (198) has a pointwise limit as $t \to \infty$. If this limit exists and (as a function of b) is a distribution function, $\{B(t)\}$ is said to *converge in distribution* in the conventional (pointwise rather than time average) sense.

From the renewal argument, it is easily shown that

$$P_b(t) = Q_o(t) + \int_0^t Q_o(t - u)\, dm_o(u), \tag{199}$$

where

$$Q_o(t) = P(B(t) \leq b,\ X_1 > t).$$

If F is nonlattice and Q_o is DRI, we can apply the KRT to (199),

$$\lim_{t \to \infty} P(B(t) \leq b) = \mu \int_0^\infty P(B(t) \leq b,\ X_1 > t)\, dt, \tag{200}$$

where, for $\mu > 0$, (200) is a distribution function. The right-hand side of (200) is time average (189), whether or not (200) exists as a pointwise limit.

How do we know Q_o is DRI? Clearly,

$$0 \leq Q_o(t) \leq P(X_1 > t) \qquad \text{for any} \quad t \geq 0,$$

where $P(X_1 > t)$ is DRI. However, this alone does not imply Q_o is DRI.

Example 2-29

To illustrate what can go wrong, suppose $\{B(t)\}$ is deterministic where for any $t \geq 0$, $B(t) = 1$ if t is rational and $B(t) = 0$ otherwise. Let the X_j be rational but nonlattice, e.g., case (e) at the end of Section 2-18. Now $\{B(t)\}$ and $\{X_j\}$ are (trivially) independent. Nevertheless, $\{B(t)\}$ is regenerative with regeneration points $X_1, X_1 + X_2, \ldots$. Clearly, $\{B(t)\}$ does *not* converge in distribution. The explanation is that $Q_o(t)$, which fluctuates wildly between 0 and $P(X_1 > t)$, is not DRI.

While Example 2-29 may seem weird, we now see the need for some condition beyond F nonlattice in order to ensure that limit (200) exists.

It turns out that there are two ways to proceed. We may impose either conditions on the sample path behavior of $\{B(t)\}$ or more stringent conditions on F. Stating these conditions requires special technical terms.

In particular, a real-valued function defined on $[0, \infty)$ is said to belong to $D[0, \infty)$ if it is right-continuous and has left-hand limits. We will require that almost all realizations of $\{B(t)\}$ have the following property.

Definition. A stochastic process $\{B(t): t \geq 0)\}$ is said to belong to $\mathscr{D}$ if

$$P\{\omega: B(t, \omega) \in D[0, \infty)\} = 1. \tag{201}$$

With regard to restrictions on F, recall that a distribution may be discrete, continuous, or mixed. A continuous distribution that may be represented as an integral of a density function is called *absolutely continuous*. Continuous distributions that do not have density functions exist. These are called *singular*.

In general, any distribution F may be (uniquely) represented as a mixture of the three possible types,

$$F = \alpha F_{\text{d}} + \beta F_{\text{ac}} + \gamma F_{\text{s}}, \tag{202}$$

where d, ac, and s denote discrete, absolutely continuous, and singular, respectively. The weights α, β, and γ are nonnegative with $\alpha + \beta + \gamma = 1$. Distribution F is said to have an *absolutely continuous component* if $\beta > 0$.

The convolution of any distribution with one that is absolutely continuous is also absolutely continuous. Hence, if $\beta > 0$ in (202), the weight on the absolutely continuous component of convolution $F^{(n)}$ is easily shown to be at least

$$1 - (1 - \beta)^n, \qquad n = 1, 2, \ldots. \tag{203}$$

For sufficiently large n, the absolutely continuous component dominates the others.

Curiously, when $\beta = 0$ but $\gamma > 0$ in (202), it is possible that some *convolution* $F^{(n)}$ will have an absolutely continuous component. [This is why "at least" preceded (203) above.] A distribution F is called *spread-out* if some convolution of F has an absolutely continuous component. Finally, we are ready to state

Theorem 20. A regenerative process $\{B(t)\}$, with finite mean cycle length $E(X_1) < \infty$, converges in distribution (in a pointwise sense) to (200) if either

(a) F is nonlattice and $\{B(t)\} \in \mathscr{D}$, or
(b) F is spread-out.

Remarks. Condition (a) is the more useful in practice. Other restrictions on the sample paths of $\{B(t)\}$ can replace those in (a); in particular, that they be left-continuous and have right-hand limits. Condition (b) implies F is nonlattice. Furthermore, it can be shown that under condition (b), Q_o in (199) is DRI, and the KRT applies. Lattice versions of Theorems 18 and 19 are known, provided that t is restricted to multiples of the span. Theorem 20 is easily extended to cover generalized regenerative processes. We require only that the first cycle ends w.p.1 and $\{B(t + X_1)\}$ *from then on* satisfy the conditions of the theorem.

2-20 ASYMPTOTIC NORMALITY AND THE REGENERATIVE METHOD*

In this book, we use analytic methods to study stochastic phenomena. For complex models, this can be difficult to do. Furthermore, it is often even more difficult to interpret our results numerically.

For example, we know that the time average of a regenerative process is the ratio of two expectations defined in terms of a single cycle. If the cycle is complex, we may not have any idea how to compute either expectation.

One way to estimate the time average (or any other property) of a "real" system is to collect data on it, and to use the data to construct an estimate.

For a stochastic model (process), we can do the same thing by simulating it over time. This means that we generate realized values of a collection of random variables, where these values determine the evolution of the process over some finite period.

Generating realized values of random variables from specified distributions is a special topic covered in many books that treat what is called "discrete event simulation." In the remainder of this section, we simply assume that this can be done.

The *regenerative method* is a way to estimate a time average $\overline{B}$ of a regenerative process $\{B(t)\}$ from a simulation of a finite number of cycles. Suppose we simulate $\{B(t)\}$ for $n \geq 1$ cycles, where $X_1, \ldots, X_n$ are the n cycle lengths, and $R_1, \ldots, R_n$ are the corresponding rewards, where for each j,

$$R_j = \int_{Z_{j-1}}^{Z_j} B(t)\, dt, \qquad j = 1, 2, \ldots.$$

Let

$$\overline{X}_n = \sum_{j=1}^{n} X_j/n \quad \text{and} \quad \overline{R}_n = \sum_{j=1}^{n} R_j/n.$$

Because the X_j and the R_j are each i.i.d., $\overline{X}_n$ converges to $E(X_j)$ and $\overline{R}_n$ converges to $E(R_j)$ w.p.1, as $n \to \infty$. (Although each sequence is i.i.d., X_j and R_j typically are dependent for the same j.) From the Renewal-Reward Theorem,

$$\overline{B} = E(R_j)/E(X_j),$$

which suggests the following *regenerative estimator* for $\overline{B}$:

$$\hat{\overline{B}}_n = \overline{R}_n/\overline{X}_n = \frac{\sum_{j=1}^{n} R_j}{\sum_{j=1}^{n} X_j} = \frac{C(Z_n)}{Z_n}. \tag{204}$$

* This is a special topic that is not needed elsewhere in this book.

where $C(Z_n)$ is the cumulative process at regeneration point Z_n. Notice that we have

$$\lim_{n\to\infty} \hat{\overline{B}}_n = \overline{B} \quad \text{w.p.1.} \tag{205}$$

Because of (205), $\hat{\overline{B}}_n$ is said to be a *consistent* estimator of $\overline{B}$.

By itself, property (205) is trivially true for most statistical estimators. For a regenerative estimator, (205) is obvious from (204) because as $n \to \infty$, $Z_n \to \infty$.

Interest in the regenerative estimator stems from the following theorem, where D denotes convergence in distribution, and $E(X_j) = 1/\mu$.

Theorem 21: Regenerative Central Limit Theorem. If $E(X_j^2)$ and $E(R_j^2)$ are finite,

$$\frac{\sqrt{n}(\hat{\overline{B}}_n - \overline{B})}{V^{1/2}} \xrightarrow[n\to\infty]{\text{D}} N(0, 1), \tag{206}$$

where

$$V = \mu^2 \operatorname{Var}(R_j - \overline{B}X_j). \tag{207}$$

Proof. Let

$$K_j = R_j - \overline{B}X_j, \qquad j = 1, 2, \ldots,$$

where the K_j are i.i.d. with $E(K_j) = 0$ and $E(K_j^2) = V(K_j) = \operatorname{Var}(R_j - \overline{B}X_j) < \infty$. We apply the standard central limit theorem to $\overline{K}_n = \sum_{j=1}^{n} K_j/n$:

$$\frac{\sqrt{n}\,\overline{K}_n}{[V(K_j)]^{1/2}} \xrightarrow[n\to\infty]{\text{D}} N(0, 1). \tag{208}$$

Because $\mu\overline{X}_n \to 1$ w.p.1, $\sqrt{n}\,\overline{K}_n$ and $\sqrt{n}\,\overline{K}_n/\mu\overline{X}_n$ have the same asymptotic (limiting) distribution. Since $\overline{K}_n/\overline{X}_n = \hat{\overline{B}}_n - \overline{B}$, we can write

$$\sqrt{n}(\hat{\overline{B}}_n - \overline{B}) = \mu\sqrt{n}\,\overline{K}_n/\mu\overline{X}_n. \tag{209}$$

Now (206) follows immediately from (208) and (209).

To apply Theorem 21, we need to estimate V. An estimate of $E(K_j^2)$ is $\sum_{j=1}^{n} K_j^2/n$; however, this estimate involves the unknown $\overline{B}$. If $\overline{B}$ is replaced by its estimator $\hat{\overline{B}}_n$, and we use $\overline{X}_n$ to estimate $1/\mu$, we have the following estimator of V:

$$\hat{V}_n = \frac{n \sum_{j=1}^{n} (R_j - \hat{\bar{B}}_n X_j)^2}{\left(\sum_{j=1}^{n} X_j \right)^2}, \tag{210}$$

where it is easily shown that $\hat{V}_n \to V$ w.p.1, as $n \to \infty$.

By the argument following (208), we can replace V by $\hat{V}_n$ in (206) without affecting the limiting distribution.

Theorem 22. If $E(X_j^2)$ and $E(R_j^2)$ are finite,

$$\frac{\sqrt{n}(\hat{\bar{B}}_n - \bar{B})}{(\hat{V}_n)^{1/2}} \xrightarrow[n\to\infty]{\text{D}} N(0, 1). \tag{211}$$

From (210) we can now estimate the variance of $\hat{\bar{B}}_n$ and, from Theorem 22, we can construct (asymptotically valid) confidence intervals. (A *confidence interval* is a technical term defined and discussed in most textbooks on mathematical statistics.)

The time average of $\{f(B(t))\}$ for some function f may be estimated in the same way simply by redefining R_j:

$$R_j = \int_{Z_{j-1}}^{Z_j} f(B(t))\, dt.$$

For example, if we define

$$\begin{aligned} f(B(t)) &= 1 \qquad \text{if } B(t) \le b, \\ &= 0 \qquad \text{otherwise}, \end{aligned}$$

we would be estimating the fraction of time $\{B(t)\}$ takes on values not exceeding b.

The regenerative method can be applied when the simulation run length is determined in a different way. The most common alternative is to simulate the process for some fixed time interval t, i.e., we generate $\{B(u)\colon 0 \le u \le t\}$, called a *run of length t*. An obvious estimator of $\bar{B}$ from this run is

$$\hat{\bar{B}}(t) = \int_0^t B(u)\, du/t = C(t)/t, \tag{212}$$

where the number of completed cycles by epoch t is $M(t)$, the number of renewals by t.

In fact, if we discard from (212) the contribution of uncompleted cycle $M(t) + 1$, (212) becomes

$$\hat{\bar{B}}_{M(t)} = \int_0^{Z_{M(t)}} B(u)\, du/Z_{M(t)}, \tag{213}$$

which is a regenerative estimator of $\overline{B}$, where the number of cycles $M(t)$ now is a random variable. Since as $t \to \infty$, $M(t)/t \to \mu$ w.p.1, where $0 < \mu < \infty$, (213) has limiting distribution (206), with $M(t)$ substituted for n there and in (204).

It is easily shown that if $B(t) \geq 0$ for all t, (212) and (213) have the same asymptotic distribution, i.e., in the limit, the contribution of the discarded cycle has no effect. For $\{B(t)\}$ unrestricted in sign, a sufficient condition is

$$E\left\{\int_0^{X_1} | B(t) | \, dt\right\} < \infty.$$

Since

$$\sqrt{t}(\hat{\overline{B}}(t) - \overline{B}) = [M(t)]^{1/2}[\hat{\overline{B}}(t) - \overline{B}][t/M(t)]^{1/2},$$

where $M(t)/t \to \mu$ w.p.1 as $t \to \infty$, we have this result:

Theorem 23: Central Limit Theorem for Cumulative Processes. If $E(X_j^2)$, $E(R_j^2)$, and $E\,\{\int_0^{X_1} | B(t) | \, dt\}$ are finite,

$$\frac{\sqrt{t}[\hat{\overline{B}}(t) - \overline{B}]}{(V/\mu)^{1/2}} = \frac{C(t) - \mu t E(R_j)}{[\mu t \operatorname{Var}(R_j - \mu E(R_j) X_j)]^{1/2}} \xrightarrow[t\to\infty]{D} N(0, 1). \tag{214}$$

To apply Theorem 23, we need to estimate V. This may be done in the same manner as before. For a run of length t, V is estimated by (210), where we substitute $M(t)$ for n. Similarly, by substituting this estimator for V in (214), we obtain a counterpart to Theorem 22 that may be used to construct confidence intervals.

Theorem 23 is of independent interest because, quite apart from statistical estimation, it provides an approximation for the distribution of $C(t)$. Furthermore, these theorems are valid for the more abstract way in which rewards and cumulative processes were defined in Section 2-3. Among other things, this permits $\{C(t)\}$ to have discontinuities corresponding to rewards collected "at the beginning" or "at the end" of a cycle. The only change in the theory is that for Theorem 23 to hold, we replace the third condition there by this: $E(P_j)$ and $E(N_j)$ are finite. (See the discussion of "partial rewards" in Section 2-3.) Under these conditions, the way in which reward accumulates during a cycle has no effect on the asymptotic distribution of $C(t)$.

In particular, by specializing Theorem 23, we can obtain asymptotic distributions for familiar quantities. For example, for $R_j = 1$ for all j, $C(t) = M(t)$ and we have that as $t \to \infty$, $M(t)$ is asymptotically normal with

$$\text{mean } \mu t \quad \text{and} \quad \text{variance } \mu^3 t V(X_j). \tag{215}$$

Remark. Compare the variance in (215) with the limit found in Problem 2-90.

Generalized Regenerative Processes and Choice of Regeneration Points

It should be clear that the first cycle can be different, i.e., our results hold for generalized regenerative processes. Furthermore, from Theorems 21 and 23, the way the run length is determined (as a number of cycles n or until epoch t) makes no difference asymptotically, provided that the run lengths are equivalent, i.e., $t \approx nE(X_j)$.

Now suppose that for the same process $\{B(t)\}$, regeneration points and cycles may be defined in more than one way. [This was the case in Example 2-11. See equation (117) and the sentence that follows.] Clearly, the asymptotic distribution of $C(t)$ in Theorem 23 cannot depend on how we happen to select regeneration points. Hence we have the following result:

Theorem 24. If the conditions of Theorem 23 hold for any choice of regeneration points, all choices of regeneration points produce asymptotically equivalent estimators of $\overline{B}$, provided that equivalent run lengths are compared.

Remark. Estimators of the *variance* of $C(t)$, based on (210), are *not* asymptotically equivalent.

This concludes an essentially complete account of the more elementary aspects of the theory of regenerative estimation. Our treatment differs somewhat from that typically found elsewhere. See the next section for a discussion of the main differences and references to more complete treatments.

Excluded from our brief account is the important topic of variance reduction techniques. Furthermore, we have presented only an asymptotic theory. The statistical properties of finite simulations (of length epoch t or of n cycles) can be quite different from what this theory predicts.

Of course, this theory applies only to regenerative processes. (This book to the contrary, there are important processes that lack this property.) For some complex regenerative processes encountered in practice, cycle lengths can be very long. Sometimes the process does not regenerate during the entire simulation run, and (210) cannot be used to estimate variance. If the number of cycles we get is small (but at least two), we can use (210), but it will be highly variable, and this variability should be taken into account.

2-21 SUMMARY AND LITERATURE GUIDE

Because virtually all the models in this book will turn out to be regenerative, this chapter provides a theoretical foundation for much that will follow. The hard part of renewal theory deals with pointwise limits. Feller [1971] provides an account of the historical development of the subject and (primarily analytic) proofs of the KRT and Blackwell's Theorem. A more probabilistic (but hardly elementary) proof is by Lindvall [1977]. Also see Asmussen [1987].

Regenerative processes were introduced and developed in several papers by Smith, see [1955] and [1958]. He coined the term KRT, proved Theorem 19 under conditions more restrictive than DRI, and proved Theorem 20(b). He also proved Theorem 23 in the context of cumulative processes, long before the regenerative method was developed. (While this method was *suggested* by Cox and Smith [1961], p. 136, formal development began about ten years later.)

As we have seen, some of the conditions for pointwise limits to exist are quite technical. Many of the proofs are delicate as well as technical. Because of the challenging nature of this theory, many well-known mathematicians were attracted to it and made significant contributions.

Unfortunately, some mistakes were made. Perhaps the most famous of these is in Section XI.8 of Feller, which mars his otherwise excellent treatment. He asserts that for regenerative processes, the only condition necessary for pointwise limits (of distribution functions) to exist is that cycle lengths have a nonlattice distribution with finite mean. From Example 2-29, this assertion clearly is false.

Miller [1972] points out the error in Feller in the 1966 edition (too late for correcting the currently available 1971 edition), relates this question to the earlier work of Smith, states Theorem 20, and proves the (a) part. Brown and Ross [1972] cover similar ground.

With all this attention to pointwise behavior, the more elementary time-average properties of regenerative processes have been neglected. This is unfortunate because time-average results

- are easier to understand,
- are much easier to prove,
- hold under weaker and less technical conditions, and
- are of paramount importance in applications.

This chapter presents a systematic treatment of the time-average behavior of regenerative processes that depends on little more than the classical strong law of large numbers, rather than the existence of pointwise limits. The main results are stated as Theorems 11 and 17. These results are not new, e.g., Brown and Ross show that $\overline{B} = E\{B_e(t)\}$. They do not state Theorem 11, which we derived without the construction of the stationary version. On the other hand, given that the limiting (time-average) distribution exists, independent of initial conditions, Theorem 11 may be regarded as a consequence of Theorem 17.

The more technical material on pointwise limits is deliberately placed near the end of this chapter. This is not to diminish either the importance of this topic or the ingenuity required for its development. Rather, the intention is to prevent it from intruding where it does not belong—in understanding and applying the time-average properties of regenerative processes.

Pointwise limits will exist in the lattice case as well. Miller has shown that when cycle lengths are lattice with span d, regenerative process $\{B_o(t)\}$ has a (pointwise) limiting distribution when t is restricted to increase by multiples of d, $t = s + d, s + 2d, \ldots$. Except in special cases however, e.g., when sample paths of $\{B_o(t)\}$ are constant between span multiples (i.e., we really have a discrete time process), this limiting distribution is *different from* our (time-average) limiting distribution, for every initial s, and its mean is *not* $\overline{B}$. There has been some confusion on this point in the literature.

In Chapter 3, we treat discrete time processes called Markov chains, where the state space is countable, and is usually taken to be the nonnegative integers. Markov chains that jump from state to state in a very general state space have been studied extensively in recent years. What are called *Harris recurrent* chains turn out to be (discrete-time) regenerative processes; they are *positive recurrent* when the mean cycle length is finite. See Nummelin [1984] and especially Asmussen, and the references therein.

Asmussen's definition of a regenerative process is more general than ours: He requires that $\{B_o(t + X_1): t \geq 0\}$ be independent of X_1, and that it be stochastically equivalent to $\{B_o(t)\}$, but *not* that it be independent of $\{B_o(t): t < X_1\}$. This implies that the cycle *lengths* X_j are i.i.d., but the cycles themselves may be dependent. This may seem odd at first, but a weakening of our definition is necessary if Harris chains are to be regenerative. He shows that the cycles are *one-dependent*, meaning that adjacent cycles are dependent, but nonadjacent cycles are independent.

If $\{B_o(t)\}$ is regenerative under our definition, $\{B_o(t), B_o(t + h): t \geq 0\}$ is regenerative under Asmussen's definition. Fortunately, the renewal argument is still valid under Asmussen's definition, a fact we used in the construction of stationary joint distributions in the proof of Theorem 17.

We usually think of queues as continuous-time processes, but we can embed discrete-time processes in them, e.g., at customer arrival epochs. Embedded processes for a remarkable number of queueing models turn out to be regenerative in the Asmussen sense, with finite mean cycle lengths; see Sigman [1988a] and [1988b]. The corresponding continuous time processes are analyzed in more detail in Sigman [1989]: They are Harris recurrent processes that are regenerative in a sense weaker than Asmussen; cycle *lengths* are also one-dependent. The renewal argument is no longer valid. Nevertheless, he shows that Theorem 11 holds for one-dependent regenerative processes. The regenerative nature of the *GI/G/c* queue is discussed in Section 11-6.

The Harris-chain approach to queueing theory is an alternative to the point process approach mentioned later in this section. It remains to be seen which approach, or perhaps a combination of the two, is better able to deal with queueing networks with general (renewal or regenerative, say) arrivals and service distributions. We will return to this question in Chapter 11.

Treatment of the Regenerative Method

Most treatments of the regenerative method require the existence of pointwise limits. Typically, the stated objective is to estimate the mean of the (pointwise) limiting distribution of some regenerative process $\{B(t)\}$. By defining a limiting distribution as a time average, the introduction of irrevelant technical details is avoided.

Why (pointwise or otherwise) $E\{B_\infty\}$ [or $E\{f(B_\infty)\}$] is chosen as a performance measure is not clear. To this author, $\overline{B}$ is more appealing, and the theory presented in the preceding section deals only with estimating $\overline{B}$. Of course, by Theorem 11, $\overline{B} = E\{B_\infty\}$, i.e., these measures turn out to be the same. Thus, the results in the preceding section apply without modification to estimating $E\{B_\infty\}$.

What is clear [from (205)] is that the regenerative estimator converges to $\overline{B}$. To assert that it also converges to $E\{B_\infty\}$ requires either Theorem 11 or a pointwise version of this result. Aside from the avoidable technical excursions mentioned earlier, Crane and Lemoine [1977] is recommended for a more complete treatment of this topic.

A regenerative estimator is a ratio of sample means, i.e., it is an example of what is called a *ratio estimator*. Ratio estimators have been studied in other contexts, e.g., survey sampling, and Theorems 21 and 22 were well known before the regenerative method was invented. Cochran [1977] has an interesting discussion of ratio estimators, with survey sampling applications. (He samples from a *finite* population, a complication that does not arise in simulation.)

There is much more to simulating stochastic processes than the regenerative method; in fact, even for regenerative processes, competing methods have been developed. A good general-purpose book on what is called *discrete event* simulation is Law and Kelton [1982]. See Rubinstein [1981] for a more theoretical treatment. One caveat: This is an active research area. Some of the advice given in these references may not hold up over time.

Stationary Point Processes

Stationary point processes include the Poisson process and the equilibrium renewal process. A brief introduction to the more technical aspects of these processes is given in the starred portion of Section 2-9. This special topic has a substantial literature. Cox and Isham [1980] provides a good intermediate-level treatment and an extensive bibliography.

Arrivals at a queue may be modeled as a stationary point process. This approach has led to highly mathematical treatments of queues, e.g., Franken et al. [1982].

The Equilibrium Distribution, Length Biasing, and the Inspection Paradox

As we shall see in later chapters, the equilibrium distribution is very important in queueing theory. For example, if G denotes a service time distribution, a remaining service (at some random time, and sometimes when observed in other ways) will be $\sim G_e$. We first met F_e and length-biased distribution F_s as the limiting distributions of the excess $\{Y(t)\}$ and spread $\{X(t)\}$ stochastic processes, respectively, for i.i.d. inter-arrival times. However, these limits hold under much more general conditions, e.g., for stationary (ergodic) point processes.

Arrival processes may be generalized in other fundamental ways. For example, suppose inter-arrival sequence $\{X_j\}$ is *regenerative* with regeneration points K_1, $K_1 + K_2$, . . . , where $E(K_1)$ and $E\left\{\sum_{j=1}^{K_1} X_j\right\}$ are finite. For every $t \geq 0$, we define $F(t)$ as the *fraction* of the X_j that are $\leq t$. Under these conditions, we show in Problem 2-91 that $\{Y(t)\}$ and $\{X(t)\}$ have limiting distributions F_e and F_s, respectively.

The fundamental reason for the generality of these results is the length biasing effect, i.e., long intervals simply take up more space than short ones. Thus we are more likely to land in a long interval (the inspection paradox). In fact, under mild assumptions about the average behavior of $\{X_j\}$, we show (Example 5-22) that these limiting distributions hold *on sample paths*. This helps explain observed real phenomena such as long waits at a bus stop. Similarly, the batch effects discussed in Section 2-5 hold under the same mild assumptions.

The Poisson Process

Our treatment of the Poisson process "in isolation" is fairly complete. Also see Çinlar [1975].

The Poisson process is an extremely important building block in stochastic modeling, frequently because of its role as an arrival process. We introduced the notion of a stopping time to characterize how a Poisson process interacts with other processes in a model.

The $M/G/\infty$ queue illustrates a typical situation. An arrival process and a service requirements process (of the arrivals) interact in a well-defined way to generate a new "number of customers in system" process.

Understanding the behavior of generated processes is the main objective of stochastic modeling. In this book, nearly all of our generated processes will be regenerative.

It is usually assumed (often implicitly) that for every (fixed) epoch t, the future increments of the Poisson process are independent of the generated process it helped create. This is called the *lack of anticipation assumption* (*LAA*). Under this assumption, we show in Section 5-16 that the (time average) limiting distribution of the generated process is the same as the distribution of the generated process obtained by averaging over the arrival epochs of the Poisson process. This turns out to be an extremely useful result, and is called *Poisson Arrivals See Time Averages* (PASTA). We discovered PASTA for a generated process in Example 2-9. Beginning in Chapter 5, we will use PASTA frequently in the analysis of generated processes.

PROBLEMS

2-1. Denote the density function of Z_2 in (4) by $g(t)$. Let X_1 and X_2 have exponential density $\lambda e^{-\lambda t}$, $t > 0$. Show that (4) becomes

$$g(t) = \lambda^2 t e^{-\lambda t}, \qquad t > 0. \tag{P.1}$$

2-2. *Continuation.* When X_1 and X_2 are uniformly distributed on (0, 1), show that (4) becomes

$$\begin{aligned} g(t) &= t && \text{for } 0 < t \le 1, \\ &= 2 - t && \text{for } 1 \le t < 2. \end{aligned}$$

2-3. Find the optimal age replacement policy, i.e., the value of T that minimizes (34), when V_j has a uniform distribution on $(0, a)$.

2-4. When V_j is ex onential, show that the optimal age replacement policy is $T = \infty$. Explain.

2-5. *Group Replacement Policies.* An alternative to age replacement policies is *group* replacement, which means that many similar or identical items are replaced simultaneously. This may be advantageous when there is some high fixed cost (whenever items are replaced) that does not depend on the number of items replaced. Examples: lights on the same fixture in a stadium, a machine that must be taken apart before any parts can be replaced. Compared with age replacement, record keeping is also reduced.

A group replacement policy for K items replaces them all at fixed intervals of length T, i.e., replacement occurs at epochs $T, 2T, \ldots$. A group replacement policy may operate *with* interim replacement, where items are replaced individually on failure during group replacement intervals, or without.

Consider a group replacement policy with group replacement interval T for K light bulbs where c_1 is the cost of each bulb and c_r is a replacement cost that is independent of the number of bulbs replaced. Let bulbs have life distribution $G(t)$.

(a) What is the cost per unit time if the policy operates without interim replacements?
(b) Would a policy of the type in (a) make sense if G were exponential?
(c) Repeat (a) and (b) for a policy that operates with interim replacements.
(d) For the policy in (a), show that the expected cost per unit time decreases with T. However, what other factor should be considered?

2-6. A machine breaks down and is repaired. The times until breakdown (up times) X_{u1}, X_{u2}, are i.i.d. Subsequent repair times (down times) $X_{d1}, X_{d2}, \ldots$ are also i.i.d. However, the time to repair a machine depends on how long it has been up, i.e., X_{dj} and X_{uj} are dependent, with some (unspecified) joint distribution. Nevertheless, the up-down cycle lengths, $X_j = X_{uj} + X_{dj}$, are i.i.d. with finite mean. Show that the fraction of time the machine is up is still (39).

2-7. For a renewal process generated by i.i.d. X_j where $P(X_j = 1) = P(X_j = 2) = 1/2$, compute $m(t)$, $E\{Z_{M(t)}\}$, and $E\{Z_{M(t)+1}\}$ for all $t \in [0, 3]$.

(a) Show that $E\{Z_{M(t)}\} \le m(t)E(X_j)$ for all $t \in [0, 3]$. On what range is this inequality strict?
(b) Show that $E\{Z_{M(t)+1}\} = [m(t) + 1]E(X_j)$ for all $t \in [0, 3]$.

2-8. For the age replacement policy T in Example 2-1, we assumed the actual time to replace each component was zero. Instead, suppose replacement times are random variables, where S_j is the time to replace the jth component. Assume S_j depends on V_j in the following way: Given $V_j \le T$, $S_j \sim G_f$, with mean s_f. Given $V_j > T$, $S_j \sim G_r$ with mean s_r. Each S_j is independent of the other V_i and S_i, $j \ne i$.

(a) In terms of the notation just given, find $E(S_j)$.

(b) Identify appropriate renewal cycles and find the mean cycle length.

(c) In addition to the costs stated in Example 2-1, suppose a reward (negative cost) is earned at the *rate* of $\$\alpha$ per unit time whenever a working component is in service. Find an expression for the cost per unit time of policy T for the model we now have.

2-9. You are planning an automobile trip during which you estimate that about half the trip (in miles) will be spent on highways and half will be spent in cities. From past experience, you estimate that you will get 40 miles per gallon on highways and 20 miles per gallon in cities.

(a) What overall mileage (miles/gal) do you expect to get on this trip?

(b) What does this question have to do with renewal theory?

2-10. *A Truncation Method*. For sequence $\{X_j\}$ that generates a renewal process, define a *truncated* sequence as follows: For arbitrary $c > 0$, let $X_j' = X_j$ if $X_j \le c$ and $X_j' = c$ if $X_j > c$, $j = 1, 2, \ldots$. In terms of $\{X_j'\}$, define Z_n', $M'(t)$, etc., with $\mu' = 1/E(X_j')$, $j \ge 2$.

(a) Show that $Z_n'(\omega) \le Z_n(\omega)$ and $M'(t, \omega) \ge M(t, \omega)$ for all n, t, and ω, i.e., these results hold on all realizations.

(b) Show that $\lim_{t\to\infty} M(t)/t \le \mu'$ with probability 1. (Technically, this is a lim sup.)

(c) Show that $\lim_{c\to\infty} E(X_j') = E(X_j)$, whether or not $E(X_j) < \infty$.

(d) Combine (b) and (c) to show that (21) holds even when $\mu = 0$.

2-11. *Continuation*. We now "truncate" in a different way to obtain another result. Let $X_j' = 0$ if $X_j \le c$ and $X_j' = c$ if $X_j > c$, $j = 1, 2, \ldots$, for any $c > 0$ such that $F(c) < 1$. To avoid minor technical details, assume $A = F$, a special case called an *ordinary* renewal process. As in Problem 2-10(a), we still have $M(t, \omega) \le M'(t, \omega)$, which implies that $M(t)$ is *stochastically smaller* than $M'(t)$.

(a) For $t \in [0, c)$ show that $M'(t)$ has a *geometric* distribution. What are p and $q = 1 - p$?

(b) For any $t \ge c$, show that $M'(t) - [t/c]$ has a *negative binomial* distribution, where $[t/c]$ is the largest integer $\le t/c$.

(c) A negative binomial distribution has finite moments of all orders. Use this fact to show that for any integer $k > 0$,

$$E\{[M(t)]^k\} < \infty.$$

2-12. *Continuation*. To generalize the preceding result let $\{M_0(t)\}$ denote the renewal process for the special case $A = F$. *For this problem only*, we represent the general case by a shift in indexing: Introduce an additional random variable $X_0 \sim A$ and define $Z_n = \sum_{j=0}^{n} X_j$, $n = 0, 1, \ldots$, where $X_1, X_2, \ldots$ are the *same* random variables as before. For any n, including X_0 in the sum can only make Z_n larger. On the other hand, because Z_n includes an additional term,

$$\{Z_n \le t,\ Z_{n+1} > t\} = \{M(t) = n + 1\}, \qquad t \ge 0.$$

Show that for all t and all realizations,

$$M(t) \le M_o(t) + 1. \tag{P.2}$$

From (P. 2) and Problem 2-11, it follows immediately that $M(t)$ has finite moments of all orders.

2-13. If X is a constant, i.e., $P(X = c) = 1$, show that X_e has a uniform distribution on $(0, c)$.

2-14. If $X \sim \exp(\mu)$, show that $X_e \sim \exp(\mu)$.

2-15. Show that $F_e = F$ if and only if F is exponential. [*Hint:* Equate the transform in (48) with $\tilde{F}(s)$.]

2-16. Higher moments of F_e: If $X \sim F$ and $X_e \sim F_e$, show that

$$E\{X_e^r\} = \mu E\{X^{r+1}\}/(r + 1), \qquad r > 0. \tag{P.3}$$

2-17. Following the relevant portions of Example 2-3, show that the limiting distribution of age is

$$\lim_{t\to\infty} \int_0^t P(U(x) \le u)\, dx/t = F_e(u). \tag{P.4}$$

2-18. Suppose you arrive "at random" at a bus stop where the frequency (number/unit time) of buses is 3 per hour. Find the distribution of the waiting time for the next bus, and its mean, if the inter-arrival times of buses are

(a) constant,

(b) exponential, or

(c) either 0 or 60 minutes, i.e., groups of 3 buses go by every hour. This unfortunate grouping of transport vehicles, particularly buses and elevators, is well known. What causes this to occur?

2-19. A subway station has both local and express service, on opposite sides of the same platform. Local trains arrive every 5 minutes (constant), and express trains arrive every 15 minutes (constant), scheduled so that every third local train arrives simultaneously with an express train. Both trains stop at your destination, with transit times of 17 minutes for a local train and 11 minutes for an express train.

You arrive "at random" on the station platform. Your *objective* is to minimize your expected travel time, $E(T)$ say, from your arrival epoch at the station until you reach your destination.

(a) What is the waiting time *distribution* until the next local train arrives? The next express train?

(b) What is the probability that the next local train arrives alone, i.e., without an express? What is the probability that the next two locals arrive alone? Briefly explain. (*Hint:* Interpret probabilities as fractions of time.)

(c) If the next local train arrives alone, should you board that train or wait for an express?

(d) Given your decision in (c), find $E(T)$.

(e) If local and express trains are boarded from different platforms, on which platform should you wait? (Running between platforms is not allowed.)

2-20. Show that

$$\lim_{t\to\infty} \int_0^t P(U(x > u,\ Y(x) > y)\ dx/t = 1 - F_e(u + y). \qquad \text{(P.5)}$$

What is the meaning of (P.5) when either $y = 0$ or $u = 0$? (No tricks are intended. It's easier to derive (P.5) directly from the joint *tail* distribution.)

2-21. An intuitive justification of (47): Suppose we arrive "at random" at a train platform. Let Y be the time until the next train arrives, U be the time since the last train, and $X = U + Y$.

(a) Find $h(y \mid x)$, the conditional density of Y given $X = x$. That is, suppose the inter-arrival times between trains were all equal to the constant x.

(b) From (a), show that the density function of Y is

$$\int_0^\infty h(y \mid x)(?)\ dx = \mu[1 - F(y)] = f_e(y),$$

where (?) is the appropriate distribution of X.

2-22. Show that for an *ordinary* renewal process ($A = F$), the spread random variable at any epoch t is *stochastically larger* than $X_j \sim F$. That is, show that for any $t > 0$,

$$P(X(t) > x) \geq 1 - F(x) \qquad \text{for all } x > 0. \qquad \text{(P.6)}$$

(*Hint:* Condition on the number of renewals in $[0, t]$ and the last renewal epoch in that interval.)

2-23. For nonnegative random variable $X \sim F$ with $E(X) < \infty$,

(a) Show that the following two properties are equivalent:

(i) $E(X - t \mid X > t) \leq E(X)$ for all $t \geq 0$, and

(ii) X_e is stochastically smaller than X, i.e.,

$$F_e(t) \geq F(t) \qquad \text{for all } t \geq 0.$$

(b) Show that if (i) holds as an *equality* for all t, X is exponential.

2-24. Verify the last step in the derivation of (48) for the case where X has a density function.

2-25. For an alternating renewal process (Example 2-2) with states 0 and 1, let the time the process spends in state i be distributed as F_i, $i = 0, 1$. If the process is in state i at epoch t, let $Y_i(t)$ be the time the process remains in state i, i.e., there is a transition from state i to state $j = 1 - i$ at epoch $t + Y_i(t)$. If the process is in state j at epoch t, define $Y_i(t) = 0$. [If there is a positive probability of one or more transitions at epoch t, by the state at epoch t, we mean at t^+, just after these transitions have occurred. Thus, realizations of $\{I(t)\}$ are continuous from the right.] Find the limiting (tail) distribution of $Y_i(t)$ and show that it can be written in form

$$\lim_{t\to\infty} \int_0^t P(Y_i(x) > y)\ dx/t = [E(X_i)/E(X)]F_{ie}^c(y), \qquad \text{(P.7)}$$

where X_i denotes an i-period duration, X denotes a cycle length, and F_{ie} is the equilibrium distribution of F_i. Interpret this result.

2-26. Show that the mean of distribution (50) [and of (51)] is

$$\mu E(X^2). \tag{P.8}$$

Why is (P.8) twice as large as (49)?

2-27. For nonnegative $X \sim F$ with mean $0 < 1/\mu < \infty$,

(a) Find $P(X > t + y \mid X > t)$.

(b) Regard your answer to (a) as a conditional distribution of the "remaining life" of X, given that it had attained an "age" of $U = t$, where U has density f_e in (47). That is, define

$$P(Y > y \mid U = t) \equiv P(X > t + y \mid X > t).$$

Uncondition the above and show that

$$P(Y > y) = F_e^c(y). \tag{P.9}$$

(c) Interpret your result in (b) in terms of observing a renewal process at random.

2-28. Derive (56) by the method used to derive (54).

2-29. Let family sizes $\nu_1, \nu_2, \ldots$ be i.i.d. with $P(\nu_i = j) = \alpha_j$, $j = 0, \ldots,$ and mean $E(\nu)$.

(a) Show that the fraction of children who are jth born *and* have l younger siblings is

$$\frac{\alpha_{j+l}}{E(\nu)}, \qquad j = 1, 2, \ldots; \quad l = 0, 1, \ldots.$$

(b) The result in (a) is a *joint* distribution. From (a) find the marginal distributions:

(i) the fraction of children who are jth born (54), and

(ii) the fraction of children who have l younger siblings.

Why are these distributions different? (Recall that the limiting distributions of excess and age are the same.)

(c) From (a) show that the distribution of the sum $k = j + l$ is (56).

2-30. If F is the distribution of an integral-valued variable, show that (50) is equivalent to (56).

2-31. Show that the mean of distribution (56) is

$$E(\nu^2)/E(\nu). \tag{P.10}$$

Show that (P.10) is equal to

$$2E(J) - 1,$$

where J has distribution (54) with mean (55). Where did the -1 come from?

2-32. One plane arrives every Thursday (or thereabouts) at a small international airport in a (small) foreign country. From past data, you estimate that the mean and variance of the number of passengers on a plane is 50 and 7450, respectively. On arrival, all passengers form a single line (queue) in front of the single passport control officer. The times to process passengers are i.i.d. with mean 15 seconds per passenger.

Let W be the waiting time of a "randomly selected" passenger, where W *includes* that passenger's processing time.

(a) Represent W as a random sum of processing times. How is the number of terms in the sum distributed? (Represent your answer in our notation; we don't have numerical values for this distribution.)

(b) From (a) and the data above, calculate

$$E(W) = 25 \text{ minutes}.$$

2-33. A sample of size 1000 adult women is drawn for a survey in which each woman is asked two questions:

(i) How many children did your mother have?

(ii) How many children did you have? (Assume their childbearing days are over.)

Suppose the distribution of number of children per woman has not changed over time and is $\alpha_j = 1/3, j = 1, 2, 3$.

(a) What is the average response (approximately) to question (i)?

(b) What is the average response (approximately) to question (ii)?

(c) If we did not know that $\{\alpha_j\}$ has not changed, the results of (a) and (b) might lead us to the conclusion that family sizes are decreasing. Instead, what is the explanation for these results? Would increasing the sample size help? [There is a subtle assumption needed in (a). What is it?]

2-34. *Continuation.* Suppose we don't know $\{\alpha_j\}$ or whether it has changed. How would you use the responses to questions (i) and (ii) to provide evidence that would either support or challenge the conclusion that the average family size is decreasing? (Assume that the proportion of female children in families does not change with family size.)

2-35. The average of the responses to the question "How many years have you been on the faculty at the University of California?" was 12.3 years. Only individuals who were on the faculty at the time of the survey were asked this question. Suppose we wish to estimate the average number of years faculty members spend on the university's faculty, θ say.

(a) Is 24.6 a reasonable estimate of θ?

(b) If 12.3 were exact, i.e., had no statistical error, would it be a lower bound on θ?

(c) Similarly, is 24.6 an upper bound on θ? Explain.

2-36. Excluding lines that begin paragraphs, the first word on a line in a manuscript tends to be longer than the average length of the words used in the manuscript (Marshall, [1979]). Why is this? Design a simple experiment to test this out yourself.

2-37. Let $\{\upsilon_i\}$ be a sequence of i.i.d. family sizes with $P(\upsilon_i = j) = \alpha_j, j = 0, 1, \ldots$.

(a) Of those children in families of constant size k, $k = 1, 2, \ldots$, find $b_j(k)$ = the fraction who are jth born, $j = 1, 2, \ldots$.

(b) Show that the mean of the distribution found in (a) is

$$\sum_j j b_j(k) = (k + 1)/2.$$

(c) Now the result in (b) is the conditional expectation of the random variable J defined in Section 2-4. Find $E(J)$, (55), by unconditioning result in (b).

2-38. For a Poisson process at rate λ, find an expression for $P(\tau_1 + \tau_2 \leq t)$ by identifying an equivalent event in terms of $\Lambda(t)$. Differentiate this expression to obtain the density function (P.1) in Problem 2-1.

2-39. *Continuation*. We obtain the density function of $\sum_{j=1}^{k} \tau_j$ by a different approach. Argue that

$$P\left(\sum_{i=1}^{k} \tau_i \in (t, t + h]\right) = P(\Lambda(t) = k - 1, \Lambda(t + h) = k) + o(h), \qquad \text{(P.11)}$$

where $o(h)$ denotes a function with the property $\lim_{h \to 0} o(h)/h = 0$. Show that the right-hand side of (P.11) can be written

$$\frac{e^{-\lambda t}(\lambda t)^{k-1}}{(k - 1)!} \lambda h e^{-\lambda h} + o(h). \qquad \text{(P.12)}$$

Divide (P.12) by h, let $h \to 0$, and obtain the density function of $\sum_{j=1}^{k} \tau_j$:

$$\lambda e^{-\lambda t}(\lambda t)^{k-1}/(k - 1)!, \qquad t > 0. \qquad \text{(P.13)}$$

The density (P.13) is that of the k-Erlang distribution. (See Chapter 5.)

2-40. Let $X \sim \exp(\lambda)$ and $Y \sim \exp(\mu)$ be independent random variables. Show that

$$\min(X, Y) \sim \exp(\lambda + \mu). \qquad \text{(P.14)}$$

2-41. Suppose that male and female applicants for a job are independent Poisson processes at rates λ and μ, respectively. Show that

(a) The composite stream of all applicants is a Poisson process at rate $\lambda + \mu$. (Don't forget to show independent increments.) Interpret (P.14) in this context.

(b) The probability that the first applicant is male is

$$\lambda/(\lambda + \mu), \qquad \text{(P.15)}$$

independent of when that applicant arrives.

2-42. In Problem 2-20, we obtained the joint distribution of the age and excess random variables when we "observe" or "enter" an ongoing renewal process "at random." Show that if these random variables are *independent*,

(a) F_e is exponential, and

(b) F is either exponential or a mixture of an exponential and a unit step at the origin. (Thus, the renewal process is either Poisson or batch Poisson.)

2-43. *Continuation of Problem 2-19*. Suppose the arrival processes of local and express trains are *independent*. Poisson processes with arrival rates 1/5 and 1/15 per minute, respectively. All other data *and your objective* remain the same.

(a) What is the waiting time distribution until the next local train arrives? The next express train?

(b) What is the probability that the next local train arrives before the express? What is the probability that the next *two* local trains arrive before the next express?

(c) If the next train that arrives is a local, should you board that train or wait for an express?

(d) Given your decision in (c), find $E(T)$.

(e) If local and express trains are boarded from different platforms, on which platform should you wait?

2-44. The arrival process of people at the ground (first)-floor elevator of an office building is Poisson at rate λ. The departure process of elevators from the ground floor is Poisson at rate μ (unrealistic, but generalizations are easy to handle). Departing elevators are large enough to hold everyone waiting.

Assume the building has $N + 1$ floors. Each person selects a floor (other than the first) randomly (with equal probability) and independently of other persons.

(a) Find the expected number of stops owing to people who enter the elevator on the ground floor.

(b) By a separate independent argument, determine the answer to (a) for $N = 1$ and as $N \to \infty$.

2-45. The arrival process of ships into San Francisco Bay is Poisson at rate of λ. The probability that any ship is foreign is q, independent of other ships. Describe the arrival process of foreign ships into the bay.

2-46. The arrival of customers at a self-service market is a Poisson process at rate λ. Customers have i.i.d. shopping times, $S_j \sim G(t)$. After completion of shopping, each customer checks himself out (it's really self-service). The checkout times $C_j \sim H(t)$ are i.i.d. and independent of shopping times. There are plenty of checkout machines; assume customers can begin checking out as soon as their shopping is completed.

At epoch t, let $N_s(t)$ be the number of customers who are shopping, and $N_c(t)$ be the number of customers who are checking out.

(a) Find the joint distribution of $N_s(t)$ and $N_c(t)$.

(b) Suppose S_j and C_j are dependent for the *same* customer, but otherwise we have the same independence assumptions. Assume we know the joint distribution of S_j and C_j, and let $K(t)$ be the distribution of $S_j + C_j$. Modify the analysis in (a). Show that $N_s(t)$ and $N_c(t)$ are still independent with

$$N_s(t) \sim P\left(\lambda \int_0^t G^c(u)\, du\right), \text{ and} \tag{P.16}$$

$$N_c(t) \sim P\left[\lambda\left(\int_0^t K^c(u)\, du - \int_0^t G^c(u)\, du\right)\right]. \tag{P.17}$$

(c) For the joint distribution in (b), let $t \to \infty$. Show that the limiting distribution is the product of marginal Poisson distributions with parameters $\lambda E(S)$ and $\lambda E(C)$. Thus, in the limit, the dependence between S_j and C_j *has no effect* on this result.

2-47. Let the arrival process of cars be a renewal process $\{M(t)\}$ generated by i.i.d. interarrival times $X_1, X_2, \ldots$, where $X_j \sim F$. Suppose the cars are randomly partitioned by make of car, where p is the probability that a car is a Ford and $1 - p$ is the probability that it is not. Let $M_1(t)$ be the number of Fords that arrive by t, and $M_2(t)$ be the number of other cars that arrive by t.

(a) What is the distribution of time between the successive arrival of Fords?

(b) Are the processes $\{M_1(t)\}$ and $\{M_2(t)\}$ (i) renewal processes, (ii) Poisson processes, (iii) or independent of each other?

2-48. *Continuation*. Suppose we reverse what we have done. Let $\{M_1(t)\}$ and $\{M_2(t)\}$ be independent renewal processes. Define the composite process $\{M(t)\}$ by $M(t) = M_1(t) + M_2(t)$, $t \geq 0$. In general, the composite process will *not* be a renewal process, i.e., inter-arrival times will not be i.i.d. Can you give an example illustrating this?

2-49. A Poisson arrival process of particles is recorded by a particle counter that operates as follows: Each recorded particle locks the counter for a period of time k, during which additional particles are not recorded. Let $R(t)$ be the number of recorded particles by epoch t. Assume the counter is initially unlocked.
(a) Is $\{R(t)\colon t \geq 0\}$ a point process? A renewal process? Explain briefly.
(b) Find

$$\lim_{t\to\infty} \frac{R(t)}{t}$$

in terms of the Poisson arrival rate, λ. For sufficiently large t, how would you estimate λ from $R(t)$?

2-50. A machine breaks down and is repaired. The times until breakdown, $X_1, X_2, \ldots$ are i.i.d. with mean $E(X)$. The machine is inspected periodically to determine whether it is still working. The times between inspections are i.i.d. $\sim \exp(\mu)$. When an inspection determines that the machine is not working, repairs begin, where repair times $Y_1, Y_2, \ldots$ are i.i.d. with mean $E(Y)$.
(a) Identify a repair "cycle" and find its mean.
(b) What fraction of time is the machine working?
(c) Suppose inspections do not occur while the machine is being repaired. What is the average number of inspections per unit time?

2-51. A component must be tested to determine whether or not it has failed, where the times between successive tests X_n are i.i.d. random variables with $P(X_n \leq t) = F(t)$ and $E(X_n) = 1/\mu$, $n = 1, 2, \ldots$. When failure is detected, the component is replaced by a new one. Component life times (in the absence of tests) are i.i.d. *exponential* with mean $1/\lambda$. Assume:

1. The test will determine whether a component has failed.
2. With probability p, the test will cause an otherwise good component to fail, independent of the age of the component and the number of prior tests. The test also detects these failures.
3. Test and replacement times are zero.

Let the cost of a test be c_t, the cost of a replacement be c_r, and the cost of failure between tests be k *times* the amount of time the failure is undetected (the time until the next test).
(a) What is the remaining life of a component that is working after a test is completed (in the absence of future tests)?
(b) Show that
(i) the expected "up" time of a component between tests is

$$\int_0^\infty e^{-\lambda t} F^c(t)\, dt, \text{ and}$$

(ii) the probability that a component survives both an interval between tests and the next test is

$$(1 - p) \int_0^\infty F(t)\lambda e^{-\lambda t}\, dt.$$

(c) Use the results in (b) to find the cost/unit time of this replacement procedure.

2-52. A machine has two parts that can fail, one of type A and the other of type B. Part lifetimes are independent, where type A parts have an *exponential* life distribution with mean $1/\mu$, and type B parts have a *general* life distribution G, with mean $E(S_B)$. Both parts are replaced on failure. In addition, whenever a type A part is replaced, the (working) type B part *is also replaced.* Type A parts cost $\$C_A$ each and type B parts cost $\$C_B$ each. In addition, for each replacement, we incur a replacement cost $\$C_r$ and a constant replacement time T, whether one or both parts are replaced.

(a) Specify appropriate regeneration cycles.

(b) Find an expression for the cost per unit time under this replacement policy.

(c) Would it make sense to also replace both parts when a type B part fails? Explain briefly.

2-53. The arrival of taxis at a taxi stand is Poisson at rate λ/hour. The arrival of customers at the stand is Poisson at rate μ/hour. Taxis will not stop (they leave empty) if another taxi is waiting at the stand. Similarly, customers will not stop if another customer is waiting at the stand. Thus, at all times, the stand is either empty, or is occupied by either a taxi or a customer.

Suppose the stand is empty at epoch 0.

(a) What is the distribution of time until the stand is occupied?

(b) Find the expected value of time from epoch 0 until, for the first time, a customer leaves in a taxi.

(c) In the long run, how many customers/hour leave in a taxi?

(d) In the long run, how many taxis/hour leave empty?

2-54. Suppose demand for an item in inventory at a warehouse is a Poisson process at rate λ, i.e. arriving customers are Poisson, where each customer demands one unit. Whenever the inventory hits 0, an order of constant size S units is placed with the factory. Delivery time (of the entire order) from the factory to the warehouse is exponential with mean $1/\mu$. During that time, demand is lost at a cost of $\$c$/unit. The inventory holding cost is $\$h$/unit/unit time, based on the *average* amount of inventory held. (These are the only relevant costs.)

(a) Define a cycle time and find its mean.

(b) Suppose the inventory hits 0 at epoch 0. Let $N(t)$ be the inventory level at epoch t, and define

$$A_1 = \int_0^{X_1} N(t)\, dt,$$

where X_1 is the duration of the first cycle and A_1 is the *area* under the inventory function over the first cycle. Show that

$$E(A_1) = (S)(S + 1)/2\lambda. \tag{P.18}$$

[*Hint:* Plot a realization of $N(t)$ as a step function and break up A_1 into rectangles.]

(c) Use (a) and (b) to show that the cost/unit time of inventory and lost sales is

$$\frac{h\mu S(S+1)/2+\lambda^2 c}{\lambda+\mu S}. \tag{P.19}$$

(d) How would the result in (c) change if delivery time had a general distribution?

2-55. If the arrival process of cars is Poisson, what can you say about the arrival process of passengers in the cars?

2-56. For the $M/G/\infty$ queue (Example 2-6), show that the departure process of completed calls $\{\Lambda(t) - N(t): t \geq 0\}$ is a *nonstationary* Poisson process.

2-57. *Continuation.* Is $\{N(t): t \geq 0\}$ a stationary or nonstationary Poisson process? What happens when a departure occurs? Is $\{N(t)\}$ a point process?

2-58. Cars arriving at a freeway entrance are Poisson at rate λ cars per hour. On entering the freeway, car i chooses velocity V_i miles per hour, $i = 1, 2, \ldots$, where the V_i are i.i.d. random variables with distribution $P(V_i \leq x) = G(x)$. (Assume cars do not interfere with each other and that each car reaches its velocity instantaneously.) Of those cars that enter the freeway between epochs 0 and t hours, let N_b, $b \geq 0$, be the number of cars that are within b miles of the freeway entrance at epoch t, where we regard t as fixed.

(a) Show that

$$N_b \sim P\left(\lambda \int_0^t G(b/x)\,dx\right). \tag{P.20}$$

(b) For $0 < a < b$, find the joint distribution of the N_a and $(N_b - N_a)$.

(c) Is $\{N_b: b \geq 0\}$ a counting process? A Poisson process? A nonstationary Poisson process? Explain.

2-59. *Continuation.* Let $t \to \infty$ in (P.20) and let $N_b(\infty)$ denote a random variable that has the corresponding limiting distribution.

(a) Show that

$$N_b(\infty) \sim P\left(\lambda b \int_0^\infty G(1/u)\,du\right), \qquad b \geq 0, \tag{P.21}$$

where the integral *may* diverge. [It obviously will diverge if $G(0) > 0$ because an infinite number of cars will pile up at the "starting gate." A sufficient condition for convergence is for velocities to be bounded away from zero, i.e., $G(x) = 0$ for some $x > 0$.]

(b) If the integral in (P.21) converges, $\{N_b(t)\}$ becomes an "asymptotically stationary" Poisson process as $t \to \infty$. Alternatively, if we choose the time interval to be $[-t, 0]$, where epoch 0 is "now," we can think of a Poisson process that began at $-\infty$, and $\{N_b(\infty): b \geq 0\}$ is a stationary Poisson process for the count of vehicles on portions of the freeway. The *arrival rate* (cars/mile) for this process is $\lambda \int_0^\infty G(1/u)\,du$. If all cars travel at the same velocity V, i.e., $P(V_i = v) = 1$, show that the arrival rate is

$$\lambda/v, \tag{P.22}$$

(c) If $P(V_i = 30) = P(V_i = 60) = 1/2$, show that the fraction of vehicles on the freeway that are traveling 30 miles per hour is 2/3. Explain. (Let b be arbitrarily large but finite.)

2-60. For a nonstationary Poisson process, are the inter-arrival times (i) exponential, (ii) independent, (iii) or identically distributed?

2-61. Suppose we have an infinite channel queue that operates like an $M/G/\infty$ queue except that the arrival process is a nonstationary Poisson process. For any $t > 0$, find the distribution of $N(t)$, the number of customers in system at epoch t.

2-62. The arrival process of buses at a service facility is Poisson at rate λ. Let υ_i be the number of customers on bus i who enter the service facility, where $\{\upsilon_i\}$ is a sequence of i.i.d. random variables with

$$P(\upsilon_i = j) = \alpha_j, \qquad j = 0, 1, \ldots,$$

and mean $E(\upsilon) < \infty$. People are served at an infinite server facility (one *person* is served at each server, not one bus), where the service times are i.i.d. with service distribution function $G(t)$. Let $N(t)$ be the number of people who arrived during $[0, t]$ that are still in service at t.

(a) Find $E\{N(t)\}$.

(b) Show that $N(t)$ has a *batch* Poisson distribution. Find the batch size distribution. (It is different from the distribution $\{\alpha_j\}$.)

(c) What can you say about the departure process $\{\Lambda(t) - N(t): t \geq 0\}$?

2-63. Let $Y_{1,n}, Y_{2,n}, \ldots, Y_{n,n}$ be n mutually independent random variables with the identical uniform distribution on $(0, t)$. Define $Z_n = \min(Y_{1,n}, \ldots, Y_{n,n})$.

(a) Find $P(Z_n > a)$.

(b) Let t be a function of n such that $\lim_{n\to\infty} n/t = \lambda$. Show that

$$\lim_{n\to\infty} P(Z_n > z) = e^{-\lambda z}.$$

2-64. Let X and Y be independent and $\sim \exp(\lambda)$. Show that $X/(X + Y) \sim$ uniform on $(0, 1)$.

2-65. *Continuation*. Let $X_1, X_2, \ldots$ be i.i.d. $\sim \exp(\lambda)$. For $j \in \{1, \ldots, n\}$ and $0 < u < 1$, show that

$$P\left(\sum_{i=1}^{j} X_i \Big/ \sum_{i=1}^{n+1} X_i \leq u\right) = \sum_{k=j}^{n} \binom{n}{k} u^k(1 - u)^{n-k}. \tag{P.23}$$

[*Hint:* Condition on the $(n + 1)$st arrival epoch from a Poisson process occurring at t. How are the n unordered arrivals in $(0, t)$ distributed?]

Remark. By this procedure, we can generate order statistics from a uniform distribution, and, by well-known techniques, order statistics from other distributions.

2-66. Suppose the arrival process at a facility is Poisson with rate that depends on the day of the week. The rate is $\lambda_1 = 10$/hour on Mondays but $\lambda_2 = 5$/hour on other weekdays. (The facility is closed on weekends.) Data were collected every weekday on the arrival process for several weeks but, unfortunately, the day-date identity of each day's data was lost. Let $\{N(t): t \geq 0\}$ denote the arrival process on a "randomly selected" weekday, where $N(t)$ is the number of arrivals during the first t hours of that day.

(a) Find the mean and variance of $N(1)$ and show that

$$V\{N(1)\}/E\{N(1)\} > 1.$$

Is $\{N(t)\}$ a Poisson process?

(b) Given $N(1) = 6$, find an expression for the probability that the data for this day were collected on a Monday. Denote your result by

$$P(M \mid N(1) = 6).$$

(Don't calculate out.)

(c) In terms of your result in (b), find $P(N(2) - N(1) = j \mid N(1) = 6)$. Does $\{N(t)\}$ have independent increments?

(d) Given $N(1) = 6$, how are the 6 (unordered) arrival epochs distributed?

Remark. This is an example of a mixture of Poisson processes, commonly called a *conditional* Poisson process. Thus, the arrival rate λ is a random variable. Given λ, the process is Poisson. As this example illustrates, however, the unconditioned process $\{N(t)\}$, where λ is random, is *not* a Poisson process.

2-67. For $g(t) = E\{X(t)\}$, $t \geq 0$,

(a) Apply Theorem 11 and earlier results to show that

$$\lim_{t\to\infty} \int_0^t g(u)\, du/t = \mu E(X^2). \tag{P.24}$$

(b) Obtain (P.24) as a direct application of renewal-reward theory.

2-68. Let $R(t)$ be the reward associated with $X(t)$ and let $H_r(t) = E\{R(t)\}$, $t \geq 0$. Assume nonnegative rewards.

(a) Show that

$$\lim_{t\to\infty} \int_0^t P(R(u) > r)\, du/t = \mu \int_0^\infty P(R_j > r, X_j > x)\, dx, \qquad j \geq 2. \tag{P.25}$$

(b) From (a) and Theorem 11, show that

$$\lim_{t\to\infty} \int_0^t H_r(u)\, du/t = \mu E(R_j X_j), \qquad j \geq 2. \tag{P.26}$$

(c) Obtain (P.26) as a direct application of renewal-reward theory.

(d) Choose R_j so that (P.24) becomes a special case of (P.26).

2-69. For Problem 2-46, use the result in part (c) there to show that

$$\lim_{t\to\infty} \int_0^t N_c(u)\, du/t = \lambda E(C). \tag{P.27}$$

2-70. For an ordinary renewal process ($A = F$) with $\mu > 0$, show that

$$E\{Z_{M(t)}\} \leq m(t)/\mu, \qquad t \geq 0. \tag{P.28}$$

Why doesn't Wald's equation apply to the sum on the left? Compare with explicit calculations made in Problem 2-7. For a Poisson process, show that the inequality in (P.28) is strict for any $t > 0$.

2-71. *Renewal-reward proof of Wald's equation.* Let $X_1, X_2, \ldots$ be i.i.d. with finite mean, and random variable N_1 be defined as a stopping time for Wald's equation, where $0 < E(N_1) < \infty$. Once N_1 occurs, let N_2 be a stopping time for the remainder of the sequence, i.e., given $N_1 = n$, N_2 is a stopping time for $X_{n+1}, X_{n+2}, \ldots$, where N_2 is independent of N_1 and has the same distribution. In this way, we define a sequence of stopping times $N_1, N_1 + N_2, \ldots$, where the N_i are i.i.d.

(a) From the theory of renewal-rewards, show that

$$\lim_{n\to\infty} Z_n/n = E\left(\sum_{j=1}^{N_1} X_j\right)\Big/E(N_1). \tag{P.29}$$

(b) Obtain Wald's equation by applying the strong law of large numbers to the left side of (P.29).

Remarks. The conditions for the previous proof are more restrictive than for the proof in the text. (In what way?) To apply the Renewal-Reward Theorem, the numerator in (P.29) needs to be finite. This is not really an assumption; it follows from the converse of the strong law. See the remark at the end of Section 1-15.

2-72. *On Example 2-17.* In Problem 2-71, we obtained a result about one stopping time by creating a sequence of stopping times. We now employ this idea in Example 2-17, where for $k \geq 1$, we define

$$N_1 + \cdots + N_k = \min\{n: S_n = k,\ n = 1, 2, \ldots\},$$

where the N_i are i.i.d., but *possibly defective* random variables. Let $P(N_1 < \infty) = \alpha$ and note that $P(N_1 + N_2 < \infty) = \alpha^2$.

(a) Argue the following:

$$\text{Given } X_1 = 1,\ N_1 = 1, \text{ and}$$

$$\text{Given } X_1 = -1,\ N_1 = 1 + N_1' + N_2',$$

where the N_i' are i.i.d. distributed as N_1.

(b) From (a), show that α is a root of the equation

$$q\alpha^2 - \alpha + p = 0, \tag{P.30}$$

and that the roots of (P.30) are p/q and 1.

(c) From (b) show that for $p \geq q$, $P(N_1 < \infty) = 1$, i.e., N_1 is a stopping time! (What range of values for α are possible?)

(d) For $p < q$, $S_n \to -\infty$. (Why?) From this, argue that $P(N_1 < \infty) < 1$. From (b), what is $P(N_1 < \infty)$?

2-73. Use a renewal argument to show that for an ordinary renewal process,

$$E\{Z_{M(t)+1}\} = [m_o(t) + 1]/\mu.$$

That is, obtain this result without direct use of Wald's equation.

2-74. For an ordinary renewal process, show that

$$P(X(t) \leq x) \equiv K_x(t) = Q_o(t) + \int_0^t Q_o(t-u)\, dm_o(u),$$

where

$$Q_o(t) = F(x) - F(t) \qquad \text{for } t \le x, \qquad \text{(P.31)}$$
$$= 0 \qquad \text{for } t > x.$$

2-75. For an ordinary renewal process, show that

$$E\{Y(t)\} = Q_o(t) + \int_0^t Q_o(t-u)\, dm_o(u),$$

where $Y(t)$ is the excess at epoch t, and

$$Q_o(t) = \int_t^\infty F^c(x)\, dx. \qquad \text{(P.32)}$$

[Argue this directly. Do *not* use the distribution of $Y(t)$.]

2-76. *Lower bound on the ordinary renewal function.*

(a) By conditioning on X_1 show that the following inequality holds between the general and ordinary renewal functions:

$$m(t) \le m_o(t) + 1, \qquad t \ge 0. \qquad \text{(P.33)}$$

(b) For $\mu > 0$, specialize (P.33) to obtain the lower bound

$$m_o(t) \ge \mu t - 1, \qquad t \ge 0. \qquad \text{(P.34)}$$

(c) Obtain (P.33) from (P.2).

2-77. *Upper bound on the ordinary renewal function.* For an ordinary renewal process with $\mu > 0$, suppose that for some finite number k,

$$E(X_j - x \mid X_j > x) \le k \qquad \text{for all } x \ge 0, \qquad \text{(P.35)}$$

where $k \ge 1/\mu$. [The property defined by (P.35) is called *bounded mean residual life* and is a generalization of the property defined in Problem 2-23.]

(a) Show that for all $t \ge 0$,

$$E\{Y(t)\} \le k. \qquad \text{(P.36)}$$

(b) From (a), show that the ordinary renewal function has the upper bound

$$m_o(t) \le \mu t - 1 + k\mu. \qquad \text{(P.37)}$$

For $k = 1/\mu$, compare this bound with the lower bound (P.34).

2-78. *Continuation.* If we do not make some assumption such as (P.35) about the tail of F, $m_o(t)$ can be arbitrarily large. Find a family of distributions with the same $\mu > 0$ which show that even $m_o(0)$ can be arbitrarily large!

2-79. Use the renewal argument to generalize the results in Problems 2-74 and 2-75 to general renewal processes. Obtain the same expressions via the direct method presented in Section 2-17.

2-80. For an ordinary renewal process with reward R_j associated with X_j, define $K_o(t) = P(R(t) > r)$.

(a) Show that

$$K_o(t) = Q_o(t) + \int_0^t Q_o(t-u)\, dm_o(u),$$

where $Q_o(t) = P_o(R_1 > r, X_1 > t)$, and P_o denotes the tail probability in the ordinary case.

(b) Find the corresponding expression for $K(t) = P(R(t) > r)$ for a general renewal process.

2-81. Equation (175) holds for an ordinary renewal process. Show that for a general renewal process,

$$E\{[M(t)]^2\} = m(t) + 2\int_0^t m(t - x)\, dm_o(x), \tag{P.38}$$

where $m_o(t)$ is the corresponding ordinary renewal function.

2-82. Apply the Key Renewal Theorem to find pointwise limits of the expressions found in Problems 2-74, 2-75, and 2-80(a). State any assumptions you find necessary for these limits to exist. Compare these limits with time averages (50), Example 2-12, and (P.25).

2-83. *Continuation of Problem 2-53*. Let state 0 denote that the stand is empty, state 1 denote "occupied by a taxi," and state 2 denote "occupied by a customer."

(a) Assuming the stand is empty at epoch 0, find an expression for $P_2(t)$, the probability that the stand is occupied by a customer at epoch t.

(b) Find the pointwise limit of $P_2(t)$. State assumptions.

(c) Independent of (a) and (b), find the *fraction of time* the system is in state 2.

2-84. For a *general* renewal process, find the pointwise limit (when it exists): $\lim_{t\to\infty} [m(t) - \mu t]$.

2-85. *Continuation*. Since the pointwise limit above depends on initial conditions (initial distribution A), so does the corresponding time average. Use $m(t) - \mu t = \mu E\{Y(t)\} - \mu E(X_1)$ to find the corresponding time average. (Where did this equation come from?)

2-86. An electronic device is composed of two independent components, each with continuous life distribution F and mean $1/\mu < \infty$. Each component is replaced on failure by a new component with independent life $\sim F$. When a component fails, the device is said to have failed. Assume downtime is negligible and does not affect residual life.

(a) Find the number of failures of the device per unit time.

(b) Averaging over the failure epochs of the component in "position 1," what is the distribution of residual life of the component in "position 2"?

(c) If $X \sim F$ and $Y \sim$ as in (b), show that

$$E\{\min(X, Y)\} = 1/2\mu.$$

(d) Generalize (c) to k components, $k \geq 2$.

2-87. Show that the nonlattice version of the Key Renewal Theorem implies the nonlattice version of Blackwell's Theorem.

2-88. For the $GI/G/\infty$ queue (Example 2-26), obtain the pointwise limit of (191) when F is nonlattice. Compare with the equivalent result for the $M/G/\infty$ queue. (The means are the same, but the limiting distributions will be different.)

2-89. For a *general* renewal process where $E(X_1) < \infty$ and F has second moment

$E(X^2) < \infty$, we may rearrange (140):

$$m(t) - \mu t = \mu E\{Y(t)\} - \mu E(X_1).$$

(a) Use this expression to find

$$\lim_{t\to\infty} \int_0^t [m(u) - \mu u]\, du/t = \mu E(Y_\infty) - \mu E(X_1) = \mu^2 E(X^2)/2 - \mu E(X_1). \qquad \text{(P.39)}$$

(b) What is the limit in (a) when F has an infinite variance?

2-90. *Continuation.* Let F have variance σ^2 and define $V(t) = \text{Var}\{M(t)\}$, $t \geq 0$.

(a) Use equation (P.38) to show that

$$V(t) = 2\int_0^t m_o(t - u)\, dm(u) + m(t) - [m(t)]^2.$$

(b) Write $V(t) = V_e(t)$ for the special case of an *equilibrium* renewal process, and show that

$$V_e(t) = 2\mu \int_0^t [m_o(u) - \mu u]\, du + \mu t.$$

(c) Show that

$$\lim_{t\to\infty} V_e(t)/t = \mu^3\sigma^2. \qquad \text{(P.40)}$$

Remark. It is not difficult to show that the limit in (c) holds for a general renewal process as well.

2-91. Let $X_1, X_2, \ldots$ be inter-arrival times of buses at a bus stop. Typically, it is *not* reasonable to assume that they are i.i.d. Instead, assume $\{X_n\}$ is a regenerative process with regeneration points $K_1, K_1 + K_2, \ldots$. (For example, $X_1, \ldots, X_{K_1}$ is the first cycle, where K_1 is the first cycle length.) Define $\{Z_n\}$, $\{M(t)\}$, and $\{Y(t)\}$ in terms of $\{X_n\}$ as was done in Section 2-2. Thus, $Y(t)$ is the waiting time for the next bus, measured from epoch t, a point on the continuous (real) time axis. Now $\{Y(t)\}$ will also be a regenerative process, with regeneration points $Z_{K_1}, Z_{K_1+K_2}, \ldots$. Assume $E(K_1)$ and $E(Z_{K_1})$ are strictly positive and finite.

(a) For $\overline{X}_n = Z_n/n$, $n = 1, 2, \ldots$, and any $x \geq 0$, define

$$\overline{X} = \lim_{n\to\infty} \overline{X}_n \quad \text{and} \quad F^c(x) = \lim_{n\to\infty} \sum_{j=1}^{n} P(X_j > x)/n.$$

Briefly explain why the following is true:

$$\overline{X} = \int_0^\infty F^c(x)\, dx \equiv 1/\mu.$$

(b) For any $y \geq 0$, let $V_n(y) = (X_n - y)^+$ and $\overline{V}_n(y) = \sum_{j=1}^{n} V_j(y)/n$, $n = 1, 2, \ldots$.

Show that

$$\overline{V}(y) \equiv \lim_{n\to\infty} \overline{V}_n(y) = \int_y^\infty F^c(u)\, du.$$

(c) Show that

$$\lim_{t\to\infty} \int_0^t P(Y(u) > y)\, du/t = E\left\{\sum_{j=1}^{K_1} (X_j - y)^+\right\} \Big/ E\{Z_{K_1}\}.$$

(d) Show that

$$E\{Z_{K_1}\} = \overline{X}E(K_1) \quad \text{and } E\left\{\sum_{j=1}^{K_1} (X_j - y)^+\right\} = \overline{V}(y)E(K_1).$$

[*Hint:* For the first result, apply the strong law of large numbers to each side of

$$\overline{X}_{T(n)} = Z_{T(n)}/T(n),$$

where $T(n) = \sum_{j=1}^{n} K_j$, $n = 1, 2, \ldots$, and let $n \to \infty$.]

(e) From (d), show that the limit in (c) is $\mu \int_y^\infty F^c(u)\, du = F_e^c(y)$!

3

DISCRETE MARKOV CHAINS

A simple notion called the *Markov property* underlies an important class of stochastic processes called *Markov processes*: Given the *present* (state of a stochastic process), the *future* (evolution of the process) is independent of the *past* (evolution of the process).

In Chapter 2, we defined four broad classes of stochastic processes in terms of time and state space, where each of them may be either discrete or continuous. Markov processes are classified in the same way.

For example, the Poisson process $\{\Lambda(t)\}$ is a discrete-state, continuous-time process. From independent increments, we also have that given $\Lambda(t) = n$ for any $t > 0$ and integer $n \geq 0$, the "future" $\{\Lambda(s): s > t \mid \Lambda(t)\}$ is independent of the "past" $\{\Lambda(s): s < t \mid \Lambda(t)\}$. That is, it has the Markov property and therefore is a Markov process. Since independence is a symmetric property, we also have that given $\Lambda(t)$, the past is independent of the future! Past and future are said to be *conditionally independent*.

In applications, the importance of the Poisson process is its frequent role as a component of a stochastic model, e.g., as an arrival process at a queue. In isolation, a Poisson process is not a particularly interesting Markov process because realizations of $\{\Lambda(t)\}$ are nondecreasing step functions. The only thing random about a Poisson process is the location of the jumps.

More complex Markov processes can make interesting stochastic models on their own. In this book, we treat in some detail discrete-state Markov processes, which are called *Markov chains*. The states of a Markov chain are usually chosen to be the nonnegative integers. Discrete-time Markov chains are treated in this chapter. Continuous-time Markov chains are treated in Chapter 4, together with some discussion of other kinds of Markov processes.

3-1 DISCRETE MARKOV CHAINS; TRANSITION PROBABILITIES

A sequence of integral-valued random variables $\{X_n: n = 0, 1, \ldots\}$ is called a *discrete Markov chain* if it has the *Markov property*:

$$P(X_{n+1} = j \mid X_n, X_{n-1}, \ldots, X_0) = P(X_{n+1} = j \mid X_n), \qquad n = 0, 1, \ldots, \quad j = 0, 1, \ldots, \tag{1}$$

where, for each value of X_n, the probabilities in (1) are called (one-step) *transition probabilities*. We will assume that transition probabilities are *stationary*, i.e., that they are independent of epoch n.

Under these assumptions, let

$$p_{ij} = P(X_{n+1} = j \mid X_n = i), \qquad i \geq 0 \quad \text{and} \quad j \geq 0,$$

and denote the corresponding (one-step) transition probability matrix by

$$P = (p_{ij}).$$

Since the rows of P are distributions, P is a square matrix of nonnegative elements with row sums equal 1. Matrices with these properties are called *stochastic*.

For $j \geq 0$, let $P(X_0 = j) = a_j$ and, for $n \geq 1$, $P(X_n = j) = a_j^{(n)}$. It will be convenient to represent these distributions as row vectors:

$$a = (a_0, a_1, \ldots) \quad \text{and} \quad a^{(n)} = (a_0^{(n)}, a_1^{(n)}, \ldots).$$

We call a and $a^{(n)}$ *probability vectors* because their elements are nonnegative and sum to 1, i.e., the corresponding distributions are proper.

Note that the number of states (values of X_j) may be either finite or infinite. In expressions that follow, we will often sum over all states, e.g., $\sum_{j=0}^{\infty} p_{ij} = 1$. In order to combine and simplify notation for the finite and infinite cases, we will delete the range of summation, e.g., $\sum_j p_{ij} = 1$, where it is understood that the range of summation, when deleted, is over all states.

Now define *n-step* transition probabilities and the corresponding matrix by

$$p_{ij}^{(n)} = P(X_n = j \mid X_0 = i),$$

$$P^{(n)} = (p_{ij}^{(n)}), \qquad n = 1, 2, \ldots.$$

We will now find $P^{(n)}$ in terms of P: For any $n \geq 1$ and $s \geq 1$,

$$\begin{aligned} p_{ij}^{(n+s)} &= P(X_{n+s} = j \mid X_0 = i) \\ &= \sum_k P(X_{n+s} = j, X_n = k \mid X_0 = i) \\ &= \sum_k P(X_{n+s} = j \mid X_n = k, X_0 = i) P(X_n = k \mid X_0 = i), \end{aligned}$$

which, from the Markov property, (1), is

$$= \sum_k P(X_{n+s} = j \mid X_n = k)P(X_n = k \mid X_0 = i). \tag{2}$$

Because transition probabilities are stationary, (2) becomes

$$p_{ij}^{(n+s)} = \sum_k p_{ik}^{(n)} p_{kj}^{(s)}, \qquad n \geq 1, \quad s \geq 1. \tag{3}$$

In matrix form, (3) is a matrix product

$$P^{(n+s)} = P^{(n)}P^{(s)}, \qquad n \geq 1, \quad s \geq 1. \tag{4}$$

Applying (4), we have $P^{(2)} = P^2$, and, in general,

$$P^{(n)} = P^n, \qquad n \geq 1, \tag{5}$$

i.e., the n-step transition probability matrix is the nth power of the one-step transition probability matrix.

For example, suppose a sequence of wet and dry days is a Markov chain where zero denotes "wet" and 1 denotes "dry." Given that today is dry, the probability of a wet day after tomorrow is $p_{10}^{(2)} = p_{10}p_{00} + p_{11}p_{10}$.

It is easy to show similar relationships between $a^{(n)}$ and P:

$$a^{(n)} = a^{(n-1)}P, \text{ and} \tag{6}$$

$$a^{(n)} = aP^n, \qquad n \geq 1. \tag{7}$$

(*Note:* We define $a^{(0)} = a$ and $P^0 = I$, the identity matrix.)

Example 3-1: Brand Switching and Market Share

Suppose we model consumer brand switching as a Markov chain, where X_n is the brand chosen on a consumer's nth purchase. In this example, the Markov property means that consumers have short memories, i.e., future purchases depend only on experience with the most recent one. The company that markets brand j may want to know how the market share captured by that brand changes as a function of brand loyalty. Specifically, if through better service and/or better product quality, p_{jj} could be increased to $p_{jj} + \delta$, how much would brand j's market share increase? (Since $\sum_k p_{jk} = 1$, we would also have to model how the p_{jk}, $k \neq j$, would change.)

To address the research issue posed in the preceding paragraph, we have to decide what is meant by *market share*. The market share of brand j in some interval on the continuous (real time) axis would be the fraction of purchases of all relevant brands that are of brand j. If all customers behave the same, this would also be the fraction of purchases made by one customer (over time) that are of brand j. To put this in mathematical terms, let

$$\begin{aligned} I_{nj} &= 1 \qquad \text{if } X_n = j, \\ &= 0 \qquad \text{otherwise,} \end{aligned}$$

for $n = 0, 1, \ldots,$ and any $j \geq 0$.

The market share of brand j is defined as the limit

$$\lim_{n\to\infty} \sum_{s=1}^{n} I_{sj}/n \quad \text{w.p.1.} \tag{8}$$

Since $E(I_{nj}) = a_j^{(n)}$, the expected-value version of (8) is

$$\lim_{n\to\infty} \sum_{s=1}^{n} a_j^{(s)}/n. \tag{9}$$

We would like to find market share in a way that is computationally easier than (9). First, suppose that for each j, the following limit holds, independent of initial vector a:

$$\lim_{n\to\infty} a_j^{(n)} = \pi_j > 0, \tag{10}$$

where $\sum_j \pi_j = 1$. By letting $n \to \infty$ in (6), it follows that the vector $\pi = (\pi_0, \pi_1, \ldots)$ satisfies

$$\pi = \pi P. \tag{11}$$

Furthermore, (10) implies that the limits (8) and (9) are also π_j. Thus, the market share of brand j is π_j, and may be found by solving the system of linear equations (11) for π.

Actually, the validity of (11) as a way of finding limits (8) and (9) is more general than the existence of limits (10). For example, consider the two-state Markov chain with transition probability matrix

$$P = \begin{pmatrix} 0 & 1 \\ 1 & 0 \end{pmatrix}, \tag{12}$$

and vector $a = (a_0, a_1)$, where $a_0 + a_1 = 1$. For n even, $a^{(n)} = (a_0, a_1)$, and for n odd, $a^{(n)} = (a_1, a_0)$. In vector form, it is easy to see that the limits (9) are $\pi = (1/2, 1/2)$, which is also the unique solution to (11), whereas the limits (10) do not exist, unless we happen to choose $a = (1/2, 1/2)$.

3-2 CONNECTION WITH RENEWAL THEORY

Finding (9) by solving (11) is an application of renewal theory. The fundamental idea is that the times (numbers of transitions) between returns to a particular state are i.i.d. random variables. This fact is a consequence of the Markov property and stationary transition probabilities. The latter property means that on each return to state j, the future evolution of the chain is governed by the same transition probability matrix.

To take advantage of renewal theory, we need to introduce the distribution of first passage times between arbitrary pairs of states, $i \geq 0$ and $j \geq 0$,

$$f_{ij}^{(n)} = P(X_n = j, X_{n-1} \neq j, \ldots, X_1 \neq j \mid X_0 = i), \quad n = 1, 2, \ldots. \tag{13}$$

In words, (13) is the probability that the first *visit* to state j occurs at epoch n, given $X_0 = i$. We can also define a *first passage time* random variable T_{ij}, where $P(T_{ij} = n) = f_{ij}^{(n)}$, $n = 1, 2, \ldots$. Thus, (13) is called a first passage time *distribution*. For $i = j$, a visit is called a *return*.

Let $f_{ij} = \sum_{n=1}^{\infty} f_{ij}^{(n)}$ be the probability that the chain *ever* visits state j, given $X_0 = i$.

To illustrate these ideas, consider the matrix in equation (14), where $abc > 0$,

$$P = \begin{pmatrix} a & b & c \\ 0 & 1 & 0 \\ 0 & 0 & 1 \end{pmatrix}. \tag{14}$$

State j is called *absorbing* if $p_{jj} = 1$. Thus states 1 and 2 are absorbing. Clearly, a chain can never leave an absorbing state. Hence, $f_{11} = f_{22} = 1$ and $f_{10} = f_{12} = f_{20} = f_{21} = 0$. If this chain is ever to return to state 0, it must do so immediately, i.e., $f_{00}^{(1)} = a$, $f_{00}^{(n)} = 0$ for $n \geq 2$, and $f_{00} = a$. Similarly, $f_{01}^{(n)} = a^{(n-1)}b$ for $n \geq 1$ (what sequence of transitions has this probability?) and $f_{01} = b/(1 - a)$. Finally, $f_{02}^{(n)} = a^{(n-1)}c$ for $n \geq 1$, and $f_{02} = c/(1 - a)$. (Why did it turn out that $f_{01} + f_{02} = 1$ here?)

In general, we say state j is *recurrent* if $f_{jj} = 1$ and *transient* if $f_{jj} < 1$. If j is recurrent, we define the *mean recurrence time* for state j, ν_j, to be

$$\nu_j = E(T_{jj}) = \sum_{n=1}^{\infty} n f_{jj}^{(n)}, \tag{15}$$

where ν_j may be either finite or infinite. State j is called *positive* if $f_{jj} = 1$ and $\nu_j < \infty$, and *null* if $f_{jj} = 1$ and $\nu_j = \infty$. (If j is transient, T_{jj} is defective, and we will set $\nu_j = \infty$.)

Suppose a Markov chain starts in state $X_0 = i$, and we are interested in counting visits to state j. given $T_{ij} = n$ for some finite n, the time (number of transitions) until the next visit to state j is distributed as T_{jj}, independent of n. Times between subsequent visits are also distributed as T_{jj}, independent of the past (provided that the earlier visits occur). Let

$$\begin{aligned} I_{ij}(n) &= 1 \qquad \text{if } X_n = j, \\ &= 0 \qquad \text{otherwise, and} \end{aligned}$$

$$M_{ij}(n) = \sum_{k=1}^{n} I_{ij}(k),$$

where $M_{ij}(n)$ is the number of visits to state j by epoch (transition) n, given $X_0 = i$. Now $\{M_{ij}(n), n = 1, 2, \ldots\}$ is a renewal process, where $\{f_{ij}^{(n)}\}$ has the role of the initial distribution A and $\{f_{jj}^{(n)}\}$ has the role of F. One or both of these distributions may be defective, i.e., $f_{ij} < 1$ and $f_{jj} < 1$ are possible. Since $E\{I_{ij}^{(n)}\} = p_{ij}^{(n)}$, the corresponding *j-renewal* function is

$$m_{ij}(n) = E\{M_{ij}(n)\} = \sum_{k=1}^{n} p_{ij}^{(k)}. \tag{16}$$

If state j is transient ($f_{jj} < 1$), the renewal process is *defective*, and [see Section 2-10 and equation (103) there]

$$\lim_{n\to\infty} m_{ij}(n) = \sum_{n=1}^{\infty} p_{ij}^{(n)} = \frac{f_{ij}}{1 - f_{jj}} < \infty, \tag{17}$$

where, for $i = j$, (17) is finite *only if* j is transient.

Remark. A weaker but conceptually important result is that if state j is transient, the number of visits to state j is finite with probability 1, i.e., $M_{ij}(\infty) < \infty$.

If state j is recurrent and $f_{ij} = 1$, the limit (17) is infinite, and, by the elementary renewal theorem,

$$\lim_{n\to\infty} m_{ij}(n)/n = \lim_{n\to\infty} \sum_{k=1}^{n} p_{ij}^{(k)}/n = 1/\nu_j \begin{cases} = 0 & \text{if } j \text{ is null,} \\ > 0 & \text{if } j \text{ is positive.} \end{cases} \tag{18}$$

It's easy to extend (18) to the case $f_{ij} < 1$, $f_{jj} = 1$; see Section 2-10.

3-3 COMMUNICATION CLASSES AND CLASS PROPERTIES

The "connectivity" of a Markov chain also plays an important role in its behavior. For $i \neq j$, we say that state i *can reach* state j if $f_{ij} > 0$. This is denoted by $i \to j$. Now $i \to j$ if and only if there is a path (finite sequence of transitions) of positive probability that starts in i and ends in j. The *length* of a path is the number of (one-step) transitions from beginning to end. Since $p_{ij}^{(n)}$ is the sum of the probabilities of all paths of length n from i to j, $p_{ij}^{(n)} > 0$ if and only if at least one of these paths has positive probability. Thus, $i \to j$ if and only if $p_{ij}^{(n)} > 0$ for some $n \geq 1$.

States i and j *communicate*, denoted by $i \leftrightarrow j$, if $f_{ij} > 0$ and $f_{ji} > 0$, i.e., if $i \to j$ *and* $j \to i$.

Communication is a relation between *pairs* of states. That is, any pair $i \neq j$ either communicates or it does not. Let's explore properties of this relation. From the definition of $\leftrightarrow$, if $i \leftrightarrow j$, then $j \leftrightarrow i$, i.e., $\leftrightarrow$ is said to be *symmetric*. Also, for $i \neq j$ and $j \neq k$, $i \leftrightarrow j$ and $j \leftrightarrow k$ together imply $i \leftrightarrow k$, i.e., $\leftrightarrow$ is said to be *transitive*. (To see this, just connect up paths of positive probability in each direction.) *By definition* we will also say that for every state i, $i \leftrightarrow i$ (even if $f_{ii} = 0$), i.e., every state i communicates with itself. This third property is called *reflexive*.

We can use the communication relation $\leftrightarrow$ to partition the set of states into subsets $A_1, A_2, \ldots$, called *communication classes* with these properties:

(a) every state i belongs to exactly one class, A_k, say,
(b) every A_k contains at least one state,
(c) states in the same class communicate with each other,
(d) states belonging to different classes do *not* communicate.

In effect, saying $i \leftrightarrow j$ is equivalent to saying that they belong to the same communication class. (Note that if we did not require $i \leftrightarrow i$ for every i, some states might be left out of this classification.)

We will also say that a set of states C is *closed* if for every pair of states $i \in C$ and $j \notin C$, $p_{ij} = 0$. Clearly, paths of positive probability starting in C must remain there, and $p_{ij}^{(n)} = 0$ for every $n \geq 1$. If C is not closed, it is said to be *open*. We use this terminology whether or not C is a communication class. (Obviously, the entire set of states is closed.)

A Markov chain is called *irreducible* if all states communicate, or, equivalently, if no proper subset (the entire set with at least one state deleted) is closed. Note that if the initial state X_0 belongs to a closed communication class, the chain will behave like an irreducible Markov chain with state space restricted to that class.

We can represent a Markov chain as a network or a graph where the states are nodes and for every pair of states (i, j) where $p_{ij} > 0$, we have a directed arc (like a one-way street) from i to j. A Markov chain is irreducible if and only if the corresponding graph is *connected*. (There is a path of one-way streets between every pair of nodes.) See Example 3-2.

Earlier, we classified individual states as to whether they were transient, null, or positive. These turn out to be (communication) class properties as well:

Theorem 1: The Class Property Theorem. The states in a communication class A are either

all positive,
all null, or
all transient.

Proof. The theorem is trivially true for communication classes containing only one state. For any pair of states $j \neq k$ such that $j \leftrightarrow k$, there are paths of positive probability of some length J from j to k and of some length K from k to j. Let $\alpha = p_{jk}^{(J)} > 0$ and $\beta = p_{kj}^{(K)} > 0$. For any $n \geq 1$, we can write

$$p_{jj}^{(J+n+K)} \geq \alpha p_{kk}^{(n)} \beta, \tag{19}$$

because the probability on the right is for a restriction on the set of paths that has the probability on the left.

Suppose state j is transient. Sum (19) from $n = 1$ to ∞. From (17) (with $i = j$) and (19), we have $\sum_{n=1}^{\infty} p_{kk}^{(n)} < \infty$, i.e., state k is transient. Since j and k are arbitrary, we have either the states in A are all transient or they are all recurrent. If j is null (and k is recurrent), (18) implies

$$\lim_{n\to\infty} \sum_{m=1}^{n} p_{jj}^{(J+m+K)}/n = 0,$$

which, together with (19), implies that k is null. Hence the states in A are either all null or all positive. This completes the proof.

Transient, null, and positive are examples of what we call class properties. A *class property* is a property of an individual state that a particular state may or may not have, such that if one state in a communication class has the property, then all states in that class do. Thus, we can without ambiguity speak of transient, null, and positive communication classes.

With this new terminology, we now state

Theorem 2. Open communication classes are transient.

Proof. Let A be an open communication class, where for some $i \in A$, $j \notin A$, $p_{ij} > 0$. Thus, $i \to j$, but $j \not\to i$, for otherwise j would belong to A. Hence, $f_{ji} = 0$. Suppose $X_0 = i$, and we condition on whether the next transition is to state j:

$$\begin{aligned} f_{ii} &= f_{ji}p_{ij} + P(\text{chain returns to state } i \mid X_1 \neq j)(1 - p_{ij}) \\ &\leq 0 + (1 - p_{ij}) < 1. \end{aligned}$$

Hence, state i is transient. But from Theorem 1, all states in A are transient, and we have Theorem 2.

Closed communication classes may be either transient, null, or positive. If $f_{ii} = 1$ and $i \leftrightarrow j$, then, by Theorem 1, $f_{jj} = 1$. It's easy to see that $f_{ij} = f_{ji} = 1$ also. (If $f_{ij} < 1$, the argument used to prove Theorem 2 can be used to show $f_{jj} < 1$, a contradiction.)

Now suppose that A is a *finite* closed communication class, i.e., the number of states in A is finite. For all $i \in A$,

$$\sum_{j\in A} p_{ij}^{(n)} = 1, \qquad n = 1, 2, \ldots, \tag{20}$$

because A is closed. If A is either transient or null, (17) and (18) imply

$$\lim_{n\to\infty} \sum_{k=1}^{n} p_{ij}^{(k)}/n = 0 \qquad \text{for all } j \in A. \tag{21}$$

Now sum (21) over all $j \in A$. Since A is finite, we can interchange limit and sum, obtaining from (20) that

$$\sum_{j \in A} \lim_{n \to \infty} \sum_{k=1}^{n} p_{ij}^{(k)}/n = \lim_{n \to \infty} \sum_{k=1}^{n} \sum_{j \in A} p_{ij}^{(k)}/n = 1, \tag{22}$$

which contradicts (21). We have proven

Theorem 3. Finite closed communication classes are positive.

An obvious consequence of these theorems we state as

Theorem 4. For a finite Markov chain, state i is transient if and only if there exists a state j such that $i \to j$ but $j \not\to i$. Otherwise, state i is positive.

Remark. Null communication classes are infinite. Thus, null states cannot occur in finite chains.

Also observe that a path has positive probability if and only if all the transition probabilities on that path are positive, regardless of their magnitudes. Thus, the communication classes of a Markov chain depend only on the location of the positive elements of the transition matrix, *not* on their magnitudes. To illustrate this, consider the following example.

Example 3-2

Consider the following transition matrix, where (*) denotes the position of each positive element, and the graph corresponding to this matrix.

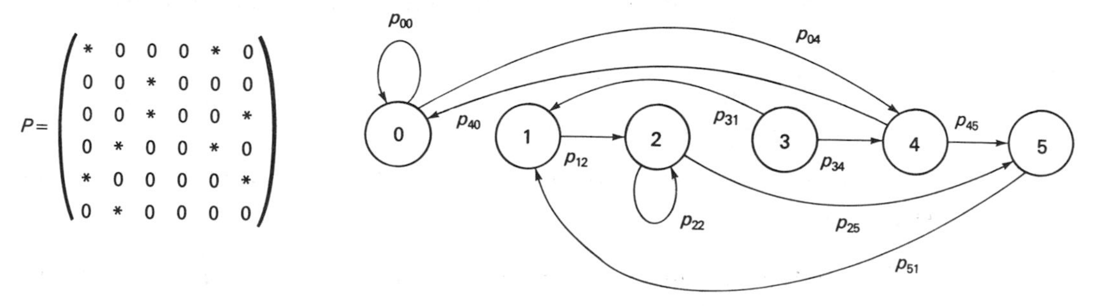

Since $0 \to 4$ and $4 \to 0$, $0 \leftrightarrow 4$. Similarly, $5 \to 1 \to 2 \to 5$ implies $5 \leftrightarrow 1 \leftrightarrow 2$. No state can reach state 3. Also, no state in {5, 1, 2} can reach {0, 4}. Hence, the communication classes are $A_1 = \{0, 4\}$, $A_2 = \{5, 1, 2\}$, and $A_3 = \{3\}$. (We will not discuss efficient ways to determine communication classes.) The "can reach" relations between the classes is

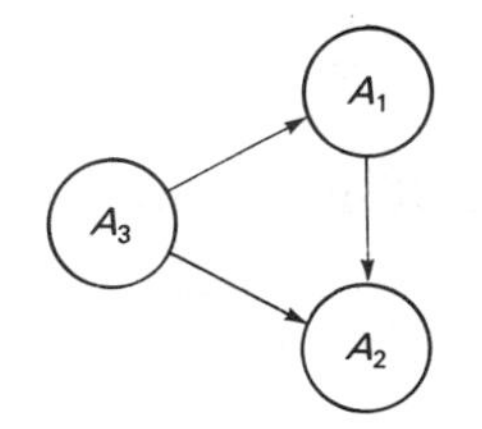

Classes A_1 and A_3 are open and therefore transient. Class A_2 is closed (finite) and therefore positive.

Remark. As the preceding example shows, finite Markov chains must contain at least one closed (and therefore positive) communication class. To see this, start with any class, A_1 say. If it is open, consider next any class it can reach, A_2 say. Continue this process. We can never return to a class considered earlier (why?). Since the number of classes is finite, we eventually will reach a class that cannot reach any other class, i.e., it is closed. (For an infinite chain, can you think of an example where there are no closed communication classes?)

3-4 IRREDUCIBLE CHAINS—POSITIVE STATES

For an arbitrary irreducible Markov chain, such as for the brand-switching example in Section 3-1, we are now ready to show how limits (9) are related to solutions π of $\pi = \pi P$. Initially, we assume all states are recurrent; in this case, $f_{ij} = 1$ for all i and j.

To begin, first define probability vectors

$$b^{(n)} = \sum_{s=1}^{n} a^{(s)}/n \quad \text{and} \quad c^{(n)} = \sum_{s=1}^{n} a^{(s-1)}/n, \qquad n = 1, 2, \ldots .$$

Now sum (7) and divide by n:

$$b^{(n)} = a \sum_{s=1}^{n} P^s/n, \tag{23}$$

where $\sum_{s=1}^{n} P^s/n$ is a matrix with elements $\sum_{s=1}^{n} p_{ij}^{(s)}/n$. From (18), the elements in column j of this matrix all converge to $1/\nu_j$. Thus, from (23),

$$\lim_{n\to\infty} b^{(n)} = \pi = (\pi_0, \pi_1, \ldots), \tag{24}$$

where $\pi_j = 1/\nu_j$. From (6), we can write

$$b^{(n)} = c^{(n)}P, \tag{25}$$

where $c^{(n)}$ also converges to vector π. Now let $n \to \infty$ in (25). For each j on the right, interchange limit and sum for $\sum_i c_i^{(n)} p_{ij}$, and we have

$$\pi = \pi P. \tag{26}$$

If all states are positive ($\nu_j < \infty$), π in (26) in a vector of strictly positive elements.

However, we do not yet know that π is a *probability* vector, i.e., that it represents a probability distribution with properties $\pi_j \geq 0$ for all j (which we already know) and $\sum_j \pi_j = 1$.

Notice that if π is a probability vector with the property $\pi = \pi P$, $\pi P = \pi P^2$, and, in general, $\pi = \pi P^n$, $n = 1, 2, \ldots$. Thus, if we choose $a = \pi$ as the distribution of X_0, the distribution of X_n is also π for every n. The stochastic process will be *stationary* and π is a *stationary* probability vector (or distribution).

Since the limit of (24) is independent of the initial vector a, this limit must be a stationary probability vector, if one exists. Obviously, if one exists, it must be unique.

We now construct a stationary probability vector. Because $b^{(n)}$ is a probability vector, the sum of the first $(k + 1)$ elements of this vector cannot exceed 1. Thus, the sum of the first $(k + 1)$ elements of the limit vector π in (26) cannot exceed 1, i.e., $\sum_{j=0}^{k} \pi_j \leq 1$ for all finite k. As $k \to \infty$, we have

$$0 < \sum_{j=0}^{\infty} \pi_j \leq 1.$$

Also note that if $\pi = \pi P$, $\alpha\pi = \alpha\pi P$, where α is any constant. Thus, if π is a vector satisfying $\pi = \pi P$, $\alpha\pi$, where $1/\alpha = \sum_j \pi_j$, is a stationary *probability* vector, i.e., such a vector exists.

Thus, if all states are positive, there is a unique stationary probability vector $\pi = \pi P$ where $\pi_j = 1/\nu_j$, $j = 0, 1, \ldots$. If all states are null, the limit of (24) is the vector $(0, 0, \ldots)$ and, from the preceding discussion, no stationary probability vector will exist. There is one other possibility that we assumed away earlier: $f_{jj} < 1$ for all j, i.e., all states are transient. In this case, (17) implies that the limit of (24) is again $(0, 0, \ldots)$, and no stationary probability vector exists. Summarizing, we have this important result:

Theorem 5. For an irreducible Markov chain, all states are positive if and only if there is a *unique* stationary probability vector π satisfying $\pi = \pi P$, where $\pi_j = 1/\nu_j > 0, j = 0, 1, \ldots$. Otherwise, either all states are transient or all states are null. In either case, no stationary probability vector exists.

Important Remarks

1. From Theorem 3, the states of an irreducible *finite* Markov chain are positive, and the relevant parts of Theorem 5 apply.
2. Limit (24) is a *time average* for discrete-time process $\{X_n\}$. For every j, π_j is the *fraction of transitions* into (or out of) state j.
3. For a Markov chain with more than one positive communication class, each class "supports" its own unique stationary probability vector, and mixtures of these vectors are also stationary. Hence, $\pi = \pi P$ has (infinitely) many solutions. (See Problems 3-21 through 3-23.) It is also easily shown that if π is a stationary probability vector for an arbitrary Markov chain, $\pi_j = 0$

for every state j that is either transient or null. Hence there is a *unique* stationary probability vector if and only if there is exactly *one* positive class.

4. From the Class Property Theorem, a recurrent irreducible Markov chain is a regenerative process. The recurrent case portions of Theorem 5 follow immediately from the treatment of regenerative processes in Chapter 2. See in particular Theorems 11 and 17 in Sections 2-12 and 2-16, respectively.

Example 3-3

Consider the brand-switching example in Section 3-1. Let $\{X_n\}$ be a Markov chain with transition probability matrix P, where for all i and j, p_{ij} is the probability that a customer's next purchase will be brand j, given the last purchase was brand i, and there are $c < \infty$ brands in all. If the chain is irreducible, the market share of brand j will be π_j, element j of the vector $\pi = (\pi_0, \pi_1, \ldots)$, where π is the unique probability vector satisfying $\pi = \pi P$.

Consider a simple symmetric case, where the probability of repeat business, $p_{jj} = 1/2$ for all j, and, for all pairs $i \neq j$, $p_{ij} = 1/2(c - 1)$. Thus, for all i, $\sum_j p_{ij} = 1$. Because of symmetry, we might expect that the market shares are equal for all brands, i.e., $\pi_j = 1/c$ for all j. It is easy to verify that this is true by showing that $\pi = (1/c, 1/c, \ldots)$ satisfies $\pi = \pi P$.

Now suppose the manufacturer of brand x wishes to increase that brand's market share by increasing p_{xx}. As p_{xx} increases, assume $p_{xj} = (1 - p_{xx})/(c - 1)$ for every $j \neq x$ and that the other rows of the matrix P are unaffected. Whether or not the assumed symmetry is present, we could solve π_x for several p_{xx} and plot the functional relationship between repeat business and market share.

Because of the symmetry in this problem, this result can be obtained analytically: Define a new two-state Markov chain, where state 0 denotes brand x and state 1 denotes *not* brand x. One can argue that this is a two-state Markov chain with transition probability matrix

$$P = \begin{pmatrix} p_{xx} & 1 - p_{xx} \\ 1/2(c - 1) & (2c - 3)/2(c - 1) \end{pmatrix}.$$

The equation for π_0 in $\pi = \pi P$ is

$$\pi_0 = \pi_0 p_{xx} + \pi_1/2(c - 1). \tag{27}$$

The equation for π_1 yields no new information. To get a unique solution we use $\pi_0 + \pi_1 = 1$ to eliminate π_1 from (27) and solve for π_0,

$$\pi_0 = [(2c - 2)(1 - p_{xx}) + 1]^{-1}, \tag{28}$$

where (28) is the market share of brand x. If $p_{xx} = 1/2$, $\pi_0 = 1/c$, as before, and π_0 increases with p_{xx}. As $p_{xx} \to 1$, $\pi_0 \to 1$. (Happiness is a satisfied customer.)

Remark. Usually, combining states as we have done destroys the Markov property, i.e., we would no longer have a Markov chain. Why does this device work here?

Example 3-4

We now illustrate the use of Theorem 5 when the number of states is infinite. Let the transition matrix be

$$\begin{pmatrix} 0 & 1 & 0 & \cdot & \cdot & \cdot \\ q & 0 & p & 0 & \cdot & \cdot \\ 0 & q & 0 & p & 0 & \cdot \\ & & \ddots & & \ddots & \end{pmatrix},$$

where $qp > 0$ and $p + q = 1$. It is easy to see that this chain is irreducible (why?). The relation $\pi = \pi P$ becomes

$$\pi_0 = q\pi_1,$$

$$\pi_1 = \pi_0 + q\pi_2,$$

$$\pi_j = p\pi_{j-1} + q\pi_{j+1}, \qquad j = 2, 3, \ldots,$$

which can be solved recursively:

$$\begin{aligned} \pi_1 &= \pi_0/q, \\ \pi_{j+1} &= p\pi_j/q, \qquad j = 1, 2, \ldots. \end{aligned} \tag{29}$$

From (29), we solve for π_j in terms of π_0:

$$\pi_j = (p/q)^{j-1}\pi_0/q, \qquad j = 1, 2, \ldots. \tag{30}$$

To find π_j, we *normalize* the distribution, which means to set

$$\sum_{j=0}^{\infty} \pi_j = \pi_0 \left[1 + \sum_{j=1}^{\infty} (p/q)^{j-1}/q \right] = 1, \tag{31}$$

which can be done if and only if $p < q$, i.e., when the infinite sum on the right converges. Thus, when $p < q$, all states are positive, where, from (31),

$$\pi_0 = (q - p)/2q, \tag{32}$$

and π_j is given by (30). It turns out that all states are transient if $p > q$, and all states are null if $p = q$.

3-5 FREQUENCIES AND RELATIVE FREQUENCIES FOR POSITIVE IRREDUCIBLE CHAINS

For positive irreducible Markov chains, the $\{\pi_j\}$ are *frequencies* (*rates*), i.e., for each j, π_j is the number of visits to state j per unit time, where by "time," we mean the discrete epochs $n = 0, 1, \ldots,$ on which the Markov chain is defined. In applications, e.g., see Problem 3-3, these epochs may be embedded as points in continuous time for some continuous-time process. When this is the case, we can remove any ambiguity and possible misunderstanding by calling π_j the *fraction of transitions* into state j.

Clearly, if the chain visits (arrives at) state j at epoch n, it must leave (depart from) state j at epoch $n + 1$. (To be consistent with our interpretation of π_j, a transition $j \to j$ counts as *both* a visit and a departure.) Now let $A_j(n)$ and $D_j(n)$ be the number of visits to and departures from state j on $\{0, \ldots, n\}$. Because visits and departures alternate,

$$|D_j(n) - A_j(n)| \leq 1, \qquad n = 1, 2, \ldots. \tag{33}$$

Now divide (33) by n and let $n \to \infty$. For every j, we have

$$\{\text{transition rate out of state } j\} = \{\text{transition rate into state } j\}. \tag{34}$$

Similarly, of those transitions that are departures from state i, p_{ij} is the fraction of them that are also visits to state j, and $\pi_i p_{ij}$ is the number of transitions of type $i \to j$ per unit time. Summing over i, $\sum_i \pi_i p_{ij}$ is the transition rate into state j. In this light, the equation $\pi = \pi P$ in Theorem 5 states that for every j,

$$\pi_j = \sum_i \pi_i p_{ij}, \tag{35}$$

which is just a particular way of writing (34). For a positive irreducible Markov chain, $\pi_j > 0$ for all j and $\sum_j \pi_j = 1$.

Similarly, transition rates *across arbitrary boundaries* must be the same. That is, suppose we partition the state space into two disjoint subsets, A and A^c. By the same argument used to obtain (33), the following holds for any such partition:

$$\underbrace{\sum_{j \in A} \sum_{i \in A^c} \pi_j p_{ji}}_{\text{rate from } A \text{ to } A^c} = \underbrace{\sum_{j \in A^c} \sum_{i \in A} \pi_j p_{ji}}_{\text{rate from } A^c \text{ to } A}. \tag{36}$$

Recurrent irreducible Markov chains are regenerative processes, where there are many choices for regeneration cycles. For arbitrary initial state i, we may choose to define cycles in terms of visits to arbitrary state j. Thus, T_{ij}, $T_{ij} + T_{jj}^{(1)}$, $T_{ij} + T_{jj}^{(1)} + T_{jj}^{(2)}$, . . . are regeneration points, where the $T_{jj}^{(k)}$ are i.i.d., distributed as T_{jj}. Alternatively, we could let returns to state i be regeneration points.

Now suppose all states are positive, and we wish to find the frequency of visiting state j in terms of regeneration cycles. From the Renewal-Reward Theorem applied to cycles defined by visits to state j, this rate is

$$1/\nu_j = \pi_j.$$

In terms of visits to state i, this rate is

$$E(V_{ij})/\nu_i = \pi_i E(V_{ij}),$$

where V_{ij} is *the number of visits to state j between returns to state i.* Since this rate cannot depend on our choice of cycles, we equate these expressions, obtaining

$$E(V_{ij}) = \pi_j/\pi_i \qquad \text{for all } i \text{ and } j. \tag{37}$$

Similarly, let V_{ijk} be *the number of transitions of type $j \rightarrow k$ between returns to state i*. Equating expressions for the rate of $j \rightarrow k$ transitions,

$$E(V_{ijk})/\nu_i = \pi_j p_{jk}, \tag{38}$$

we have

$$E(V_{ijk}) = \pi_j p_{jk}/\pi_i \qquad \text{for all } i, j, \text{ and } k. \tag{39}$$

Clearly, we could have anticipated (37) and (39), e.g., for (37), π_j/π_i is the number of visits to state j per visit to state i, which is a *relative frequency*.

The distribution of V_{ij} is found in Problem 3-26, where it is also shown that for a recurrent irreducible chain,

$$E(V_{ij}) < \infty \qquad \text{for all } i \text{ and } j, \tag{40}$$

even when all states are null!

Now (34) will hold for any process, Markov or not. Hence, it will hold for a process that combines the states of a positive irreducible chain into new states (see Problem 3-24), where the process obtained in this manner (usually) is not a Markov chain (see Problem 3-20).

Modeling Implications

Markov chains are sometimes used to model phenomena where they may not be appropriate. For example, in models of brand switching (Example 3-3), do we really believe that consumer behavior is not affected by experience with products prior to the most recent one?

Nevertheless, if the purpose of these models is to summarize aggregate behavior, little harm is done. Thus, in Problem 3-12, the mix of vehicles on the highway is determined by the given data, whether or not the sequence of cars and trucks is a Markov chain. On the other hand, it is inappropriate to conclude that a stochastic process is a Markov chain simply because estimates of π and P satisfy (with negligible error) $\pi = \pi P$. (See Problem 3-25.)

What difference does it make? When we question the appropriateness of a Markov chain model, we usually are referring to the validity of the Markov property (conditional independence) for the chosen state space. Transient results, such as the distribution of the V_{ij} and of first passage times are directly affected, as are all of the results in Section 3-7.

For chains generated by coin tossing and in other gambling situations, conditional independence is a direct consequence of independent trials, which in this context is an assumption we are quite willing to accept.

In other contexts, caution is advised, particularly when a chain model is used for *predictive* purposes. For brand switching in Example 3-3, we assumed that improving brand x would affect only *row* x of the transition matrix. If consumers have longer memories, the p_{ij} would be meaningful only as frequencies, and *all* of them may change.

3-6 COSTS AND REWARDS FOR POSITIVE IRREDUCIBLE CHAINS

In applications, costs and rewards may accumulate as a Markov chain evolves, in a manner consistent with Section 2-3. Let R_n be the reward associated with the state visited at epoch n, where R_n may depend on *both* X_n and X_{n-1}. Let the cumulative reward over the first n transitions be

$$C(n) = \sum_{s=1}^{n} R_s(X_{s-1}, X_s), \tag{41}$$

with expected value $c(n) = E(C(n))$.

Given X_0 and any sequence of transitions $X_1, \ldots, X_n$, we assume the rewards R_s, $s = 1, \ldots, n$ are independent, where for each s, the distribution of R_s depends only on X_{s-1} and X_s. For any (i, j) such that $(X_{s-1}, X_s) = (i, j)$, let

$$r_{ij} = E\{R_s(i, j)\}, \tag{42}$$

where the r_{ij} are assumed to be finite, and

$$r_i = \sum_j r_{ij} p_{ij}, \tag{43}$$

where, owing to assumptions made below, the r_i are well defined and finite. Thus, r_{ij} is the expected reward associated with a transition of type $i \to j$ and r_i is the expected reward associated with a departure from state i.

We would like to find a computationally convenient expression for the reward rate (reward per transition or unit time),

$$\lim_{n\to\infty} C(n)/n. \tag{44}$$

To do this, suppose $X_0 = i$ (arbitrary) and we consider cycles defined by returns to state i. Thus T_{ii} is the length of the first cycle and $C(T_{ii})$ is the total reward for this cycle. For a positive irreducible chain,

$$0 < \nu_i = 1/\pi_i < \infty. \tag{45}$$

By renumbering the R_s appropriately, we can rewrite $C(T_{ii})$ in terms of the number of transitions between each pair of states,

$$C(T_{ii}) = \sum_j \sum_k \sum_{s=1}^{V_{ijk}} R_s(j, k), \tag{46}$$

where the V_{ijk} are defined prior to (38). Provided

$$E\{ \sum_{s=1}^{T_{ii}} | R_s(X_{s-1}, X_s) | \} < \infty,$$

we can find the expected value of (46) by the interchanging expectation and sum,

$E\sum\sum\sum = \sum\sum E\sum$. The expected value of the inner sum may be found by conditioning† and (39) to be

$$E\{\sum_{s=1}^{V_{ijk}} R_s(j, k)\} = r_{jk}E(V_{ijk}) = r_{jk}p_{jk}\pi_j/\pi_i. \tag{47}$$

Hence, the expected value of (46) is

$$E\{C(T_{ii})\} = \sum_j \sum_k r_{jk}p_{jk}\pi_j/\pi_i = \sum_j r_j\pi_j/\pi_i. \tag{48}$$

From the Renewal-Reward Theorem, the limit (44) is $\pi_i E\{C(T_{ii})\}$, and we have the random-variable version of

Theorem 6: Reward Rates for Markov Chains. For a positive irreducible Markov chain, with $E\{\sum_{s=1}^{T_{ii}} | R_s(X_{s-1}, X_s) | \} < \infty$ for any initial state $X_0 = i$,

(Random-variable version) $$\lim_{n\to\infty} C(n)/n = \sum_j r_j\pi_j \quad \text{w.p.1, and} \tag{49}$$

(Expected-value version) $$\lim_{n\to\infty} c(n)/n = \sum_j r_j\pi_j. \tag{50}$$

The expected-value version of Theorem 6 follows from the statement of the Renewal-Reward Theorem because we know limits (49) and (50) exist and are the same. For $R_n = X_n$, Theorem 6 is a special case of Theorem 11 in Chapter 2.

Remark. We defined the R_n as though rewards are "collected at the end" of a transition. However, as was shown in Section 2-3, the reward rate does not depend on how reward accumulates during a cycle.

We now illustrate how to use Theorem 6.

Example 3-5

Partition a metropolitan area into three states (zones), where 0 denotes downtown, 1 denotes suburbs, and 2 denotes the airport. Let $\{X_n\}$ be a Markov chain with these states, where X_n is the location of a taxi after n trips. Let p_{ij} be the probability that a customer boarding in state i wishes to go to state j, and r_{ij} be the (expected) fare for this trip. For this example, the corresponding matrices are

$$P = (p_{ij}) = \begin{pmatrix} 1/2 & 1/4 & 1/4 \\ 1/2 & 1/4 & 1/4 \\ 3/4 & 1/4 & 0 \end{pmatrix}, \qquad r = (r_{ij}) = \begin{pmatrix} 1.0 & 2.0 & 4.0 \\ 2.0 & 2.0 & 4.0 \\ 4.0 & 4.0 & — \end{pmatrix}.$$

To use the preceding development, we need to define the expected fare per visit to each state. To do this, let R_{n+1} be the fare of the ride that originates at X_n

† We could use Wald's equation (Section 2-13) instead of conditioning on V_{ijk} here, but it is not necessary to do so.

and terminates at X_{n+1}. Thus, $r_i = \sum_j r_{ij}p_{ij}$, which, from the above matrices, give $r_0 = 2.0$, $r_1 = 2.5$, and $r_2 = 4.0$.

To apply (50), we need to solve $\pi = \pi P$, with $\sum_i \pi_i = 1$. For the above P matrix, the solution is: $\pi_0 = 11/20$, $\pi_1 = 5/20$, and $\pi_2 = 4/20$. From (50)

$$\lim_{n\to\infty} c(n)/n = \pi_0 r_0 + \pi_1 r_1 + \pi_2 r_2 = 101/40,$$

which is the expected fare/trip.

3-7 TRANSIENT BEHAVIOR

So far in this chapter, we have been concerned about the long-run behavior of positive, irreducible chains. State j is called *absorbing* if $p_{jj} = 1$. A chain with one or more absorbing states is clearly not irreducible. Instead of returning over and over to each state, the chain may (sooner or later) visit an absorbing state and be trapped (absorbed) there forever. For example, in Problem 3-1, the rat will eventually (with probability 1) leave the maze.

In *ruin games*, usually between two players, one player wins when the other(s) are ruined (go broke). Markov chain models of ruin games have an absorbing state for each player, where entering an absorbing state means the corresponding player wins the game. Of central interest is the probability of absorption into each absorbing state. We will also be interested in the expected duration of the game, which is the expected number of transitions until one player wins. More generally, instead of absorbing states, we may have closed (positive or otherwise) communication classes.

Sometimes the sequence of transitions is important, even in irreducible chains. We may want to find the probability that one state or set of states is visited before some other state or set of states. It is important to realize that for distinct states j, k, and l,

$$P(\text{chain visits } k \text{ before } l \mid X_0 = j) \tag{51}$$

does *not* depend on rows k and l of the transition matrix P. Thus (51) for an irreducible chain is the same as it would be if k and l were absorbing states. On the other hand, we would expect (51) to depend on row j.

With these considerations in mind, let T and C be disjoint subsets of states, and, for every $j \in T$, define

$$x_j(n) = P(X_n \in C, X_{n-1} \in T, \ldots, X_1 \in T \mid X_0 = j \in T), \tag{52}$$

$$y_j(n) = P(X_n \in T, \ldots, X_1 \in T \mid X_0 = j \in T), \tag{53}$$

$$x_j = \sum_{n=1}^{\infty} x_j(n), \text{ and} \tag{54}$$

$$y_j = \lim_{n\to\infty} y_j(n). \tag{55}$$

Now y_j is simply the probability that the chain *never* leaves T, given $X_0 = j \in T$. The meaning of x_j depends on the application. If $T \cup C$ is the entire set of states, x_j is the probability that the chain *ever* visits C, given $X_0 = j \in T$. More generally, x_j is the probability that the chain visits C *before* any state not in either C or T, given $X_0 = j \in T$. In this context, x_j is sometimes called a *taboo probability* where the states in neither C nor T are "taboo" states. In some applications, T will be the set of transient states and C will be a closed communication class. It is important to realize that while our notation suppresses C, x_j depends on the choice of both C and T.

Rather than finding x_j or y_j for some fixed j, we will find a system of equations for $\{x_j, j \in T\}$ and $\{y_j, j \in T\}$.

We find a system of equations for $\{y_j\}$ from recursion relations for the $\{y_j(n)\}$. First,

$$y_j(1) = P(X_1 \in T \mid X_0 = j \in T) = \sum_{i \in T} p_{ji} \qquad \text{for all } j \in T. \tag{56}$$

To find $y_j(n + 1)$, condition on X_1 and observe that *given* $X_1 = i \in T$, the conditional probability that the chain remains in T through the first $n + 1$ transitions is $y_i(n)$. Hence, for $n = 1, 2, \ldots,$

$$y_j(n + 1) = \sum_{i \in T} p_{ji} y_i(n) \qquad \text{for all } j \in T. \tag{57}$$

Now let $n \to \infty$ in (57). From (55), we have

$$y_j = \sum_{i \in T} p_{ji} y_i \qquad \text{for all } j \in T. \tag{58}$$

[We could have derived (58) directly, without (56) and (57), but these intermediate results turn out to be important.]

Clearly, $\{y_j = 0 \text{ for all } j \in T\}$ is a solution to (58). Does this imply $\lim_{n \to \infty} y_j(n) = 0$ for all j? Not necessarily! If not, there must be more than one solution to (58). How are we to determine which solution to (58) corresponds to the "true" probabilities (55)?

Since probabilities are bounded by 1, we can restrict our attention to solutions with this property. In particular, let $\{z_j\}$ be any solution to (58) that is bounded by 1, i.e., $|z_j| \le 1$ for all j. Hence, from (56) and $|\sum| \le \sum |\cdot|$,

$$|z_j| \le \sum_{i \in T} p_{ji} |z_i| \le \sum_{i \in T} p_{ji} = y_j(1) \qquad \text{for all } j \in T, \tag{59}$$

or (term by term) $\{z_j\}$ is bounded by $\{y_j(1)\}$. In fact, suppose $\{z_j\}$ is bounded by $\{y_j(n)\}$. From (57),

$$|z_j| \le \sum_{i \in T} p_{ji} y_i(n) = y_j(n + 1) \qquad \text{for all } j \in T, \tag{60}$$

from which we conclude that for $n = 1, 2, \ldots,$

$$|z_j| \le y_j(n) \qquad \text{for all } j \in T. \tag{61}$$

For each j, $y_j(n)$ is monotone nonincreasing with limit y_j. Hence, (61) holds in the limit and we have proven

Theorem 7. The set of probabilities $\{y_j\}$ defined in (55) is the *maximal* solution to (58) that is bounded by 1; $\lim_{n\to\infty} y_j(n) = 0$ for all $j \in T$ if and only if $\{y_j = 0 \text{ for all } j \in T\}$ is the *unique* bounded solution to (58).

Remarks. Theorem 7 is the key to proving the other theorems in this section. Compare the proofs of Theorems 7 and 8 with that of Theorem 15 in Chapter 2.

To obtain a system of equations for the $\{x_j\}$ directly, condition on X_1. On leaving T, the chain must visit C immediately; otherwise, it would visit taboo states. Let I_j be the indicator of the union of the events with probability (52), so that $E(I_j) = x_j$. We have

$$E(I_j \mid X_1 \in C, X_0 = j) = 1,$$

$$E(I_j \mid X_1 = i \in T, X_0 = j) = x_i, \text{ and}$$

$$x_j = \sum_{i\in C} p_{ji} + \sum_{i\in T} p_{ji}x_i \qquad \text{for all } j \in T. \tag{62}$$

As with (58), (62) may have more than one solution. If $\{u_j\}$ and $\{v_j\}$ are *distinct bounded* solutions to (62), $\{u_j - v_j\}$ is a nonzero bounded solution to (58). From Theorem 7, we have

Theorem 8. The set of probabilities (54) is the *unique* bounded solution to (62) if and only if $\lim_{n\to\infty} y_j(n) = 0$ for all $j \in T$, i.e., if and only if for every initial state $X_0 = j \in T$, the probability that the chain remains in T forever is zero.

When (62) does not have a unique solution, the next theorem characterizes the probabilistic solution.

Theorem 9. The set of probabilities $\{x_j\}$ defined by (54) is the *minimal* nonnegative solution to (62).

Proof. For every $j \in T$, let

$$z_j(n) = \sum_{s=1}^{n} x_j(s) = P(\text{chain visits } C \text{ by epoch } n \mid X_0 = j \in T), \; n = 1, 2, \ldots.$$

By conditioning on X_1, it is easy to see that for every $j \in T$,

$$z_j(n+1) = \sum_{i\in C} p_{ji} + \sum_{i\in T} p_{ji}z_i(n), \qquad n = 1, 2, \ldots. \tag{63}$$

Now let $\{v_j\}$ be *any* nonnegative solution to (62). Clearly, for every $j \in T$,

$$v_j \geq \sum_{i \in C} p_{ji} = z_j(1) \text{ and, from (63),}$$

$$v_j \geq \sum_{i \in C} p_{ji} + \sum_{i \in T} p_{ji} z_i(1) = z_j(2).$$

Continuing in this manner, we have

$$v_j \geq z_j(n) \qquad \text{for every } j \in T \text{ and } n = 1, 2, \ldots. \tag{64}$$

Now let $n \to \infty$ in (64), and we have the theorem,

$$v_j \geq x_j \qquad \text{for every } j \in T.$$

In applications, the conditions for uniqueness in Theorem 8 are often satisfied. For example, because transient states are visited only a finite number of times, the chain will leave T with probability 1 if T is a finite number of transient states. Similarly, if T is a proper subset of a recurrent communication class, the chain will leave T with probability 1 (because for $i \in T$, $j \notin T$, where $i \leftrightarrow j$, $f_{ij} = 1$). Summarizing, we have

Theorem 10. The set of probabilities (54) is the unique bounded solution to (62) if either (i) T is a finite set of transient states or (ii) T is a proper subset of a recurrent communication class.

Remark. The first condition covers the next two examples. The second suggests an approach for determining whether the states of an irreducible chain are transient. The result will be Theorem 11.

Example 3-6

For the Markov chain with transition matrix (14), let $T = \{0\}$ and $C = \{1\}$. For this choice of C and T, $x_0 = f_{01}$, and (62) becomes

$$f_{01} = b + af_{01},$$

$$f_{01} = b/(1 - a).$$

On the other hand, for $T = \{0\}$, $C = \{2\}$, $x_0 = f_{02}$, and (62) becomes

$$f_{02} = c + af_{02},$$

$$f_{02} = c/(1 - a).$$

Example 3-7: A Ruin Game

Two players are tossing (possibly biased) coins where, on each toss, the probability player 1 wins 1 cent is p, and the probability player 1 loses 1 cent is $q = 1 - p$, where c is the combined financial assets (total number of pennies) of both players. Define a Markov chain $\{X_n\}$, where $X_n = j$ means that player 1 has j cents after the nth toss. The game continues until one player goes broke (the other player wins).

Thus, the states are $0, 1, \ldots, c$, where states 0 and c are absorbing. Absorption in state 0 means player 1 goes broke (player 2 wins), and absorption in state c means player 1 wins. The transition probability matrix is

$$P = \begin{pmatrix} 1 & 0 & 0 & \cdot & \cdot & \cdot & \cdot \\ q & 0 & p & 0 & \cdot & \cdot & \cdot \\ 0 & q & 0 & p & \cdot & \cdot & \cdot \\ \vdots & & \ddots & & \ddots & & \\ 0 & \cdot & \cdot & \cdot & q & 0 & p \\ 0 & \cdot & \cdot & \cdot & \cdot & \cdot & 1 \end{pmatrix}$$

Now $T = \{1, \ldots, c - 1\}$ is a finite set of transient states. For $C = \{c\}$, x_j is the probability that player 1 wins, given $X_0 = j \in T$. From Theorem 10 and (62), $\{x_j\}$ is the unique solution to

$$\begin{aligned} x_1 &= px_2, \\ x_j &= qx_{j-1} + px_{j+1}, \qquad j = 2, \ldots, c - 2, \\ x_{c-1} &= qx_{c-2} + p. \end{aligned}$$

If the game is fair ($p = q = 1/2$), it is easy to verify that

$$x_j = j/c, \qquad j = 1, \ldots, c - 1,$$

is the solution to these equations.

For a problem where we wish to find probabilities of type (51), consider the following example.

Example 3-8

Suppose the sequence of weather types is a 4-state Markov chain with the following states:

state 0: cool and clear
state 1: cool and foggy
state 2: hot and clear
state 3: rain

The transition probability matrix is

$$P = \begin{pmatrix} .2 & .4 & .3 & .1 \\ .3 & .4 & .1 & .2 \\ .1 & .6 & .2 & .1 \\ .4 & .2 & .1 & .3 \end{pmatrix}.$$

It is now cool and clear in the Napa Valley; the grapes are not quite ripe. Another hot spell would ripen them, but rain before a hot spell would ruin them.

The grower must decide whether to pick less than ideal grapes now or gamble on the weather.

To help the grower decide, we will compute the probability that a hot spell will occur before rain. That is, we want to find x_0 = the probability that the chain visits state 2 before state 3, given $X_0 = 0$. This probability will be found together with the corresponding probability, given $X_0 = 1$.

Thus, for $T = \{0, 1\}$ and $C = \{2\}$, (62) becomes

$$x_0 = .3 + .2x_0 + .4x_1,$$

$$x_1 = .1 + .3x_0 + .4x_1,$$

which has the solution $x_0 = 11/18$, $x_1 = 17/36$. [While the solution is obviously unique, notice that condition (ii) of Theorem 10 applies, which guarantees uniqueness.]

Conditions for the States of an Irreducible Chain to Be Transient

For an irreducible Markov chain, let $T = \{1, 2, \ldots\}$ and $C = \{0\}$. Let $X_0 = 0$ and consider the probability of returning to state 0. For any $j \in T$, let E_j be the event that the chain visits state j prior to returning to state 0, where $P(E_j) > 0$. If E_j occurs, the probability of returning to state 0 is $1 - y_j$, the probability that the chain leaves T. Conditioning on E_j, we have

$$\begin{aligned} f_{00} &= (1 - y_j)P(E_j) + P(\text{chain returns to state } 0 \mid E_j^c)P(E_j^c) \\ &< 1 \qquad \text{if } y_j > 0. \end{aligned}$$

Thus $f_{00} = 1$ only if $y_j = 0$ for all $j \in T$. On the other hand, $y_j = 0$ for all $j \in T$ implies $f_{00} = 1$. From Theorem 7, we have

Theorem 11. For an irreducible Markov chain with state space $\{0, 1, \ldots\}$, all states are transient if and only if

$$y_j = \sum_{i=1}^{\infty} p_{ji} y_i, \qquad j = 1, 2, \ldots \tag{65}$$

has a nonzero bounded solution.

Remark. By applying Theorem 11 together with Theorem 5, we can determine when the states of an irreducible Markov chain are null (when they are neither transient nor positive).

The form of Theorem 11 is the result of the choice $C = \{0\}$. As the Example 3-9 shows, this is often a convenient choice. However, a minor modification of Theorem 11 is valid for any choice of state $C = \{j\}$. In fact, C can be any (nonempty) *finite* subset of states.

Example 3-9

We now apply Theorem 11 to a Markov chain with transition matrix

$$P = \begin{pmatrix} 0 & 1 & 0 & \cdot & \cdot & \cdot \\ q & 0 & p & 0 & \cdot & \cdot \\ 0 & q & 0 & p & \cdot & \cdot \\ & & \ddots & & \ddots & \end{pmatrix},$$

where $pq > 0$, $p + q = 1$. Equation (65) becomes

$$y_1 = py_2, \tag{66}$$
$$y_n = py_{n+1} + qy_{n-1}, \qquad n = 2, 3, \ldots.$$

Before proceeding, let's see if we can deduce the form of the *probabilistic* solution to (66). First, observe that a path from state j to state $j - 1$, e.g., $j \to j + 1 \to j \to j - 1$, has the same probability (pq^2 in this case) for any j. Because of this correspondence between paths, $f_{j,j-1} = \alpha$, $\alpha \in (0, 1]$, for some α *that is independent* of j. Also observe that because transitions are to adjacent states, a path from state j to state 0 must pass through states $j - 1, j - 2, \ldots, 1$. Hence,

$$f_{j0} = f_{j,j-1} \cdot f_{j-1,j-2} \cdot \ldots \cdot f_{10} = \alpha^j, \text{ and} \tag{67}$$

$$y_j = 1 - f_{j0} = 1 - \alpha^j \qquad \text{for some } \alpha \in (0, 1]. \tag{68}$$

Thus, (66) will have a nonzero bounded solution if and only if it has a solution of form (68) for some $\alpha \in (0, 1]$.

Now plug (68) into *any* equation in (66) (they all will yield the same information), where we have chosen $y_1 = py_2$ below:

$$1 - \alpha = p(1 - \alpha^2), \tag{69}$$

a quadratic equation with two roots. Obviously, one root is $\alpha = 1$.

Since the transient case corresponds to finding a root $\alpha \in (0, 1)$, we can find the other root by factoring and canceling $1 - \alpha$:

$$1 = p(1 + \alpha), \text{ or}$$
$$\alpha = q/p. \tag{70}$$

Hence, there exists a root $\alpha \in (0, 1)$, namely (70), if and only if $p > q$, *and all states are transient if and only if* $p > q$.

Our probabilistic argument determined the y_j directly. Instead, we could use (66) to solve for y_j in terms of any $y_1 > 0$. Notice that when $p > q$, $\lim_{j\to\infty} y_j = 1$ for our solution,

$$y_j = 1 - (q/p)^j, \qquad j = 1, 2, \ldots, \tag{71}$$

i.e., (71) is the largest solution to (66) that is bounded by 1. Hence, either by our direct argument or Theorem 7, we have for $p > q$,

$$f_{j0} = 1 - y_j = (q/p)^j, \qquad j = 1, 2, \ldots, \tag{72}$$

but, for $p \leq q$,

$$f_{j0} = 1, \qquad j = 1, 2, \ldots .$$

Combining these results with those of Example 3-4, we have that all states are null if and only if $p = q$.

Example 3-10

We now apply the preceding results to the Markov chain in Example 2-17. In that example, $\{S_n\}$ is a Markov chain with state space $\{\ldots, -1, 0, 1, \ldots\}$, and transition probabilities

$$p_{j,j+1} = p, \qquad p_{j,j-1} = q, \qquad j = 0, \pm 1, \ldots . \tag{73}$$

N is a stopping time if and only if $f_{01} = 1$. Since we are interested in visiting a *higher* state, the roles of p and q are reversed, and $f_{01} = 1$ if and only if $p \geq q$. It is also easy to show that $E(N) < \infty$ if and only if $p > q$. (See Problem 3-30.)

The methods of this section may also be used to find moments of first passage times. For an irreducible *positive* Markov chain and any fixed i, partition the state space into $C = \{i\}$ and $T = \{j: j \geq 0\} - C$. Given $X_0 = j$ and conditioning on X_1, the mean first passage time to state i, $E\{T_{ji}\}$, satisfy:

$$E(T_{ji}) = 1 + \sum_{k \neq i} p_{jk} E(T_{ki}) \qquad \text{for all } j \in T, \tag{74}$$

$$E(T_{ii}) = 1 + \sum_{k \neq i} p_{ik} E(T_{ki}), \tag{75}$$

where (74) is a form similar to (62). Now the $E\{T_{ji}\}$ are finite. If $\{E\{T_{ji}\}: j \neq i\}$ is also known to be a *bounded* function of j, the same theory applies. For example, this will be true if the number of states is finite. Since $f_{ji} = 1$ for all $j \in T$, we have

Theorem 12. For an irreducible positive Markov chain and any fixed state i, the set of mean first passage times $\{E\{T_{ji}\}: j \neq i\}$ is the unique bounded solution to (74), whenever it is bounded. This solution is bounded if the chain is finite. Equation (75) is determined by the unique solution to (74).

The same methods can be used to find higher moments, e.g., $E\{T_{ij}^2\}$. See Problem 3-33. Other problems can be formulated and solved in a similar manner, e.g., the mean duration of the game in Example 3-7.

3-8 PERIODIC AND APERIODIC STATES; POINTWISE LIMITS

The distribution $\{f_{jj}^{(n)}\}$ is obviously lattice with span $d \geq 1$. Because Markov chains are defined on discrete time, $n = 0, 1, \ldots$, the distinction between $d = 1$ and $d > 1$ is important. If $\{f_{jj}^{(n)}\}$ has span $d > 1$, state j is called *periodic* with *period* (span) $d > 1$, where d must be an integer. If $\{f_{jj}^{(n)}\}$ has span $d = 1$, state j is called *aperiodic*. We will use this terminology whenever $0 < f_{jj} \leq 1$.

We now extend the class property theorem.

Theorem 13. All states in the same communication class have the same span, i.e., span d is a class property.

Proof. Refer back to equation (19) for any pair of states $j \neq k$ such that $j \leftrightarrow k$. Let j and k have spans d_j and d_k, respectively. By the definition of J and K, there is a path of positive probability of length $J + K$ both from j to j and from k to k. Hence, $J + K$ is a multiple of both d_j and d_k. Any n for which $p_{kk}^{(n)} > 0$ must be a multiple of d_k. In fact, d_k is the greatest common divisor of the set $\{n: n \geq 1, p_{kk}^{(n)} > 0\}$. From (19), $p_{jj}^{(J+n+K)}$ is positive whenever $p_{kk}^{(n)} > 0$, and *may* be positive for values of n that are not multiples of d_k. Hence, $d_j \leq d_k$. However, because j and k are arbitrary, $d_k \leq d_j$ as well, i.e., $d_j = d_k$.

Periodic states and (irreducible) chains are common. In Example 3-4, transitions occur only between adjacent states ($j \to j + 1$ or $j \to j - 1$). Hence, a *return* to a state can occur only after an even number of transitions. Thus, this chain is periodic with period $d = 2$. On the other hand, if $p_{jj} > 0$ for some state in an irreducible Markov chain, the chain is aperiodic.

From Blackwell's theorem,

$$\lim_{n\to\infty} p_{jj}^{(n)} = \lim_{n\to\infty} [m_{jj}(n) - m_{jj}(n - 1)] = 1/\nu_j, \tag{76}$$

if state j is aperiodic, where this limit is zero if j is either transient or null. If j is periodic and positive, the limit (76) does not exist; instead, if j has period $d > 1$, we have

$$\lim_{n\to\infty} p_{jj}^{(nd)} = \lim_{n\to\infty} [m_{jj}(nd) - m_{jj}((n - 1)d)] = d/\nu_j. \tag{77}$$

In either case, of course,

$$\lim_{n\to\infty} \sum_{k=1}^{n} p_{jj}^{(k)}/n = 1/\nu_j, \tag{78}$$

where we will call (78) a time (or *transition*) average, and (76) is a *pointwise* limit. Thus, for Markov chains, periodic plays the role of lattice in renewal theory.

For irreducible aperiodic Markov chains, we can somewhat strengthen Theorem 5. By conditioning on T_{ij}, we can write

$$p_{ij}^{(n)} = \sum_{k=1}^{n} f_{ij}^{(k)} p_{jj}^{(n-k)}. \tag{79}$$

If all states are positive, $f_{ij} = 1$ and $p_{jj}^{(n)}$ has limit (76). From (79), it follows that for all i and j,

$$\lim_{n\to\infty} p_{ij}^{(n)} = 1/\nu_j = \pi_j > 0, \tag{80}$$

where $\pi = \pi P$.

We can also characterize the behavior of periodic chains in more detail, should that be desired. In particular, consider an arbitrary but fixed j in an irreducible periodic Markov chain. For any $k \in \{1, \ldots, d\}$, let S_k be the set of states that can be reached only from state j in k, $d + k$, $2d + k, \ldots$ transitions. Obviously, $j \in S_d$. In fact, it can be shown that every state belongs to exactly one S_k and that the S_k are not empty. Furthermore, it can be shown from (79) that for any $i \in S_k$,

$$\lim_{n\to\infty} p_{ij}^{(nd+k)} = d/\nu_j. \tag{81}$$

Remark. It is important to observe that neither the time-average results in Sections 3-4 through 3-6 nor the transient results in Section 3-7 depend on the distinction between periodic and aperiodic states.

3-9 BRANCHING PROCESSES

Our treatment of this special topic is independent of the rest of this chapter.

Branching processes were introduced over 100 years ago to treat the problem of the extinction of family names. More generally, they can be used to study populations of individuals that reproduce and die. The population may be biological or physical, e.g., neutrons or cosmic rays. There is also a connection between branching processes and busy periods for the $M/G/1$ queue. (See Chapter 8 for a detailed discussion.)

Let Y_r be the number of *progeny* (in the context of family names, male children) of (male) individual r belonging to some population, $r = 1, 2, \ldots$. Assume the Y_r are i.i.d. with distribution

$$P(Y_r = j) = \alpha_j, \qquad j = 0, 1, \ldots, \tag{82}$$

where $0 < \alpha_0 < 1$, and (82) has finite mean and variance

$$E(Y_r) = b, \qquad V(Y_r) = \sigma^2. \tag{83}$$

Let $K_n - 1$ be the number of descendants of a particular individual after n generations, $n = 0, 1, \ldots$. Thus, $K_0 = 1$, and K_n includes the original individual.

Now define $X_0 = 1$, $X_n = K_n - K_{n-1}$, $n = 1, 2, \ldots$, i.e., for $n \geq 1$, X_n is the number of descendants *on the* nth *generation*. By appropriately numbering the $\{Y_r\}$, we have

$$X_1 = Y_1, \text{ and } X_{n+1} = \sum_{K_{n-1}+1}^{K_n} Y_r, \qquad n = 2, 3, \ldots, \tag{84}$$

where the number of terms in the sum (84) is X_n.

The sequence $\{X_n\}$ is a Markov chain with specified initial conditions ($X_0 = 1$). Let's find the transition matrix P for this chain.

Clearly, if $X_n = 0$ for some $n \geq 1$, $X_s = 0$ for all $s > n$, and we say *extinction* has occurred. Thus, $p_{00} = 1$, i.e., state 0 is *absorbing*. Furthermore,

$$P(X_{n+1} = j \mid X_n = 1) = p_{1j} = \alpha_j, \qquad j = 0, 1, \ldots,$$

and, from the independence of $\{Y_r\}$, row i of P is the *i-fold convolution* of (82),

$$\{p_{ij}; j \geq 0\} = \{\alpha_j\}^{*i}, \qquad i = 0, 1, \ldots. \tag{85}$$

We include $i = 0$ in (85) for convenient reference later. This is consistent with earlier conventions: The 0-fold convolution of a distribution is the unit step at the origin, i.e., $\{\alpha_j\}^{*0} = \{1, 0, \ldots\}$.

Let the probability of extinction by the nth generation be

$$q_n = P(X_n = 0 \mid X_0 = 1), \qquad n = 1, 2, \ldots, \tag{86}$$

and the *probability of extinction* be

$$q = \lim_{n \to \infty} q_n, \tag{87}$$

where, because the q_n are monotone (why?) the limit (87) clearly exists.

Our first problem is to find q. To do this, define the generating functions

$$A_n(z) = E\{z^{X_n} \mid X_0 = 1\}, \qquad n = 0, 1, \ldots, \qquad z \in [0, 1], \tag{88}$$

where $A_0(z) = z$. For $n = 1$, we delete the subscript on A,

$$A_1(z) \equiv A(z) = \sum_{j=0}^{\infty} \alpha_j z^j. \tag{89}$$

In terms of generating functions, (87) becomes

$$q = \lim_{n \to \infty} A_n(0). \tag{90}$$

We will find (90) from a recursion relation obtained by conditioning on $X_1 = j$, and observing that X_{n+1} is the *sum* of the descendants of the j individuals at epoch 1, *n generations later*. (Each of these individuals begins his own branching process, independent of the others.)

$$E\{z^{X_{n+1}} \mid X_1 = j, X_0 = 1\} = E\{z^{X_{n+1}} \mid X_1 = j\}$$

$$= [A_n(z)]^j, \qquad n = 1, 2, \ldots, \tag{91}$$

$$A_{n+1}(z) = E\{[A_n(z)]^{X_1} \mid X_0 = 1\} = A[A_n(z)], \quad n = 1, 2, \ldots. \tag{92}$$

Before proceeding, we sketch the shape of $A(z)$ as a function of $z \in [0, 1]$. We have $A(0) = \alpha_0$ and $A(1) = 1$. If $\alpha_0 + \alpha_1 = 1$, $A(z)$ is linear. Otherwise, the second derivative $A''(z) > 0$ for $z \in (0, 1)$. Thus, $A(z)$ is increasing with *strictly increasing* first derivative, and is said to be *strictly convex*. We distinguish two cases, where the linear case is included in Figure 3-1(a).

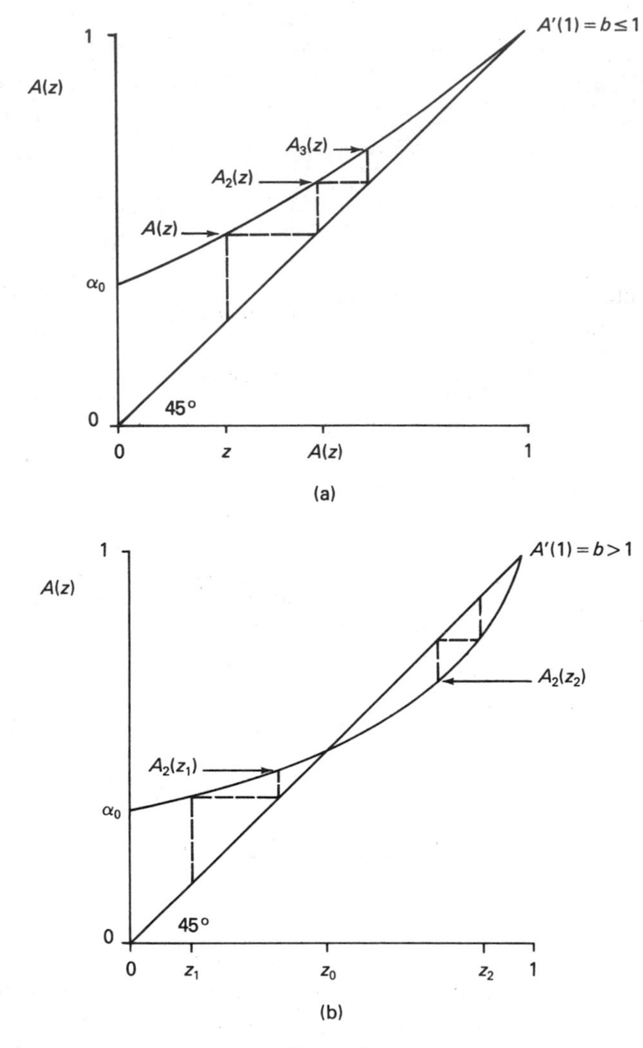

Figure 3-1

In both cases, the straight 45° line is the function $f(z) = z$. The dashed lines will be explained later. For $b > 1$, we have Figure 3-1(b) because $A(z)$ must lie below $f(z)$ on some interval including the point $z = 1$. In this case it is easy to see that for some $z_0 \in (0, 1)$, we must have $A(z) = f(z)$. Because $A'(z)$ is strictly increasing, z_0 must be *unique*. On the other hand, for $b \leq 1$, $A'(z) < 1$ for all $z < 1$, and $A(z)$ must lie strictly above $f(z)$ for all $z \in [0, 1)$, which is Figure 3-1(a). We state some of these results as

Theorem 14. For $0 < \alpha_0 < 1$, the equation $A(z) = z$ has a *unique* root (solution) $z_0 \in (0, 1)$ if and only if $A'(1) = b > 1$. Otherwise, this equation has no root in (0, 1). The value $z = 1$ is always a root.

We now investigate the behavior of $A_n(z)$ as a function of n for fixed $z \in [0, 1)$. For Figure 3-1(a), $z < A(z) < 1$, and the dashed lines show geometrically how the values $A(z), A_2(z), \ldots$ are determined. In this case, $A_n(z)$ is a strictly increasing sequence, bounded above by 1. Hence,

$$\lim_{n\to\infty} A_n(z) = 1 \tag{93}$$

for any $z \in [0, 1)$ in Figure 3-1(a). (Why can't the limit be < 1?) For Figure 3-1(b), $A_n(z)$ is monotone increasing with upper bound z_0 for any $z \in [0, z_0)$, but monotone *decreasing* with *lower bound* z_0 for any $z \in (z_0, 1)$. In either case,

$$\lim_{n\to\infty} A_n(z) = z_0, \tag{94}$$

for any $z \in [0, 1)$ in Figure 3-1(b). From (90), we see that $q = 1$ if $b \le 1$ and $q = z_0$ if $b > 1$. Summarizing, we have

Theorem 15. For any $z \in [0, 1)$, $A_n(z)$ is a monotone function of n. For $b \le 1$, the sequence is increasing with limit (93). For $b > 1$, the sequence is increasing for $z < z_0$ and decreasing for $z > z_0$, where z_0 is the unique root of $A(z) = z$ in (0, 1). In either case, $A_n(z)$ has limit (94). The probability of extinction is

$$\begin{aligned} q &= 1 & & \text{for } b \le 1, \\ &= z_0, \quad 0 < z_0 < 1 & & \text{for } b > 1. \end{aligned} \tag{95}$$

We now interpret Theorem 15 in terms of the sequence $\{X_n\}$. For $b \le 1$,

$$\lim_{n\to\infty} (X_n \mid X_0 = 1) = 0 \quad \text{w.p.1}, \tag{96}$$

as well as in distribution. [Recall that if $X_n(\omega) = 0$ at some sample point ω, $X_s(\omega) = 0$ for all $s > n$.]

For $b > 1$, $\{X_n\}$ does not converge in the sense of (96). However, we can say something about its limiting distribution. Clearly,

$$\lim_{n\to\infty} P(X_n = 0 \mid X_0 = 1) = z_0. \tag{97}$$

What happens if extinction does not occur? From (94), we have a limiting generating function $A_\infty(z)$, say,

$$A_\infty(z) = z_0 < 1, \qquad \text{for } z \in [0, 1), \tag{98}$$

a generating function of a *defective* random variable X_∞, say, where

$$\begin{aligned} P(X_\infty = 0) &= z_0, \\ P(X_\infty = \infty) &= 1 - z_0. \end{aligned} \tag{99}$$

In this extended sense, we say that $\{X_n \mid X_0 = 1\}$ converges in distribution to X_∞. Thus, either extinction occurs, or the population becomes (in the limit) infinitely large.

By conditioning on X_1, we may also find moments of X_n, e.g., by the same argument used to obtain (92),

$$E(X_{n+1} \mid X_1 = j, X_0 = 1) = jE(X_n \mid X_0 = 1),$$

$$E(X_{n+1} \mid X_0 = 1) = bE(X_n \mid X_0 = 1), \qquad n = 0, 1, \ldots,$$

$$E(X_n \mid X_0 = 1) = b^n, \qquad n = 1, 2, \ldots. \tag{100}$$

With a bit more effort, the variance may be found as a function of $n = 1, 2, \ldots$:

$$V(X_n \mid X_0 = 1) = \begin{cases} \dfrac{\sigma^2 b^n(b^n - 1)}{b^2 - b} & \text{for } b \neq 1, \\ n\sigma^2 & \text{for } b = 1. \end{cases} \tag{101}$$

The case $b = 1$ is interesting. We have $\lim_{n\to\infty} E(X_n \mid X_0 = 1) = 1 \neq E(X_\infty)$. As discussed in Section 2-12, convergence in distribution does not necessarily imply convergence of moments to the corresponding moments of the limiting distribution. Observe that in this case, $V(X_n \mid X_0 = 1) \to \infty$ as $n \to \infty$.

Total Number of Descendants of an Individual

For $q = 1$ $(b \leq 1)$,

$$K = \sum_{n=0}^{\infty} X_n \tag{102}$$

is a proper random variable, where $K - 1$ is the total number of descendants of an individual. Given $x_1 = j$, let $K^{(i)} - 1$ be the number of descendants of each of the i individuals in the first generation, $i = 1, \ldots, j$, where the $K^{(i)}$ are i.i.d. distributed as K. With this notation,

$$K = 1 + \sum_{i=1}^{Y_1} K^{(i)}, \tag{103}$$

where, given $X_0 = 1$, we have $X_1 = Y_1$, where Y_1 has distribution (82).

Let K have generating function $\Gamma(z) = E\{z^K \mid X_0 = 1\}$. From (103) and that the $K^{(i)}$ are i.i.d. $\sim K$,

$$E\{z^K \mid Y_1, X_0 = 1\} = z[\Gamma(z)]^{Y_1}, \tag{104}$$

$$\Gamma(z) = zE\{[\Gamma(z)]^{Y_1} \mid X_0 = 1\},$$

$$\Gamma(z) = zA[\Gamma(z)], \tag{105}$$

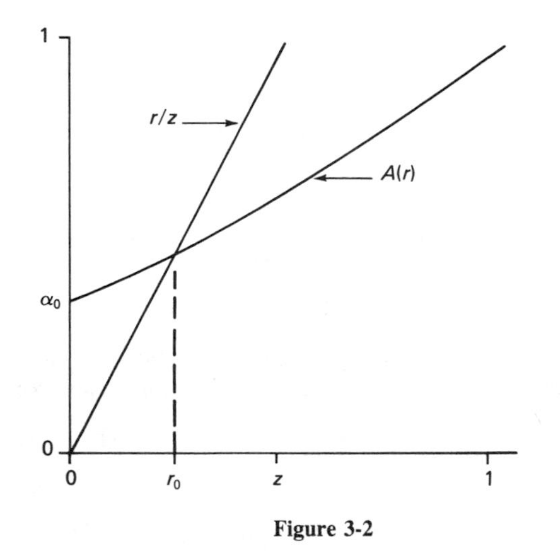

Figure 3-2

where (105) determines $\Gamma(z)$ only implicitly. Nevertheless, we now show that (105) determines $\Gamma(z)$ *uniquely*. In particular, for any $z \in (0, 1)$, there is a unique $r_0 \in (0, 1)$ that satisfies $r_0 = zA(r_0)$, where $r_0 = \Gamma(z)$. To see this, first observe that for $b \leq 1$, $A(z)$ has the shape in Figure 3-1(a). For fixed z, $A(r)$ and r/z are plotted in Figure 3-2. Since r/z has slope > 1, there is clearly a unique solution to $A(r) = r/z$. Of course, $\Gamma(z)$ in turn determines the distribution of K uniquely.

Summarizing, we have proven

Theorem 16. Let Y and K have generating functions $A(z)$ and $\Gamma(z)$, respectively, where Y is the number of progeny and $K - 1$ is the total number of descendants of an individual that initiates a branching process. K is a proper random variable if and only if $b = E(Y) \leq 1$, in which case, $\Gamma(z)$ and the distribution of K are uniquely determined by (105).

Example 3-11

For $A(z) = \alpha_0 + \alpha_1 z$, i.e., $\alpha_1 = 1 - \alpha_0$, (105) becomes

$$\Gamma(z) = z(\alpha_0 + \alpha_1\Gamma(z)), \qquad \Gamma(z) = z\alpha_0/(1 - z\alpha_1).$$

To find the distribution of K, we expand the denominator of $\Gamma(z)$ in a (geometric) power series and equate coefficients on z^j, $j = 0, 1, \ldots$.

$$E(z^K) = \sum_{j=0}^{\infty} P(K = j)z^j = \Gamma(z) = \sum_{j=1}^{\infty} \alpha_0\alpha_1^{(j-1)}z^j, \text{ or}$$

$$P(K = j) = \alpha_0\alpha_1^{(j-1)}, \qquad j = 1, 2, \ldots,$$

i.e. $K - 1$ has a geometric distribution! Can you give an elementary explanation for this? (Derive the distribution of K by a direct argument for this special case.)

Example 3-12

For $\alpha_0 = \alpha_2 = 1/2$,

$$A(z) = (1 + z^2)/2,$$
$$\Gamma(z) = z[1 + (\Gamma(z))^2]/2, \text{ or}$$
$$z\Gamma^2(z) - 2\Gamma(z) + z = 0,$$

which, for each fixed z, is a quadratic equation for $\Gamma(z)$, where only for the *smaller* of the two roots is $0 \leq \Gamma(z) \leq 1$. Hence,

$$\Gamma(z) = (1 - \sqrt{1 - z^2})/z.$$

Finding the distribution of K is tedious but can be done by representing $\Gamma(z)$ as a power series. To do this we can use a Taylor series expansion of $\sqrt{1 + x}$:

$$\sqrt{1 + x} = 1 + x/2 - x^2/8 + x^3/16 - 5x^4/128 + \cdots.$$

Substituting $x = -z^2$ into this expansion, $\Gamma(z)$ becomes

$$\Gamma(z) = z/2 + z^3/8 + z^5/16 + 5z^7/128 + \cdots,$$

from which we get

$P(K = 1) = 1/2, \quad P(K = 3) = 1/8, \quad P(K = 5) = 1/16, \quad P(K = 7) = 5/128,$

and so on. (Why must K be an odd number here?)

As evident in Example 3-12, the distribution of K is usually complicated and messy to find explicitly. Moments may easily be found, however, from either (103) or (105). The mean and variance of K are

$$E(K) = 1/(1 - b), \tag{106}$$
$$V(K) = \sigma^2/(1 - b)^3, \tag{107}$$

for $b < 1$, where $E(K) = \infty$ when $b = 1$, even though K is a proper random variable.

3-10 REVERSED AND REVERSIBLE POSITIVE IRREDUCIBLE CHAINS

Consider the market-share problem introduced in Section 3-1 and treated in Example 3-3. Typically, we would not know the transition probability matrix. To estimate this matrix, suppose we collect data at "point of sale" locations. As customers make purchases, they are asked to name the brand they had been using. (This could be done by including this question on a warranty registration card.) Of those responses that bought brand i, let b_{ij} be the fraction who said they had been using brand j, for all $i, j \geq 0$. How can information of this kind be used to estimate transition probabilities?

If the data were collected at epoch n, b_{ij} would be an estimate of

$$P(X_{n-1} = j \mid X_n = i) = \frac{P(X_{n-1} = j, X_n = i)}{P(X_n = i)}$$

$$= \frac{P(X_n = i \mid X_{n-1} = j)P(X_{n-1} = j)}{P(X_n = i)}$$

$$= p_{ji}a_j^{(n-1)}/a_i^{(n)} \qquad \text{for all } i \text{ and } j. \tag{108}$$

Now (108) holds for an arbitrary Markov chain where, in effect, we have reversed time. It was observed earlier that given the present state of a Markov chain, the past is independent of the future. Thus, a Markov chain with reversed time is also a Markov chain! It will be called a *reversed* chain. To keep these chains distinct, we will sometimes refer to the original chain as the *forward* chain.

The probabilities (108) are one-step transition probabilities in the reversed direction. Unless the Markov chain is stationary ($a = \pi$ for some probability vector satisfying $\pi = \pi P$), (108) depends on n, i.e., the reversed chain does *not* have stationary transition probabilities.

In the remainder of this section, we assume the chain is irreducible and positive, in which case there exists a unique stationary π.

In the context of the market-share problem, what do we mean by epoch n? Clearly, we would be collecting data on many customers over some limited (real) time interval. The observed fraction of purchases of brand j would be an estimate of π_j. In effect, we are observing customers "at random," and the Markov chain $\{X_n\}$ is stationary. In this case (108) becomes

$$P(X_{n-1} = j \mid X_n = i) = p_{ji}\pi_j/\pi_i \equiv q_{ij} \qquad \text{for all } i \text{ and } j. \tag{109}$$

From the b_{ij}, estimates of the π_j, and (109), the matrix P may be estimated as follows:

$$\hat{p}_{ji} = b_{ij}\hat{\pi}_i/\hat{\pi}_j \qquad \text{for all } i \text{ and } j. \tag{110}$$

Let

$$Q = (q_{ij}). \tag{111}$$

When $\{X_n\}$ is stationary, Q is the stationary one-step transition probability matrix for the reversed chain. It follows immediately (from the joint probability) that

$$q_{ij}^{(n)} \equiv P(X_s = j \mid X_{s+n} = i) = p_{ji}^{(n)}\pi_j/\pi_i \qquad \text{for all } i \text{ and } j, \tag{112}$$

and the corresponding n-step transition probability matrix for the reversed chain is

$$Q^{(n)} = (q_{ij}^{(n)}) = Q^n. \tag{113}$$

For all i and j, $q_{ij} > 0$ if and only if $p_{ji} > 0$. Hence, a chain with transition matrix Q is irreducible. Furthermore, from (112),

$$\lim_{n\to\infty} \sum_{s=1}^{n} q_{ij}^{(s)}/n = \pi_j > 0,$$

so that this chain is positive with stationary vector $\pi = \pi Q$. Alternatively, we can verify $\pi = \pi Q$ algebraically, or from the frequency interpretation of π applied to the reversed chain. That is, let $\{X_n: n = 0, \pm 1, \ldots\}$ be a positive irreducible Markov chain. (In effect, the chain started at $-\infty$ rather than epoch 0.) The fraction of transitions into state j must be the same, whether we are looking forward ($n = 0, 1, \ldots$) or in reverse ($n = 0, -1, \ldots$).

Reversible Chains

A reversible stochastic process is a process that, when time is reversed, has the same stochastic properties that it had before. The forward and reversed processes are *stochastically equivalent*. The formal definition is as follows.

Definition. A stochastic process $\{X(t)\}$ is called *reversible* if for every n, $t_1, \ldots, t_n$, and τ, $(X(t_1 + \tau), \ldots, X(t_n + \tau))$ has the same distribution as $(X(-t_1), \ldots, X(-t_n))$.

Since τ is arbitrary, a reversible process is stationary. Also notice that we are allowing time to be negative. Thus, whether t is discrete or continuous, the process "starts" at $-\infty$ rather than epoch 0.

For a stationary Markov chain, transition matrix P for the forward chain and Q for the reversed chain, together with (the same) stationary probability vector π, completely determine joint distributions. Hence, a stationary Markov chain is *reversible*† if and only if

$$Q = P, \tag{114}$$

which is equivalent to

$$\pi_i p_{ij} = \pi_j p_{ji} \qquad \text{for all } i \text{ and } j, \tag{115}$$

i.e., for every pair of states, the transition rate in the forward chain from i to j is equal to the rate from j to i. In fact, the following is also true:

Theorem 17. If there exists a set of positive $\{\pi_j\}$ that satisfy (115) and sum to 1, and the chain is irreducible, then the chain is positive and the stationary version of the chain is reversible.

† Thus a reversed chain need not be reversible. This unfortunate terminology has become standard.

Proof. Summing (115) on i, we have

$$\pi_j = \sum_i \pi_i p_{ij} \qquad \text{for all } j, \tag{116}$$

i.e., π is a stationary probability vector, and the chain is positive. The reversed chain is reversible because (115) holds where the π_j are now frequencies.

We now present some examples of reversible chains.

Example 3-13

Refer back to Example 3-4. We need verify (115) only for positive elements of P. We have

$$\pi_0 = q\pi_1,$$

$$\pi_i p = q\pi_{i+1}, \qquad i = 1, 2, \ldots .$$

Thus, the solution obtained in Example 3-4 satisfies (115), and the chain is reversible in this case ($q < p$).

This example can easily be generalized: Let $p_{i,i+1} = p_i$, $p_{i,i-1} = 1 - p_i = q_i$, where $p_i q_i > 0$, $i = 1, 2, \ldots$. We have

$$\pi_0 = q_1\pi_1 \tag{117}$$

$$\pi_i p_i = q_{i+1}\pi_{i+1}, \qquad i = 1, 2, \ldots .$$

Hence, the chain is reversible under the same normalization conditions for it to be positive.

The key to this example is that because one-step transitions are all between adjacent states, the nonzero rates in (115) are transition rates across the boundary between sets of form $\{0, \ldots, i\}$ and $\{i + 1, i + 2, \ldots\}$. From (36), these rates must be equal.

Can Reversible Chains Be Periodic?

Because state changes are either $+1$ or -1, in the preceding example, the chain can return to a state only after an even number of transitions. In fact, it is periodic with period $d = 2$. (See Section 3-8.) By permitting $p_{ii} > 0$ for some i, the chain would be aperiodic, yet it would still be reversible. On the other hand, let $p_{ij} > 0$ for some pair of states i and j in a positive chain with period $d \geq 3$. Clearly, the transition rate from i to j is $\pi_i p_{ij} > 0$. However, the transition rate in the reverse direction would have to be zero, for otherwise, $p_{ji} > 0$ and the chain would have period $d \leq 2$. Hence, a periodic chain with period $d \geq 3$ *cannot be reversible*. Nevertheless, as illustrated in Example 3-13, reversible chains with period $d = 2$ are important special cases.

Example 3-14: Trees

A Markov chain can be represented as a graph, where the states are nodes and for each $p_{ij} > 0$, we have a directed arc from node i to node j. The chain is irreducible if every pair of nodes is connected by a path of (one-way) arcs in each direction. The graph of an irreducible chain is a *tree* if it has no loops, e.g.,

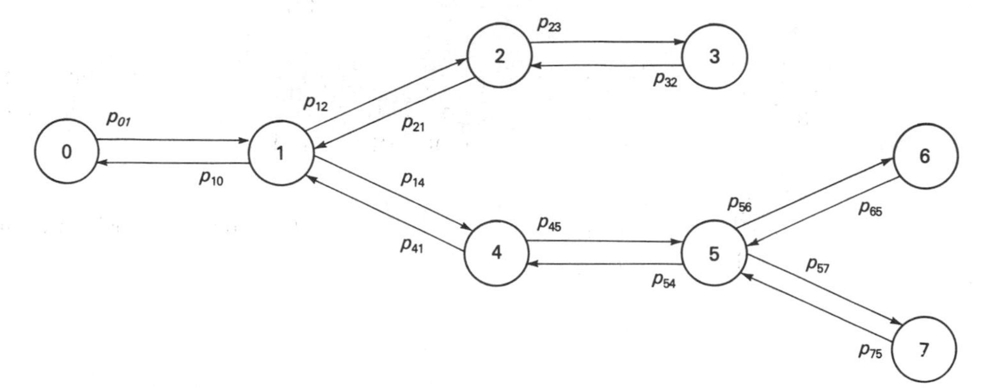

By considering transitions across boundaries, we see immediately that a positive Markov chain with tree structure is reversible. (Incidentally, a tree is also periodic.) The chain in Example 3-13 is also a tree, where the graph of states forms a single line.

Example 3-15: Transition Probabilities Based on Weights

For a positive irreducible Markov chain, suppose we assign a *weight* $w_{ij} = w_{ji}$ to each pair of states and define transition probabilities in terms of these weights: For all i and j, let $p_{ij} = w_{ij}/w_i$, where, for all i, $w_i = \sum_j w_{ij} < \infty$. These weights can be thought of as unnormalized transition rates. Since $w_{ij} = w_{ji}$ for all i and j, this suggests (115) may hold. In fact, if $w = \sum_i w_i < \infty$, we have a solution to (115),

$$\pi_i w_{ij}/w_i = \pi_j w_{ji}/w_j \qquad \text{for all } i \text{ and } j, \tag{118}$$

where for all i, $\pi_i = w_i/w$. Since $\sum_i \pi_i = 1$, we have by Theorem 17 that the chain is reversible. (Interpretation: The "message rate" between cities i and j is proportional to the product of weighting factors such as population.)

Kolmogorov's Criterion

To establish that a chain is reversible from (115), we need to be able to find or at least represent appropriately the stationary probability vector π. Can reversibility be established without doing this? First, observe that when $Q = P$ holds, paths in the forward and reverse direction have the same probability, e.g.,

$$\begin{aligned} p_{ij}p_{jk}p_{kl} &= q_{ij}q_{jk}q_{kl} \\ &= (p_{ji}\pi_j/\pi_i)(p_{kj}\pi_k/\pi_j)(p_{lk}\pi_l/\pi_k), \text{ or} \\ \pi_i p_{ij}p_{jk}p_{kl} &= \pi_l p_{lk}p_{kj}p_{ji}. \end{aligned} \tag{119}$$

When $l = i$ in (119), we can cancel π_i, and we have

$$p_{ij}p_{jk}p_{ki} = p_{ik}p_{kj}p_{ji}. \tag{120}$$

When the initial and terminal states of a path are the same, a path is called a *loop*. We are now ready to state

Theorem 18: Kolmogorov's Criterion. A stationary Markov chain is reversible if and only if the probability of traversing any loop is equal to the probability of traversing the same loop in the reverse direction.

Proof. The necessity of the criterion follows from the argument leading to (120). To prove sufficiency, the criterion implies that for any pair of states $i \neq j$, and any $n \geq 1$,

$$p_{ij}p_{ji}^{(n)} = p_{ij}^{(n)}p_{ji}. \tag{121}$$

To see this, observe that there is a one-to-one correspondence between loops with probability included on the left side of (121) and loops traversed in the reverse direction with probability on the right. From (121),

$$\left(\sum_{s=1}^{n} p_{ji}^{(s)}/n\right) p_{ij} = \left(\sum_{s=1}^{n} p_{ij}^{(s)}/n\right) p_{ji}, \qquad n = 1, 2, \ldots. \tag{122}$$

Now let $n \to \infty$ in (122), and we have (115).

While obviously a special case, reversible chains occur often enough in applications to warrant special attention. This is particularly true for continuous-time Markov chains, where certain queueing models turn out to be reversible. We will treat this topic in more detail in Chapter 6.

3-11 STATIONARY MEASURES FOR RECURRENT IRREDUCIBLE CHAINS*

A vector of real numbers $u = (u_0, u_1, \ldots)$ is called a *stationary measure* for a Markov chain with transition matrix P if $u_j \geq 0$ for all j, $u_j > 0$ for some j, and

$$u = uP. \tag{123}$$

If we also have

$$\sum_j u_j = 1, \tag{124}$$

u is a probability vector. For the remainder of this section, we deal only with irreducible chains.

* This is an advanced topic that is needed occasionally in starred portions of subsequent chapters.

If u is a stationary measure, $u = uP^n$ for every $n \geq 1$. For any pair of states i and j, irreducibility implies $p_{ij}^{(n)} > 0$ for some $n \geq 1$. It follows that if $u_i > 0$ for some i,

$$u_j \geq u_i p_{ij}^{(n)} > 0,$$

for any j. Hence, u is a stationary measure only if $u_i > 0$ for all i.

In this section, we explore existence and uniqueness properties of stationary measures without imposing normalizing condition (124). (If $\sum_j u_j < \infty$, a stationary probability vector exists, and we have the conclusions of Theorem 5.)

We will show that for recurrent chains, a stationary measure exists that is *unique up to a multiplicative constant*. That is, if u and v are measures that satisfy (123), $u = cv$ for some constant $c > 0$.

We begin by defining the following for all states i, j, and k:

$$_ip_{jk}^{(n)} = P(X_n = k, X_{n-1} \neq i, \ldots, X_1 \neq i \mid X_0 = j), \quad n = 1, 2, \ldots, \tag{125}$$

which we called a *taboo* probability in Section 3-7, where i is now the taboo state. In terms of visits to state k (k-renewals), define

$$_im_{jk}^{(n)} = \sum_{s=1}^{n} {}_ip_{jk}^{(s)}, \qquad n = 1, 2, \ldots, \tag{126}$$

which is a renewal function, where the epoch of first visit to k (first passage time) has distribution

$$_if_{jk}^{(n)} = P(X_n = k, X_{n-1} \notin \{i, k\}, \ldots, X_1 \notin \{i, k\} \mid X_0 = j), \qquad n = 1, 2, \ldots, \tag{127}$$

and the distribution of times between returns to k is

$$_if_{kk}^{(n)} = P(X_n = k, X_{n-1} \notin \{i, k\}, \ldots, X_1 \notin \{i, k\} \mid X_0 = k), \qquad n = 1, 2, \ldots. \tag{128}$$

For $i \neq k$, the probability of visiting i between returns to k is strictly positive for an irreducible chain. This implies that (128) is a defective distribution, and (126) is the renewal function for a defective renewal process. Hence, for any i, j, and k such that $i \neq k$,

$$_im_{jk}(\infty) < \infty. \tag{129}$$

To construct a stationary measure for a recurrent chain, suppose $X_0 = i$ and, for any $k \neq i$, define

$$\begin{aligned} I_n(i, k) &= 1 \qquad \text{if } X_n = k, \quad X_{n-1} \neq i, \ldots, X_1 \neq i, \\ &= 0 \qquad \text{otherwise.} \end{aligned}$$

The number of visits to state k between returns to state i has representation

$$V_{ik} = \sum_{n=1}^{\infty} I_n(i, k),$$

with [from (129)] finite expectation

$$0 < E(V_{ik}) = {}_im_{ik}(\infty) < \infty.$$

[See also Problem 3-26 for a proof of the finiteness of $E(V_{ik})$.]

By conditioning on X_n, it is easy to see that

$$_ip_{ik}^{(n+1)} = \sum_{j \neq i} {}_ip_{ij}^{(n)} p_{jk}, \qquad n = 1, 2, \ldots . \tag{130}$$

Summing (130) on n and observing that $_ip_{ik}^{(1)} = p_{ik}$, we have that for $k \neq i$,

$$E(V_{ik}) - p_{ik} = \sum_{j \neq i} E(V_{ij}) p_{jk}. \tag{131}$$

For a *recurrent* chain, $E(V_{ii}) = 1$ (formally, $\sum_{n=1}^{\infty} {}_ip_{ik}^{(n)} = f_{ii} = 1$), and, combining this with (131), we have that for every i,

$$E(V_{ik}) = \sum_j E(V_{ij}) p_{jk}, \qquad k = 0, 1, \ldots , \tag{132}$$

i.e., $\{E(V_{ik}), k = 0, 1, \ldots\}$ is a *stationary measure*.

Remark. If the chain is *transient*, $E(V_{ii}) < 1$, and instead of (132), we have that for any i,

$$E(V_{ik}) > \sum_j E(V_{ij}) p_{jk}, \qquad k = 0, 1, \ldots , \tag{133}$$

where these inequalities are strict for those k where $p_{ik} > 0$.

To prove uniqueness of stationary measures, we need to refine our knowledge of the renewal functions introduced in Section 3-2, e.g., equation (18). In particular, for a null recurrent Markov chain, $m_{ij}(\infty) = \infty$ but $\lim_{n \to \infty} m_{ij}(n)/n = 0$.

To go beyond these results, we condition on the first passage time T_{ij} and write

$$m_{ij}(n) = \sum_{s=1}^{n} f_{ij}^{(s)} [m_{jj}(n - s) + 1]. \tag{134}$$

For a recurrent chain, T_{ij} is proper, which implies that for arbitrary $\varepsilon > 0$, we can find a k such that $P(T_{ij} \leq k) > 1 - \varepsilon$. From this and (134), we can bound $m_{ij}(n)$ for any $n \geq k$:

$$[m_{jj}(n - k) + 1](1 - \varepsilon) \leq m_{ij}(n) \leq m_{jj}(n) + 1. \tag{135}$$

From (135) it follows immediately that for a recurrent Markov chain,

$$\lim_{n\to\infty} m_{ij}(n)/m_{jj}(n) = 1 \qquad \text{for all } i \text{ and } j. \tag{136}$$

Remarks. For both $m_{jj}(n)$ and $m_{ij}(n)$, cycles are times between returns to state j, and these are examples of *ordinary* and *general* renewal functions. The right-hand inequality in (135) holds for all n and for renewal processes in general. (See Problems 2-12 and 2-76.) The left-hand inequality in (135) is also true, and, for any *proper* renewal process, we have, in the notation of Chapter 2,

$$\lim_{t\to\infty} m(t)/m_o(t) = 1, \tag{137}$$

even when $\mu = 0$.

For arbitrary $i \neq j$, visits to state j can also be represented in terms of i-renewals:

$$M_{ij}(n) \leq \sum_{s=1}^{M_{ii}(n)+1} V_{ij}^{(s)}, \qquad n = 1, 2, \ldots, \tag{138}$$

where the $V_{ij}^{(s)}$ are the number of visits to state j between successive returns to state i. From (138), we can write

$$[m_{ii}(n) + 1]E(V_{ij}) - \sum_k p_{ik}^{(n)} \cdot {}_im_{kj}(\infty)$$

$$= m_{ij}(n) \leq [m_{ii}(n) + 1]E(V_{ij}), \; n = 1, 2, \ldots, \tag{139}$$

where the right-hand inequality is simply an application of Wald's equation. The equality on the left is obtained by subtracting the visits to state j that occur between epochs n and the end of the i-renewal cycle in progress at epoch n. The second term on the far left involves taboo-type renewal functions introduced earlier in this section. By the same argument used to obtain the right-hand inequality in (135),

$$\sum_k p_{ik}^{(n)} \cdot {}_im_{kj}(\infty) \leq {}_im_{jj}(\infty) + 1 < \infty. \tag{140}$$

From (139) and (140), we have that for a *recurrent* chain,

$$\lim_{n\to\infty} m_{ij}(n)/m_{ii}(n) = E(V_{ij}) \qquad \text{for all } i \text{ and } j. \tag{141}$$

Remarks. Limits (136) and (141) are called *ratio limit theorems*. The interpretation of $E(V_{ij})$ as a *relative* frequency is valid for null as well as positive chains. It is easily shown that when random variables $M_{ij}(n)$ replace the $m_{ij}(n)$ in (141), the limit remains the same.

We are now ready to deal with the uniqueness question. Suppose P has a stationary measure u, i.e., $u = uP$, where, because the chain is irreducible, $u_j >$

0 for all j. In terms of u, we can define a reversed chain analogous to (109), where matrix $Q = (q_{ij})$ is now defined by

$$q_{ij} = u_j p_{ji}/u_i \qquad \text{for all } i \text{ and } j, \tag{142}$$

where Q is a legitimate transition probability matrix, i.e., the q_{ij} are nonnegative and have row sums $\sum_j q_{ij} = 1$ for all i. It is immediate that u is also a stationary measure for Q.

While (142) may appear artificial when u is not a probability vector, Q has the same algebraic properties found in Section 3-10, i.e., it may be shown that for $Q^n = (q_{ij}^{(n)})$, the following holds for all i and j:

$$q_{ij}^{(n)} = u_j p_{ji}^{(n)}/u_i, \qquad n = 1, 2, \ldots. \tag{143}$$

Clearly, the reversed chain is irreducible and, from (143), the forward and reversed chains are of the same type: transient, null, or positive. (They also have the same period or span.)

In the recurrent case, where we have shown that a stationary measure exists, we use (143) to write

$$u_i \sum_{s=1}^{n} q_{ij}^{(s)} \Big/ \sum_{s=1}^{n} q_{jj}^{(s)} = u_j \sum_{s=1}^{n} p_{ji}^{(s)} \Big/ \sum_{s=1}^{n} p_{jj}^{(s)}, \qquad n = 1, 2, \ldots. \tag{144}$$

Now let $n \to \infty$. From (136) and (141), we have

$$u_i = u_j E(V_{ji}) \qquad \text{for all } i \text{ and } j. \tag{145}$$

Since u is arbitrary, we have proven

Theorem 19. For a recurrent irreducible Markov chain, there exists a stationary measure $u = uP$, where $u_j > 0$ for all j, and u is unique up to a multiplicative constant. For all i and j, the ratios u_i/u_j are uniquely determined by (145).

Remarks. As illustrated in the following examples, a transient Markov chain may also have a stationary measure, which may or may not be unique in the sense of Theorem 19, or it may not have any stationary measure. Thus, finding a stationary measure u, such that $\sum_j u_j = \infty$, implies the chain is *not* positive. The chain will be transient if we are able to show either that no stationary measure exists or that uniqueness fails.

Example 3-16

A transition probability matrix P and the corresponding Markov chain are called *doubly stochastic* if $\sum_i p_{ij} = 1$ for all j. It is easy to show that $u = (1, 1, \ldots)$ is a stationary measure for any doubly stochastic matrix. If the chain is irreducible and has an infinite number of states, it *cannot* be positive. (See Problems 3-16 and 3-17 for finite-state examples.)

Example 3-17

A special case of a doubly stochastic matrix is: $p_{i,i+1} = p$, $p_{i,i-1} = q$, for $i = 0, \pm 1, \ldots$, where $pq > 0$ and $p + q = 1$. Setting $u = uP$, we have that for all i,

$$u_i = pu_{i-1} + qu_{i+1}.$$

We know from earlier examples that the chain is null when $p = q$, and $u_i = 1$ for all i is the unique stationary measure for this case. For $p \neq q$, $u_i = (p/q)^i$ for all i is also a stationary measure, but now the chain is transient.

Example 3-18

In this example, we show that in the transient case, a stationary measure may not exist. Let the transition probabilities be $p_{i,i+1} = p_i$, $p_{i,0} = q_i$, for $i = 0, 1, \ldots$, where $p_i q_i > 0$ and $p_i + q_i = 1$. This chain is irreducible, and it is easy to see that all states are transient if and only if $\prod_{i=0}^{\infty} p_i > 0$. (While not needed here, it may be shown that this condition is equivalent to $\sum_i q_i < \infty$; the equivalence of these expressions is the mathematical content of Problem 3-36.)

By equating $u = uP$, we have

$$u_0 = \sum_{j=0}^{\infty} q_j u_j \quad \text{and} \quad u_{j+1} = p_j u_j \qquad \text{for } j \geq 0,$$

from which we obtain

$$u_j = \alpha_j u_0 \qquad \text{for } j \geq 1,$$

where $\alpha_j = \prod_{i=0}^{j-1} p_i$ for $j \geq 1$, and we define $\alpha_0 = 1$ for use below.

By substituting $u_j = \alpha_j u_0$ into the equation for u_0, we have

$$u_0 = u_0 \sum_{j=0}^{\infty} q_j \alpha_j.$$

Hence, there will exist a (unique) stationary measure if and only if $\sum_{j=0}^{\infty} q_j \alpha_j = 1$. Now $q_0 \alpha_0 = (1 - p_0)\alpha_0 = \alpha_0 - \alpha_1$, and

$$q_j \alpha_j = (1 - p_j) \prod_{i=0}^{j-1} p_i = \alpha_j - \alpha_{j+1} \qquad \text{for } j \geq 1.$$

Hence,

$$\sum_{j=0}^{\infty} q_j \alpha_j = \lim_{n \to \infty} \sum_{j=0}^{n} q_j \alpha_j = \lim_{n \to \infty} \sum_{j=0}^{n} (\alpha_j - \alpha_{j+1}) = \lim_{n \to \infty} (1 - \alpha_{n+1}) = 1 - \prod_{i=0}^{\infty} p_i,$$

and we conclude that a stationary measure exists if and only if $\prod_{i=0}^{\infty} p_i = 0$, which is the condition for all states to be recurrent. In the transient case, a stationary measure does not exist.

PROBLEMS

3-1. A rat is placed in the maze below consisting of 6 cells numbered 1 through 6.

1	2	3
4	5	6

The rat moves from cell to cell at random and is equally likely to leave a cell through any of the openings available, independent of the sequence of previously visited cells. This includes the opening in cell 6 to the outside, where "outside" is denoted by state 0. Once outside, the rat will not (of its own volition) return to the maze.

Let X_0 be the cell in which the rat is initially placed and X_n be the position (cell number or state 0) of the rat after n moves, $n = 1, 2, \ldots$.

(a) Is $\{X_n\}$ a Markov chain? Does it have stationary transition probabilities?

(b) If the rat leaves the maze on the third move, so that $X_3 = 0$, what do we mean by X_n, $n > 3$?

(c) Determine the transition probability matrix P for this problem.

(d) Find $P(X_2 = 2 \mid X_0 = 2)$.

3-2. Let $\{X_n, n = 0, 1, \ldots\}$ be a sequence of i.i.d. random variables with $P(X_n = j) = a_j > 0$, $j = 0, 1, \ldots$. Define sequences $\{S_n\}$, $\{M_n\}$, $\{m_n\}$ by $S_n = \sum_{j=0}^{n} X_j$, $M_n = \max\{X_0, \ldots, X_n\}$, and $m_n = \min\{X_0, \ldots, X_n\}$, $n = 0, 1, \ldots$.

(a) Which of the sequences $\{X_n\}$, $\{S_n\}$, $\{M_n\}$, and $\{m_n\}$ are Markov chains?

(b) For each sequence in (a) that is a Markov chain, determine the transition probability matrix P.

3-3. In Problem 2-53, we will say that the system is in state 0 if the stand is empty, in state 1 if the stand is occupied by a taxi, and in state 2 if the stand is occupied by a customer. Model the sequence of state changes as a Markov chain and determine the corresponding transition probability matrix.

3-4. In terms of $\{X_n\}$ in Problem 3-2, define $K_0 = 0$, $K_n = X_n + X_{n-1}$, $n = 1, 2, \ldots$. Is $\{K_n\}$ a Markov chain?

3-5. Derive (6) and (7).

3-6. Apply a renewal-type argument to find f_{01} for the Markov chain with transition matrix P given by (14).

3-7. For which of the following transition probability matrices is the corresponding Markov chain irreducible?

$$\text{(a)} \begin{pmatrix} 0 & 1 & 0 \\ 0 & 0 & 1 \\ 1 & 0 & 0 \end{pmatrix} \quad \text{(b)} \begin{pmatrix} .2 & 0 & .8 \\ .6 & .2 & .2 \\ .8 & 0 & .2 \end{pmatrix} \quad \text{(c)} \begin{pmatrix} 1 & 0 & 0 \\ 0 & 1 & 0 \\ 0 & 0 & 1 \end{pmatrix}$$

3-8. *Continuation.* For each Markov chain, find *all* stationary probability vectors, i.e., all probability vectors π that satisfy $\pi = \pi P$. (Work out directly, without use of Markov-chain theorems.)

3-9. Which of the Markov chains in Problem 3-2 are irreducible?

3-10. For recurrent states i and j, $i \neq j$, such that $i \leftrightarrow j$, show that $f_{ij} = f_{ji} = 1$.

3-11. For an infinite Markov chain, give an example where all communication classes are open.

3-12. Suppose that 2/3 of the trucks on a highway are followed by a car but only 1/5 of the cars are followed by a truck. What fraction of vehicles on the highway are cars?

3-13. Suppose the probability of rain today is .6 if rain fell yesterday, but only .2 if it did not, where these probabilities do not depend on the weather on previous days.

(a) Given rain fell today, what is the probability of rain on the day after tomorrow?

(b) What is the average duration (number of days) of a rainy period?

(c) On what fraction of days does rain fall?

3-14. *Continuation.* Repeat (b) for dry periods and verify (c) by viewing the sequence of rainy and dry periods as an alternating renewal process (Example 2-2).

3-15. Consider a Markov chain with transition probability matrix

$$P = \begin{pmatrix} p_0 & p_1 & p_2 & p_3 & \cdots \\ 1 & 0 & 0 & 0 & \cdots \\ 0 & 1 & 0 & 0 & \cdots \\ 0 & 0 & 1 & 0 & \cdots \\ & & & \ddots & \end{pmatrix},$$

where $p_j > 0$ for all j and $\sum_{j=0}^{\infty} p_j = 1$.

(a) Is the chain irreducible?

(b) What is the probability of returning to state 0? Are all states recurrent?

(c) From definitions and elementary considerations, under what conditions is $\nu_0 < \infty$? When are all states positive?

(d) When are all states null?

(e) Verify (c) and (d) by determining conditions for which a stationary probability vector exists. Under these conditions, find π.

(f) Find the mean recurrence time for state 23.

3-16. A square matrix $P = (p_{ij})$ of nonnegative elements is called *stochastic* if for all i, $\sum_j p_{ij} = 1$. It is called doubly stochastic if it also has the property: for all j, $\sum_i p_{ij} = 1$. For a *finite* Markov chain with a doubly stochastic transition probability matrix, "guess" the form of a stationary probability vector π. Verify that $\pi = \pi P$. (*Hint:* Look at special cases, e.g. the two-state case, and generalize what you find.) Under what conditions will your guess be unique?

3-17. *Continuation.* Stocks are usually traded in lots of 100 shares. Sales of less than 100 shares are called *odd lot* sales. These can be handled separately, but sometimes 100 share lots are broken down (or conversely, odd lots are combined) to complete transactions. To simplify matters, we consider customers who wish to buy odd lots only, and that these sales are made by breaking down 100 share lots. Whatever is left over from a lot is used to satisfy demand before a new lot is broken down. Suppose odd lot demands $D_1, D_2, \ldots$ are i.i.d., with $P(D_j = i \text{ shares}) = d_i > 0$, $i = 1, \ldots, 99$. What fraction of sales are completed such that they leave no shares in an odd lot?

3-18. S units of an item are stocked in inventory. Demand for the item is Poisson with mean λ units/day. If the inventory hits zero by the end of a day, S units are reordered; they are available by the beginning of the next day. Otherwise, no reorder is placed. Demand during a day in excess of available inventory that day is lost. Let X_n be the inventory level at the end of day n and

$$p_{ij} = p(X_{n+1} = j \mid X_n = i); \qquad i, j = 0, 1, \ldots, S.$$

(a) Construct the transition probability matrix $P = (p_{ij})$.
(b) In terms of P, state how you would find the fraction of days on which reorders are placed (no manipulations required).
(c) In terms of your answer in (b), at what rate (units/day) are orders filled? At what rate are orders lost? What fraction of demand is lost?

3-19. A company has three machines. On any day, each working machine breaks down with probability p, independent of other machines. At the end of each day, the machines that have broken down are sent to a repairman who can work on only one machine at a time. When the repairman has one or more machines to repair at the beginning of a day, he repairs and returns exactly one at the end of that day.

Let X_n = the number of working machines at the end of day n (beginning of day $n + 1$), after breakdowns and repair that day are included.

(a) What fraction of days begin with j working machines, $j = 0, 1, 2, 3$?
(b) What is the average number of working machines at the beginning of a typical day?
(c) If a working machine at the beginning of a day brings in \$$r$ of revenue that day and each machine costs \$$c$ to repair, what is the net reward (revenue − cost) per day?

3-20. In Problem 3-1, suppose we combine states 5 and 6 into a new state $5'$. (There is no change in the maze; we are just not distinguishing whether the rat is in cell 5 or 6.) By considering one or more particular sequences of transitions, argue that the sequence $\{X_n\}$ for new state space $\{0, \ldots, 4, 5'\}$ is *not* a Markov chain. When we combined states in Example 3-3, we still had a Markov chain. Why?

3-21. For an irreducible Markov chain with $c < \infty$ states, suppose there exists a vector υ with arbitrary (possibly negative) real elements such that $\upsilon = \upsilon P$.
(a) Show that this is impossible if υ is independent of the unique probability vector $\pi = \pi P$. (*Hint:* Construct a new stationary probability vector.)
(b) From (a), show that the rank of the matrix $(P - I)$ is $c - 1$.

3-22. *Continuation.* Let $\pi^{(s)}$ be a stationary probability vector for an otherwise arbitrary Markov chain. Show that $\pi_j^{(s)} = 0$ if state j is either transient or null. (Hence, the probability mass of the vector is concentrated on the positive states.)

3-23. *Continuation.* For a Markov chain with $c < \infty$ states and k *positive* communication classes, show that the rank of $(P - I)$ is $c - k$.

3-24. As discussed in Section 3-5, the equation $\pi = \pi P$ for an irreducible, positive Markov chain simply equates transition rates into and out of each state. These results do not depend on the Markov property. To illustrate, suppose $\{X_n\}$ is an irreducible, positive

Markov chain with matrix P and unique vector $\pi = \pi P$. Let us collapse the state space into two states, $0'$ and $1'$,

$$0' = \{0\}, \qquad 1' = \{1, 2, \ldots\},$$

with "transition matrix" $P' = (p'_{ij})$, where

$$p'_{00} = p_{00}, \qquad p'_{01} = 1 - p_{00},$$

$$p'_{10} = \sum_{i \geq 1} \pi_i p_{i0} / \sum_{i \geq 1} \pi_i, \quad \text{and} \quad p'_{11} = 1 - p'_{10}.$$

(a) In terms of the original Markov chain, justify P', and, in particular, p'_{10} above. That is, interpret p'_{10} as a frequency.

(b) Show that the solution to $\pi' = \pi' P'$ is $\pi'_0 = \pi_0$, $\pi'_1 = 1 - \pi_0$.

3-25. Suppose each day's weather is categorized into four types:

0: primarily sunny,
1: primarily cloudy but dry,
2: primarily wet (rain), and
3: primarily wet (snow).

Let X_n be the weather type on day n, where it has been conjectured that $\{X_n\}$ is a Markov chain. To verify this, data were collected over an $(n + 1)$-day period, i.e., $X_0, \ldots, X_n$ were observed, for some large $n > 100$.

For all states i and j, let N_{ij} be the number of observed transitions of type $i \to j$, and $N_j = \sum_i N_{ij}$ be the number of observed visits to state j, where $\sum_j N_j = n$. It was decided to estimate π and P as follows:

$$\hat{p}_{ij} = N_{ij} / \sum_j N_{ij} \qquad \text{for all } i \text{ and } j,$$

$$\hat{\pi}_j = N_j / n \qquad \text{for all } j.$$

(a) Show that if $X_0 = X_n = k$, say, for any state k, $\hat{\pi} = \hat{\pi}\hat{P}$ *exactly*! (*Hint:* Argue that under this assumption, $\sum_i N_{ij} = \sum_i N_{ji}$ for all j.)

(b) Actually, we would not expect $X_0 = X_n$. Suppose that for the data collected above $X_0 \neq X_n$, but the elements of $|\hat{\pi} - \hat{\pi}\hat{P}|$ were all less than 1/100. The conclusion was that "the data fit a Markov chain model remarkably well." Do you agree?

3-26. For any pair of recurrent states $i \neq j$ such that $i \leftrightarrow j$, let

$${}_if_{ij} = P(\text{chain visits } j \text{ before returning to } i \mid X_0 = i).$$

Show that

(a) ${}_if_{ij} > 0$.

(b) $P(V_{ij} > k) = {}_if_{ij}(1 - {}_jf_{ji})^k$, $k = 0, 1, \ldots$.

(c) $E(V_{ij}) < \infty$.

(d) V_{ij} has finite moments of all orders.

3-27. **(a)** For an irreducible *positive* Markov chain, show that the mean first passage times are finite:

$$E(T_{ij}) < \infty \qquad \text{for all } i \text{ and } j. \tag{P.1}$$

(*Hint:* For $i \neq j$, consider the event: {chain visits state j prior to state $i \mid X_0 = i$}.)

(b) Let P and P' be transition probability matrices of irreducible chains that are *identical* except for a finite number of elements in the top row:

$$p_{0j} \neq p'_{0j} \qquad \text{for some } j, \quad j \leq k, \quad k < \infty.$$

Show that all states are positive for P if and only if all states are positive for P'.

(c) Apply (b) and what you know about another Markov chain to determine conditions under which all states are positive for the following:

$$\begin{pmatrix} .2 & .4 & .1 & .3 & 0 & \cdots \\ q & 0 & p & 0 & & \cdots \\ 0 & q & 0 & p & & \\ & & \ddots & & \ddots & \end{pmatrix},$$

where $pq > 0$, $p + q = 1$.

3-28. For a Markov chain with transition matrix,

$$P = \begin{pmatrix} .1 & .2 & 0 & .3 & .4 \\ .4 & 0 & .3 & .1 & .2 \\ .5 & .1 & .1 & .2 & .1 \\ .4 & 0 & 0 & 0 & .6 \\ .1 & .6 & .1 & .1 & .1 \end{pmatrix},$$

find the probability that the chain visits state 3 before state 4, given

$$X_0 = i, \qquad i = 0, 1, 2, 3, 4.$$

3-29. As an application of Theorem 12, find $\{E(T_{ij})\}$ for the Markov chain in Problem 3-13 and verify that $E(T_{ii}) = 1/\pi_i$, $i = 0, 1$.

3-30. With reference to Example 3-10, we want to show that for $p > q$, $E(N) < \infty$. Refer now to Example 3-4, where we found that all states are positive if and only if $p < q$. In particular, $E(T_{00}) = 1/\pi_0 = 2q/(q - p) < \infty$.

(a) Use the top row of P in Example 3-4 to show that for $p < q$, $E(T_{10}) < \infty$.

(b) *Reversing the roles of p and q*, we have that $E(T_{10})$ in Problem 3-33 is equal to $E(N) = E(T_{01})$ in Example 3-10. Find $E(T_{10})$ in (a) and, reversing p and q, show that for $p > q$,

$$E(N) = 1/(p - q) < \infty,$$

in agreement with equation (123) in Chapter 2.

3-31. In Problem 3-1, find a system of linear equations for $E(T_{i0})$, $i = 1, \ldots, 6$.

3-32. Consider a Markov chain with transition probability matrix

$$P = \begin{pmatrix} 0 & 1 & 0 & \cdots & & \\ a & b & c & 0 & \cdots & \\ 0 & a & b & c & & \\ & \ddots & \ddots & \ddots & & \end{pmatrix},$$

where $abc > 0$, $a + b + c = 1$.

(a) Is the chain irreducible?

(b) Show that all states are positive if and only if $a > c$.

(c) Argue that for $i \geq 1$, f_{i0} has form $f_{i0} = \beta^i$, $0 < \beta \leq 1$. Use this information to show that all states are transient if and only if $a < c$.

(d) When are all states null?

3-33. For a Markov chain with a *finite* set T of transient states and a set C of recurrent states, define the following for each $i \in T$: T_i = the first passage time to C, given $X_0 = i \in T$; $\mu_i = E(T_i)$, and $\gamma_i = E(T_i^2)$.

(a) Show that $\{\gamma_i\}$ is a solution to

$$\gamma_i = 1 + 2 \sum_{j \in T} p_{ij}\mu_j + \sum_{j \in T} p_{ij}\gamma_j \qquad \text{for all } i \in T. \tag{P.2}$$

(b) Assuming the γ_i are finite (they are!), show that the solution to (P.2) is unique.

(c) Apply the result (P.2) in (a) to find the γ_i for the chain with transition matrix

$$P = \begin{pmatrix} 1 & 0 & 0 & 0 & 0 \\ 1/2 & 0 & 1/2 & 0 & 0 \\ 0 & 1/2 & 0 & 1/2 & 0 \\ 0 & 0 & 1/2 & 0 & 1/2 \\ 0 & 0 & 0 & 0 & 1 \end{pmatrix}.$$

(Find μ_i, $i = 1, 2, 3$ first.)

3-34. In Example 3-8, we found $x_{02} = 11/18$ and $x_{12} = 17/36$, where now

$$x_{i2} = P(\text{chain visits state 2 before state 3} \mid X_0 = i),\ i = 0, 1.$$

(a) Use the same method to find

$$x_{i3} = P(\text{chain visits state 3 before state 2} \mid X_0 = i),\ i = 0, 1.$$

Verify that $x_{i2} + x_{i3} = 1$ for $i = 0, 1$.

(b) Suppose grapes can be sold for:

$300/ton if picked before chain reaches either states 2 or 3,
$500/ton if picked after chain reaches state 2 (before 3),
$100/ton if picked after chain reaches state 3 (before 2).

Our objective is to *maximize expected selling price per ton.*

Given the chain is now in state 0, show that waiting until the chain reaches either states 2 or 3 is better than picking immediately.

(c) A third policy is this: Wait until the chain leaves state 0 and pick during the next weather type (state 1, 2, or 3), whatever it turns out to be. If the chain is now in state 0, find the expected selling price per ton under this policy, and show that this policy is the best of the three. (*Hint:* Condition on the next transition.)

(d) If the chain initially is in state 1, show that the best policy is to pick immediately. (Can you do this without any calculation?)

3-35. Let T be the set of transient states in a Markov chain, and for all $i, j \in T$, let $M_{ij}(\infty)$ be the total number of visits to state j given $X_0 = i$, and $c_{ij} = E\{M_{ij}(\infty)\}$.

(a) Show that $\{c_{ij}\}$ satisfies

$$c_{ij} = p_{ij} + \sum_{k \in T} p_{ik}c_{kj} \qquad \text{for all } i, j \in T.$$

(b) Let T be a finite set. For each fixed $j \in T$, explain why the solution to the equations in (a), $\{c_{ij}: i \in T\}$, is unique.

(c) For the Markov chain with state space {0, 1, 2, 3, 4} and transition matrix

$$p = \begin{pmatrix} 1 & 0 & 0 & 0 & 0 \\ q & 0 & p & 0 & 0 \\ 0 & q & 0 & p & 0 \\ 0 & 0 & q & 0 & p \\ 0 & 0 & 0 & 0 & 1 \end{pmatrix},$$

where $pq > 0$, find an explicit expression for c_{22}.

3-36. Let a Markov chain have transition probabilities $p_{i0} = q_i$, $p_{i,i+1} = 1 - q_i$, $i = 0, 1, \ldots$, $p_{ij} = 0$ otherwise, where for each i, $0 < q_i < 1$. Show that

(a) All states are transient if and only if

$$\sum_{i=0}^{\infty} q_i < \infty, \text{ and}$$

(b) All states are null if and only if

$$\sum_{i=0}^{\infty} q_i = \infty \quad \text{and} \quad \sum_{n=1}^{\infty} \prod_{i=0}^{n-1} (1 - q_i) = \infty.$$

3-37. Let P be a transition probability matrix and $Q = (P + I)/2$, where I is the identity matrix, and P and I are the same size (possibly infinite). Show the following:

(a) Q is a transition probability matrix. (In what follows, when we refer to P and Q, we mean Markov chains with these transition probability matrices.)

(b) A set of states is a communication class in P if and only if it is a communication class in Q.

(c) A probability vector π satisfies $\pi = \pi P$ if and only if it satisfies $\pi = \pi Q$.

(d) State j is positive in P if and only if it is positive in Q.

(e) State j is transient in P if and only if it is transient in Q.

(f) State j is null in P if and only if it is null in Q.

(g) For any state j, $f_{jj}(Q) = [f_{jj}(P) + 1]/2$, where Q and P simply denote each chain.

(h) For any state j, $E\{T_{jj}(P)\} = E\{T_{jj}(Q)\}$.

3-38. For branching processes, find (101).

3-39. Find (106) and (107).

3-40. A branching process is a Markov chain with transition probability matrix (85). Thus, the general theory we have developed should apply.

(a) First, suppose that the initial conditions of a branching process are changed to $X_0 = i$ for any $i \geq 1$. Argue that $P(\text{extinction} \mid X_0 = i) = q^i$, $i = 1, 2, \ldots$, where $P(\text{extinction} \mid X_0 = 1) = q$, as before.

(b) Apply Theorem 11 to (85) in order to derive (95). [If you insist on dealing with an irreducible chain, change the top row of (85) to (0, 1, 0, . . .). Does this change really matter?]

3-41. For a Markov chain with transition probability matrix

$$P = \begin{pmatrix} \sum_{i>0} a_i & a_0 & 0 & \cdots \\ \sum_{i>1} a_i & a_1 & a_0 & \cdots \\ \vdots & & \ddots & \ddots \end{pmatrix},$$

where $a_i > 0$ for all i, it can be shown that a stationary probability vector, when it exists, has form

$$\pi_j = (1 - \beta)\beta^j, \qquad j = 0, 1, \ldots, \qquad \text{for some } \beta \in (0, 1).$$

Use this information to show that all states are positive if and only if $\sum_{i=0}^{\infty} ia_i > 1$.

3-42. Your company manufactures brand X, which competes in a certain market with brand Y. Laboratory tests have shown that the mean life of items of each brand are $\mu_x = 100$ days and $\mu_y = 150$ days, respectively. (Lifetimes are independent of each other and of customer purchasing decisions. The life distribution depends only on the brand.)

It is assumed that customers buy one item at a time and replace on failure. The sequence of purchases that a customer makes is assumed to be an irreducible two-state Markov chain, with states X and Y, and an *unknown* transition probability matrix.

Let V_{xy} be the number of purchases a customer makes of brand Y between purchases of brand X, D_x be the number of *days* between a customer's successive purchase of brand X, and f_x be the fraction of time (days) that a customer spends using brand X.

(a) Show that $E(D_x) = \mu_x + E(V_{xy})\mu_y$.

(b) Using (a), show that

$$f_x = \frac{\mu_x \pi_x}{\mu_x \pi_x + \mu_y \pi_y},$$

where π_x is the market share (fraction of purchases) of brand X, and $\pi_y = 1 - \pi_x$.

(c) A survey of customers found that 50 percent of the customers were using brand X (when asked), and the rest were using brand Y. Treating these estimates as exact, what is the market share of brand X?

(d) For all $i, j \in \{X, Y\}$, let

f_{ij} = fraction of time (days) for which it is true that a customer is "now" (i.e., on a randomly selected day) using brand j *and* bought brand i on the preceding purchase.

Use the cycle defined in (a) to find expressions for f_{xx} and f_{xy}. (These results should involve transition probabilities.) How would you use estimates of the f_{ij} to estimate the transition probabilities?

3-43. Consider a branching process with $X_0 = 1$.

(a) Condition on X_1 to find an expression for q in terms of the result in Problem 3-40(a).

(b) Show that (a) is equivalent to $q = A(q)$, where

$$A(z) = \sum_{j=0}^{\infty} \alpha_j z^j.$$

(c) From this analysis and Theorem 14, but *not* the more detailed analysis in Section 3-9, why must $q = 1$ if $A'(1) = b \leq 1$?

3-44. Let the probability that a machine breaks down on its jth day of service be $P(\nu = j) = \alpha_j > 0, j = 1, 2, 3, \ldots$, where $E(\nu) = \sum_{j=1}^{\infty} j\alpha_j < \infty$. When the machine breaks down, it is repaired overnight. The process is repeated, where the times (number of days) between failure are i.i.d. with distribution $\{\alpha_j\}$.

For $n = 0, 1, \ldots$, let $X_n = j$ if on day n, the machine is beginning its jth day of service since the last repair, $j = 1, 2, \ldots$. Let the state space of $\{X_n\}$ be $\{1, 2, \ldots\}$, i.e., $X_n = 0$ cannot occur.

(a) Briefly argue that $\{X_n\}$ is a Markov chain and find the transition probability matrix.

(b) Is the chain irreducible?

(c) From what you know about renewal theory, find π_1 *without* solving $\pi = \pi P$.

(d) Determine whether the states in this chain are positive, transient, or null.

(e) Show that $b = (b_1, b_2, \ldots)$ satisfies $b = bP$, where $b_j = P(\nu \geq j), j = 1, 2, \ldots$. What is π?

3-45. *Continuation*. For the same model, we now define a *different* chain. By $X_n = j$ we mean that a working machine at the beginning of day n breaks down on its jth day of additional use (on day $n + j - 1$), i.e., j is the *remaining life* until breakdown.

(a) Find transition matrix P for this chain and show that it has the same stationary probability vector π that we found in Problem 3-44.

(b) Show that the chain in this problem and in the preceding one are reversed versions of each other.

(c) Interpret (b).

3-46. Let $\{X_n\}$ and $\{Y_n\}$ be independent Markov chains; it is easily shown that $\{(X_n, Y_n)\}$ is a Markov chain with a two-dimensional state space. Show that if $\{X_n\}$ and $\{Y_n\}$ are both reversible, then so is $\{(X_n, Y_n)\}$.

3-47. *Truncation*. Let $\{X_n\}$ be a reversible Markov chain with matrix $P = (p_{ij})$ and stationary probability vector π. Partition the state space into disjoint subsets A and A^c. We now define a new Markov chain $\{X'_n\}$ over the truncated state space A, with the following transition probabilities (p'_{ij}):

$$p'_{ij} = p_{ij} \qquad \text{for all } i, j \in A, \quad i \neq j,$$

$$p'_{ii} = p_{ii} + \sum_{j \in A^c} p_{ij} \qquad \text{for all } i \in A,$$

$$p'_{ij} = 0 \qquad \text{otherwise.}$$

Show that if $\{X'_n\}$ is irreducible, it is reversible with stationary probability vector π', where

$$\pi'_j = \pi_j / \sum_{j \in A} \pi_j \qquad \text{for all } j \in A.$$

3-48.* *Stationary measures and reversed chains*. Let $u = (u_0, u_1, \ldots)$ be a stationary measure for an irreducible chain, where we show in Section 3-11 that $u_j > 0$ for all j. Corresponding to joint distributions, joint measures are defined in an obvious way, e.g., the measure for $\{X_n = i, X_{n+1} = j\}$ is $u_i p_{ij}$. Define transition matrix $Q = (q_{ij})$

* This problem is dependent upon material in Section 3-11*.

for a reversed chain analogous to (109),

$$q_{ij} = P(X_{n-1} = j \mid X_n = i) = p_{ji}u_j/u_i \qquad \text{for all } i \text{ and } j,$$

and note that the analog of (112) holds.

(a) Show that u is a stationary measure for the reversed chain.

(b) Show that the reversed chain and the original chain are either both positive, both null, or both transient.

3-49.* *Continuation. Reversible recurrent chains; Kolmogorov's Criterion.* Now suppose the chain is recurrent, so that u is unique up to a multiplicative constant, and the reversed chain is uniquely determined. The reversed chain is called *reversible* if and only if it has the same joint measures (up to a multiplicative constant). It is easy to see that the reversed chain is reversible if and only if $Q = P$, which is equivalent to

$$u_i p_{ij} = u_j p_{ji} \qquad \text{for all } i \text{ and } j.$$

Kolmogorov's Criterion (Theorem 18) also holds. The proof of necessity is the same as that given in Section 3-10 for the positive case. Use (136) and (141) to prove sufficiency.

* This problem is dependent upon material in Section 3-11*.

4

CONTINUOUS-TIME MARKOV CHAINS

In Chapter 3, we treated discrete-time Markov chains, where transitions occur on the integers 1, 2, ... In continuous time, we can think of these chains as processes that jump from state to state, and remain in each state a fixed length of time. A natural generalization is to consider processes that jump from state to state, but remain in each state a *random* length of time. In this chapter, we consider discrete-state Markov processes of this type.

As we will find in Chapters 5 through 7, continuous-time Markov chains are widely used for modeling stochastic phenomena. The goal of this chapter is to present enough of the underlying theory so that you can build and analyze your own Markov chain models with confidence.

The basic theory is presented in Sections 4-1 through 4-6. We treat what are called uniformizable chains in Section 4-7, semi-Markov processes in Section 4-8, and some more technical issues in Section 4-9. The important topic of reversible and quasi-reversible Markov chains is treated in Chapter 6.

In this chapter, "Markov chain" or just "chain" means the continuous-time process.

4-1 PURE-JUMP CONTINUOUS-TIME CHAINS

Let $\{X(t): t \geq 0\}$ be a continuous-time stochastic process with finite or countable state space $\mathcal{S}$; usually $\mathcal{S}$ is $\{0, 1, \ldots\}$, or a subset thereof. We say $\{X(t)\}$ is a *continuous-time Markov chain* if the *transition probabilities* have the following property: For every $t, s \geq 0$ and $j \in \mathcal{S}$,

$$P\{X(s + t) = j \mid X(u); u \leq s\} = P\{X(s + t) = j \mid X(s)\}. \tag{1}$$

It is easy to show that (1) implies the *conditional independence of* $\{X(s); s < t\}$ and $\{X(s); s > t\}$, given $X(t)$, for every $t \geq 0$. This is what we call the *Markov property*.

We assume *stationary* transition probabilities, which means that (1) is independent of s. For every pair of states (i, j), denote these probabilities by

$$P_{ij}(t) = P\{X(s + t) = j \mid X(s) = i\}, \tag{2}$$

where $P_{ij}(t) \geq 0$. From (1), (2), and the fact that $X(t)$ is proper,

$$P_{ij}(t + s) = \sum_k P_{ik}(t)P_{kj}(s), \text{ and} \tag{3}$$

$$\sum_j P_{ij}(t) = 1, \tag{4}$$

for all $t, s \geq 0$, where (3) is known as the *Chapman-Kolmogorov equation*. [As in Chapter 3, when the range of summation for the sums such as in (3) and (4) is not specified, the sum is over all states.] When we use matrix notation for these continuous-time quantities, boldface will denote a matrix, e.g.,

$$\mathbf{P}(t) = (P_{ij}(t)).$$

Unlike discrete-time chains, there are no "one-step" or shortest step transitions. This leads to new mathematical problems that we will briefly explore and partially resolve in Sections 4-1 through 4-3.

For any initial state $X(0) = i$ [or distribution of $X(0)$], joint distributions of $\{X(t)\}$ are readily determined from (3). For later use, we introduce compact notation for the dependence of joint probabilities on initial conditions, where this notation suppresses everything else: For arbitrary integer $n \geq 1$, $s_1, \ldots, s_n$, and $j_1, \ldots, j_n$, let

$$h(i) = P\{X(s_1) = j_1, \ldots, X(s_n) = j_n \mid X(0) = i\}. \tag{5}$$

For any fixed t, the evolution of a Markov chain after t is independent of prior history. We now consider what happens when t is replaced by a random variable. To do this, we define a nonnegative *possibly defective* random variable $\mathcal{T}$ to be a *stopping time* for $\{X(t)\}$ if for every $t \geq 0$, the event $\{\mathcal{T} \leq t\}$ depends only on $\{X(s); s \leq t\}$.

Remark. Compare this definition with how we defined stopping times for the Poisson process and for Wald's equation. In those cases, we relied on independence (of increments or of an i.i.d. sequence), an option not available here. The advantage of these earlier definitions is that they allow stopping times to depend not only on the process, but also on events completely independent of the process; they may be thought of as *randomized* stopping times. Also note that we now permit stopping times to be defective. The reason is convenience; e.g., with this definition, a quantity such as the first passage time from state i to state j turns out to be a stopping time, even when there is a positive probability that state j will never be reached.

We are now ready for an important

Definition. A Markov chain $\{X(t)\}$ is said to be *strong Markov* if for every stopping time $\mathcal{T}$,

$$P\{X(\mathcal{T} + s_1) = j_1, \ldots, X(\mathcal{T} + s_n) = j_n \mid X(s); s \leq \mathcal{T}\} = h[X(\mathcal{T})], \tag{6}$$

for $\{\mathcal{T} < \infty\}$, where h is defined in (5).

Now (6) means that the process after $\mathcal{T}$ has joint distributions that depend only on the state of the chain at $\mathcal{T}$, and these distributions are the same as those of the original chain. It follows that the process after $\mathcal{T}$ is a Markov chain with the same structure as before, and that it is independent of $\mathcal{T}$ and the past history of the chain.

For any stopping time that takes on at most a countable set of possible values, it is easily shown that (6) follows from (3). It follows immediately that *discrete-time Markov chains are strong Markov*. In fact, we used this property many times in Chapter 3 without comment, particularly with regard to first-passage-time distributions.

One might expect (6) to hold for continuous-time chains as well, but it turns out that without some further restriction, this need not be the case. Throughout this book, we impose the following restriction:

Assumption. For every initial state $X(0) = i$, $\{X(t)\}$ is a *pure-jump process*, which means that the process jumps from state to state, and remains in each state a *strictly positive* sojourn time.

To illustrate, let $t_0 = 0$, and successive jumps occur at random epochs $t_1 < t_2 < \cdots$, with *sojourn times* $\tau_n = t_{n+1} - t_n > 0$, $n = 0, 1, \ldots$; see Figure 4-1. By convention, $X(t_n)$ is the new state visited at the nth jump, so that sample paths are continuous from the right. We have ruled out what are called *instantaneous states*, and because of the regularity of sample paths under this assumption, we state without proof

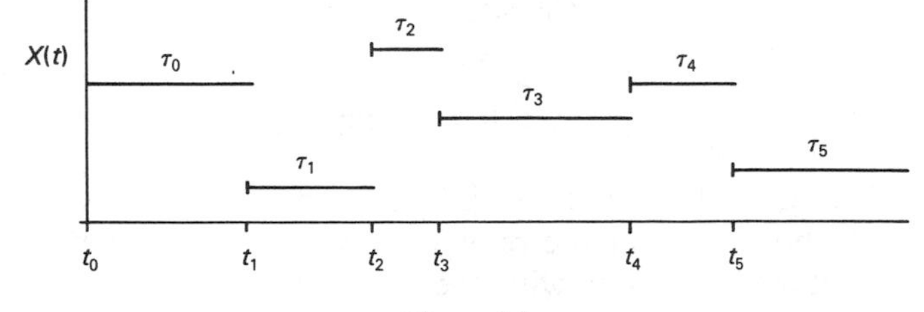

Figure 4-1

Theorem 1. A pure-jump continuous-time Markov chain is strong Markov.

For every $i \in \mathcal{S}$, let $g_i(t) = P(\tau_0 > t \mid X(0) = i)$ for all $t \geq 0$, which is the *tail* distribution of τ_0, given initial state i. For all $s, t > 0$,

$$\begin{aligned} &g_i(s + t) \\ &= P(\tau_0 > s + t \mid X(0) = i) = P(X(u) = i; u \leq s + t \mid X(0) = i) \\ &= P(X(u) = i; u \in (s, s + t] \mid X(u) = i; u \leq s)P(X(u) = i; u \leq s \mid X(0) = i) \\ &= P(X(u) = i; u \in (s, s + t] \mid X(s) = i)P(X(u) = i; u \leq s \mid X(0) = i), \end{aligned}$$

which becomes

$$g_i(s + t) = g_i(s)g_i(t) \qquad \text{for all } s, t > 0. \tag{7}$$

We know (e.g., see Problem 1-29) that (7) is the *defining* property of an exponential distribution, i.e., g_i is of form $e^{-a_i t}$, indicating dependence on initial conditions, where $0 \leq a_i \leq \infty$. The pure-jump assumption rules out $a_i = \infty$. For $a_i = 0$, $P_{ii}(t) = 1$ for all $t \geq 0$, and state i is *absorbing*; in this case, we set $\tau_0 = \infty$. Combining these cases,

$$(\tau_0 \mid X(0) = i) \sim \exp(a_i), \qquad \text{where } 0 \leq a_i < \infty, \quad i \in \mathcal{S}. \tag{8}$$

For $a_i > 0$, the first jump occurs, and for $j \neq i$, we define

$$p_{ij} = P(X(\tau_0) = j \mid X(0) = i) \qquad \text{for all } i, j \in \mathcal{S}, \tag{9}$$

where $\sum_{j \neq i} p_{ij} = 1$, $p_{ii} = 0$. From the Markov property, it follows that

$$P(X(\tau_0) = j \mid X(0) = i, \tau_0 > t) = p_{ij} \qquad \text{for all } t > 0, \tag{10}$$

i.e., given $X(0) = i$, the next state visited is independent of the sojourn time in state i. For $a_i = 0$, define $p_{ii} = 1$.

Now $\tau_0 = t_1, t_2, \ldots$ are stopping times for $\{X(t)\}$, and we define $X_0 = X(0)$, and the sequence of states visited by $X_n = X(t_n)$, $n = 1, 2, \ldots$. From the strong Markov property, the argument for (8) and (10) can be repeated for each t_n, and the conclusion is

Theorem 2. The sequence $\{X_n\}$ is a *discrete-time* Markov chain with transition probability matrix $P = (p_{ij})$ given by (9). Given $\{X_n\}$, $\{\tau_n\}$ is a sequence of *independent exponential* random variables. For every n, the distribution of τ_n depends *only* on X_n;

$$(\tau_n \mid X_n = i) \sim \exp(a_i) \qquad \text{for } i \in \mathcal{S}.$$

The discrete-time chain $\{X_n\}$ is said to be *embedded* in $\{X(t)\}$. We could quibble about whether the remainder of $\{X_n\}$ is well defined after the chain hits an absorbing state, but with the convention $p_{ii} = 1$ for absorbing states, the behavior of $\{X_n\}$ is consistent with $\{X(t)\}$.

It will be convenient to have abbreviations for each of these chains: Denote the continuous-time chain $\{X(t)\}$ by *CTMC* and the corresponding embedded chain $\{X_n\}$ by *EMC*.

For every initial state $X(0) = i$, t_1 is exponential, and it is easy to show that

$$\lim_{t \to 0} P(t_2 \le t \mid X(0) = i)/t = 0, \tag{11}$$

which is the *orderliness* property first introduced in Section 2-6.

A function $g(t)$ with the property that $g(t)/t \to 0$, as $t \to 0$, is said to be $o(t)$, and this is denoted by setting

$$g(t) = o(t).$$

Thus $P(t_2 \le t \mid X(0) = i) = o(t)$ for every i. This compact way of expressing information about limits will be useful in Section 4-6.

From Theorem 2 and (11), the following limits are easy to establish for every nonabsorbing state i:

$$\lim_{t \to 0} [1 - P_{ii}(t)]/t = a_i \equiv -a_{ii},\ 0 < a_i < \infty, \text{ and for every } j \ne i, \tag{12}$$

$$\lim_{t \to 0} P_{ij}(t)/t = a_i p_{ij} \equiv a_{ij} \ge 0, \tag{13}$$

where $\sum_{j \ne i} a_{ij} = -a_{ii}$. We can think of a_i as an "infinitesimal" rate out of state i, and a_{ij} as the corresponding rate from i to j. This terminology is unfortunate because infinitesimal does not mean the a_i are small, but rather that they are (approximate) rates over short time intervals. For this reason, we will call the a_i and a_{ij}, $j \ne i$, *conditional* rates, meaning that they are rates conditioned on the chain being in state i.

When all the states are nonabsorbing, we define a *conditional-rate matrix* $\mathbf{A} = (a_{ij})$. Notice that the row sums of $\mathbf{A}$ all equal 0.

For any $i \ne j$, we say state i *can reach* state j in the CTMC if for some $t \ge 0$, $P_{ij}(t) > 0$. Clearly, i can reach j in the CTMC if and only if there is a path of positive probability from i to j in the EMC, i.e., i can reach j in the EMC. Define communication classes and the terms open, closed, and irreducible as we did in Chapter 3. We have

Theorem 3. The CTMC and the EMC have the same communication classes. In particular, the CTMC is irreducible if and only if the EMC is irreducible.

Remarks. Communication classes are determined by the location of the positive elements in the transition probability matrix of the EMC, or equivalently, by the location of the positive elements in the matrix $\mathbf{A}$. Because sojourn times can be arbitrarily small, it is easy to see that if i can reach j, $P_{ij}(t) > 0$ for *all* $t > 0$.

One drawback to our approach is that for nonabsorbing states i, $p_{ii} = 0$. We defined a jump as an actual change of state. We did not allow jumps from i to i. The primary reason for this is so that limits (11) and (12) are unambiguous. As illustrated in the next section, we will sometimes need to relax this restriction. There are important theoretical reasons for permitting $i \to i$ transitions as well. It turns out that we will be able to "bend this rule" when it is convenient to do so.

4-2 SOME EXAMPLES

When continuous-time chains are used in modeling, it is rare that we start with transition probabilities such as (2). Instead, the chain usually is the result of the interaction of more elementary processes. This information is then used to determine **A** or equivalent quantities in Theorem 2.

Example 4-1: The Poisson Process

Let $\{\Lambda(t)\}$ be a Poisson process at rate λ. Independent increments gives us (1). That and stationary increments give us (2). In this case, the definition of a Poisson process gives us (2) explicitly for $i = 0$ and $s = 0$, and with little effort, for other values. It is also easy to identify the quantities in Theorem 2: $a_i = \lambda$ and $p_{i,i+1} = 1$ for all i.

Example 4-2: The M/M/1 Queue

The arrival of customers at a service facility (called the *system*) is a Poisson process at rate λ. A single server serves one customer at a time; service times are i.i.d. $\sim$ exp (μ). On completion of service, customers depart from the system. Let $N(t)$ be the number of customers in system at epoch $t \geq 0$. If $N(t) = 0$, the next jump is an arrival, a time $\sim$ exp (λ). If $N(t) = i > 0$, the *remaining* service of the customer in service (from t on) is $\sim$ exp (μ), and the time until the next jump, the minimum of two independent exponential random variables, is $\sim$ exp ($\lambda + \mu$). The probability of an arrival before service completion is $p_{i,i+1} = \lambda/(\lambda+\mu)$. It follows that $\{N(t)\}$ is an irreducible Markov chain with transition rates $a_0 = \lambda$, $a_i = \lambda + \mu$ for $i \geq 1$, $a_{i,i+1} = \lambda$ for $i \geq 0$, $a_{i,i-1} = \mu$ for $i \geq 1$, and $a_{ij} = 0$ otherwise.

Suppose we put an upper limit on the number of customers in system, e.g., $N(t) \leq l$ for all t and some integer l. This can be done by defining $a_l = a_{l,l-1} = \mu$. On the other hand, suppose that arrivals occur when the number of customers in system reaches l, but these customers depart immediately and are lost. This suggests that we let the transition rate out of state l be $a_l' = \lambda + \mu$, where we have $l \to l$ transition rate $a_{ll}' = \lambda$.

Example 4-3: The M/M/∞ Queue

This is like the preceding example except that now we have an infinite number of servers, each with service rate μ. Given $N(0) = i$, each of the i remaining service times is $\sim$ exp (μ), $\tau_0 \sim$ exp ($\lambda + i\mu$), and $p_{i,i+1} = \lambda/(\lambda + i\mu)$. It follows that for all i, $a_i = \lambda + i\mu$, $a_{i,i+1} = \lambda$, and $a_{i,i-1} = i\mu$.

Example 4-4: The Pure-Birth Process

This is like a Poisson process except that $a_i = a_{i,i+1} \equiv \lambda_i$, i.e., the sojourn times depend on the state of the process. The term "pure birth" comes from the study of biological populations, where λ_i may increase in i. In fact, one can construct examples where λ_i increases so rapidly that

$$P\left(\sum_{n=0}^{\infty} \tau_n < \infty \mid X(0) = i\right) > 0,$$

i.e., with positive probability, there is an infinite number of jumps in finite time, which is called an *explosion*.

4-3 REGULAR CHAINS

In Example 4-4, we have identified another technical problem—the possibility of an explosion. This problem is interesting mathematically (what happens after an explosion?) and is of some practical interest as well, e.g., branching process models of a nuclear explosion. For the applications we have in mind, however, explosions will not occur.

Definition. A pure-jump Markov chain is called *regular* if for every initial state $X(0) = i$, the number of transitions in finite time is finite with probability 1.

To explore the possibility of an explosion further, let

$$S(i) = \left(\sum_{n=0}^{\infty} \tau_n \mid X(0) = i\right) \quad \text{and} \quad y_i = E(e^{-S(i)}) \qquad \text{for every } i,$$

where $S(i)$ is a defective random variable, $y_i = 0$ is equivalent to $P(S(i) = \infty) = 1$, and $y_i \in (0, 1)$ means that $P(S(i) < \infty) > 0$. Thus a chain is regular if and only if $y_i = 0$ for all i. Using the notation in (12), which assumes no $i \to i$-type transitions, we are now ready to state

Theorem 4. A pure-jump Markov chain is regular if and only if the only bounded solution to

$$y_i = \sum_j a_{ij} y_j, \qquad i = 0, 1, \ldots, \text{ is} \tag{14}$$

$\{y_i = 0 \text{ for all } i\}$.

Theorem 4, which we prove in Section 4-9, is used primarily in the development of the theory (e.g., in Section 6-7); it can be difficult to apply to particular models. We give some simple sufficient conditions for regularity at the end of this section. Fortunately, these conditions will cover virtually all the Markov chain models of queues in this book.

When a chain is regular, the occurrence of an event $\{X(t) = j \mid X(0) = i\}$ is completely determined by the countable sequence of jumps and sojourn times in Theorem 2. Starting with this sequence and some specified matrix $\mathbf{A}$, sample paths $X(t, \omega)$ and transition matrix $\mathbf{P}(t)$ with property (4) are determined for every $t \geq 0$. Thus the renewal theory methods used on discrete-time chains in Chapter 3 apply to regular CTMCs as well. We also use the same terminology.

For all states i and j in a CTMC, define $\mathcal{T}_{ij}$ as the first passage time (epoch of the first visit) to state j, given $X(0) = i$. (For $i = j$, the chain must first leave j.) By this, we mean that that $\mathcal{T}_{ij}$ is the sum of T_{ij} sojourn times, where T_{ij} is the corresponding first passage time (*number of jumps*) in the EMC. If T_{ij} is defective, define $\mathcal{T}_{ij} = T_{ij} = \infty$ on the set of points $\{\omega\}$ where T_{ij} does not occur.

It follows that $\mathcal{T}_{ij}$ is well defined *whether or not* the CTMC is regular, and $\mathcal{T}_{ij}$ is proper if and only if T_{ij} is proper. We say state i *can reach* state j if $P(\mathcal{T}_{ij} < \infty) > 0$. Now $P(\mathcal{T}_{ij} < \infty) = P(T_{ij} < \infty)$, and we again have that the CTMC and the EMC have the same communication classes, and the same communication structure between classes, e.g., $i \rightarrow j$, but $j \nrightarrow i$.

State j in the CTMC is called *transient* if $\mathcal{T}_{jj}$ is defective, and *recurrent* if $\mathcal{T}_{jj}$ is proper; if recurrent, state j is *positive* if $E(\mathcal{T}_{jj}) < \infty$, and *null* if $E(\mathcal{T}_{jj}) = \infty$. Clearly, state j is recurrent in the CTMC if and only if it is recurrent in the EMC. However, it is possible for state j to be positive in the CTMC and null in the EMC, and also the converse.

Finally, it is easily shown that for a CTMC, the states in a communication class are either (i) all transient, (ii) all positive, or (iii) all null.

For a regular CTMC and every i and j, we define a j-renewal process of visits to state j, given $X(0) = i$. Let $\mathcal{T}_{ij} \sim F_{ij}$, where in the terminology of Chapter 2, F_{ij} is the distribution of time until the first renewal, and F_{jj} is the distribution of time between subsequent renewals.

By conditioning on $\mathcal{T}_{jj}$ and τ_0, it is easy to find a renewal equation for $P_{jj}(t)$,

$$P_{jj}(t) = e^{-a_j t} + \int_0^t P_{jj}(t - u)\, dF_{jj}(u), \tag{15}$$

which has solution

$$P_{jj}(t) = e^{-a_j t} + \int_0^t e^{-a_j(t-u)}\, dm_{jj}(u) \qquad \text{for all } j, \tag{16}$$

where $m_{jj}(t) = \sum_{n=1}^{\infty} F_{jj}^{(n)}(t)$. For $i \neq j$, we have the general renewal process case, and by the method in Section 2-16, it is easy to show that

$$P_{ij}(t) = \int_0^t e^{-a_j(t-u)}\, dm_{ij}(u) \qquad \text{for } i \neq j. \tag{17}$$

where $m_{ij}(t)$ is the corresponding general renewal function.

From (16), (17), and the Key Renewal Theorem, it follows immediately that pointwise limits exist and have the following form:

$$\lim_{t\to\infty} P_{ij}(t) = 0 \qquad \text{if } j \text{ is transient or null,} \tag{18}$$

and, if j is positive and $\mathcal{T}_{ij}$ is proper,

$$\lim_{t\to\infty} P_{ij}(t) = 1/a_j E(\mathcal{T}_{jj}) \equiv p_j > 0, \tag{19}$$

where p_j is also the *fraction of time* the chain spends in state j.

Remark. It is not necessary to use pointwise limits here. The time-average versions of (18) and (19) may be obtained from the Renewal-Reward Theorem. Equation (20) below may also be deduced in this manner.

If the CTMC is irreducible and positive, the $\mathcal{T}_{ij}$ are proper, and (19) will hold for all i and j. In addition,

$$\sum_j p_j = 1. \tag{20}$$

In this case, certain quantities have important frequency interpretations:

$$a_j p_j = 1/E(\mathcal{T}_{jj}), \tag{21}$$

which is the *transition rate* of visits to state j, i.e., the number of transitions either into or out of state j per unit time (rate in = rate out). Every time there is a visit to state j, there is an opportunity to visit some state k next, where p_{jk} is the probability of such a visit. In the long run, p_{jk} is the fraction of visits to state j for which state k is visited next. It follows that the transition rate of type $j \to k$ is

$$p_j a_j p_{jk} = p_j a_{jk}. \tag{22}$$

Equating rates into and out of state k, it follows that

$$\overset{\text{out}}{a_k p_k} = \overset{\text{in}}{\sum_{j\neq k} a_{jk} p_j} \qquad \text{for all } k \in \mathcal{S}. \tag{23}$$

Equations (23) are called *balance equations* because they balance (equate) transition rates into and out of every state.

For an irreducible regular CTMC $\{X(t)\}$, suppose probability distribution $\{p_j\}$ satisfies balance equations (23). We show in Section 4-9 that $\{p_j\}$ also satisfies

$$\sum_i p_i P_{ij}(t) = p_j \qquad \text{for every } j \in \mathcal{S} \text{ and } t > 0, \tag{24}$$

i.e., $[X(t) \mid X(0) \sim \{p_j\}] \sim \{p_j\}$, which means that with these initial conditions, $\{X(t)\}$ is a *stationary* stochastic process. Now $p_j > 0$ for some j, and (24) implies that state j (and hence every state) is positive. Clearly, if $\{p_j\}$ satisfies {24}, it also satisfies (23), and (23) has a unique solution. We have shown

Theorem 5. An irreducible regular Markov chain is *positive* if and only if there is a *unique* probability distribution $\{p_j\}$ that satisfies the balance equations (23). $\{p_j\}$ satisfies (23) if and only if it satisfies (24). This solution has the properties $p_j > 0$ for all j, and $\sum_j p_j = 1$. If the chain is either transient or null, there is no probability distribution that satisfies the balance equations.

Because of (24), we call the p_j *steady-state* probabilities and $\{p_j\}$ either the *stationary* or *steady-state* distribution. Clearly, no stationary distribution exists in either the transient or null cases.

Remark. When the chain is explosive, a probability distribution that satisfies (23) need not satisfy (24). In fact, $\{P_{ij}(t)\}$ is not uniquely determined in this case. If we restrict ourselves to the *minimal* solution $\{P_{ij}^{(\infty)}(t)\}$ (see Section 4-9), (24) will not hold.

As will be shown, finite-state chains are regular, and it is easily shown that *irreducible finite-state* chains are positive.

As with discrete-time chains, or for that matter general stochastic processes, we can arbitrarily partition state space $\mathscr{S}$ of a positive irreducible Markov chain into two subsets, E and E^c, and equate transition rates from one subset to the other:

$$\underset{\text{rate of } E \to E^c \text{ transitions}}{\sum_{i \in E} \sum_{j \in E^c} p_i a_{ij}} = \underset{\text{rate of } E^c \to E \text{ transitions}}{\sum_{i \in E^c} \sum_{j \in E} p_i a_{ij}}. \tag{25}$$

By choosing E appropriately, we can sometimes use (25) to find $\{p_j\}$ quickly. For the $M/M/1$ queue in Example 4-2, we can partition $\mathscr{S}$ into $E = \{0, \ldots, j\}$ and $E^c = \{j + 1, j + 2, \ldots\}$ for any state j, where in Figure 4-2, states are represented as nodes, and possible transitions as arcs. The boundary between E and E^c is represented by a vertical dashed line. The only transitions across this boundary are of type $j \to (j + 1)$ and $(j + 1) \to j$.

It follows that a solution to (23) for this model must also satisfy

$$\lambda p_j = \mu p_{j+1}, j = 0, 1, \ldots, \text{ which is equivalent to} \tag{26}$$

$$p_j = \rho^j p_0, \quad j = 0, 1, \ldots, \tag{27}$$

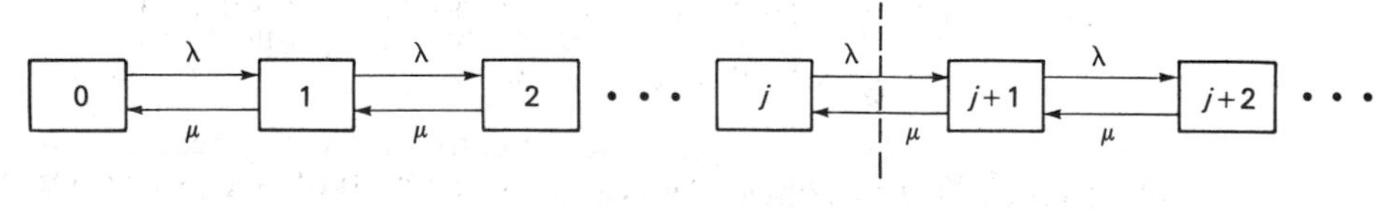

Figure 4-2

where $\rho = \lambda/\mu$. (The converse is also true.) A solution will exist provided we can normalize (27). Setting

$$\sum_j p_j = p_0 \sum_j \rho^j = 1,$$

we have that all states are positive if and only if $\rho < 1$, in which case,

$$p_j = (1 - \rho)\rho^j, \qquad j = 0, 1, \ldots. \tag{28}$$

Sometimes it is convenient to permit $k \to k$ transitions at some specified rate a'_{kk} for each state k. [Given $X(0) = k$, we can introduce a random variable $Y_0 \sim \exp(a'_{kk})$, and define the time until the first transition to be $\tau'_0 = \min(\tau_0, Y_0)$.] We have introduced transitions that are not real jumps because the state doesn't change. In fact, between each real jump, there is a geometric number of these transitions. A chain that is regular in terms of real jumps will remain regular when these additional transitions are counted.

A transition of type $k \to k$ is both out of and into state k, i.e., *both* rates increase by $a'_{kk}p_k$. For an irreducible chain, balance equations may be altered to include these transitions:

$$(a_k + a'_{kk})p_k = \sum_{j \neq k} a_{jk}p_k + a'_{kk}p_k \quad \text{for all } k \in \mathcal{S}. \tag{29}$$

Clearly, adding a term to both sides of one or more balance equations has no effect on their solution, and *Theorem 5 still applies*.

Now suppose that a Poisson process at rate λ is a component of a model that generates a positive irreducible Markov chain, as was the case in Examples 4-2 and 4-3. For each state j, let $a_j = \lambda + (a_j - \lambda)$, where $a_j - \lambda$ is the transition rate generated by "everything else." Every time the chain visits state j, there is an opportunity for a Poisson "event" to "find" the chain there. (When these events are arrivals, as in Examples 4-2 and 4-3, we say that an arrival observes or "sees" the chain in state j.) For example, given $X(0) = j$, let $Y_0 \sim \exp(\lambda)$ be the time until the next Poisson event, and $Z_0 \sim \exp(a_j - \lambda)$ be the time until something else happens, where $\tau_0 = \min(Y_0, Z_0)$. The Poisson event will find the chain in state j with probability $P(Y_0 < Z_0) = \lambda/a_j$.

Now λ/a_j is also the long-run *fraction* of visits to state j for which a Poisson event finds the chain in that state. Since state j is visited at rate a_jp_j, the rate at which Poisson events find the chain in state j is

$$a_jp_j \cdot \lambda/a_j = \lambda p_j.$$

Poisson events occur at rate λ, and it follows that the fraction of Poisson events that *find* the chain in state j is

$$\lambda p_j/\lambda = p_j \qquad \text{for all } j \in \mathcal{S}, \tag{30}$$

which is the fraction of *time* the chain is in state j.

Important Remark. Property (30) is called "Poisson Arrivals See Time Averages," or *PASTA*. Applications of this result are often to Poisson arrivals, but in general, the Poisson event process can play any role. This property holds for general stochastic processes, Markov or not, where state j is replaced by an arbitrary set of states B that is visited a fraction of time $p(B)$. The independent-increments property turns out to be of crucial importance; arrival processes lacking this property generally will not have the PASTA property (30). See Sections 5-16 and 5-17 for a treatment of general stochastic processes and further discussion.

We will make frequent use of the PASTA property in later chapters. For an elementary application, consider the $M/M/1$ queue in Example 4-2 with upper limit l on the number of customers in system. Equation (26) still holds for $j = 0, \ldots, l-1$, and normalization now gives

$$p_j = (1-\rho)\rho^j/(1-\rho^{(l+1)}), \qquad j = 0, \ldots, l. \tag{31}$$

Being finite state, this chain is positive for any $\rho > 0$. From PASTA, the fraction of arrivals lost is p_l in (31). For this result, we need to define $l \to l$ transition rate $a'_{ll} = \lambda$.

We will also associate costs and rewards with positive irreducible chains. Typically, rewards will accumulate in either or both of two ways:

(i) There is a reward rate function $f(j)$ of the state of the chain such that the cumulative reward during the interval $[0, t]$ is

$$C(t) = \int_0^t f[X(u)]\,du.$$

Provided that $\sum_j |f(j)|\,p_j < \infty$, the reward rate is

$$\lim_{t\to\infty} C(t)/t = \sum_j f(j)p_j. \tag{32}$$

(ii) Rewards depend only on the sequence of states visited, not the sojourn times, where $r(j)$ is the reward received for each visit to state j. The cumulative reward on $[0, t]$ now is

$$C(t) = \sum_{\{n: t_n \le t\}} f[X(t_n)].$$

Provided that $\sum_j |r(j)|\,a_j p_j < \infty$, the reward rate now is

$$\lim_{t\to\infty} C(t)/t = \sum_j r(j)a_j p_j. \tag{33}$$

We now give some elementary conditions, each of which ensures that a Markov chain is regular. We will return to this topic later in the chapter.

Condition 1. A Markov chain is regular if there is a finite upper bound on a_i for all $i \in \mathcal{S}$.

To see that Condition 1 is sufficient, let $a_i \leq \nu < \infty$ for all i and define a Poisson "arrival" process at rate ν. Let $\{X(t)\}$ be a pure-jump process, where jumps are permitted to occur only at the arrival epochs of the Poisson process $t_1' < t_2' < \cdots$. If $X(t_n') = i$, a jump occurs at epoch t_{n+1}' with probability a_i/ν. Otherwise, an $i \to i$ transition occurs. On the subsequence $\{t_n\}$ where actual jumps occur, let $\{X(t_n)\}$ be a discrete-time Markov chain with transition probability matrix P. It is easy to see that $\{X(t)\}$ is a continuous-time Markov chain with fewer jumps than a Poisson process, and therefore is regular. The Poisson process here is said to *uniformize* the transition rates of the chain; see Section 4-7.

Remark. From Condition 1, *finite-state chains* are regular.

Condition 1 is too strong to cover our needs. In Example 4-3, where $\{a_i\}$ is unbounded, notice that the arrival and departure epochs of every customer are jumps in the chain; at most two jumps are associated with any customer in any interval. Because the arrival process is regular, so is the chain.

More generally, each arrival in the queueing network models of Chapter 6 induces a finite (with probability 1) number of jumps in the chain. For these models, we state

Condition 2. A queueing network that has a regular arrival process is regular.

If a chain begins in a recurrent state, it will return over and over, with i.i.d. times between returns, and a finite number of jumps between returns. The number of returns in a finite interval is finite w.p.1, and we have

Condition 3. A Markov chain is regular if for every initial state $X(0) = i$, it will (with probability 1) visit a recurrent state in a finite number of jumps. In particular, a recurrent irreducible chain is regular.

An irreducible CTMC is recurrent if and only if the corresponding EMC is recurrent. This idea, together with balance equations (23), can be used to determine conditions for a chain to be positive. See Section 4-4.

We close this section with the observation that all of the methods and most of the results for the transient behavior of discrete-time chains in Section 3-7 apply to regular continuous-time chains.

4-4 THE BIRTH AND DEATH PROCESS

The birth and death process is a Markov chain with state space $\{0, 1, \ldots\}$, where all jumps are between adjacent states. Thus we have $a_{j,j+1} \equiv \lambda_j > 0$, $a_{j+1,j} \equiv \mu_{j+1} > 0$, $\mu_0 \equiv 0$, and $a_j = \lambda_j + \mu_j$, for $j = 0, 1, \ldots$. This is Figure 4-2 except that now we have state-dependent subscripts on λ and μ. In fact, Examples 4-2 and 4-3 are special cases of a birth and death process. Sometimes we put an upper limit on the state space, l say, $0 < l < \infty$, by setting $\lambda_l = 0$, but in this section, we deal with the infinite state case. Birth and death processes are irreducible.

As with the pure-birth process, this terminology arises from the study of biological populations, where births occur at "up" jumps, and deaths occur at "down" jumps. If λ_j increases in j (not necessarily monotone), the possibility of an explosion looms again.

Under what conditions is a birth and death process positive? To find out, we first try to solve the balance equations. From (25) and Figure 4-2,

$$\lambda_j p_j = \mu_{j+1} p_{j+1}, \text{ and} \tag{34}$$

$$p_j = b_j p_0, \text{ where} \tag{35}$$

$$b_0 \equiv 1, \text{ and } b_j = \frac{\lambda_0 \cdots \lambda_{j-1}}{\mu_1 \cdots \mu_j}, \qquad j = 1, 2, \ldots. \tag{36}$$

Thus a solution to the balance equations exists if we can normalize (35).

Simply solving the balance equations is not enough, however, because we don't know whether the process is regular. An irreducible CTMC is regular if it and the corresponding EMC are recurrent.

The transition probability matrix for the EMC is

$$P = \begin{pmatrix} 0 & 1 & 0 & & \cdots \\ d_1 & 0 & u_1 & & \cdots \\ 0 & d_2 & 0 & u_2 & \cdots \\ \vdots & & \ddots & & \ddots \end{pmatrix},$$

where

$$u_j = \lambda_j/(\lambda_j + \mu_j) \text{ and } d_j = \mu_j/(\lambda_j + \mu_j), \qquad j = 1, 2, \ldots. \tag{37}$$

We apply Theorem 11 in Chapter 3, which for this chain becomes: The chain is transient if and only if there is a nonzero bounded solution $\{y_j\}$ to

$$y_1 = u_1 y_2, \tag{38}$$

$$y_j = u_j y_{j+1} + d_j y_{j-1}, \qquad j = 2, 3, \ldots, \tag{39}$$

where (39) can be written $d_j(y_j - y_{j-1}) = u_j(y_{j+1} - y_j)$, and hence

$$y_{j+1} - y_j = \left(\frac{d_2 \cdots d_j}{u_2 \cdots u_j}\right)(y_2 - y_1), \qquad j = 2, 3, \ldots. \tag{40}$$

For $y_1 > 0$, the y_j are increasing in j, and there is a bounded solution if and only if the sum on j of (40) converges, and conversely, the chain is *recurrent* if and only if this sum *diverges*.

From (37), we can substitute for d_i and u_i in (40) and cancel $\lambda_i + \mu_i$, $i = 2$,

$\ldots, j$. The divergence condition is usually stated in terms of the b_j. To this end, multiply each term in (40) by the constant $\mu_1/(\lambda_0\lambda_1)$. By combining the divergence and normalization conditions, we have

Theorem 6. A birth and death process is *recurrent* if and only if

$$\sum_j (\lambda_j b_j)^{-1} = \infty. \tag{41}$$

A birth and death process is *positive* if and only if we also have

$$\sum_j b_j < \infty, \tag{42}$$

in which case

$$p_0 = \left(\sum_{j=0}^{\infty} b_j\right)^{-1} \quad \text{and} \quad p_j = b_j p_0, \qquad j = 0, 1, \ldots, \tag{43}$$

where the p_j are well defined as limits (19) and as fractions of time.

It is easy to verify that (41) and (42) are satisfied for Example 4-3, and, when $\rho = \lambda/\mu < 1$, for the unbounded $M/M/1$ queue in Example 4-2. However, because we already know these chains are regular, it is necessary only to verify (42). When $\rho = 1$ for the $M/M/1$ queue, (41) is satisfied, but (42) is not. This, of course, is the null case! When $\rho > 1$, the $M/M/1$ queue is transient but still regular.

From Condition 2, a birth and death process is regular if for some λ, $\lambda_j < \lambda < \infty$ for all j. (We can uniformize the up jumps only with a Poisson process at rate λ.) In this case, it is easy to see that (42) implies (41).

4-5 TIME AND TRANSITION AVERAGES FOR RECURRENT IRREDUCIBLE CHAINS

When both the CTMC and the EMC are positive irreducible chains, $\{p_j\}$ is the unique solution to (23) and $\{\pi_j\}$ is the unique solution $\pi = \pi P$, where P is the transition probability matrix for the embedded chain. Both $\{p_j\}$ and $\{\pi_j\}$ are probability distributions that have frequency interpretations: p_j is the fraction of time that the CTMC spends in state j, and π_j is the fraction of transitions that are visits to state j. How are these distributions related?

We also know that $a_j p_j$ is the rate that state j is visited, which implies that

$$\pi_j \propto a_j p_j. \tag{44}$$

From normalization, we have

$$\pi_j = a_j p_j / [\sum_j a_j p_j] \quad \text{and} \quad p_j = (\pi_j/a_j)/[\sum_j \pi_j/a_j]. \tag{45}$$

An interesting interpretation of the right-hand equation in (45) is that the p_j are *length biased* π_j, where length is the mean time in state.

Remark. If you wish to estimate $\{p_j\}$ from a simulation, this can be done by simulating the EMC, estimating $\{\pi_j\}$, and then using (45). This avoids having to generate the exponential sojourn times.

In elementary models such as in Examples 4-2 and 4-3, conditions for the CTMC and the EMC to be positive are the same. However, this need not be so. For example, consider a birth and death process where $\lambda_j = \mu_j$ for all $j \geq 1$, so that $u_j = d_j = 1/2$ for all j, and the EMC is null. If we also have $\lambda_j/\mu_{j+1} = r < 1$ for all j, (42) is satisfied, and the CTMC is positive. Examples for the converse case are also easy to construct.

Stationary Measures*

If the CTMC is positive and the EMC is null, $a_j p_j > 0$ is still the rate that the chain visits state j. The expected number of visits to state j between returns to state i,

$$E(V_{ij}) = \pi_j/\pi_i = a_j p_j/a_i p_i \qquad \text{for all } i \text{ and } j, \tag{46}$$

has physical meaning as a ratio of rates. In this case, $\{\pi_j\}$ satisfies $\pi = \pi P$, but $\sum_j \pi_j = \infty$. In Section 3.11 we called $\{\pi_j\}$ a stationary measure and showed that in the recurrent case, this measure is unique up to a multiplicative constant.

For a recurrent CTMC, let $C_{ij}(t)$ be the cumulative amount of time during $[0, t]$ that the chain spends in state j, given $X(0) = i$. As an application of the strong law of large numbers, it is easy to show that

$$\lim_{t\to\infty} C_{ij}(t)/C_{ii}(t) = a_i E(V_{ij})/a_j \equiv p_j/p_i \qquad \text{for all } i \text{ and } j, \tag{47}$$

i.e., the ratios p_j/p_i have meaning as relative amounts of time, even when the CTMC is null. From (46), we still have $\{p_j\} = \{\pi_j/a_j\}$.

If π is a stationary measure for the EMC, it is easy to see that $\{p_j\} = \{\pi_j/a_j\}$ satisfies balance equations (23) [or (29)], and conversely. For irreducible chains, $\pi_j > 0$ for all j; hence, $p_j > 0$ for all j. π is unique up to a multiplicative constant if and only if $\{p_j\}$ is. If the CTMC is positive, $\{p_j\}$ is stationary in the sense of (24). We now show that this is true in the null case as well.

Suppose $\{X(t)\}$ is regular, and define a discrete-time Markov chain $\{Y_n\}$ as follows: For arbitrary *fixed* $\delta > 0$, let

$$Y_n = X(n\delta), \qquad n = 0, 1, \ldots,$$

where $\{Y_n\}$ is called the *discrete skeleton* of $\{X(t)\}$, and has one-step transition

* This is the continuous-time analog of Section 3-11*, an advanced topic.

probability matrix $\mathbf{P}(\delta)$. It is immediate that $\{Y_n\}$ is irreducible if and only if $\{X(t)\}$ is. It is easily shown that (irreducible) $\{Y_n\}$ is recurrent if and only if $\{X(t)\}$ is. Furthermore, in contrast with the EMC, it is easily shown (as an application of the Renewal-Reward Theorem) that $\{Y_n\}$ is *positive* if and only if $\{X(t)\}$ is.

Now suppose that $\{Y_n\}$ is recurrent with stationary measure $\{p_j'\}$. Thus

$$[X(n\delta) \mid X(0) \sim \{p_j'\}] \sim \{p_j'\}, \qquad n \geq 1,$$

i.e., $\{X(t)\}$ is stationary is least for those t that are multiples of δ. By subdividing δ, this is easily extended to t of form $m\delta/n$ (the rationals). We now extend this to all t:

$$\sum_i p_i' P_{ij}(t) = p_j' \qquad \text{for every } j \in \mathcal{S} \text{ and } t > 0. \tag{48}$$

Let $\delta = 1$, and $t_n \to t$, where the t_n are rational and t is irrational. The $P_{ij}(t)$ are continuous, and we have that for every n,

$$\sum_{i=0}^{n} p_i' P_{ij}(t_n) \to \sum_{i=0}^{n} p_i' P_{ij}(t) \leq p_j', \text{ and as } n \to \infty, \text{ we get}$$

$$\sum_i p_i' P_{ij}(t) \leq p_j' \qquad \text{for every } j.$$

Now apply the Chapman-Kolmogorov equation (3) for *rational* $t + s$:

$$p_j' = \sum_i p_i' P_{ij}(t + s) = \sum_k \sum_i p_i' P_{ik}(t) P_{kj}(s).$$

If the preceding inequality is strict for some j, we have a contradiction, so that (48) holds. Notice that $\{p_j'\}$ is independent of the choice of δ.

For chain $\{Y_n\}$, p_j'/p_i' is the number of visits to j per visit to i. For fixed i and j, let $\delta \to 0$, and we see that $p_j'/p_i' = p_j/p_i$, as defined in (47), so that up to a multiplicative constant, $\{p_j'\} = \{p_j\}$, and

$$\sum_i p_i P_{ij}(t) = p_j \qquad \text{for every } j \in \mathcal{S} \text{ and } t > 0. \tag{49}$$

We have shown

Theorem 7. For a recurrent irreducible CTMC, there exists a stationary measure $\{p_j\}$ that satisfies (23) and (49), where $p_j > 0$ for all j, and $\{p_j\}$ is unique up to a multiplicative constant. For all i and j, the ratios p_j/p_i are determined by (46) and have interpretation (47).

If we find a solution to (23) with the property $\sum_j^{\infty} p_j = \infty$, we know from Theorem 7 that the CTMC is *not* positive. Notice that when $\{p_j\}$ is a probability distribution, we get the conclusions of Theorem 5 without having to assume or verify separately that the CTMC is recurrent.

For example, $\{p_j = \rho^j, j \geq 0\}$ is a stationary measure for an $M/M/1$ queue for every $\rho > 0$, and normalization determines that $\rho < 1$ is the positive case. From the analysis of the EMC in Section 4-4, $\rho = 1$ is the null case.

4-6 THE BACKWARD AND FORWARD EQUATIONS

We now derive two systems of linear differential equations for the transition probabilities $P_{ij}(t)$. This is done by breaking the interval $(0, t + h]$ into subintervals of length h and t, and then letting $h \to 0$. When h is the left-hand interval, which is "backward" in time, we will get what are called the *backward equations*. When h is the right-hand "forward" interval, we will get the *forward equations*. We exclude $i \to i$-type transitions.

First break $(0, t + h]$ into $(0, h]$ and $(h, t + h]$. From limits (12) and (13), which in turn depended on orderliness property (11), we can write

$$P_{ii}(h) = 1 - a_i h + o(h) \qquad \text{for all } i, \text{ and}$$

$$P_{ij}(h) = a_{ij} h + o(h) \qquad \text{for all } i \neq j,$$

where the notation $o(h)$ is defined in Section 4-1. Recalling the Chapman-Kolmogorov equation (3), we have for all i and j that

$$P_{ij}(t + h) = \sum_k P_{ik}(h) P_{kj}(t) = (1 - a_i h) P_{ij}(t) + \sum_{k \neq i} a_{ik} h P_{kj}(t) + o(h), \tag{50}$$

$$[P_{ij}(t + h) - P_{ij}(t)]/h = -a_i P_{ij}(t) + \sum_{k \neq i} a_{ik} P_{kj}(t) + o(h)/h. \tag{51}$$

Letting $h \to 0$ in (51), we have the *backward equations*:

$$P'_{ij}(t) = -a_i P_{ij}(t) + \sum_{k \neq i} a_{ik} P_{kj}(t) \qquad \text{for all } i, j \in \mathscr{S}. \tag{52}$$

Notice that in (50), we combined $o(h)$ terms into a single term on the right. In fact, what we are actually doing is interchanging limit and sum, and taking the limit term by term inside the summation. For any initial state i, the orderliness property (11) can be used to justify this interchange.

Now break $(0, t + h]$ into $(0, t]$ and $(t, t + h]$, and write for all i and j that

$$P_{ij}(t + h) = \sum_k P_{ik}(t) P_{kj}(h) = P_{ij}(t)(1 - a_j h) + \sum_{k \neq j} P_{ik}(t) a_{kj} h + o(h), \tag{53}$$

$$[P_{ij}(t + h) - P_{ij}(t)]/h = - a_j P_{ij}(t) + \sum_{k \neq j} a_{kj} P_{ik}(t) + o(h)/h. \tag{54}$$

Letting $h \to 0$ in (54), we have the *forward equations*:

$$P'_{ij}(t) = -a_j P_{ij}(t) + \sum_{k \neq j} a_{kj} P_{ik}(t) \qquad \text{for all } i, j \in \mathscr{S}. \tag{55}$$

As with the derivation of the backward equations, combining $o(h)$ terms in (53) is equivalent to interchanging limit and sum. Unfortunately, because the initial state $(X(t) \mid X(0) = i)$ in interval $(t, t + h]$ is now random, this interchange cannot be justified directly without making some additional assumption. For example, it would be sufficient to have an upper bound on the a_i, but this condition is too restrictive. In Section 4-9, we will justify the forward equations indirectly by use of the backward equations. An easy consequence of the theory developed there is

Theorem 8. For a regular Markov chain, the forward and backward equations have the same unique solution $\{P_{ij}(t);\ t \geq 0\}$, where for all i and t, $\sum_j P_{ij}(t) = 1$.

Thus the forward equations are valid, i.e., they give the correct solution. To illustrate one use of them, let $t \to \infty$ in (55). For a regular irreducible chain, limits of the $P_{ij}(t)$ exist, which implies that

$$\lim_{t \to \infty} P'_{ij}(t) = 0 \qquad \text{for all } i \text{ and } j,$$

and the limit of (55) turns out to be the balance equations (23)! [What do we get when $t \to \infty$ in (52)?]

In particular cases, either the forward or the backward equations can be solved to find $\{P_{ij}(t)\}$; usually, the forward equations are easier. There is a formal representation for the solution to the backward and forward equations, but it is of little help computationally.

4-7 UNIFORMIZABLE CHAINS

A Markov chain $\{X(t)\}$ is called *uniformizable* if for some $\nu < \infty$, $a_i \leq \nu$ for all i. This idea was used in Section 4-3 to show that a chain is regular when $\{a_i\}$ is bounded. Obviously, finite-state chains are uniformizable. We now present some other applications of this idea; all chains in this section are assumed to be uniformizable.

Notice that we can choose any ν such that $\nu \geq \sup(a_i)$. Jumps of $\{X(t)\}$ are permitted to occur only at the arrival epochs $t'_1 < t'_2 < \cdots$ of a Poisson process $\{\Lambda(t)\}$ at rate ν. For all states i and j, define

$$r_{ij} = P(X(t'_{n+1}) = j \mid X(t'_n) = i) = \begin{cases} a_{ij}/\nu & \text{for } j \neq i, \text{ and} \\ 1 - a_i/\nu & \text{for } j = i, \end{cases} \tag{56}$$

where $R = (r_{ij})$ is the transition probability matrix of discrete-time chain

$\{X(t_n')\}$, a stochastic process independent of $\{\Lambda(t)\}$. It is easy to see that for all i and j,

$$P_{ij}(t) = \sum_{n=0}^{\infty} r_{ij}^{(n)} e^{-\nu t}(\nu t)^n/n!, \quad \text{or in matrix notation}$$

$$\mathbf{P}(t) = \sum_{n=0}^{\infty} R^n e^{-\nu t}(\nu t)^n/n! = e^{-\nu t[I-R]}, \tag{57}$$

where I is the identity matrix.

Remark. We assumed above that transitions of type $i \rightarrow i$ were excluded in the original chain. If they are permitted, it is easy to modify (56) to include this possibility.

For a uniformized positive irreducible chain, let $\{\pi_j'\}$ satisfy $\pi' = \pi' R$. Either from PASTA or (45) (with $a_j = \nu$ for all j) it is obvious that

$$\{\pi_j'\} = \{p_j\}. \tag{58}$$

There are often computational advantages to (57) for computing transient probabilities, particularly for finite-state chains, but we shall not explore this idea.

4-8 SEMI-MARKOV PROCESSES

We now generalize continuous-time chains to processes that jump from state to state, where the sequence of states visited is a discrete-time chain, but the sojourn times in each state now have an arbitrary distribution. The state space is the nonnegative integers.

Let X_0 be the initial state and X_n be the state entered on the nth jump, $n \geq 1$, where $\{X_n\}$ is a discrete-time Markov chain with transition probability matrix $P = (p_{ij})$. Let τ_0 be the sojourn time in the initial state and τ_n be the sojourn time in the state entered on the nth jump, $n \geq 1$, where, given $\{X_n\}$, the τ_n are independent nonnegative random variables. For each $n \geq 0$, the distribution of τ_n depends *only* on the values of X_n and X_{n+1}; for all states i and j, let

$$H_{ij}(t) \equiv P(\tau_n \leq t \mid X_n = i \text{ and } X_{n+1} = j), \qquad t \geq 0.$$

The distribution of a sojourn time in state i is

$$H_i(t) \equiv \sum_j p_{ij} H_{ij}(t), \qquad t \geq 0.$$

Let $H_{ij}(t)$ have mean η_{ij} and $H_i(t)$ have mean $1/a_i \equiv \sum_j \eta_{ij} p_{ij}$, where we assume that $0 < a_i < \infty$ for all i, which rules out instantaneous states. We permit both $p_{ii} > 0$ and $H_i(0) > 0$, and no special notation is needed to include these possibilities. An $i \rightarrow i$ transition will also be called a jump.

Example 4-5

To help visualize the processes we are about to define, consider a taxi driver who picks up passengers at various locations (states) and takes them to other locations, where X_n is the location of the taxi at the end of the nth trip, and p_{ij} is the probability that a passenger picked up at location i is taken to location j. Notice that the duration of a trip from i to j (sojourn time) is permitted to depend on both i and j. We will be interested in representing such quantities as the number of visits to various locations by epoch t, and the current state of the taxi at epoch t, which will mean the location of the taxi at the time that the current trip began. (Trip times include idle times of the taxi prior to the beginning of each trip.)

Now let $t_0 = 0$ and $t_n = \sum_{i=0}^{n-1} \tau_i$, $n = 1, 2, \ldots$, be the successive epochs at which jumps occur, and for all $t \geq 0$, let

$$\mathcal{N}(t) = \max\{n: t_n \leq t\}, \tag{59}$$

which is the number of jumps by epoch t. As with continuous-time chains, the number of jumps in finite time can be infinite (an explosion), i.e., $\mathcal{N}(t)$ may be a defective random variable. For every $t \geq 0$ and sample point ω such that $\mathcal{N}(t, \omega) < \infty$, define

$$X(t) \equiv X_{\mathcal{N}(t)}. \tag{60}$$

Thus, aside from the possibility of an explosion, $\{X(t): t \geq 0\}$ is a well-defined stochastic process called a *semi-Markov process (SMP)*, where it is the process that jumps from state to state at random times.

The sample paths of $\{X(t)\}$ will look like Figure 4-1, except that we permit multiple jumps at the same epoch; they are right continuous and have left-hand limits. If more than one jump occurs at epoch t, it is easy to see from (60) that $X(t) = X(t^+)$, the state entered on the last jump.

We call the sequence $\{(X_n, t_n)\}$ a *Markov renewal process (MRP)*, and, as with continuous-time chains, we call $\{X_n\}$ the corresponding *embedded Markov chain (EMC)*. An MRP differs from an EMC simply by associating a real-time epoch with each jump.

Example 4-6: The Alternating Renewal Process

This process, originally introduced as Example 2-2, is a trivial example of a two-state SMP with EMC transition probabilities $p_{01} = p_{10} = 1$.

We introduce the MRP in order to discuss elementary properties of continuous-time processes before dealing with explosions. Clearly, state i can reach state j in the MRP if and only if i can reach j in the EMC, and the processes have the same communication classes. First passage times $\mathcal{T}_{ij}$ for the MRP are defined as they were for the CTMC in Section 4-3, and we use the same notation and terminology. It follows that j is recurrent in the MRP if and only if it is recurrent in the EMC, and that for the MRP, the states in a communication class are either (i) all transient, (ii) all positive, or (iii) all null.

For irreducible processes, we can distinguish transient from recurrent by analyzing the EMC. In the recurrent case, we write

$$\mathcal{T}_{jj} = \tau_0 + \sum_{i \neq j} \sum_{k}^{V_{ji}} \tau_{ik}, \tag{61}$$

where V_{ji} is the number of visits to i between returns to j, and τ_{ik} is the sojourn time in state i on the kth visit. Recall that

$$E(V_{ji}) = \pi_i/\pi_j, \tag{62}$$

where $\{\pi_j\}$ is the stationary distribution of the EMC when it is positive, and is the stationary measure (see Section 3-11) when the EMC is null. In either case, (62) is determined uniquely. By taking the expected value of (61) (interchange E and first $\sum$, and apply Wald's equation), we get

$$E(\mathcal{T}_{jj}) = 1/a_j + \sum_{i \neq j} (\pi_i/a_i)/\pi_j. \tag{63}$$

Thus state j in the MRP (and the irreducible MRP itself) is positive if and only if

$$\sum_i \pi_i/a_i < \infty, \tag{64}$$

which is the required condition for determining $\{p_j\}$ in (45).

Given $X_0 = i$, let $M_{ij}(t)$ be the number of visits of the MRP to state j by epoch $t \geq 0$. For each fixed i and j, we have defined a (possibly defective) renewal process with renewal function $m_{ij}(t) = E\{M_{ij}(t)\}$, $t \geq 0$.

A new feature of the MRP is the possibility that $\mathcal{T}_{jj}$ has a lattice distribution, and it is easily shown that the states in a communication class all have the same span. Notice that lattice is a property of first passage time distributions $F_{jj}(t)$, and need have no relation to the periodicity of the EMC. For applications in this book, this feature is of minor importance.

To apply renewal theory to SMPs, we need to ensure that explosions do not occur, so that the state of an SMP at any epoch t is determined (w.p.1) by a finite number of jumps. Following the terminology of Section 4-3, we have the

Definition. An SMP is called *regular* if for every initial state $X_0 = i$, the number of jumps in finite time is finite with probability 1.

For brevity, we apply irreducible and other terms describing an MRP to the corresponding SMP. From the preceding discussion, an SMP will be regular if the EMC is irreducible and recurrent. More generally, Condition 3 in Section 4-3 is also a sufficient condition for an SMP to be regular. In place of Condition 1, another sufficient condition for an SMP to be regular is that there exists a distribution function $H(t)$ with $H(0) < 1$ such that for all states i,

$$H(t) \geq H_i(t), \qquad t \geq 0. \tag{65}$$

In particular, *finite-state* SMPs are regular. Condition (65) is a stochastic ordering; as with uniformization, we can define another SMP with sojourn time distribution $H(t)$ that "runs faster" than the original one.

For the remainder of this section, we deal only with regular SMPs and employ the same terminology and notation used for regular CTMCs in Section 4-3. In fact, aside from nonexponential sojourn times and that pointwise limits may not exist, much of the analysis is identical.

For all i and j, we now define

$$P_{ij}(t) = P(X(t) = j \mid X_0 = i), \qquad t \geq 0. \tag{66}$$

One technical difference to observe here is that we are conditioning on X_0 rather than $X(0)$; we will treat the SMP as though it "just entered" state i at epoch 0.

Defining renewals as visits to state j, the renewal equation for $P_{jj}(t)$ is easily shown to be

$$P_{jj}(t) = H_j^c(t) + \int_0^t P_{jj}(t - u)\, dF_{jj}(u), \tag{67}$$

which has solution

$$P_{jj}(t) = H_j^c(t) + \int_0^t H_j^c(t - u)\, dm_{jj}(u) \qquad \text{for all } j. \tag{68}$$

For $i \neq j$, we have

$$P_{ij}(t) = \int_0^t H_j^c(t - u)\, dm_{ij}(u). \tag{69}$$

If j is transient or null, $\lim_{t \to \infty} P_{ij}(t) = 0$. If j is positive, $\mathcal{T}_{ij}$ is proper, *and* F_{jj} is nonlattice, we have by the Key Renewal Theorem that the limit of (68) and (69) exists as $t \to \infty$, and is given by (19), where p_j is also the fraction of time the SMP spends in state j. As remarked in Section 4-3, the interpretation of p_j as a fraction of time depends only on the Renewal-Reward Theorem and does not require $F_{jj}(t)$ to be nonlattice.

Now suppose $\{X(t)\}$ is irreducible. If positive recurrent, $\{X(t)\}$ is a regenerative process with finite mean cycle length in terms of returns to arbitrary state j, and Sections 2-11 and 2-16 apply. From the preceding analysis, we also have

Theorem 9. An irreducible SMP is positive recurrent if and only if the EMC is recurrent and (64) holds, in which case

$$p_j = (\pi_j/a_j)/[\sum_j \pi_j/a_j] > 0, \qquad j = 0, 1, \ldots, \tag{70}$$

is the fraction of time the SMP spends in state j.

Remarks. Distribution $\{p_j\}$ depends only on the *mean* sojourn times, i.e., for fixed EMC transition matrix P, it is *insensitive* to the sojourn time distributions. Furthermore, if we *define* $a_{ij} = a_i p_{ij}$ for all i and j, (70) is a solution to (23). While the a_{ij} are no longer conditional rates, the $p_i a_{ij}$ are long-run rates.

We can easily find other quantities either as fractions of time or as pointwise limits. For example, suppose we define the "excess" $Y(t) = t_{\mathcal{N}(t)+1} - t$, the time until the first jump after t, and the corresponding "age" $U(t) = t - t_{\mathcal{N}(t)}$. For a positive, irreducible, nonlattice SMP,

$$\lim_{t\to\infty} P[X(t) = j,\ Y(t) \le y] = p_j H_{je}(y) \qquad \text{for all } j, \tag{71}$$

and arbitrary initial state, where H_{je} is the equilibrium distribution of H_j.

Embedded Markov chains have been used to analyze queues for a long time; for important examples, see Chapters 7 and 8. These examples may also be viewed as applications of the theory developed here.

Example 4-7: The M/G/1 Queue

Poisson arrivals (of customers) at rate λ are served in their order of arrival by a single server. Let S_n be the service time of the nth arrival, where $\{S_n\}$ is an i.i.d. sequence with some general distribution G, and finite mean $E(S_n) = 1/\mu$. This model is called the $M/G/1$ queue.

Let X_n be the number of customers who are left behind by the nth departure, Y_n be the number of arrivals during S_n, and t_n be the nth departure epoch, $n = 1, 2, \ldots$. It is easy to see that

$$X_{n+1} = X_n - 1 + Y_n \quad \text{if } X_n > 0, \qquad X_{n+1} = Y_n \quad \text{if } X_n = 0,$$

and $\{X_n\}$ is an irreducible Markov chain. This chain is analyzed in Section 8-1, where it is shown that $\{X_n\}$ is positive if and only if $\rho \equiv \lambda/\mu < 1$.

Now $\{X_n, t_n\}$ is also an MRP. To see this, notice that the sojourn time following t_n is S_{n+1} if $X_n > 0$, and is S_{n+1} plus an independent *idle period* if $X_n = 0$. Thus, $H_i = G$ for $i \ge 1$, and $H_0 = \exp(\lambda) * G$. From Theorem 9, we see that the MRP is also positive if and only if $\rho < 1$.

Let $N(t)$ be the number of customers in system (being served or not) at epoch $t \ge 0$. It follows that when $\rho < 1$, $\{N(t)\}$ regenerates at those t_n for which $X_n = 0$, and has finite mean cycle length.

Important Remarks. The jumps of $\{N(t)\}$ occur at arrival epochs as well as departure epochs; it is a different stochastic process from SMP $\{X(t)\}$, and p_j in (70) is *not* the fraction of time $\{N(t)\}$ spends in state j. It turns out (see Section 8-3) that for the $M/G/1$ queue, $\{N(t)\}$ has stationary distribution $\{\pi_j\}$. This distribution is *not* insensitive to sojourn time distributions because transition matrix P depends on service distribution G. The primary advantage of introducing an MRP in an embedded chain analysis is access to Theorem 9.

4-9 MINIMAL SOLUTIONS TO THE BACKWARD AND FORWARD EQUATIONS*

We now return to the possibility of an explosion in a CTMC. For all states i and j, and $t \geq 0$, let

$$P_{ij}^{(n)}(t) = P[X(t) = j, t_{n+1} > t \mid X(0) = i], \qquad n = 0, 1, \ldots, \tag{72}$$

where t_n is the epoch of the nth jump. As n increases, (72) is nondecreasing and bounded by 1. Let

$$\lim_{n\to\infty} P_{ij}^{(n)}(t) = P_{ij}^{(\infty)}(t) \qquad \text{for all } i, j, \text{ and } t, \tag{73}$$

which is the probability that starting in state i, the chain is in state j at epoch t, and the number of jumps by t is finite. We have

$$P_{ij}^{(0)}(t) = \delta_{ij}e^{-a_i t}, \tag{74}$$

where $\delta_{ij} = 1$ if $i = j$, $\delta_{ij} = 0$ otherwise.

By conditioning on the first transition, it is easy to see that for all i, j, and t,

$$P_{ij}^{(n+1)}(t) = \delta_{ij}e^{-a_i t} + \sum_{k\neq i} \int_0^t P_{kj}^{(n)}(t-u)a_{ik}e^{-a_i u}\,du, \qquad n = 0, 1, \ldots. \tag{75}$$

Now let $n \to \infty$ in (75) and change the integration variable to $x = t - u$. We have that the limits (73) satisfy

$$P_{ij}^{(\infty)}(t) = \delta_{ij}e^{-a_i t} + \sum_{k\neq i} \int_0^t P_{kj}^{(\infty)}(x)a_{ik}e^{-a_i(t-x)}\,dx. \tag{76}$$

Differentiating (76) with respect to t, it is a few lines of algebra to show that $\{P_{ij}^{(\infty)}(t)\}$ satisfies the backward equations (52).

In fact, (76) is just the integral-equation version of the backward equations, i.e., (76) and (52) [together with boundary conditions $P_{ii}(0) = 1$ for all i] are equivalent, and we now explore the question of uniqueness of solutions to the backward equations.

Let $\{P_{ij}(t)\}$ be any (possibly defective) probability distribution that satisfies (76). From (74), we see that

$$P_{ij}(t) \geq P_{ij}^{(0)}(t),$$

and from (75),

$$P_{ij}(t) \geq P_{ij}^{(n)}(t), \text{ for } n = 1, 2, \ldots,$$

and, by letting $n \to \infty$,

$$P_{ij}(t) \geq P_{ij}^{(\infty)}(t) \qquad \text{for all } i, j, \text{ and } t. \tag{77}$$

* This is an advanced topic.

Thus $\{P_{ij}^{(\infty)}(t)\}$ is what we call the *minimal* solution to the backward equations.

If the CTMC is regular, the minimal solution is proper, and therefore is the *unique* solution to the backward equations. When the CTMC is not regular,

$$1 - \sum_j P_{ij}^{(\infty)}(t) \tag{78}$$

is the probability of infinitely many jumps by epoch t, when the chain starts in state i.

Remark. Similar methods were used in Section 3-7 to establish uniqueness of solutions to certain linear equations.

In the same manner, we now show that the minimal solution to the backward equations is also the minimal solution to the forward equations (55). This is done with use of a different recursion formula for the $P_{ij}^{(n)}(t)$ that is found by conditioning on the *last* transition in $[0, t]$.

let $\mathcal{N}(t)$ be the number of transitions in $[0, t]$, and for all i, j, and t, and let

$$R_{ij}^{(n)}(t) = P[X(t) = j, \mathcal{N}(t) = n \mid X(0) = i], \qquad n = 0, 1, \ldots,$$

where

$$R_{ij}^{(0)}(t) = \delta_{ij} e^{-a_i t}.$$

Suppose the $(n + 1)$st transition occurs at $x \in [0, t]$ and is of the type $k \to j$. In order for the event $\{X(t) = j, \mathcal{N}(t) = n + 1\}$ to occur, the sojourn time in state j must exceed $t - x$, an event with probability $e^{-a_j(t-x)}$. The probability that this type of $(n + 1)$st transition occurs in the interval $(x, x + h)$ is $R_{ik}^{(n)}(x) a_{kj} h + o(h)$, i.e., we are defining a density function. It follows that for all i, j, and t,

$$R_{ij}^{(n+1)}(t) = \sum_{k \neq j} \int_0^t R_{ik}^{(n)}(x) a_{kj} e^{-a_j(t-x)}\, dx, \qquad n = 0, 1, \ldots. \tag{79}$$

Now

$$P_{ij}^{(n)}(t) = \sum_{l=0}^{n} R_{ij}^{(l)}(t),$$

and it follows from (79) that for all i, j, and t,

$$P_{ij}^{(n+1)}(t) = \delta_{ij} e^{-a_i t} + \sum_{k \neq j} \int_0^t P_{ik}^{(n)}(x) a_{kj} e^{-a_j(t-x)}\, dx, \qquad n = 0, 1, \ldots. \tag{80}$$

Now let $n \to \infty$ in (80), and we have that $\{P_{ij}^{(\infty)}(t)\}$ also satisfies

$$P_{ij}^{(\infty)}(t) = \delta_{ij} e^{-a_i t} + \sum_{k \neq j} \int_0^t P_{ik}^{(\infty)}(x) a_{kj} e^{-a_j(t-x)}\, dx, \qquad \text{for all } i, j, \text{ and } t. \tag{81}$$

By differentiating (81), we have that the minimal solution to the backward equations is a solution to the forward equations (55), and in fact that (81) and (55) are equivalent. By the argument used above, this solution is also the minimal solution to the forward equations, and is unique and proper when the chain is regular. We have completed the proof of Theorem 8 in Section 4-6.

When the chain is not regular, it may be shown that $\{P_{ij}^{(\infty)}(t)\}$ satisfies the Chapman-Kolmogorov equation (3), but not the normalizing condition (4). In this case, there can exist many proper solutions $\{P_{ij}(t)\}$ to the backward equations that also satisfy (3), some of which may (or may not) also satisfy the forward equations. We illustrate some of these possibilities in Example 4-8, but a treatment of this topic is beyond our scope and the needs of this book.

Remark. The conclusions above hold for pure-jump Markov processes with (essentially) an arbitrary uncountable state space.

Proof of Theorem 4-4. We now prove a necessary and sufficient condition for a chain to be regular. In Section 4-3, we defined $S(i) = (\sum \tau_n \mid X(0) = i)$ and, conditioning on the first transition, we have that for every i,

$$y_i = E(e^{-S(i)}) = E(e^{-\tau_0}e^{-(S(i)-\tau_0)} \mid X(0) = i)$$

$$= E(e^{-\tau_0} \mid X(0) = i)E(e^{-(S(i)-\tau_0)} \mid X(0) = i) = \sum_{j \neq i} a_i p_{ij} y_j/(1 + a_i).$$

From this, $a_{ij} = a_i p_{ij}$, $a_{ii} = -a_i$, and $\{y_i\}$ satisfies (14). If the chain is not regular, $y_i > 0$ for some i; i.e., the condition is sufficient. Now suppose the chain is regular, and let $\{y_i\}$ be a bounded solution to (14), where, by rescaling, we have $|y_i| \leq 1$ for all i. Now let $y_i^{(0)} = 1$ and

$$y_i^{(n)} = E\{e^{-(\tau_0 + \cdots + \tau_{n-1})}\}, \qquad n \geq 1,$$

where $y_i^{(n)} \downarrow n$, and $y_i^{(n)} \to 0$ as $n \to \infty$. By conditioning on the first transition, we have that for all i,

$$y_i^{(n+1)} = \sum_{j \neq i} a_{ij} y_j^{(n)}/(1 + a_i) \quad \text{and} \quad y_i^{(n+1)} \geq \sum_j a_{ij} y_j^{(n)}, \qquad n \geq 0. \tag{82}$$

From $|y_i| \leq y_i^{(0)}$ for all i, and the inequality in (82), we have by induction that $|y_i| \leq y_i^{(n)}$ for all n and i. Now let $n \to \infty$, and $y_i = 0$ for all i.

Remarks. Compare this proof with that of Theorem 7 in Chapter 3. If $y_i > 0$ for some i, it can be shown that for any positive t, the probability of an explosion by epoch t is strictly positive, given $X_0 = i$, and the corresponding minimal solution to the forward and backward equations is defective.

Example 4-8: The Pure-Birth Process

This process was introduced as Example 4-4. We have $a_i = a_{i,i+1} \equiv \lambda_i$, $i = 1, 2, \ldots$. The forward equations are

$$P'_{i0}(t) = -\lambda_0 P_{i0}(t),$$

$$P'_{ij}(t) = -\lambda_j P_{ij}(t) + \lambda_{j-1} P_{j-1,j}(t), \qquad j = 1, 2, \ldots,$$

where $P_{ij}(t) = 0$ for $i > j$. Because these equations can be solved recursively, the solution is unique for any specified initial conditions, and therefore must be the minimal solution $\{P_{ij}^{(\infty)}(t)\}$.

If the λ_i increase rapidly, the minimal solution may not be proper, which can occur only if the chain is explosive. From Theorem 4, it can be shown that a pure-birth process is regular if and only if $\sum_j \lambda_j^{-1} = \infty$. [The mathematics of showing this is the same as in Problem 3-36(a).] Hence

$$\sum_j P_{ij}^{(\infty)}(t) < 1,\ t > 0 \quad \text{if and only if} \quad \sum_j \lambda_j^{-1} < \infty. \tag{83}$$

[The condition is clearly sufficient, because it implies $E\{S(i)\} < \infty$.]

When a pure-birth process is explosive, the solution to the backward equations is not unique. One solution, for which $\{P_{ij}(t)\}$ is proper, can be constructed by starting the process over in some fixed state, 0 say, each time an explosion occurs. Let $S_i(0)$ be the times between successive explosions. The process (by definition) regenerates at explosion epochs $S_1(0)$, $S_1(0) + S_2(0)$, ...

Proof That a Solution to the Balance Equations Is Stationary

For an irreducible regular chain, let $\{p_j\}$ satisfy (20) and (23). From (80),

$$\sum_i p_i P_{ij}^{(n+1)}(t) = p_j e^{-a_j t} + \sum_{k \neq j} \int_0^t \sum_i p_i P_{ik}^{(n)}(x) a_{kj} e^{-a_j(t-x)}\, dx.$$

From (74), $\sum_i p_i P_{ij}^{(0)} = p_j e^{-a_j t} \leq p_j$, for all j and t. By induction on n,

$$\begin{aligned} \sum_i p_i P_{ij}^{(n+1)}(t) &\leq p_j e^{-a_j t} + \sum_{k \neq j} \int_0^t p_k a_{kj} e^{-a_j(t-x)}\, dx \\ &\leq p_j e^{-a_j t} + p_j(1 - e^{-a_j t}) = p_j. \end{aligned}$$

Now let $n \to \infty$, and we have

$$\sum_i p_i P_{ij}^{(\infty)}(t) \leq p_j \qquad \text{for all } j \text{ and } t. \tag{84}$$

From regularity, $\sum_j P_{ij}^{(\infty)}(t) = 1$. Consequently, summing (84) on j, both sides sum to 1. Thus the inequalities in (84) must be strict, which is (24).

Remarks. The preceding proof, which does not use either the forward or backward equations, is from Miller [1963]. Because the results in this chapter are well known, a literature review will not be provided. However, some of our results may be hard to find in various "complete" treatments. See Asmussen [1987] for a readable and more complete development of the theory of CTMCs and more general pure-jump chains, and other important references. Similarly, for stationary measures, see Kelly [1983] for a recent account and some new results. Chung [1967] is a frequently-referred-to general reference on CTMCs.

PROBLEMS

4-1. For an alternating renewal process, let the time spent in state i be distributed exp (a_i), $i = 0, 1$. Use balance equations to find p_i and verify that this is consistent with the Renewal-Reward Theorem.

4-2. Derive (17).

4-3. Apply the Renewal-Reward Theorem to obtain (19) as a time average.

4-4. Consider an $M/M/1$ queue where customers are impatient. A customer who finds the server busy will wait a time $\sim$ exp (γ) for service to begin, and will leave without service if service has not begun by then. Determine the transition rate matrix **A** and the balance equations for this model.

4-5. Formulate Problem 2-53 as a CTMC. Solve the balance equations for the steady-state probabilities, and answer (c) and (d).

4-6. Formulate Problem 2-54 as a CTMC. Solve for the steady-state probabilities and use them to find equation (P.20).

4-7. A company has $k \geq 2$ trucks, where the time until breakdown of a working truck is $\sim$ exp (λ). Trucks are repaired at a facility with c repairmen, $1 \leq c \leq k$. Each truck is repaired by one repairman; each repair time is $\sim$ exp (μ). Breakdown and repair times are independent; repaired trucks will eventually break down again in the same manner.

(a) Formulate as a CTMC. State whether or not (i) the chain is irreducible, and (ii) the balance equations have a unique solution.
(b) Write out the balance equations for $\{p_j\}$.
(c) Suppose $c = k$ (think of assigning one man permanently to each truck).
  (i) What fraction of time is each truck working?
  (ii) Without algebra, deduce the solution $\{p_j\}$ in (b).
(d) Does $\{p_j\}$ in (c) depend on having exponential up and repair times?
(e) Suppose we have $c = 1$, $k = 2$, exponential up times, and *constant* repair times $(= 1/\mu)$.
  (i) Formulate as a regenerative process—define your cycles.
  (ii) Find $q \equiv$ the probability that a working truck fails during a repair.
  (iii) Find the fraction of time there are j working trucks, $j = 0, 1, \ldots, k$.
(f) Do your answers in (e) depend on having constant repair times?

4-8. *The linear birth and death process.* Let $X(t)$ be the size of a population at epoch t, where each individual in the population has a life that is $\sim$ exp (μ). Until death, each individual gives birth to new individuals in accordance with a Poisson process at rate λ.

(a) Formulate as a birth and death process. What are the λ_i and μ_i? (To make irreducible, temporarily define $\lambda_0 = \lambda$.)
(b) Show that the irreducible chain is recurrent if and only if $\lambda \leq \mu$. When is it

positive? What does this mean as far as the probability of extinction (absorption in state 0) of the real process, where $\lambda_0 = 0$?

(c) Given $X(0) = i \geq 1$, the i individuals operate independently. What does this tell us about the probability of extinction as a function of i? Given $X(0) = i$, find the probability of extinction.

(d) How is $\{P_{ij}(t)\}$, $i \geq 2$, related to $\{P_{1j}(t)\}$?

[*Remark.* Equation (83) in Section 4-9 implies this process is regular; see Problem 4-14.]

(e) This model is a disguised branching process (Section 3-9) in continuous time. Remove the disguise and relate to (b).

4-9. *The $M/M/\infty$ queue.* (Example 4-3)

(a) Write down the forward equations.

(b) From the $M/G/\infty$ queue, deduce $\{P_{0j}(t)\}$ and verify that this distribution satisfies the forward equations.

(c) From (b), deduce $\{P_{ij}(t)\}$ for $i \geq 1$.

4-10. Derive equations (67) and (69).

4-11. Derive (71) by first showing that for all i and j,

$$P[X(t) = j,\ Y(t) > y \mid X_0 = i] = \delta_{ij} H_j^c(t + y) + \int_0^t H_j^c(t + y - u)\, dm_{ij}(u).$$

where $\delta_{ij} = 1$ if $i = j$, $\delta_{ij} = 0$ otherwise.

4-12. Consider the model in Problem 4-7 with $k \geq 2$, $c = 1$, exponential breakdown times, and a *general* repair time distribution G with mean $1/\mu$.

(a) For this model, define an MRP and the sojourn time distributions.

(b) Is the SMP positive?

(c) Let $N(t)$ be the number of working trucks at epoch $t \geq 0$, and p_j be the fraction of time there are j working trucks, $j = 0, \ldots, k$. Why is $\{N(t)\}$ a different stochastic process from the SMP? Nevertheless, why is $\{N(t)\}$ regenerative with finite mean cycle length? It follows that the (time) average number of working trucks, denoted by L, is

$$L = \sum_j j p_j.$$

(d) Suppose we can work out the stationary distribution π for the EMC (the transition probabilities are complicated), where $1/\pi_k$ is the mean number of service completions between visits of the EMC to state k (all trucks working). Let $\mathcal{T}_{kk}$ be the time between returns to state k in the SMP. Represent $\mathcal{T}_{kk}$ as a sum of random variables; show that

$$E(\mathcal{T}_{kk}) = 1/k\lambda + 1/\mu\pi_k,$$

and therefore that the *number of service completions per unit time* is

$$1/[\pi_k/k\lambda + 1/\mu].$$

(e) For each $j = 0, \ldots, k$, define a Poisson process at rate λj, where an "arrival" in this process results in a truck breakdown when there are j working trucks, and no change otherwise. From PASTA, argue that the *number of breakdowns per unit time* is

$$\sum_j \lambda j p_j = \lambda L.$$

(f) Why must the breakdown and repair rates be equal? Consequently,

$$L = 1/[\pi_k/k + \lambda/\mu].$$

Remark. The combination of ideas used here is typical of what frequently will be done in later chapters. This application of PASTA goes beyond CTMCs and is justified in Section 5-16.

4-13*. For the linear birth and death process in Problem 4-8, find stationary measure $\{p_j\}$ when $\lambda = \mu$.

4-14*. For the linear birth and death process in Problem 4-8, bound the number of jumps that occur by epoch t in terms of a pure-birth process that results from setting $\mu = 0$. Show that this birth and death process is regular.

* This is an advanced topic.

5

INTRODUCTION TO QUEUEING THEORY

Incoming orders at a warehouse and telephone calls at a switchboard are examples of *arrival streams* of what we shall call *customers*. Customers arrive at a *service facility,* such as a warehouse or a switchboard. A customer may or may not wait at the facility, and the corresponding waiting time may or may not depend on other customers. Usually, waiting time does depend on other customers. These are problems of *congestion,* and the primary theoretical tool for studying these problems is known as *queueing theory*.

In queueing theory, customers are usually treated as discrete (customer 1, 2, etc.) and the corresponding queue size (number of customers in queue) is integral valued. Events such as the arrival or departure of a customer are permitted to occur at arbitrary epochs on the continuous-time axis. This will be our approach except in Section 5-4, where queue size is treated as a continuous variable.

Inventory theory and the theory of dams are related theoretical topics. Inventory in a warehouse waiting to fill orders and water in a reservoir waiting to be released may be viewed as queues. In the case of a dam, it is natural to treat queue size as continuous.

Most operational problems to which queueing theory is applied consider the arrival stream as having a known and fixed stochastic structure and treat the service facility as subject to some control with respect to operation and/or design. Regardless of whether this is so, a mathematical description of both is necessary before analysis is possible.

The Poisson, renewal, and general point processes discussed in Chapter 2 are often used as models of arrival streams.

5-1 CONCEPTS, DEFINITIONS, AND NOTATION

Consider an arbitrary arrival stream of customers $C_j, j = 1, 2, \ldots,$ where

$$t_j = \text{arrival epoch of } C_j, \text{ with } t_0 \equiv 0;\ t_0 \leq t_1 \leq \cdots < \infty, \text{ and}$$

$$T_j = t_{j+1} - t_j = \text{inter-arrival time between } C_{j+1} \text{ and } C_j.$$

Notice that customers are numbered in their order of arrival.

Each customer has a service requirement; on completion of service, the customer departs. A customer who has arrived but not yet departed is said to be *in the system*. While in the system, a customer is either *in service,* i.e., is receiving service, or is delayed, owing to the presence of other customers, and is said to be *in queue*. With these definitions in mind, let

$$S_j = \textit{service time} \text{ (time spent in service) of } C_j,$$

$$D_j = \textit{delay} \text{ in } \textit{queue} \text{ of } C_j \text{ (not including } S_j),$$

$$W_j = D_j + S_j = \textit{waiting time} \text{ in } \textit{system} \text{ of } C_j, \text{ and}$$

$$t_j + W_j = \textit{departure epoch} \text{ of } C_j \text{ (from the system).}$$

We would also like to represent the *number* of customers in system, service, and queue. To do this, we define an indicator function for every $t \geq 0$,

$$I_j(t) = 1 \qquad \text{if } t_j \leq t < t_j + W_j, \text{ i.e., if } C_j \text{ is in the system at epoch } t,$$

$$= 0 \qquad \text{otherwise} \left[\textit{Note: } \int_0^\infty I_j(t)\,dt = W_j\right], \text{ and}$$

$$N(t) = \sum_{j=1}^{\infty} I_j(t) = \textit{number} \text{ (of customers) } \textit{in system} \text{ at epoch } t.$$

Similarly, for $t \geq 0$, define

$$N_q(t) = \textit{number} \text{ (of customers) } \textit{in queue} \text{ at epoch } t,$$

$$N_s(t) = \textit{number} \text{ (of customers) } \textit{in service} \text{ at epoch } t,$$

where $N(t) = N_q(t) + N_s(t)$,

$$\Lambda(t) = \max\{j\colon t_j \leq t\}$$

$$= \text{number of arrivals (of customers) by epoch } t, \text{ and}$$

$$\Omega(t) = \Lambda(t) - N(t)$$

$$= \text{number of departures (of customers from the system) by epoch } t.$$

We will regard all quantities defined in this section to be random variables.

For any point ω in the sample space, all quantities such as $t_j(\omega)$ are real numbers. For any model we consider, properties of the arrival process (in terms of either $\{t_j\}$ or $\{T_j\}$) and service requirements (in terms of $\{S_j\}$) will be specified. Together with properties of the service facility, e.g., there is a single server, the arrival and service properties will determine all other quantities. For example, $\{t_j(\omega)\}$ and $\{S_j(\omega)\}$ at some point ω determine $\{N(t, \omega)\}$ as an ordinary (deterministic) function of t.

The *performance* of a queueing model will be defined in terms of properties of one or more of the following: $\{N(t)\}$, $\{N_q(t)\}$, $\{W_j\}$, and $\{D_j\}$. We will be concerned primarily with long-run behavior where *measures* of performance are defined as limits that are averages over either *time* or *customers*. The limits are defined for an arbitrary point ω, but the functional dependence on ω will be suppressed, e.g., $N(t, \omega) \equiv N(t)$. Furthermore, it is assumed that these limits are independent of ω. That is, each limit is a constant for all ω, or, at worst, it is a constant on a set of ω that has probability 1. In the language of previous chapters, all limits are assumed to hold with probability 1.

Define L, λ, and w as the following limits:

$$\lim_{t\to\infty} \int_0^t N(u)\,du/t = L, \text{ \textit{the average number} (of customers) \textit{in system},} \tag{1}$$

$$\lim_{t\to\infty} \Lambda(t)/t = \lambda, \text{ \textit{the arrival rate}, and} \tag{2}$$

$$\lim_{n\to\infty} \sum_{j=1}^{n} W_j/n = w, \text{ \textit{the average waiting time in system} (per customer).} \tag{3}$$

Similarly, define

$$\lim_{t\to\infty} \int_0^t N_q(u)\,du/t = Q, \text{ \textit{the average number in queue},} \tag{4}$$

$$\lim_{n\to\infty} \sum_{j=1}^{n} D_j/n = d, \text{ \textit{the average delay in queue},} \tag{5}$$

$$\lim_{t\to\infty} \int_0^t N_s(u)\,du/t = L_s, \text{ \textit{the average number in service}, and} \tag{6}$$

$$\lim_{n\to\infty} \sum_{j=1}^{n} S_j/n = 1/\mu, \text{ \textit{the average time in service}, where} \tag{7}$$

μ is called the *service rate*. Notice that (1), (2), (4), and (6) are averages over *time*, while (3), (5), and (7) are averages over *customers*. Also observe that in our terminology, *delay* always refers to customers in *queue*, whereas *waiting time* always refers to customers in *system*.

5-2 THE BASIC RELATION: $L = \lambda w$

The averages L and w defined in (1) and (3) are the two most common measures of performance of a queue. It is of fundamental importance that these measures satisfy the simple relation $L = \lambda w$. Here we will be content to demonstrate the plausibility of this result. A proof is presented in Section 5-15.

Consider a realization of $\Lambda(t)$ and the W_j on some interval $[0, T]$ in Figure 5-1, where the step function represents $\Lambda(t)$, and W_j is the length of rectangle j. Note that rectangle j has height 1 and *area* W_j.

For t in Figure 5-1, $N(t) = 2$; i.e., C_2 and C_4 are in the system at epoch t, but C_1 and C_3 have already departed. At T, $N(T) = 0$; the first four customers have departed but C_5 has not yet arrived. Notice that C_3 departs before C_2; it is not required that customers depart in their order of arrival.

Because the first four customers in Figure 5-1 have departed by T, $W_j = \int_0^\infty I_j(t)\,dt = \int_0^T I_j(t)\,dt$ for these customers. Thus we can write

$$\int_0^T N(t)\,dt = \int_0^T \sum_{j=1}^{\Lambda(T)} I_j(t)\,dt = \sum \int \cdots = \sum_{j=1}^{\Lambda(T)} W_j, \tag{8}$$

i.e., the *area* under the function $N(t)$ between 0 and T is the *sum* of the waiting times of all arrivals in $[0, T]$. Equality holds in (8) only because $N(T) = 0$. Otherwise, the sum would be greater than the integral.

For arbitrary $t \geq 0$, where $N(t) \geq 0$,

$$\int_0^t N(u)\,du = \sum_{j=1}^{\Lambda(t)} W_j + \text{an error term}. \tag{9}$$

If $N(t) > 0$, $\int_0^t I_j(u)\,du < W_j$ for some $j \leq \Lambda(t)$, and the error term is strictly negative. Dividing by t, we have

$$\int_0^t N(u)\,du/t = [\Lambda(t)/t] \cdot \left[\sum_{j=1}^{\Lambda(t)} W_j/\Lambda(t)\right] + (\text{error term})/t. \tag{10}$$

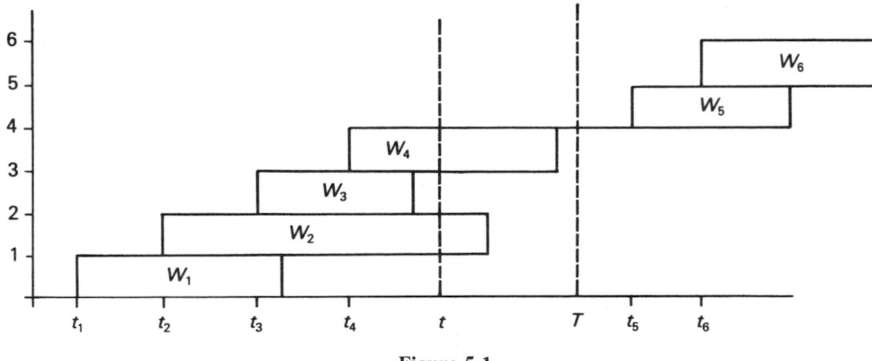

Figure 5-1

Now let $t \to \infty$ [and $\Lambda(t) \to \infty$] in (10). From (1), (2), and (3), we have

$$L = \lambda w, \tag{11}$$

provided that (error term)/$t \to 0$. In fact, the hard part of any proof of $L = \lambda w$ is to show that the error term does not contribute to the limit. In Section 5-15, we prove Theorem 1, which is also known as *Little's formula*.

Theorem 1. If the limits (2) and (3) exist with $\lambda < \infty$ and $w < \infty$, then limit (1) exists where $L < \infty$ and $L = \lambda w$.

The theorem means that for practical purposes, (11) is *always* true. The same approach clearly works for delay: If (2) and (5) exist and are finite,

$$Q = \lambda d. \tag{12}$$

Remark. Even when service is split up, with portions performed at separate times, (12) holds provided that we also have (3) with $w < \infty$. Similarly, customer waiting times may be split up, where customers leave the system and return later. This is discussed under "Multiple Visits" in Section 5-17.

Similarly,

$$L_s = \lambda/\mu. \tag{13}$$

If there is one server who serves only one customer at a time, L_s is the *proportion* of time the server is busy. If there is more than one server, where each server serves one customer at a time, L_s is the *expected number* of busy servers. The right-hand side of (13) is also called the *offered load*. In terms of λ and μ (characteristics of the arrival stream of customers), the offered load determines the minimum number of servers required to handle that load.

The only precaution we have in using these results is that each limit in (11), (12), and (13) must refer to the *same* stream of customers. For example, in models where not all customers are served, the offered load would still be defined as λ/μ, where λ is the arrival rate of *all* customers. However, L_s does not equal the offered load because the arrival rate of those served, λ_s say, is less than λ. Thus $L_s = \lambda_s/\mu$.

In applications, we will usually find either L or w from a corresponding limiting distribution. For example, let

$$P_n(t) = P\{N(t) = n\}.$$

In this book (e.g., see Section 2-3 for the corresponding expression in terms of indicator variables), the limit

$$\lim_{t\to\infty} \int_0^t P_n(u)\, du/t = p_n, \qquad n = 0, 1, \ldots, \tag{14}$$

is the *proportion of time* the system is in "state" n, i.e., that the number of customers in system is n. Similarly, for $t \geq 0$,

$$\lim_{n\to\infty} \sum_{j=1}^{n} P(W_j \leq t)/n = W(t) \tag{15}$$

is a waiting time distribution, where for every t, (15) is the *proportion of customers* who have waiting time $\leq t$. Limiting distributions for the number of customers in queue and the delay in queue may also be defined.

In terms of these distributions, we will find L and w as follows:

$$L = \sum_{n=0}^{\infty} np_n, \text{ and} \tag{16}$$

$$w = \int_0^\infty W^c(t)\, dt. \tag{17}$$

It requires proof that L and w can be found in this way; for regenerative processes, (16) and (17) are shown in Section 2-12.

We call a queue *stable* if $\{N(t)\}$ and $\{W_j\}$ have proper limiting distributions in a time-average sense, which will be true if these processes are regenerative with finite mean cycle length, e.g., when $\{N(t)\}$ is an irreducible continuous-time Markov chain that is *positive recurrent*. In the regenerative case, *unstable* will correspond to having either cycle lengths with infinite mean (the *null* case in Markov chains), or cycle lengths that are defective random variables (the *transient* case).

Example 5-1: The M/G/∞ Queue

In Section 2-6, we introduced as Example 2-6 the queueing model with Poisson arrivals, an infinite number of servers (so that no queueing occurs), and independent service times with distribution function G and mean $1/\mu$. In this chapter, the number of calls in progress at epoch t is the number of customers in system, $N(t)$. We showed that $N(t)$ has the Poisson distribution

$$N(t) \sim P\left[\lambda \int_0^t G^c(u)\, du\right]. \tag{18}$$

It follows that $N(t)$ converges in distribution as $t \to \infty$ to $P(\lambda/\mu)$, i.e.,

$$\lim_{t\to\infty} P\{N(t) = n\} = e^{-\lambda/\mu}(\lambda/\mu)^n/n! = p_n, \qquad n = 0, 1, \ldots, \text{ and} \tag{19}$$

$$L = \lambda/\mu. \tag{20}$$

Because no queueing occurs, $W_j = S_j, j = 1, 2, \ldots$, and

$$w = \lim_{n\to\infty} \sum_{j=1}^{n} S_j/n = 1/\mu. \tag{21}$$

Notice that we have an independent verification of $L = \lambda w$ for this model.

Remark. Pointwise limit (19) implies (14), but otherwise is not needed for (20).

There are many applications of $L = \lambda w$. Obviously, if we are able to find either L or w by an independent analysis, the other is easily obtained. In some cases, this equation will be one member of a set of simultaneous equations that determine both quantities. At other times, we will be concerned about whether (and in what direction) L or w change if the operation of the service facility changes. An example follows.

Example 5-2

There was a time when entering a bank was quite a challenge. A separate queue formed for each teller. Trying to choose the shortest line (in terms of service times, not simply number of customers) would add interest to an otherwise boring task. Most banks now have a single queue that feeds all tellers. (Excluded from consideration here are special-function tellers such as check-cashing-only.) This change makes taking scheduled breaks easier for tellers. Are there any advantages to customers?

To consider this question, let us fix everything else, i.e., assume that the arrival stream of customers is unaltered by the change, the service times are the same random variables, and the number of tellers available is also fixed. The change really amounts to changing the *order* of service. Under a single queue, customers begin service (leave the queue) in their order of arrival. In the separate queue case, this need not occur. If the service times have the same distribution, and, for each j, the service time of C_j is independent of other service times *and* of C_j's order of service, the change will not affect the distribution of the number of customers in system, $N(t)$. Thus L and $w = L/\lambda$ are also unaffected, i.e., the *expected* waiting time is the same. (Whether *your* expected waiting time is the same depends on whether you are better or worse than average at choosing the shortest line.)

The equation $L = \lambda w$ is a relation between means of distributions (14) and (15). In Example 5-2, a change in operation of a facility did not affect $\{p_n\}$. However, in that example, the waiting time distribution, (15), *does* change. In fact, in Section 5-14, we will show that the change to a single queue in this example has the effect of reducing the *variance* of (15). More generally, operating with a single queue makes the waiting times more predictable. This is usually regarded as desirable, but not always. For example, if you have 10 minutes to cash a check, the probability of your doing so depends on the waiting time distribution. If $w > 10$, increasing predictability can *decrease* this probability.

5-3 SERVICE FACILITIES; THE RULES OF THE GAME

The preceding notation and results were presented in an abstract setting because they hold in general, whenever the queue is *stable,* i.e., when $\sum_{n=0}^{\infty} p_n = 1$, independent of how the service facility actually operates. We now introduce terminology for the operation of service facilities.

A *channel* is a portion of a facility capable of providing service to one customer at a time. A single ticket window at a box office is an example of a *single-channel* facility, and may be pictured as shown in Figure 5-2. At the epoch this "picture" was taken, there were five customers in system, of which four were in queue, and one was in the service facility (in service). Usually, there is no limit on the number of customers allowed in system or queue. We have a *queue limit* model if there is a limit, i.e., if there is an upper bound on the number in system. Ordinarily, arrivals that would violate this bound are assumed to depart instantaneously without receiving service. We may think of them as representing lost sales or unsatisfied demand.

Series, parallel, and more complex configurations of service facilities occur. The term *multichannel* facility will refer to the parallel configuration. (The terms *single server* and *multiserver* are also used.) For example, c toll booths across a highway is a c-channel facility. In this case, it is usually assumed that any customer may be completely served by any channel. Unless otherwise specified, we will assume that a single queue feeds all channels. Alternatively, each channel could have its own queue (see Example 5-2). In this case, rules by which arriving customers choose which queue to join must be specified. Under certain conditions, customers in one queue may be allowed to switch to another. This is known as *jockeying*.

The term "queues in *tandem*" will be used to denote the series configuration. For example, consider a production line where every customer receives a set of operations performed in a fixed sequence. *Station* j will denote the facility at which operation j is performed, where $j = 1$ denotes the first operation performed, etc. A queue may be allowed to form at every station or possibly only at the first. A buildup of the queue at station j, $j \geq 2$, may interfere with the operation of station $(j - 1)$, forcing it to stop working. This is known as *blocking*.

More generally, the sequence of operations may be different from one customer to another, e.g., in a machine shop. There is no natural ordering of the operations, and the stations may be numbered in an arbitrary manner. A model of this kind is called a *network* of queues. Complex communication and data transmission systems may be viewed as networks of queues.

Return now to a single-channel or multichannel arrangement (in parallel).

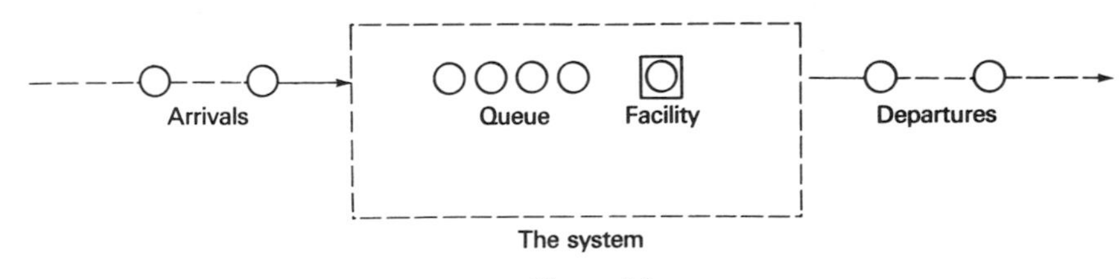

Figure 5-2

Unless otherwise specified, we will assume that arrivals join the end of the queue and begin service in the order of their arrival. This is known as a first-in-first-out (*FIFO*) order of service (also called first-come-first-served or *FCFS*). Note that FIFO refers to the order of departures *from the queue*. Since service times vary, the order of departures *from the system* may be different. One alternative to FIFO is last-in-first-out (*LIFO*) order of service (from the queue). LIFO may occur inadvertently, e.g., when a stock clerk in a supermarket places new stock in front of old, and customers take stock from the front.

Other alternatives to FIFO may be based on factors other than location in the arrival stream. Customers in queue may be ranked in some fashion, i.e., assigned priority numbers. The topic of *priority queues* is broad. Important factors in assigning priority numbers are waiting time costs and differences in service times. See Section 5-14 and Chapter 10.

5-4 DETERMINISTIC QUEUES—CONTINUOUS APPROXIMATION

Queueing theory is concerned primarily with congestion caused by stochastic effects, e.g., more arrivals than expected or longer service times than "usual." However, there are circumstances where congestion and long delays are predictable. For example, during the morning and evening rush hours, the arrival rate is (predictably) very high, and long queues form at certain locations every day.

A first approximation to rush hour phenomena is to ignore stochastic fluctuations and treat the problem deterministically. Because there are many arrivals and long queues, it is convenient to treat number of customers as a *continuous* variable. In this section, we assume that $\Lambda(t)$ is a continuous differentiable function, where, for $t \geq 0$,

$$\lambda(t) = \frac{d}{dt}\Lambda(t) = \text{arrival rate at epoch } t.$$

To illustrate, assume that customers are served at a single- or multiple-channel facility. Since the queue will be large (relative to the number of servers) we will not make any distinction between the number of customers in queue and in system. That is,

$$N(t) \approx N_q(t) \equiv Q(t), \text{ number in queue at epoch } t,$$

where the notation $Q(t)$ will be used only for deterministic queues. Let

$$\mu(t) = \text{service capacity at epoch } t\text{, i.e., maximum rate customers can be served.}$$

For example, for the toll booths at the San Francisco Bay Bridge, the service capacity would be proportional to the number of toll booths in operation.

When $\lambda(t) > \mu(t)$, the queue increases, and when $\lambda(t) < \mu(t)$, the queue (if positive) decreases. The *departure* rate is assumed to be the service capacity whenever the queue (size) is positive. When the queue is zero and $\lambda(t) \leq \mu(t)$, the departure rate equals the arrival rate. Under a deterministic approximation, no queue would form if $\lambda(t) \leq \mu(t)$ for all t. Thus, the departure function has the properties

$$\begin{aligned} \frac{d}{dt}\Omega(t) &= \mu(t) && \text{if } Q(t) > 0, \\ &= \min(\lambda(t), \mu(t)) && \text{if } Q(t) = 0. \end{aligned} \tag{22}$$

For convenient graphical representation, we assume FIFO, i.e., that customers depart (at least from the queue) in their order of arrival. Under FIFO, let

$$D(t) = \text{delay in queue of an arrival at epoch } t,$$

$$\mathscr{D}(\Lambda) = \text{delay in queue of arrival number } \Lambda.$$

These are different functions (of t and Λ), but are related, i.e.,

$$\mathscr{D}(\Lambda(t)) = D(t). \tag{23}$$

The function $D(t)$ is particularly easy to represent on a graph because under FIFO, a customer who is arrival number Λ is also departure number Λ,

$$\Lambda(t) = \Omega(t + D(t)), \tag{24}$$

i.e., $D(t)$ is the horizontal distance between the two functions.

In a typical application, we are given $\Lambda(t)$ and $Q(0)$. These and $\mu(t)$ determine all other quantities. The function $\mu(t)$ may also be specified, or we may analyze the consequences of different functions under certain constraints. For the Bay Bridge example, we may want to schedule the number of open toll booths under the constraint that the total service capacity $\int_0^T \mu(t)\,dt$ is fixed. In Figure 5-3, we illustrate the case $\mu(t) = \mu$, a constant.

Let our reference interval, e.g., morning rush hour, be $[0, T]$, where

$$L \approx Q = \int_0^T Q(t)\,dt/T, \text{ and} \tag{25}$$

$$\lambda = \frac{\Lambda(T)}{T}. \tag{26}$$

How do we define d? The continuous analog of (5) is

$$d = \int_0^{\Lambda(T)} \mathscr{D}(\Lambda)\,d\Lambda/\Lambda(T), \tag{27}$$

or, changing variables and using (23), we have

$$d = \int_0^T D(t)\lambda(t)\,dt/\Lambda(T). \tag{28}$$

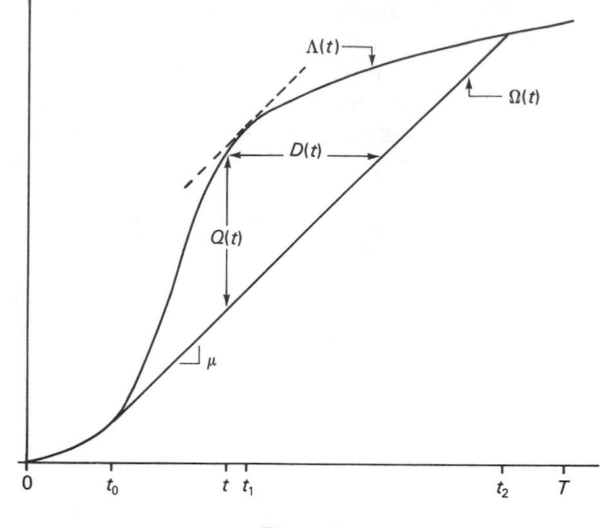

Figure 5-3

The numerators in (25), (27), and (28) are just different representations for the area between the arrival and departure functions, Λ and Ω. Consequently, from (25), (26), and (28), we again have

$$Q = \lambda d. \tag{29}$$

It is important to observe that d is an average over *customers,* i.e., customers are given equal weight. This accounts for the *weighting factor* $\lambda(t)$ in (28).

LIFO Delays

How would we represent LIFO delays? When $\lambda(t) > \mu(t)$, the departure rate is $\mu(t)$, even under LIFO. Hence, even though the most recent arrivals are served first, not all of them can be served immediately. In effect, the arrival stream is split. At rate $\mu(t)$, arrivals depart immediately, and at rate $\lambda(t) - \mu(t)$ they join the queue. [Alternatively, the *proportion* of arrivals that depart immediately is $\mu(t)/\lambda(t)$.]

When do those who queue depart? Under LIFO, customers join the *front* of the queue. When C_j departs, the customers left behind are the same ones who were there when C_j arrived. Hence, the queue *lengths* at a customer's arrival and departure epochs must be equal. With this observation, it is a simple geometric construction in Figure 5-3 to represent

$D_L(t)$ = delay in queue under LIFO of those arrivals at epoch t who join the queue,

where $D_L(t)$ may be either greater or less than $D(t)$.

Let the queue begin to form at t_0, reach its maximum at t, and vanish at t_2. For those unfortunate ones who join the queue at t_0,

$$D_L(t_0) = t_2 - t_0, \tag{30}$$

which is much greater than the maximum of $D(t)$,

$$\max_t D(t) = D(t_1). \tag{31}$$

However, the average delay under LIFO,

$$d_L = \int_{t_0}^{t_1} D_L(t)[\lambda(t) - \mu(t)]\, dt/\Lambda(T), \tag{32}$$

is the *same* as under FIFO, (28), because the numerator in (32) is yet another representation of the area between Λ and Ω.

As in Example 5-2, LIFO merely changes the order of service in a way unrelated to customer service times. For stochastic models, we will show in Section 5-14 that changing from FIFO to LIFO *increases the variance* of the waiting time distribution.

Tandem Queues

An interesting generalization of the model in Figure 5-3 is to process arrivals in series, where μ_i is the service capacity of station i. The departure function from station 1, $\Omega_1(t)$, becomes the arrival function at station 2, and so forth.

For example, letters at a post office are sorted several times before they depart for their eventual destinations. This problem has been studied for an arbitrary number, k say, of queues in tandem under the constraint

$$\sum_{i=1}^{k} \mu_i = \text{a constant}. \tag{33}$$

If we allocate personnel to service tasks (successive sorts), where the service rates of each person on each task is the same, (33) represents the constraint that the work force is constant. We now illustrate this problem for the case $k = 2$.

Example 5-3: Queues in Tandem with Constant Work Force

Let the respective service rates be μ_1 and μ_2 at stations 1 and 2, respectively, where $\mu_1 + \mu_2 = \mu$, a constant. The problem is to split μ into μ_1 and μ_2 in a way that minimizes *overall* average delay from arrival until departure from the *second* station. From $L = \lambda w$, we know that this is equivalent to minimizing the area between $\Lambda(t)$ and $\Omega_2(t)$. Clearly, $\mu_1 < \mu_2$ is wasteful because there will be idle personnel at the second station. In Figure 5-4, we show the case $\mu_1 > \mu_2$. The notation in Figure 5-4 is an obvious extension of that in Figure 5-3, e.g., $D_i(t)$ is the delay in queue at station i, $i = 1, 2$, of an arrival at epoch t. When $\mu_1 > \mu_2$, we can lower $\Omega_1(t)$ and raise $\Omega_2(t)$ for *every* t by moving μ_1 down closer to $\mu/2$. This reduces the area between

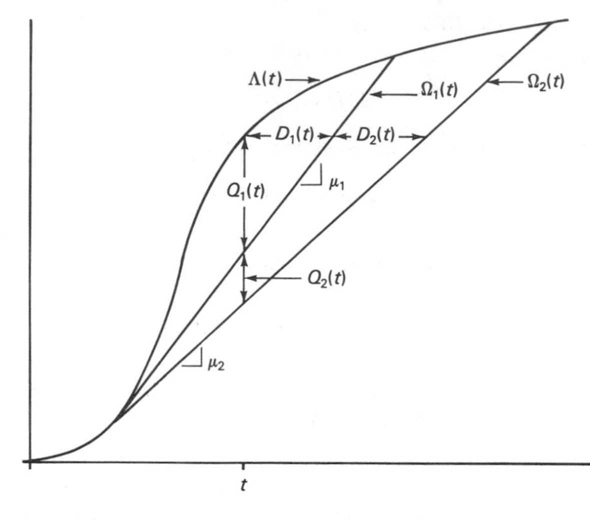

Figure 5-4

Λ and Ω_2. In fact, the *optimal* solution (the one that minimizes this area) is

$$\mu_1 = \mu_2 = \mu/2, \tag{34}$$

i.e., keep *all* of the queue at the first station.

Perhaps the easiest way to understand this result is to suppose that we have m workers to allocate to the stations, where a worker can process (serve) r_i customers per unit time (per hour, say) at station i, $i = 1, 2$. An allocation of m_i workers to station i, where $m_1 + m_2 = m$, results in a service capacity of $\mu_i = m_i r_i$ there. For the special case $r_1 = r_2$, fixing $\mu_1 + \mu_2 = \mu$ is equivalent to fixing m. If m is large, we can, at least as an approximation, ignore the constraint that m is an integer, and treat the allocations as continuous variables.

For every $t \geq 0$, define

$V(t)$ = total amount of *work in system,* in worker hours, that remains to be done on all customers in the system at epoch t.

For our two-station model,

$$V(t) = (1/r_1 + 1/r_2)Q_1(t) + Q_2(t)/r_2. \tag{35}$$

An allocation is *efficient* if all workers are fully utilized at every epoch t where $V(t) > 0$. The function $\{V(t);\ t \geq 0\}$ is the same for *every* efficient allocation. An allocation that is not efficient can only increase $V(t)$ for some t.

Before a typical rush hour, $V(t) = 0$ for t on some interval $0 \leq t \leq t_0$, say. On some interval to the right of t_0,

$$V(t) = (1/r_1 + 1/r_2)[\Lambda(t) - \Lambda(t_0)] - m(t - t_0), \tag{36}$$

i.e., (36) simply states that remaining work is the difference between the cumulative amount of work that has come in and the amount of work performed, where work is performed at rate m, when m workers are fully utilized.

Since $\{V(t)\}$ is fixed, we can minimize $Q(t) = Q_1(t) + Q_2(t)$ for *all* t by *maximizing* the remaining work/customer. This is accomplished by keeping all of the queue *at the first station,* which in turn is accomplished (efficiently) by having equal processing rates, $\mu_1 = \mu_2$. It is easy to see that this result holds for an arbitrary number of tandem stations, whether or not the processing rates are equal.

One must not carry this message too far, however. This conclusion depends on having a tandem arrangement of stations, where all customers are served at all stations, in some particular order. If only a fraction of arrivals at the first station are served at the second station, for example, it may *not* be optimal to keep all of the queue at the first station. See Problem 5-4.

Remark. The concept of work in system is very important and will appear again in Section 5-13 and in Chapters 10 and 11.

Dispatching Policies

Another variation on the model in Figure 5-3 is to empty the queue periodically with an "instantaneous" service. This would correspond to trucks picking up all sorted mail at a post office at certain times. For arrival function $\Lambda(t)$ and dispatch epochs $t_1 < t_2 < \cdots$, the departure function $\Omega(t)$ is a step function with jumps at $t_1, t_2, \ldots,$ where

$$\Omega(t_i) = \Lambda(t_i), \qquad i = 1, 2, \ldots . \tag{37}$$

Given a dispatch occurs at T (the end of the day, say), we would like to schedule an intermediate dispatch at epoch t, $0 < t < T$, to minimize the average delay. Again, this is equivalent to minimizing the area between Λ and Ω on the interval $[0, T]$.

5-5 THE $M/M/1$ QUEUE

We now return to treating queues as stochastic phenomena. To begin, consider a single-channel (server) queue with Poisson arrivals at rate λ and exponential service at rate μ: the $M/M/1$† queue. The footnote describes the usual shorthand

† The notation $A/B/c$ is used to classify certain common single- and multiple-channel queueing models, where c is the number of channels, A denotes the inter-arrival time distribution, and B denotes the service time distribution. Implicit with this notation is the assumption that the inter-arrival times $\{T_j\}$ and the service times $\{S_j\}$ are independent and that the random variables in each sequence are identically distributed. The symbols in positions A and B denote specific families of distributions, where M denotes Poisson (exponential); D, constant (deterministic); G, general; E_k, k-Erlang; and H, hyper-exponential. Erlang and hyper-exponential distributions are defined in Section 5-10. Unless the contrary is stated, we assume FIFO and, when $c > 1$, that all channels are fed by a single queue. Thus, the single-channel queue with Poisson arrivals and exponential service is denoted $M/M/1$. Similarly, $M/G/\infty$ denotes an infinite-channel queue with Poisson arrivals and general service (see Example 5-1). A notational anomaly is the use of GI rather than G to denote a general inter-arrival distribution, e.g., the $GI/M/c$ queue.

notation for denoting a variety of single- and multiple-channel queues, together with implicit assumptions that are made when this notation is used.

The number-in-system process $\{N(t): t \geq 0\}$ is a simple example of a continuous-time Markov chain (Chapter 4), where only two types of transitions are possible, an arrival, which increases $N(t)$ by one customer, and a departure (service completion), which decreases $N(t)$ by one.

Let $P(2, t)$ be the probability of two or more transitions in an interval of length t (at least two arrivals, or two departures, or one of each). As $t \to 0$, we show in Chapter 4 that $P(2, t)$ gets small in the following sense:

$$\lim_{t \to 0} \frac{P(2, t)}{t} = 0. \tag{38}$$

This is a generalization of the *orderliness* property of a Poisson process (Section 2-6). Any function of t with property (38) will be denoted by $o(t)$, i.e., $o(t)$ denotes any function of "smaller order" than t, meaning that

$$\lim_{t \to 0} \frac{o(t)}{t} = 0. \tag{39}$$

Given $N(t) = 0$, the next transition must be an arrival. Thus, the time spent in state 0 has exponential density $\lambda e^{-\lambda x}$, $x > 0$, i.e., is $\sim \exp(\lambda)$. From (38) and (39), we can write

$$P(N(t) = 0 \mid N(0) = 0) = e^{-\lambda t} + o(t), \tag{40}$$

where the first term on the right is the probability of *no* transitions in $(0, t)$, and the second term is the probability of an event that involves at least two transitions. From this it is simple to verify

$$\lim_{t \to 0} \frac{P(N(t) > 0 \mid N(0) = 0)}{t} \equiv a_0 = \lambda. \tag{41}$$

This is the transition rate *out* of state 0, given that the chain is in state 0, i.e., it is a conditional transition rate, and will be called a *conditional rate*. Transitions out of state 0 are into state 1, i.e.,

$$\lim_{t \to 0} \frac{P(N(t) = 1 \mid N(0) = 0)}{t} \equiv a_{01} = \lambda, \tag{42}$$

the conditional rate from state 0 to state 1.

Given $N(t) = n > 0$, the next transition may be either an arrival or a departure. The time until the next arrival is $\sim \exp(\lambda)$. From the memoryless property of the exponential, the time until the next departure is $\sim \exp(\mu)$. The time until the next *transition* is the minimum of these two independent random variables and is $\sim \exp(\lambda + \mu)$. Furthermore, the probability that the next transition is an arrival (i.e., the next arrival occurs before the next departure) is $\lambda/(\lambda + \mu)$. With these considerations in mind, the corresponding conditional rates are

$$\begin{aligned} a_n &= \lambda + \mu, \text{ rate out of state } n, \\ a_{n,n+1} &= \lambda, \text{ rate from state } n \text{ to state } n+1, \\ a_{n,n-1} &= \mu, \text{ rate from state } n \text{ to state } n-1. \end{aligned}$$

In terms of these rates, the limits

$$\lim_{t \to \infty} P(N(t) = n) = p_n \tag{43}$$

may be found as the solution to a system of equations called *balance* equations. See Section 4-3 and equation (23) there.

$$\begin{aligned} &\textit{out} \quad \textit{in} \\ \lambda p_0 &= \mu p_1 \\ (\lambda + \mu)p_n &= \mu p_{n+1} + \lambda p_{n-1}, \qquad n = 1, 2, \ldots . \end{aligned} \tag{44}$$

For each n, the limit p_n in (43) is also the proportion of time there are n customers in system; in Markov chain terminology, this is the proportion of time the chain (or system) is in *state n*. Balance equations, e.g., (44), have the intuitive interpretation of equating *unconditional* transition rates (number of transitions per unit time) into and out of each state. Thus, the transition rate out of state n is given on the left as the conditional rate out when there times the proportion of time there, $(\lambda + \mu)p_n$. The transition rate into state n is the sum of rates into state n from other states, where each term in the sum is the product of an appropriate rate and proportion of time, $\mu p_{n+1} + \lambda p_{n-1}$. The corresponding equation is the balance equation *about* state n.

Fortunately, (44) is easy to solve. In terms of $\rho = \lambda/\mu$, $p_1 = \rho p_0$. The balance equation about state 1, $(\lambda + \mu)p_1 = \mu p_2 + \lambda p_0$, contains the terms in the balance equation about state 0. Canceling these terms, we have $\lambda p_1 = \mu p_2$. Recursively, it is easy to show that $\lambda p_n = \mu p_{n+1}$, $n = 1, 2, \ldots$, and that

$$p_n = \rho p_{n-1} = \rho^n p_0. \tag{45}$$

We still have $\sum_{n=0}^{\infty} p_n = 1$ to use, which, with (45), gives

$$p_0 \sum_{n=0}^{\infty} \rho^n = 1. \tag{46}$$

Thus, we are able to solve for p_0 (and p_n) if and only if the series in (46) converges. This occurs when $\rho < 1$, where $\sum_{n=0}^{\infty} \rho^n = 1/(1 - \rho)$. Thus, when $\rho < 1$, $p_0 =$

$1 - \rho$ and

$$p_n = (1 - \rho)\rho^n, \qquad n = 0, 1, \ldots . \tag{47}$$

The average number of customers in system, L, is simply the mean of the distribution (47), which is a geometric distribution:

$$L = \sum_{n=0}^{\infty} np_n = \frac{\rho}{1 - \rho}. \tag{48}$$

The average number of customers in the queue is

$$Q = \sum_{n=1}^{\infty} (n - 1)p_n = \frac{\rho^2}{1 - \rho}. \tag{49}$$

An easy check on what we have done is to use (13), i.e., the average number of customers in *service* is

$$L_s = \lambda/\mu = \rho = 1 - p_0, \tag{50}$$

check! We could find Q from (48) and (50), $Q = L - L_s$.

Equations (48) to (50) are valid when $\rho < 1$. When $\rho \geq 1$, the series in (46) diverges and no solution of form $p_n > 0$, $\sum_{n=0}^{\infty} p_n = 1$ exists. This means that when $\rho \geq 1$,

$$\lim_{t \to \infty} P(N(t) = n) = 0, \qquad n = 0, 1, \ldots . \tag{51}$$

A more meaningful way of writing (51) is

$$\lim_{t \to \infty} P(N(t) > n) = 1 \tag{52}$$

for every finite n. Thus, the queue grows without bound, and both L and Q are infinite.

In the terminology of Section 5-2, the $M/M/1$ queue is *stable* for $\rho < 1$, and *unstable* for $\rho \geq 1$. The stability condition $\rho = \lambda/\mu < 1$ simply means that the arrival rate λ must be *less* than the service rate μ. Even when the queue is stable, queues can be long. For example, we plot L vs. ρ in Figure 5-5.

As $\rho \to 1$, L grows without bound.†

The average waiting time in system, w, may be found from (48),

$$w = L/\lambda = 1/(\mu - \lambda). \tag{53}$$

Similarly, the average delay in queue is

$$d = Q/\lambda = w - 1/\mu = 1/(\mu - \lambda) - 1/\mu. \tag{54}$$

† The behavior of the queue when $\lambda = \mu$ is quite different from when $\lambda > \mu$. In the former case, the Markov chain is *null*; in the latter, the chain is *transient*.

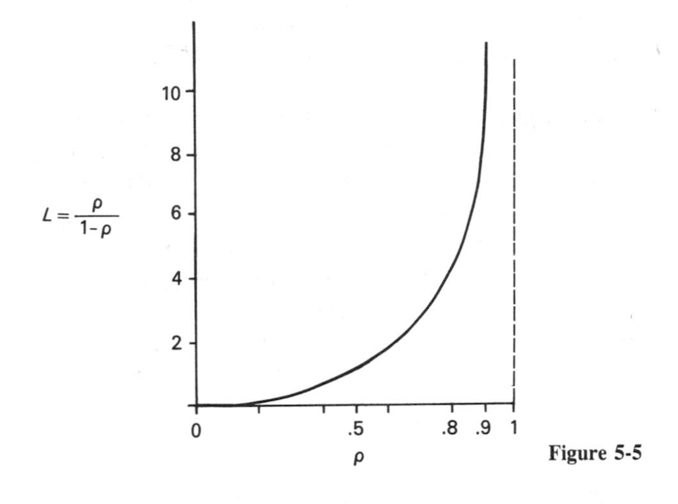

Figure 5-5

The behavior of the $M/M/1$ queue is typical of many queueing models: For stability, the arrival rate must be less than the service capacity. The behavior of L and w for queues that operate close to capacity will look (roughly) like Figure 5-5.

As Example 5-4 illustrates, it is easy to incorporate cost considerations into queueing models for decison-making purposes.

Example 5-4

You are considering renting some data processing equipment. Model 1 has service rate $\mu_1 = 100$ customers/hour and rents for $c_1 = \$1000$/month. Model 2 has service rate $\mu_2 = 200$ customers/hour and rents for $c_2 = \$1800$/month. The arrival rate of customers (orders to be processed) is $\lambda = 80$ customers per hour. The waiting time cost is judged to be \$1.00/customer-hour. Assume Poisson arrivals during normal working hours (about 200 hours/month), exponential service, and that the equipment operates only during these hours. Which model should you rent? (Ignore carryover effects, e.g., over a weekend.)

To decide, let us compare the total expected cost per hour for each model. The rental cost per hour for Model i is $c_i/200$, $i = 1, 2$. Should the waiting time cost/hour be figured in terms of L or w? In terms of L, it is simply $(\$1.00)L = \$L$/hour. In terms of w, we would first calculate the average cost per customer, $(\$1.00)w = \$w$. To get the average cost per hour, multiply by λ = average number of customers per hour, $\lambda \cdot \$w = \$\lambda w$. From $L = \lambda w$, it *doesn't make any difference* which way we perform the calculation. Hence, the total expected cost/hour of Model i is

$$L_i + c_i/200, \qquad i = 1, 2, \tag{55}$$

where L_i is given by (48) and $\rho_i = \lambda_i/\mu_i$. Computing (55), we obtain \$9/hour for Model 1 and \$9.67 per hour for Model 2. You should rent Model 1. Notice, however, that $\rho_1 = .8$ and results are quite sensitive to small changes in λ. Don't sign any long-term contracts!

5-6 ARRIVAL AND TIME AVERAGES; FIFO WAITING TIMES

For the $M/M/1$ queue, we found the *distribution* of number of customers in system, $\{p_n\}$, but only the average waiting time, w. How would we find the corresponding waiting time distribution (15) under FIFO? This calculation depends on the congestion of the queue *as found by an arrival.*

Let a_n be the proportion of arrivals who find n other customers in system on arrival, $n = 0, 1, \ldots,$ and let N_a denote a random variable with distribution $\{a_n\}$. Thus, N_a is the number of customers in system found by an arrival. Let W be a FIFO waiting time of a "typical" customer, i.e., one who has distribution defined by (15). We can represent W in terms of N_a as follows:

$$W = \sum_{j=1}^{N_a+1} S_j, \tag{56}$$

i.e., for a single-server FIFO queue, a customer's waiting time is the sum of: the service times of those customers found in queue, the remaining service time of the customer (if any) found in service, and the service time of the arriving customer. Because service is exponential, the remaining service time is $\sim$ exp (μ). Thus, (56) is the sum of $N_a + 1$ i.i.d. exponential random variables. The distribution of W may be found in terms of the distribution of N_a, $\{a_n\}$.

Because arrivals are Poisson,

$$\{a_n\} = \{p_n\}, \tag{57}$$

i.e., for each n, the proportion of *arrivals* that find n customers in system is equal to the proportion of *time* there are n customers in system.

It is important to realize that contrary to $L = \lambda w$, (57) depends on properties of the Poisson arrival process. When arrivals are not Poisson, results such as (57) are usually false. For example, consider the $D/D/1$ queue with constant interarrival times of 10 minutes each and constant service times of 9 minutes each. Suppose an arrival occurs at epoch 0 to find the system empty. Now plot $N(t)$ as a function of t as shown in Figure 5-6. Clearly, the fraction of time the number in system equals 1, $p_1 = .9$. The corresponding fraction of arrivals $a_1 = 0$. Since by definition, p_1 and a_1 must fall in the interval [0, 1], they could not be much farther apart!

Result (57) is an example of what is known as "Poisson Arrivals See Time Averages" or *PASTA*. In Chapter 4 we showed the following: The probability that a Poisson event (not necessarily an arrival) finds a continuous-time Markov chain in some state is equal to the fraction of time it is there. More generally, an arrival may observe other aspects of a system such as remaining service or work in system, and the system may be a general stochastic process, rather than a Markov chain. PASTA will hold under these generalizations as well. See Section 5-16.

In Chapter 4, we also showed that the terms in balance equations such as

(44) are indeed transition rates. In particular, for our *M*/*M*/1 model,

$$\lambda p_n \tag{58}$$

is the rate (number of events per unit time) that arrivals find n customers in system.

The equality of transition rates into and out of each state, or across other boundaries, is of course true for arbitrary stochastic processes, not just Markov chains. Thus, for the process in Figure 5-6, 6 customers per hour find the system empty and 6 customers per hour leave the system empty. However, for arbitrary processes, there is no direct connection between transition rates and the proportion of time a process is in various states.

Now that we know the distribution of N_a in (56), we may find the distribution of W. This will be done by finding $E\{e^{-sW}\}$. First observe that the exponential S_j each have transform $E\{e^{-sS_j}\} = \mu/(\mu + s)$. Hence,

$$E\{e^{-sW} \mid N_a\} = [\mu/(\mu + s)]^{(N_a+1)}, \tag{59}$$

because the transform of a sum of independent random variables is the product of their transforms. By unconditioning (59), we have

$$\begin{aligned} E\{e^{-sW}\} &= E\{[\mu/\mu + s]^{(N_a+1)}\} \\ &= \sum_{n=0}^{\infty} \left(\frac{\mu}{\mu + s}\right)^{(n+1)} (1 - \rho)\rho^n \\ &= \frac{\mu - \lambda}{\mu + s} \sum_{n=0}^{\infty} \left(\frac{\lambda}{\mu + s}\right)^n = \frac{(\mu - \lambda)/(\mu + s)}{1 - \lambda/(\mu + s)} \\ &= \frac{\mu - \lambda}{\mu - \lambda + s}. \end{aligned} \tag{60}$$

Notice that (60) is also the transform of an exponential distribution, now with parameter $(\mu - \lambda)$, i.e.,

$$W \sim \exp(\mu - \lambda). \tag{61}$$

The mean of (61) is $w = 1/(\mu - \lambda)$, previously found in (53).

While (57) may be used to obtain the waiting time distribution for other models, results are usually complicated. [An exception is (84).] It is also very

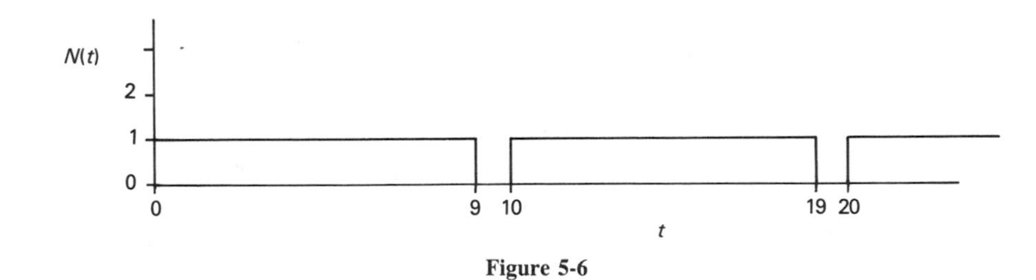

Figure 5-6

useful for obtaining means, e.g., L and w. We now illustrate this for the $M/M/1$ queue. The expected value of (56) is

$$w = (E(N_a) + 1)/\mu. \tag{62}$$

From (57), $E(N_a) = \sum_{n=0}^{\infty} np_n = L$, or

$$w = (L + 1)/\mu. \tag{63}$$

With $L = \lambda w$ and (63), we have two equations and two unknowns, from which L *and* w may be found, in agreement with (48) and (53). Notice that to do this, we did not have to solve the balance equations!

5.7 THE *M/M/*1 QUEUE WITH QUEUE LIMIT

For stable $M/M/1$ queues, $p_n > 0$ for all n. Sooner or later, the number of customers in system will reach any finite n. In practical situations, this may be physically impossible. For example, a gas station may not have room for more than l cars, for some $l > 0$.

For an $M/M/1$ queue, assume that arrivals who find l customers in system depart immediately without receiving service. This is a *queue limit* model, where the limit on the queue size is $(l - 1)$. In terms of transition rates, this is equivalent to letting λ depend on the number of customers in system, n say, where $\lambda_n = \lambda$ for $n < l$ and $\lambda_l = 0$.

There are only a finite number of states, and the balance equations for this model are

$$\begin{aligned} \lambda p_0 &= \mu p_1, \\ (\lambda + \mu)p_n &= \mu p_{n+1} + \lambda p_{n-1}, \qquad n = 1, \ldots, l - 1, \\ \mu p_l &= \lambda p_{l-1}. \end{aligned} \tag{64}$$

As was done in Section 5-5, (64) may be solved recursively for p_n in terms of p_0,

$$p_n = \rho^n p_0, \tag{65}$$

where $\rho = \lambda/\mu$. Now solve for p_0,

$$\sum_{n=0}^{l} p_n = p_0 \sum_{n=0}^{l} \rho^n = p_0(1 - \rho^{l+1})/(1 - \rho) = 1,$$

$$p_0 = (1 - \rho)/(1 - \rho^{l+1}). \tag{66}$$

Note that since the number of states is finite, a solution always exists, regardless of the magnitude of ρ. (What happens when $\rho = 1$?)

From (65) and (66), finding L is a straightforward algebraic exercise:

$$L = \sum_{n=0}^{l} np_n = p_0 \sum_{n=0}^{l} n\rho^n = \frac{\rho[1 - (l + 1)\rho^l + l\rho^{l+1}]}{(1 - \rho)(1 - \rho^{l+1})}. \tag{67}$$

To find w, we may apply $L = \lambda w$. However, not all arrivals are served. First, let us agree to define w as the average waiting time in system *of those who are served*. Since customers who are not served depart immediately and are not counted in L, we may find w from

$$w = L/\lambda_s, \tag{68}$$

where λ_s is the arrival rate of those served.

In Section 5-6, we observed that for an $M/M/1$ queue, which has Poisson arrivals, the fraction of arrivals who find the system in state n is p_n, the corresponding fraction of time. That result applies here as well. For this purpose, let $\lambda_l = \lambda$. An arrival who finds l customers in system generates a transition from l to l. The effect on (64) is to add λp_l to *both* sides of the balance equation about state l. [The solution to (64) is unchanged.] The fraction of arrivals who are served is $1 - p_l$. The corresponding arrival rate of those who are served is

$$\lambda_s = \lambda(1 - p_l). \tag{69}$$

If customers who are not served represent lost sales at a cost of $\$C$ each, the average cost of lost sales per unit time is

$$\lambda p_l C, \tag{70}$$

which decreases as l increases. This would help us determine whether increasing the size of the customer waiting room is worthwhile.

5-8 BIRTH AND DEATH MODELS

A *birth and death process* is a continuous-time Markov chain where only transitions between adjacent states are permitted. For each state n, let the transition rates be

$$a_n = \lambda_n + \mu_n,$$

$$a_{n,n+1} = \lambda_n,$$

$$a_{n,n-1} = \mu_n, \qquad n = 0, 1, \ldots,$$

where $\mu_0 = 0$. The terminology arises from the study of biological populations where arrivals represent births and departures represent deaths. The birth and death rates are permitted to be functions of the population size.

The $M/M/1$ queue is an example of a birth and death queue where $\lambda_n = \lambda$ for all n, $\mu_0 = 0$, and $\mu_n = \mu$ for $n \geq 1$. With the addition of a queue limit, $\lambda = 0$ for $n \geq l$.

Now let λ_n and μ_n be arbitrary strictly positive functions of n, except for $\mu_0 = 0$ and, in the queue limit case, $\lambda_n = 0$ for $n \geq l$. The balance equations are

$$\begin{aligned} \lambda_0 p_0 &= \mu_1 p_1, \\ (\lambda_n + \mu_n)p_n &= \mu_{n+1}p_{n+1} + \lambda_{n-1}p_{n-1}, \qquad n = 1, 2, \ldots . \end{aligned} \tag{71}$$

By solving (71) recursively in the same manner as we did for the $M/M/1$ queue, we have

$$\lambda_n p_n = \mu_{n+1}p_{n+1}, \qquad n = 0, 1, \ldots . \tag{72}$$

From (72),

$$p_n = b_n p_0, \qquad n = 0, 1, \ldots, \tag{73}$$

where $b_0 = 1$ and

$$b_n = \prod_{i=1}^{n} (\lambda_{i-1}/\mu_i), \qquad n = 1, 2, \ldots . \tag{74}$$

If $\sum_{n=0}^{\infty} b_n$ converges,

$$p_0 = 1 \Big/ \sum_{n=0}^{\infty} b_n, \tag{75}$$

and we have found the unique probability distribution $\{p_n\}$ that satisfies the balance equations. If $\sum_{n=0}^{\infty} b_n$ diverges, no solution exists.

Some constraint on λ_n and μ_n as functions of n is needed to prevent the number of transitions in finite time from growing too rapidly. We will not go into details here; see Section 4-4 for a theoretical treatment of these processes. As shown there, an upper bound on λ_n is sufficient to prevent bizarre bahavior. When this condition is satisfied, then either

1. $\sum_{n=0}^{\infty} b_n$ converges, the queue is *stable*, and, for every n,

$$\lim_{t \to \infty} P(N(t) = n) = p_n > 0, \qquad \sum_{n=0}^{\infty} p_n = 1, \text{ or}$$

2. $\sum_{n=0}^{\infty} b_n$ diverges,

$$\lim_{t \to \infty} P(N(t) = n) = 0 \qquad \text{for every } n, \text{ and}$$

the queue increases without bound.

Before presenting some examples, we give an intuitive explanation of (72). In Figure 5-7, we represent the states as nodes in a transition diagram, where arcs connecting the nodes represent the possible transitions between pairs of states. The balance equation about state n equates transition rates in and out of the dashed circle around state n. We may also equate transition rates across other boundaries, in particular, the vertical dashed line between adjacent states. Transition rates across this boundary have a very simple form: The transition rate from left to right is the left-hand side of (72). The transition rate from right to left is the right-hand side of (72).

Example 5-5: The $M/M/\infty$ Queue

Consider an infinite channel queue with Poisson arrivals at rate λ and exponential service at rate μ per channel. Clearly, $\lambda_n = \lambda$ for all n. When there are n customers in system, the time until the next service completion is the minimum of n exponential service times. The corresponding service rate is $\mu_n = n\mu$. Therefore, from (72),

$$\lambda p_n = (n + 1)\mu p_{n+1}, \text{ and} \tag{76}$$

$$p_n = (\lambda/\mu)^n p_0/n!, \qquad n = 0, 1, \ldots . \tag{77}$$

Setting $\sum_{n=0}^{\infty} p_n = 1$, we find $p_0 = e^{-\lambda/\mu}$, and

$$p_n = e^{-\lambda/\mu}(\lambda/\mu)^n/n!, \qquad n = 0, 1, \ldots , \tag{78}$$

a Poisson distribution in agreement with Example 5-1.

Example 5-6: Finite Source Models

Suppose we have m machines where the time until breakdown of each working machine is exponential with rate λ (mean $1/\lambda$). If the breakdown of a machine constitutes an arrival at a repair facility, we have a *finite source* model. When there are n machines at the facility, the arrival rate of the next machine is $(m - n)\lambda$.

A finite source model is one component of a queueing model; it describes the nature of the arrival process. We also have to model the facility. Suppose there is a single repairman at the facility, and that repair times are exponential at rate μ. Let state n denote the number of machines at the repair facility. From (72),

$$\lambda(m - n)p_n = \mu p_{n+1}, \qquad n = 0, 1, \ldots , m - 1, \tag{79}$$

which is easily solved for the $\{p_n\}$.

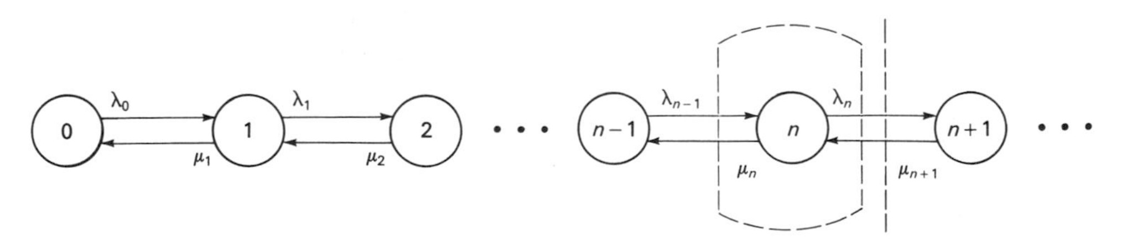

Figure 5-7

If m is very large and $m\lambda/\mu$ is less than (and not too close to) 1, results with the finite source model will differ little from an $M/M/1$ queue with Poisson arrival rate $m\lambda$. In fact, in most applications, a finite population generates the arrivals. The assumption of Poisson arrivals is an approximation (often a very good one).

5-9 THE *M/M/c* QUEUE; COMPARISON WITH SINGLE-CHANNEL QUEUES

This is an example of a birth and death model, but it deserves special attention because of its relationship to the $M/M/1$ queue and to more general multichannel queues. For arrival rate λ and service rate μ/server, (72) becomes

$$\begin{aligned} \lambda p_n &= (n+1)\mu p_{n+1}, \qquad n = 0, 1, \ldots, c-1, \\ \lambda p_n &= c\mu p_{n+1}, \qquad n = c, c+1, \ldots . \end{aligned} \tag{80}$$

For $n \geq c - 1$,

$$p_{n+1} = \rho p_n \tag{81}$$

where $\rho = \lambda/c\mu$, from which it is easy to see that the queue is stable when $\rho < 1$ and unstable otherwise. As defined in this way, ρ is called the *utilization factor*, i.e., it is the fraction of the service capacity that is used. Thus, if each channel (server) is busy the same fraction of time, that fraction is ρ. An alternative and equally meaningful way of stating the stability condition is that the *offered load*, λ/μ, must be less than the number of servers assigned to handle that load, c. We stress this result because it turns out to hold for $GI/G/c$ queues and even more general arrival and service processes.

For $\rho < 1$ it is easy to solve (80) for

$$p_n = \begin{cases} p_0(c\rho)^n/n! & \text{for } n \leq c, \\ p_0\rho^n c^c/c! & \text{for } n > c, \end{cases} \tag{82}$$

where

$$p_0 = \left[\frac{(c\rho)^c}{(1-\rho)c!} + \sum_{j=0}^{c-1} \frac{(c\rho)^j}{j!}\right]^{-1} . \tag{83}$$

From $\{p_n\}$, it is easy to write an expression for L. Because $\{p_n\}$ has a geometric tail, i.e., (81), we can also easily find the distribution of delay in queue. Let D be a random variable having the limiting delay in queue distribution, analogous to (15). From PASTA and the methods used to obtain (61), it can be shown that

$$(D \mid D > 0) \sim \exp\,(c\mu - \lambda). \tag{84}$$

From (84),

$$d = P(D > 0)/(c\mu - \lambda), \tag{85}$$

where $P(D > 0)$ is the corresponding fraction of time that all channels are busy,

$$P(D > 0) = 1 - \sum_{j=0}^{c-1} p_j. \tag{86}$$

Now we can find w from d,

$$w = d + 1/\mu, \tag{87}$$

and either L or Q by dividing the appropriate quantity by λ.

We now compare the performance of single- and multiple-channel queues. One issue is whether a fast single server (channel) is better than two or more slow servers (channels). The other has to do with *pooling*, i.e., is it better to serve separate arrival streams of customers at separate single-channel facilities or to pool them and serve them at a single multichannel facility?

In Example 5-7, measures of performance L and Q for a single-channel queue with service rate 2μ are compared with those for a two-channel queue with service rate μ/channel. The stability condition $\rho = \lambda/2\mu < 1$ is the same for both, and formulas are at hand for the computation. Before proceeding, it is important to reflect on this question in the abstract because we may want to apply what we learn in a more general setting.

What are the advantages of each arrangement? When there is one customer in the system, that customer receives service rate 2μ in the single-channel case but only μ in the two-channel case, i.e., in the latter case, the full capacity of the service facility is not being used. This would seem to favor the single-channel case. On the other hand, a customer with an unusually long service time would "plug up" a single channel, allowing a (possibly long) queue to form, while this would be bypassed in the two-channel case. This would seem to favor the two-channel case. Let's see what actually happens in the example and resume this discussion later.

Example 5-7: One Fast Versus Two Slow Servers

We now compare measures of performance of an $M/M/1$ queue with service rate 2μ and an $M/M/2$ queue with service rate μ/channel, where both queues have arrival rate λ. Let L_i, Q_i, and $\{p_n^{(i)}\}$ denote results for each queue where i is the number of channels, $i = 1, 2$. The transition diagrams for the two models are shown in Figure 5-8. The vertical dashed lines show correspondence between states in the two models that have the same number of customers in *system*. The diagonal dashed lines show correspondences between states that have the same number of customers in *queue*.

The ratios of successive terms in each sequence $\{p_n^{(i)}\}$, $i = 1, 2, \ldots$ are determined by the transition rates on the arcs, in accordance with (72). Thus $p_{n+1}^{(i)}/p_n^{(i)} = \rho$, $n = 1, 2, \ldots$, for both systems. Only for $n = 0$ are the ratios different, i.e.,

$$p_1^{(1)}/p_0^{(1)} < p_1^{(2)}/p_0^{(2)}. \tag{88}$$

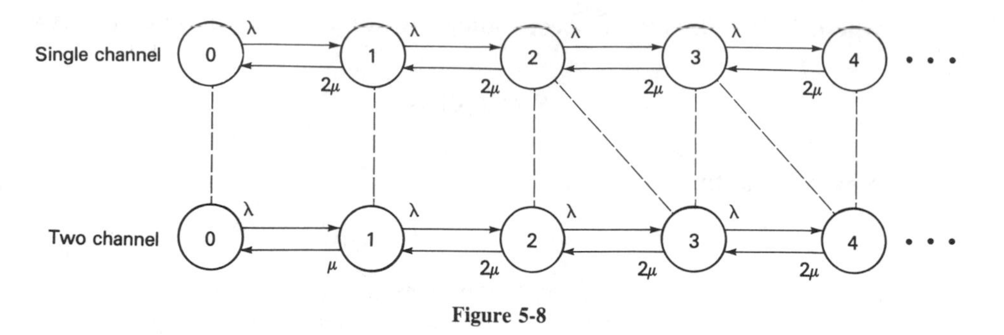

Figure 5-8

Since $\sum_{n=0}^{\infty} p_n^{(i)} = 1$, $i = 1, 2$, (88) implies

$$p_0^{(1)} > p_0^{(2)} \quad \text{and} \quad p_n^{(1)} < p_n^{(2)}, \qquad n = 1, 2, \ldots . \tag{89}$$

On the other hand, let's pretend for the moment that $p_0^{(2)} = 0$, i.e., that state 0 is deleted from the two-channel model. The two models would simply differ in the labeling of state numbers, where corresponding states are indicated by the diagonal dashed lines, and we would have $p_n^{(1)} = p_{n+1}^{(2)}$, $n = 0, 1, \ldots$. However, $p_0^{(2)} > 0$. The effect of this is to reduce the size of all the other terms since their ratios are fixed. Hence,

$$p_n^{(1)} > p_{n+1}^{(2)}, \qquad n = 1, 2, \ldots . \tag{90}$$

From (89)

$$L_1 = \sum_{n=1}^{\infty} np_n^{(1)} < \sum_{n=1}^{\infty} np_n^{(2)} = L_2, \tag{91}$$

but from (90)

$$Q_1 = \sum_{n=1}^{\infty} (n-1)p_n^{(1)} > \sum_{n=2}^{\infty} (n-2)p_n^{(2)} = Q_2. \tag{92}$$

Thus, in terms of the usual measure L, the fast single server is better, but in terms of Q, the reverse is true!

In fact, we have shown more: (89) implies that the distribution of number in system is *stochastically smaller* in the single-channel case, whereas (90) implies the reverse for the number in queue.

How general are the results of Example 5-7? It is easy to generalize the number of servers. In fact, if we compare r channels at rate c each with c channels at rate r each, where $r < c$, it can be shown that

$$L_r < L_c, \quad \text{but} \quad Q_r > Q_c, \tag{93}$$

where the subscripts denote the number of channels.

To what extent do these results depend on the stochastic assumptions of Poisson arrivals and exponential service? It is difficult to answer this question fully, but for (91) to hold, it turns out that the arrival process is irrelevant, provided that the service distribution is sufficiently *regular*; in particular, (91) holds whenever $V(S)/E^2(S) \leq 1$. (What is this ratio for exponential service?) A derivation may be found in Chapter 11, where the notion of full versus partial use of service capacity results in an inequality on *work in system*.

The hooker is that service must be sufficiently regular; i.e., the notion of plugging up the single channel is also relevant. The importance of this effect should grow with service time dispersion. This is the basis for Example 5-8.

Example 5-8: Plugging Up a Single-Channel Queue

Consider two single-channel queues:

$$A/G/1 \quad \text{where } A = \exp(\lambda),$$
$$G = (1 - \beta)U_0 + \beta \exp(2\alpha), \text{ and}$$

$$B/F/1 \quad \text{where } B = \exp(\beta\lambda),$$
$$F = \exp(2\alpha),$$

where U_0 denotes the distribution with all its mass at 0. In the $A/G/1$ queue, an arrival has exponential service with probability β and zero service time with probability $(1 - \beta)$; the arrival streams of each type of customer are Poisson at rates $\beta\lambda$ and $(1 - \beta)\lambda$, respectively. Thus, the $B/F/1$ queue is simply the $A/G/1$ queue where the customers with zero service time have been deleted. The presence of customers with zero service time has no effect on the delay of other customers, and, since the arriving zero-service-time customers are Poisson, they experience the same congestion. Consequently, the expected delay in queue, d_1, must be the same for these two queues. Now consider the corresponding $A/G/2$ and $B/F/2$ queues with 2α replaced by α, i.e., service times take twice as long. Again, the expected delay in queue, d_2, must be the same for these two queues.

Since the $B/F/1$ and $B/F/2$ queues have the same arrival rates and are of the type considered in Example 5-7, (92) implies

$$d_2 < d_1. \tag{94}$$

For the $A/G/1$ and $A/G/2$ queues, the respective average waiting times are

$$w_1 = d_1 + \beta/2\alpha \quad \text{and} \quad w_2 = d_2 + \beta/\alpha. \tag{95}$$

Now fix α and $\beta\lambda$ but let $\beta \to 0$. Since the $B/F/1$ and $B/F/2$ queues are unchanged, so are d_1 and d_2. However, $w_1 \to d_1$ and $w_2 \to d_2$. Hence, for sufficiently small β, (94) implies

$$w_2 < w_1, \tag{96}$$

which reverses the inequality (91)! Small β means that the $A/G/1$ and $A/G/2$ queues are flooded with zero service time customers, and $V(S)/E^2(S)$ can be arbitrarily large.

Example 5-9: Pooling

Now take up the question of pooling. Consider c $M/M/1$ queues, each with arrival rate λ/c and service rate μ. If we pool the arrivals and place the c service channels in parallel, we have a single $M/M/c$ queue with arrival rate λ and service rate μ/channel. The single- and multichannel queues all have utilization factor $\rho = \lambda/c\mu$. Now L and Q depend only on ρ, not on λ and μ individually. For each of the single-channel queues, the expected number in queue, Q_1, is the same as for an $M/M/1$ queue with arrival rate λ and service rate $c\mu$, a fast single server. From (93) with $r = 1$,

$$Q_1 > Q_c, \tag{97}$$

where the left-hand side of (97) is only for *one* of the $M/M/1$ queues. For the collection of c $M/M/1$ queues, the total expected number in queue is c times as large, and the corresponding relation between the average delays [from $d_1 = Q_1/(\lambda/c) = cQ_1/\lambda$ and $d_c = Q_c/\lambda$] is

$$d_1 > cd_c. \tag{98}$$

Thus, pooling is better. For higher utilization, where ρ is near 1 and "most" of w is d, pooling is much better.

The superiority of pooling can be shown to be a very general result independent of the nature of the arrival process and the distribution of service, although the degree of superiority is not necessarily as great as (98). For airline reservation systems, c can be several hundred. Under these circumstances, one can provide good service and have high utilization at the same time. Of course, our results do not cover special-purpose queues or servers with different, specialized training.

5-10 THE SCOPE OF MARKOVIAN QUEUES

The birth and death models we have considered are Markov processes where only transitions between adjacent states are permitted. There was little question about the choice of an appropriate state space; state n meant n customers in system. Consequently, the balance equations reduced to a simple recursive relation for the steady-state probabilities. We have also been restricted to dealing with exponential distributions, which is often unrealistic. Through a series of examples, we will show how to extend the balance equation technique to cover more general situations.

A key consideration is that the state space be valid, i.e., do we have a Markov process? The fundamental test is the Markov property: Given the present state, is the future independent of the past?

Writing balance equations can become a chore, even when the state space is obvious and simple. The main difficulty seems to be remembering how each state can be entered, i.e., the terms on the right-hand side of the balance equations. To keep track of this information, it is often helpful to represent states as nodes and transitions as arcs in a diagram, e.g., see Figures 5-7 and 5-8. When in a state, various transitions can occur, e.g., an arrival or service completion. Each of these possible transitions is represented by an arc from that state to the state that would be entered if a transition of that type occurred. Each arc is labeled with the appropriate transition (arrival or service) rate. In Figure 5-7, the balance equations follow by equating rates into and out of each state, e.g., across the dashed circle.

Example 5-10: Two Queues in Tandem with Blocking and Queue Limit

Tandem refers to a series arrangement of servers. Consider two single-channel queues in tandem with Poisson arrivals at rate λ at the first station and exponential service at rate μ_i at station i, $i = 1, 2$.† Assume the following operating rules:

1. No queue is allowed in front of either station.
2. Both stations may operate simultaneously except that if a customer completes service at station 1 while station 2 is busy, station 1 is blocked from accepting another customer. The blocking customer will not begin service at station 2 until the customer already there departs.

A valid state space for this model is $\mathscr{S} = \{(0, 0), (0, 1), (1, 0), (1, 1), (b, 1)\}$, where the first element of (i, j) refers to the first station and j the second, where 0, 1, and b mean empty, working, and blocked, respectively. The transition diagram is shown in Figure 5-9.

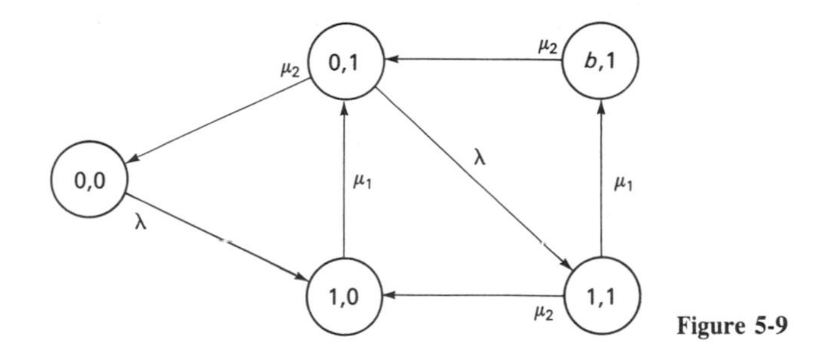

Figure 5-9

The balance equations are

$$\lambda p_{00} = \mu_2 p_{01},$$

$$\mu_1 p_{10} = \mu_2 p_{11} + \lambda p_{00},$$

$$(\lambda + \mu_2) p_{01} = \mu_2 p_{b1} + \mu_1 p_{10},$$

$$(\mu_1 + \mu_2) p_{11} = \lambda p_{01},$$

$$\mu_2 p_{b1} = \mu_1 p_{11},$$

which, together with $\sum_{ij} p_{ij} = 1$, have a unique solution. In terms of these proba-

† Implicit in the analysis is the additional assumption that the service times of the *same* customer at different stations are independent.

bilities, the fraction of arrivals lost is

$$p_{10} + p_{11} + p_{b1}.$$

Other quantities of interest follow immediately:

$$L = p_{01} + p_{10} + 2(p_{11} + p_{b1}),$$

$$w = L/\lambda(p_{00} + p_{01}).$$

Example 5-11: Two Queues in Tandem with No Queue Limits

This is the same model as in Example 5-10 except that unlimited queues are allowed at each station. Obviously, no blocking occurs. For state space $\{(m\ n)\}$ where m is the number of customers at the first station and n is the number at the second, the transition diagram is shown in Figure 5-10.

The balance equations are straightforward and are left as an exercise. It is easy to verify that the balance equations have the solution

$$p_{mn} = (1 - \rho_1)\rho_1^m(1 - \rho_2)\rho_2^n, \qquad n, m \geq 0, \tag{99}$$

where $\rho_i = \lambda/\mu_i$, $i = 1, 2$. Let N_i denote the number of customers in system at station i. From (99), the N_i are independent random variables with marginal distributions of the form of the $M/M/1$, (47).

The *product-form* of solution (99) generalizes to networks of queues, but, with some notable exceptions, breaks down if we fail to have both Poisson arrivals and exponential service. See Chapter 6.

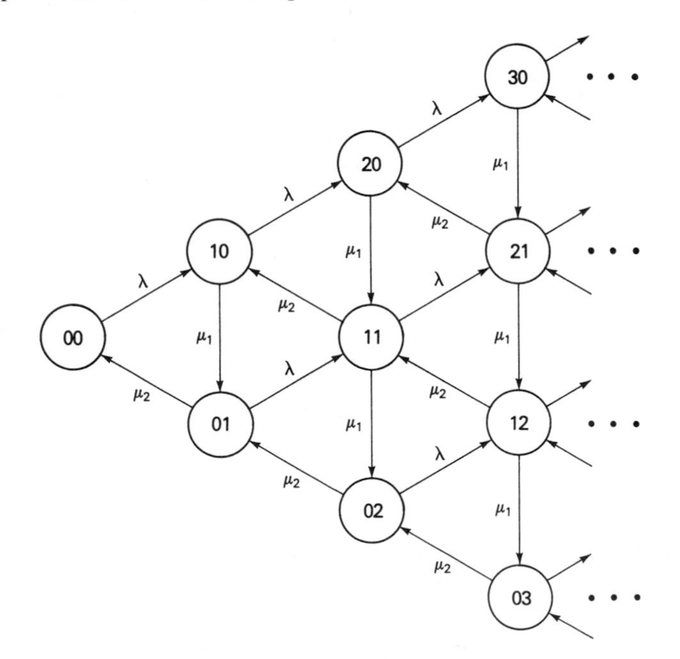

Figure 5-10

Batch arrivals and service can also be handled by these techniques. For example, suppose we have a batch Poisson arrival process of customers (Section 2-9). That is, the arrival process of *batches* of customers is Poisson at rate λ, where each batch contains a random number of customers with distribution $\{\alpha_j\}$. If customers are served individually at a single-channel queue with exponential service, a valid state space is $\mathcal{S} = \{n\}$, where n is the number of customers in the system. An arrival of a batch of size j finding n customers in system would result in a transition to state $n + j$. The corresponding transition rate is $\lambda\alpha_j$. We will treat batch arrivals using a different approach in Section 5-13.

Relaxing the Exponential Assumption

The time spent in each state of a continuous-time Markov chain has an exponential distribution. If in an application, data clearly show that inter-arrival times and/or service times are more (or less) regular than exponential, e.g., $V(S)/E^2(S) << 1$, what do we do? If we are to use the techniques of this section, we have to fabricate other distributions from the exponential. There are two ways to do this: convolution and mixture.

A *convolution* is a distribution obtained by adding independent random variables, as was done in Section 1-10. A *mixture* is a distribution defined as a weighted sum of distributions; e.g., the distribution G in Example 5-8 is a mixture. A convolution makes the resulting distribution more regular, e.g., $V(S)/E^2(S)$ gets smaller, whereas a mixture makes the resulting distribution less regular (more disperse). A more refined fit to data can be obtained by taking mixtures of convolutions.

A *k-Erlang* distribution, E_k, is a k-fold convolution of an exponential distribution, that is, it is the distribution of the sum of k i.i.d. exponential random variables. For example, a k-Erlang service time, S, can be represented as the sum

$$S = \sum_{i=1}^{k} Y_i, \tag{100}$$

where the Y_i are i.i.d. $\sim$ exp $(k\mu)$. Thus,

$$E(S) = kE(Y_i) = 1/\mu, \tag{101}$$

$$V(S) = kV(Y_i) = 1/k\mu^2, \text{ and} \tag{102}$$

$$V(S)/E^2(S) = 1/k. \tag{103}$$

Notice that (103) can be made as small as desired (short of 0) by increasing k. We will still call μ the service rate even though S is not exponential and μ is *not* a conditional rate in a Markov chain.

To incorporate Erlang service into a Markov chain model of a queue, we need to know not only the number of customers in system but also "enough" about the remaining service of the customer in service, where enough means that the Markov property holds. A useful notion is to think of the Y_i as service phases that are performed in sequence. After a time Y_1, the first phase is completed and there are now $k - 1$ phases remaining, and so forth. Thus, if we knew the number of phases completed, $k - j$ say, there would be j phases remaining, and the remaining service distribution would be j-Erlang.

Example 5-12: The $M/E_k/1$ Queue

Let the arrival rate be λ and the service times be defined as in (100). From the preceding discussion a valid state space is $\mathcal{S} = \{(n, j)\}$, where n is the number of customers in the system and, for $n > 0$, j is the number of remaining phases associated with the customer in service, $j = 1, 2, \ldots, k$. For $n = 0$, we define $j = 0$. It turns out to be convenient to define the number of *phases in system* by

$$m = k \max(n - 1, 0) + j. \tag{104}$$

From (104), the arrival of a customer increases the number of phases in system by k. For $m > 0$, the completion of a phase (at *exponential* phase rate $k\mu$) decreases the number of phases in system by one. For state space $\mathcal{S} = \{m;\ m = 0, 1, \ldots\}$, the transition diagram is shown in Figure 5-11.

It is straightforward to write out the balance equations for $\{q_m\}$, the distribution of the number of *phases* in system. It is not difficult to relate $\{q_m\}$ to the distribution of the number of customers in system, $\{p_n\}$. Similarly, L can be obtained from $L_p = \sum_{m=0}^{\infty} mq_m$. In Section 5-12, we show that

$$L = \rho^2(k + 1)/2k(1 - \rho) + \rho, \tag{105}$$

where $\rho = \lambda/\mu$. However, because the $M/E_k/1$ queue is a special case of the $M/G/1$ queue in Section 5-13, we will not go into these details here.

For the first time, we are now able to explore the effect of the service distribution on system performance. From (105), L decreases as k increases, i.e., as the service distribution becomes more regular. For single-channel queues, this behavior is quite general. It is usually true for multichannel queues as well.

Erlang arrivals can be formulated in a similar way, where we would have to keep track of the status of the next arrival. For multichannel Erlang service, we would have to keep track (in terms of the number of remaining phases) of the status of each customer in service.

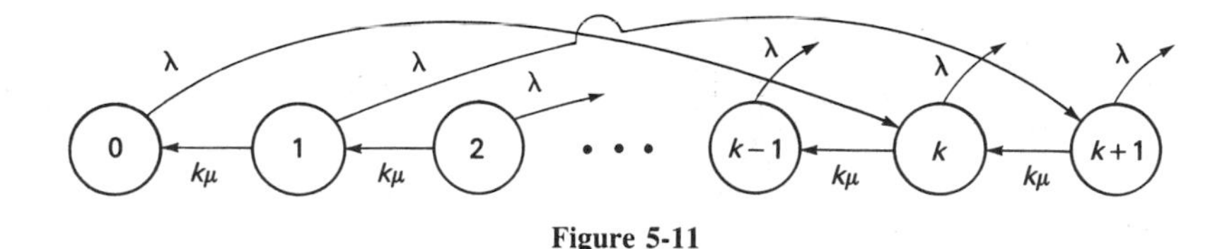

Figure 5-11

Example 5-13: The $E_k/M/1$ Queue

Now suppose that an inter-arrival time T has the representation

$$T = \sum_{i=1}^{k} Y_i, \tag{106}$$

where the Y_i are i.i.d. $\sim$ exp $(k\lambda)$. Thus, T has a k-Erlang distribution with mean $E(T) = 1/\lambda$, i.e., the arrival rate is λ. Service times are distributed exp (μ).

As with Erlang service, we can think of arrivals passing through a sequence of k exponential phases, where now we choose to number the phases $0, \ldots, k - 1$. By phase j, we mean that the next arrival has completed j phases. Thus, we can imagine a system that looks like the one in Figure 5-12.

The elongated next-arrival box is simply a device for keeping track of the status of the next arrival. The next arrival spends a time $\sim$ exp $(k\lambda)$ in each phase and actually arrives at the system on leaving phase $k - 1$. When this occurs, the "new" next arrival enters phase 0. Thus, the next arrival box is never empty. As pictured in Figure 5-12, the next arrival is in phase j, and there are $n = 4$ customers in system.

The state space for this Markov process is $\{(n, j)\}$, where $n \geq 0$ and $j \in \{0, \ldots, k - 1\}$. As in Example 5-12, this two-dimensional state space can be converted to a one-dimensional space,

$$m = kn + j, \qquad m = 0, 1, \ldots, \tag{107}$$

where m is called the number of *phases* in the system. (Notice that m includes phases associated with the next arrival, who is not yet in the system.)

Let $\{q_m\}$ be the distribution of the number of phases in the system. The completion of an arrival phase (at rate $k\lambda$) increases m by one. From (107), the completion of a customer service time decreases m by k. Hence, the balance equations for $\{q_m\}$ are

$$\begin{aligned} k\lambda q_0 &= \mu q_k, \\ k\lambda q_m &= \mu q_{m+k} + k\lambda q_{m-1}, \qquad m = 1, 2, \ldots, k - 1, \\ (k\lambda + \mu) q_m &= \mu q_{m+k} + k\lambda q_{m-1}, \qquad m = k, k + 1, \ldots. \end{aligned} \tag{108}$$

Notice that terms μq_m appear on the left only for $m \geq k$; for $m < k$, $n = 0$.

Let $\{p_n\}$ be the distribution of the number of customers in system. Thus, $p_0 = q_0 + \cdots + q_{k-1}$, and, from (107), $p_n = \sum_{m=nk}^{(n+1)k-1} q_m$, $n \geq 1$.

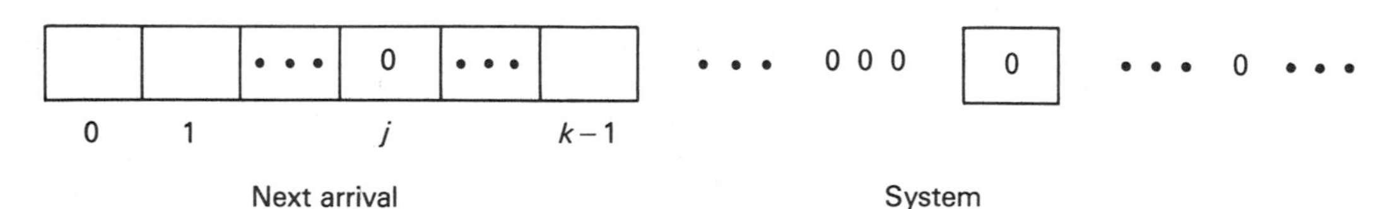

Figure 5-12

The $E_k/M/1$ queue is a special case of the $GI/M/1$ queue treated in Chapter 8. Nevertheless, in order to illustrate a useful technique, we will solve (108) in Section 5-12.

To represent distributions that are *less regular* than the exponential, we define the *hyper-exponential distribution, H,* to be a mixture of exponentials. That is, a hyper-exponential service time S has distribution of form

$$P(S \leq t) \equiv H(t) = \beta(1 - e^{-\mu_1 t}) + (1 - \beta)(1 - e^{-\mu_2 t}), \tag{109}$$

where $0 < \beta < 1$. We define the service rate μ by

$$1/\mu \equiv E(S) = \beta/\mu_1 + (1 - \beta)/\mu_2. \tag{110}$$

More generally, a mixture may involve more than two component distributions.

The easiest way to think about hyper-exponential service is to suppose that there are two types of customers, type 1 with service time distribution exp (μ_1), and type 2 with distribution exp (μ_2), where β is the proportion of customers who are of type 1. The service distribution averaged over both types of customers is the mixture (109).

Mixtures of distributions are less regular than the original distributions in the sense that the coefficient of variation is increased. In particular, for S with distribution (109), it is easily shown that

$$V(S)/E^2(S) > 1, \tag{111}$$

whereas this ratio is equal to 1 for the component exponential distributions. In fact, the ratio in (111) can be made arbitrarily large. This was the idea behind Example 5-8, where one of the distributions was the unit step function at the origin, U_0. By letting $\mu_2 \to \infty$ in (109), exp $(\mu_2) \to U_0$, and the limit of (109) is a distribution of form G in Example 5-8.

Example 5-14: The $M/H/1$ Queue

We have Poisson arrivals at rate λ and hyper-exponential service distribution (109). In order to have a Markov process, we would need to know not only the number of customers in system n, but also enough about the customer in service. In particular, if we know that the customer in service is of type i, $i = 1, 2$, then the remaining service time distribution would be exp (μ_i). Thus, an appropriate state space would be $\{(n, i); n = 0, 1, \ldots; i = 1, 2\}$. The transition diagram for this state space is shown in Figure 5-13. The coefficients β and $1 - \beta$ on the μ_i are the probabilities that the next customer to enter service is of type i, $i = 1, 2$. The balance equations are now straightforward.

We could keep track of the customer type of each customer in queue. This would result in a different state space that, while also correct, is more detailed and complicated. (For the simpler formulation, we could pretend that we don't ask a customer "What type are you?" until that customer is about to enter service.)

For hyper-exponential arrivals, the inter-arrival distribution would have a form analogous to (109). The state space would have to include knowledge of the next arrival "type." After each arrival, the type of the next arrival may switch, with probabilities specified in the definition of H.

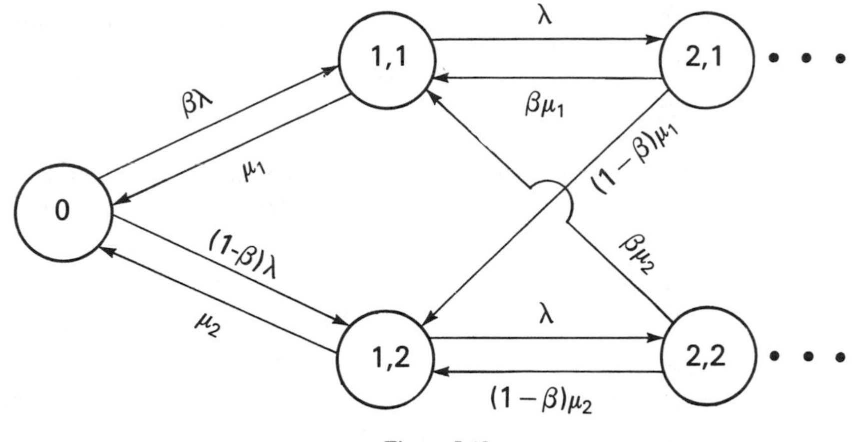

Figure 5-13

Remark. Examples 5-12 through 5-14 illustrate how to formulate queueing models that involve certain nonexponential distributions. The notions of phases and customer type were introduced to facilitate understanding of these models. It is not necessary that the real system have physical characteristics of this nature. What is important is that inter-arrival and service time distributions be correctly represented (or approximated). Of course, the physical nature of a system may suggest formulations of the type illustrated by these examples.

Batch Effects—Connection with the Erlang Distribution

Suppose the arrival process of customers is batch Poisson, where batches are Poisson at rate λ, and each batch contains a constant $k \geq 1$ customers. Customers are served at a single-channel facility (one *customer* at a time) with exponential service at rate μ. An arriving batch (at rate λ) increases the number of customers in system by k. A service completion (at rate μ) decreases the number of customers in system by one.

Inspection of Example 5-12 will show that it is the same model as that described earlier, where what we were calling phases are now customers. There is also a scale change: the service rate is now μ rather than $k\mu$, and the arrival rate (of customers) is now $k\lambda$. For ρ defined by $\rho = k\lambda/\mu$, the solution for $\{q_m\}$ is the same, where $\{q_m\}$ is now the distribution of the number of customers in system. Similarly, $L = L_p$, where L_p is given by equation (132) in Section 5-12. Finally, notice that the service time of a batch (i.e., the sum of customer service times in that batch) has a k-Erlang distribution.

More generally, suppose the batches in a batch Poisson arrival process are i.i.d. random variables, where the size of batch j has distribution

$$P(\nu_j = n) = \alpha_n, \qquad n = 0, 1, \ldots, \tag{112}$$

and customers are served at a single-channel facility with exponential service. An equivalent model is to regard batches as customers, where the arrival process of batches is Poisson, and the service time distribution of a batch is a *mixture* of Erlangs. (For $\alpha_0 > 0$, one of the components in this mixture is the 0-Erlang, which is just the unit step at the origin, U_0.)

Similarly, there is a connection between batch service and Erlang arrivals. Suppose arriving customers are Poisson, where customers are served in constant batches of size $k \geq 1$, i.e., k customers are served together, where the time to serve all k customers is exponential, and there is a single server. Suppose further that the server will not begin service unless a full batch is present, i.e., the number in system must be at least k. It is readily seen that what were called phases in Example 5-13 are now customers in the model described here.

Transit vehicles often provide batch service. For the model previously described, each vehicle has fixed capacity and will not begin a trip until it is full. Random capacity (e.g., arriving vehicles may be partially full) is an obvious generalization analogous to batch arrivals with random batch size. Alternatively, if vehicles are permitted to begin service while partially empty, there is no equivalent Erlang arrival model. No matter. This variation can be modeled and solved as a separate special case. Batch arrivals and/or service can of course occur in more general situations. For example, in Section 5-13 we generalize an $M/G/1$ queue to batch Poisson arrivals.

Because many models are easily generalized to the corresponding models with i.i.d. batch arrivals, we will employ a special *batch arrivals notation*. For example, the $M/M/1$ queue with batch arrivals discussed before will be denoted $BM/M/1$. Similarly, we will have $BM/G/1$ and even $BGI/G/c$ queues.

Approximating General Distributions

We now discuss how to approximate the distribution of a nonnegative random variable *arbitrarily closely* in the sense described as follows. The context is service times, but the technique obviously may be applied to distributions of other quantities, e.g., inter-arrival times.

In Example 5-12, we introduced the k-Erlang distribution. As $k \to \infty$, (102) and (103) imply that E_k converges in distribution to the unit step function at $1/\mu$, $U_{1/\mu}$, i.e., in the limit, we have constant service with mean $1/\mu$.

Now suppose that we want to approximate the distribution of a service time S where, for $0 < a < b < \infty$,

$$\begin{aligned} P(S = a) &= 1 - \beta, \text{ and} \\ P(S = b) &= \beta \qquad \text{for } \beta \in (0, 1). \end{aligned} \tag{113}$$

Another way of expressing (113) is to write

$$S = a + (b - a)I, \tag{114}$$

where $P(I = 1) = \beta$, $P(I = 0) = 1 - \beta$. This suggests the following approximation for the distribution of S:

$$(1 - \beta)E_k + \beta E_k * E_l, \tag{115}$$

where E_k has phase rate k/a and E_l has phase rate $l/(b - a)$. As $k \to \infty$ and $l \to \infty$, (115) converges in distribution to S.

Clearly, an arbitrary discrete distribution may be approximated in this manner. Since discrete distributions may be used to approximate continuous distributions, any distribution may be approximated in this manner. Furthermore, (115) is a special case of service with $k + l$ exponential service phases, where, on completion of phase j, there is a probability β_j of going to phase $j + 1$, and a probability $1 - \beta_j$ that service is completed with that phase.

Thus, to approximate an arbitrary distribution G, we need only consider phase-type distributions.

Definition. A distribution is said to be of *phase-type* (with *k phases*) if for some interger $k \geq 1$, it is of form

$$(1 - \beta_1) \exp(\mu_1) + (1 - \beta_2)\beta_1 \exp(\mu_1) * \exp(\mu_2) + \cdots + \prod_{j=1}^{k-1} \beta_j[\exp(\mu_1) * \cdots * \exp(\mu_k)], \tag{116}$$

where $\beta_j \in (0, 1)$, $j = 1, \ldots, k - 1$, and $\beta_k = 0$.

One way to represent phase-type service is to think of a service channel as an elongated box (see Figure 5-14) where we have shown a customer in phase j. Customers enter service in phase 1. On completion of phase j, either service is complete (with probability $1 - \beta_j$), or the customer enters phase $j + 1$. The time to complete phase j has distribution $\exp(\mu_j)$. We are assuming independence of phase service times and of events that terminate service on the completion of each phase.

From the definition of phase-type service, we can think of a service time as a first passage time in a continuous-state Markov chain from phase 1 to some new state, $k + 1$ say, that represents completion of service.

The k-Erlang distribution is obviously a special case of a phase-type distribution. Although not so obvious, the hyper-exponential is also of phase type, i.e., as shown in Problem 5-36, a hyper-exponential distribution has a representation of form (116) for $k = 2$. If $G(0) > 0$, this can be approximated by setting $G(0) = 1 - \beta_1$ and letting μ_1 be arbitrarily large.

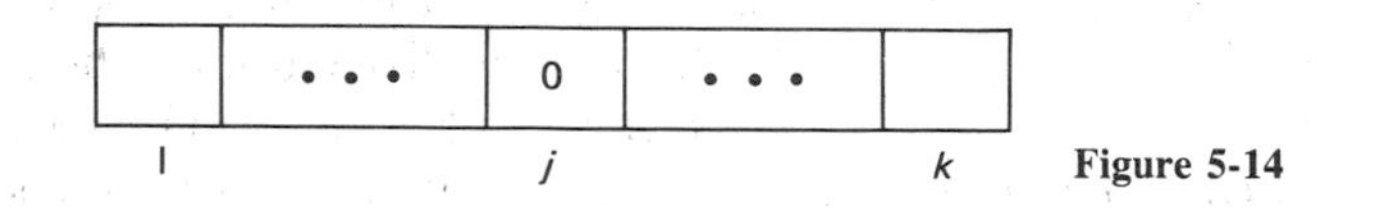

Figure 5-14

Let $\mathscr{H}$ be the collection of phase-type distributions as defined by (116). What we argued explicitly for discrete distributions is now stated as

Theorem 2. For an arbitrary distribution G of a nonnegative random variable, there exists a sequence of distributions $H_n \in \mathscr{H}$, $n = 1, 2, \ldots$, such that $\{H_n\}$ converges in distribution to G.

Because of this theorem, we say that G can be approximated *arbitrarily closely* by phase-type distributions. In particular, for any $\varepsilon > 0$, we can find an $H \in \mathscr{H}$ such that

$$| G(t) - H(t) | < \varepsilon$$

for all t, except for those t that belong to (arbitrarily) small neighborhoods about any points where G is discontinuous. It follows that if G has finite rth moment $1 \leq r < \infty$, we can find an $H \in \mathscr{H}$ for which the first r moments are arbitrarily close to those of G.

Applications

A queue with phase-type inter-arrival and/or service distributions is a continuous-time Markov chain with state space composed of the number of customers in system (or a more detailed description of the current location and possibly the type and past locations of every customer in system) and, for each phase-type distribution in progress, the current phase of that distribution.

Many interesting questions, such as how performance measures vary with the regularity of various distributions, can be investigated. Conjectures may be supported or refuted by simple examples. Several problems at the end of this chapter are of this type.

Usually, performance measures get worse (e.g., L increases) as distributions become "less regular" or "more variable" in some sense. For stable queues, the variability of the arrival process and service times is the primary reason that queues form at all. (Rush hour behavior is different. Queues form at predictable times because during certain periods, the arrival rate exceeds the service capacity. Of course, the queue size will vary from day to day. Similarly, if the server breaks down, or (in the highway context) an accident occurs, long queues may form.)

The most common measure of "more variable" is increasing coefficient of variation, e.g., increasing $V(S)/E^2(S)$. However, most queueing models are too complex for this ratio to be a reliable guide to comparative system performance. Other measures of variability are introduced in Chapter 11.

In some cases, the variability of the service distribution (by any measure) has no effect on important measures of system performance. Systems for which this is true are said to be *insensitive* to the service distribution. The most famous result of this kind, known as Erlang's loss formula, is discussed in the next section. We will show there that some performance measures for a loss system are insensitive to service distributions of phase-type. In the next chapter, insensitivity is discussed in the context of queueing networks.

For obvious mathematical and practical reasons, we will want to take advantage of insensitivity when it occurs. However, results of this kind should be regarded as "exceptions to the rule."

The use of phase-type distributions to approximate distributions for which performance measures are sensitive (i.e., *not* insensitive) can lead to computational difficulties. The balance equations may be intractable to solve. To approximate a distribution "arbitrarily closely" may require an arbitrarily large k. Furthermore, it is not clear how closely a performance measure is approximated by replacing a general distribution by a phase-type distribution that is close in some specified way.

5-11 THE $M/G/c$ LOSS SYSTEM—INSENSITIVITY

The $M/G/\infty$ queue, Example 5-1, is the most elementary example illustrating what is called insensitivity. The limiting distribution of the number of customers in system $\{p_n\}$ is a Poisson distribution with mean $\lambda E(S)$, *independent of the form of the service distribution*. Results that depend only on the mean of a distribution are said to be *insensitive* to that distribution. Thus, $\{p_n\}$ is insensitive to the service distribution G. It is important to observe that no queueing occurs in an infinite server queue.

By an $M/G/c$ *loss* system, we mean an $M/G/c$ queue with no queue allowed. Arrivals who find all servers busy are "lost," i.e., they leave immediately without being served, and they do not return. One of the most famous discoveries in the queueing literature is that the limiting (or stationary) distribution $\{p_n\}$ for an $M/G/c$ loss system is also insensitive to G. This remarkable result has a long history. The earliest proofs were for service distributions of phase-type. Many proofs for arbitrary G have appeared in the literature over the years.

We now illustrate this result for an elementary phase-type distribution.

Example 5-15: The $M/E_2/c$ Loss System

Let the arrival rate be λ and the exponential phase rate of the 2-Erlang distribution be 2μ for each phase. Let the state space be $\{(n, j); 0 \le j \le n \le c\}$, where n is the number of customers in system and j is the number of these customers who are undergoing their first service phase. The balance equations are

$$\begin{aligned}
\lambda p_{00} &= 2\mu p_{10}, \\
(\lambda + 2n\mu)p_{nj} &= 2(j + 1)\mu p_{n,j+1} \\
&\quad + 2(n - j + 1)\mu p_{n+1,j} + \lambda p_{n-1,j-1}, \qquad 0 \le j \le n < c, \\
2c\mu p_{cj} &= 2(j + 1)\mu p_{c,j+1} + \lambda p_{c-1,j-1}, \qquad 0 \le j \le c.
\end{aligned} \tag{117}$$

Notice that to save space, we have combined certain cases. Define $p_{nj} = 0$ for any state (n, j) not in the state space. With this definition, appropriate terms on the right of (117) drop out if either $j = n$ or $j = 0$.

Since this system is stable for any choice of λ and μ, there is a unique solution to the balance equations. To obtain this solution, it is enough to verify that an "educated guess" for the solution in fact satisfies the balance equations. Our guess is

$$p_{nj} = [\rho^n p_{00}/n!]\left[\binom{n}{j}(1/2)^n\right], \qquad 0 \le j \le n \le c, \tag{118}$$

where

$$p_{00} = \left[\sum_{n=0}^{c} \rho^n/n!\right]^{-1}, \tag{119}$$

where $\rho = \lambda/\mu$ here. It is readily verified that (118) satisfies the balance equations (plug in) and that (119) is just the normalizing condition. Hence, our guess is correct!

From (118), the distribution $p_n = \sum_{j=0}^{n} p_{nj}$, $n = 0, \ldots, c$, is

$$p_n = \rho^n p_0/n!, \tag{120}$$

where $p_0 = p_{00}$, and $\{p_n\}$ is the distribution of number of customers in system. Notice that (120) satisfies

$$\lambda p_n = (n + 1)\mu p_{n+1}, \qquad n = 0, \ldots, c - 1, \tag{121}$$

which is the birth and death recursion for the $M/M/c$ loss system. Hence, the distribution of $\{p_n\}$ for these two systems is the same!

Plugging in a guess is a rather unsatisfactory way of proving a result such as (118). It is not constructive, nor does it provide insight as to why (118) is true. Nevertheless, as is illustrated in Chapter 6, most insensitivity results are proven in this manner.

Now (118) is a joint distribution of a pair of random variables, (N, J) say. Given $N = n$, J has a binomial distribution. One way to obtain this binomial distribution is to suppose that each of the n customers in system is equally likely to be in each service phase, independent of other customers. The *remaining* service time distribution of any of these customers is

$$\exp(2\mu)/2 + \exp(2\mu) * \exp(2\mu)/2. \tag{122}$$

It is easily shown (either by transform methods or by a direct renewal-reward argument applied to E_2) that (122) is the *equilibrium distribution* of E_2! This observation suggests (i.e., leads to an educated guess about) the form of the solution for the general case.

Now consider the $M/G/c$ loss system. We need to define a state space for this model so that it becomes a Markov process. Obviously, at any epoch t, we need to know the number of customers in system, N. Given $N = n$, suppose we know the *remaining service times* $S_{r1}, \ldots, S_{rn}$ of each of these customers. [An alternative would be to know the *attained* service (amount of service received) of each of these customers.] The state space is the possible values of $(N, S_{r1}, \ldots, S_{rN})$. We now state without proof Theorem 3. (See Section 6-8 for a discussion of two methods of proof.)

Theorem 3: Erlang's Loss Formula. For an $M/G/c$ loss system with arrival rate λ and service rate μ, $0 < \mu < \infty$, and $\rho = \lambda/\mu$, the joint (limiting or stationary) distribution of $(N, S_{r1}, \ldots, S_{rN})$ is

$$P(N = n, S_{ri} \le x_i, i = 1, \ldots, n) = p_n \prod_{i=1}^{n} G_e(x_i), \tag{123}$$

where G_e is the equilibrium distribution of G, and $\{p_n\}$ is the probability distribution that satisfies (121).

The most important measure of performance for a loss system is the fraction of customers lost. From PASTA, this fraction is p_c. Thus, this measure of performance is insensitive to G. (We don't need PASTA here. The service rate of those served can be found from $L = \lambda w$, i.e., $L_s = \lambda_s/\mu$, or $\lambda_s = \mu L_s$, where $L_s = \sum_{n=0}^{c} nP_n$. The fraction of customers served is $f_s = \lambda_s/\lambda$. The fraction lost is $f_l = 1 - f_s$. A few lines of algebra will show that $f_l = p_c$.)

It is important to realize that not *every* performance measure is insensitive to G. In particular, the remaining service time distribution, G_e, depends on G. Of course, the waiting time distribution is G for this model.

Furthermore, insensitivity with respect to G depends on having Poisson arrivals. This result can be generalized to a "Poisson-like" arrival process where the arrival rate λ_n is a function of the number of customers in system, $n = 0, \ldots, c$. This is the arrival process for the "birth" part of a birth and death process. In this case, (123) is still valid, where $\{p_n\}$ now satisfies the recursion

$$\lambda_n p_n = (n + 1)\mu p_{n+1}, \qquad n = 0, \ldots, c - 1. \tag{124}$$

5-12 SOLVING BALANCE EQUATIONS—GENERATING FUNCTIONS

Generating functions were introduced in Section 1-12 to find properties of integral-valued random variables. In this section, we illustrate how to find the generating function of the distribution that is the solution to a set of balance equations. While the generating function of a distribution determines that distribution, inverting a generating function to find the corresponding distribution can be tedious. Even when this is the case, however, moments are often easy to obtain.

Example 5-16: The $M/E_k/1$ Queue

We now return to Example 5-12 and write out balance equations for the distribution of the number of phases in the system, $\{q_m\}$. From Figure 5-11, they are

$$\begin{aligned} \lambda q_0 &= k\mu q_1 \\ (\lambda + k\mu)q_m &= k\mu q_{m+1}, \qquad m = 1, \ldots, k-1, \\ (\lambda + k\mu)q_m &= k\mu q_{m+1} + \lambda q_{m-k}, \qquad m = k, k+1, \ldots. \end{aligned} \tag{125}$$

Now define the generating function

$$G(z) = \sum_{m=0}^{\infty} q_m z^m. \tag{126}$$

To find $G(z)$, notice the systematic form of (125). We will multiply the equations in (125) by successively higher powers of z and then add the equations *column by column*. We then factor out $G(z)$ for each column sum and solve for $G(z)$.

From the form of (126), it saves a little algebra to multiply successive equations in (125) by powers of z high enough so that for every term, the power of z is at least as high as the subscript on q. This is accomplished by multiplying the top equation by z, the next by z^2, and so on. Adding column by column, we obtain

$$(\lambda + k\mu)zG(z) - k\mu z q_0 = k\mu G(z) - k\mu q_0 + \lambda z^{k+1}G(z). \tag{127}$$

To explain (127), the second column on the right of (125) produces the sum $\lambda q_0 z^{k+1} + \lambda q_1 z^{k+2} + \ldots.$ By factoring out λz^{k+1}, we have the term on the far right in (127). The first column on the right of (125) has a factor $k\mu$ and a missing term, $k\mu q_0$. By summing, we have the first two terms on the right in (127). Obtaining the left side of (127) is similar. To solve for $G(z)$, first divide by μ and set $\rho = \lambda/\mu$. We obtain

$$G(z) = \frac{kq_0(1 - z)}{\rho z^{k+1} - (\rho + k)z + k} \equiv \frac{N(z)}{D(z)}, \tag{128}$$

where $N(z)$ and $D(z)$ denote the numerator and denominator of this expression.

To find q_0, we use normalization. However, simply setting $G(1) = 1$ will not work here because (128) has indeterminant form 0/0 at $z = 1$. On the other hand, a generating function of a distribution, e.g. (126), is a power series that converges at least for $|z| \leq 1$, and is continuous on that region. Hence, normalization may be performed by setting

$$\lim_{z \to 1} G(z) = 1.$$

[Technically, $z = 1$ is approached from inside (the left) of the region. This may be denoted by $z \to 1^-$.] Because the expression is indeterminant, we differentiate numerator and denominator (employing L'Hospital's rule),

$$\lim_{z \to 1} G(z) = \lim_{z \to 1} N'(z)/D'(z) = 1, \tag{129}$$

from which we obtain

$$q_0 = 1 - \rho, \tag{130}$$

and

$$G(z) = \frac{k(1 - \rho)(1 - z)}{\rho z^{k+1} - (\rho + k)z + k}. \tag{131}$$

From (130), we have stability, i.e., a limiting *probability* distribution $\{q_m\}$ will exist if and only if $q_0 > 0$ $(\rho < 1)$. This result and (130) could have been anticipated from the observation that $m = 0$ if and only if $n = 0$, meaning that the system is empty of *customers*. However, from $L = \lambda w$, the fraction of time a stable single-channel queue is busy is $\lambda E(S) = \rho$. Thus, (130) can be deduced without formally going through the steps in (129). (Nevertheless, going through these steps is a useful check for algebraic errors.)

We can find the average number of phases in system, $L_p = \sum_{m=0}^{\infty} mq_m$, by differentiating (131). We again have an indeterminant expression in terms of $N(z)$, $D(z)$, and their derivatives.

$$\begin{aligned} L_p = \lim_{z\to 1} G'(z) &= \lim_{z\to 1} (N'D - D'N)D^{-2} \\ &= \lim_{z\to 1} (N''D - D''N)/2DD' \\ &= 0 - \lim_{z\to 1} N/D \lim_{z\to 1} D''/2D' \\ &= -D''(1)/2D'(1) = \rho(k + 1)/2(1 - \rho). \end{aligned} \tag{132}$$

To find the average number of *customers* in system, $L = \sum_{n=0}^{\infty} np_n$, write (104) in terms of random variables,

$$M = kN_q + J, \tag{133}$$

where $L_p = E(M)$, $Q = E(N_q)$, and

$$L = Q + \rho = (L_p - E(J))/k + \rho. \tag{134}$$

To find $E(J)$, observe that every customer eventually passes through every service phase, where $P(J = j)$ is the fraction of time there is a customer in phase $k - j + 1$.

From $L = \lambda w$,

$$P(J = j) = \lambda E(\text{time in phase } (k - j + 1)) = \rho/k, \qquad j = 1, \ldots, k,$$

and

$$E(J) = \rho(k + 1)/2. \tag{135}$$

By combining (132), (134), and (135), we get (105).

Example 5-17: The $E_k/M/1$ Queue

We now return to Example 5-13. Define the generating function for the number of phases in system

$$H(z) = \sum_{m=0}^{\infty} q_m z^m,$$

where now $\{q_m\}$ is the solution to (108). Starting with z^k, multiply the balance equations in (108) by successively higher powers of z and sum:

$$(k\lambda + \mu)z^k H(z) - \mu z^k \sum_{j=0}^{k-1} q_j z^j = \mu H(z) - \mu \sum_{j=0}^{k-1} q_j z^j + k\lambda z^{k+1} H(z),$$

$$H(z) = \frac{(1 - z^k) \sum_{j=0}^{k-1} q_j z^j}{k\rho z^{(k+1)} - (k\rho + 1)z^k + 1}, \tag{136}$$

where $\rho = \lambda/\mu$.

Unfortunately, (136) has k unknowns in the numerator, $q_0, \ldots, q_{k-1}$. Normalization provides only one condition for determining them. To obtain more information, observe again that the generating function of a distribution converges for $|z| \leq 1$. (If we allow z to be complex, this means that z must lie inside or on the boundary of the unit circle in the complex plane.) Hence, any root z_r of the denominator of $H(z)$ for which $|z_r| \leq 1$ *must* also be a root of the numerator. This is the essential observation. To make use of it, we need to locate the roots of the denominator.

To do this, we apply a result from the theory of complex variables known as Rouché's Theorem.* Applications of this result to the analysis of queues primarily occur when, as here, transform methods are used. Rouché's Theorem is not used elsewhere in this book.

To apply this theorem, let g be the denominator of (136) and let $f(z) = 1 - (k\rho + 1)z^k$. For $|z| = 1 + \delta$, $\delta > 0$, we have

$$|f(z) - g(z)| = |k\rho z^{k+1}| = k\rho(1 + \delta)^{k+1} = k\rho(1 + k\delta) + k\rho\delta + o(\delta), \tag{137}$$

$$|f(z)| = |1 - (k\rho + 1)z^k| \geq (k\rho + 1)(1 + \delta)^k - 1 = k\rho(1 + k\delta) + k\delta + o(\delta). \tag{138}$$

* Let z be a complex variable, Ω be an open set on the complex plane, a be an arbitrary complex number, and r be an arbitrary positive (real) number. An open disk with center a and radius r is defined as the set $D(a, r) = \{z: |z - a| < r\}$. Similarly, $\overline{D}(a, r) = \{z: |z - a| \leq r\}$. A function f is said to be *analytic* on Ω if its derivative $f'(z)$ exists $\forall\, z \in \Omega$. A *zero* of an analytic function f on Ω is a value of $z \in \Omega$ such that $f(z) = 0$. Polynomials are analytic on the entire complex plane. The zeros of a polynomial are customarily called roots. As a generalization of the notion of the order of a multiple root, a zero of an analytic function may be of order greater than one (it is always a positive integer). The following is the statement of Rouché's Theorem: Let f and g be analytic on Ω, $\overline{D}(a, r) \subset \Omega$, and

$$|f(z) - g(z)| < |f(z)| \text{ on } |z - a| = r.$$

Then f and g have the same number of zeros in $D(a, r)$ (if they are counted according to their multiplicities).

For $\rho < 1$ and sufficiently small positive δ, (137) and (138) imply

$$|f(z) - g(z)| < |f(z)| \text{ on } |z| = 1 + \delta. \tag{139}$$

Since the k roots of $f(z)$ all have magnitude $[1/(k\rho + 1)]^{1/k} < 1 + \delta$, g has exactly k roots in $D(0, 1 + \delta)$, where δ is arbitrarily small. This means that g has exactly one root, which we denote by z_0, with magnitude greater than 1. Since z_0 must be real and is obviously not negative, $z_0 > 1$. We already know one of the roots in $D(0, 1 + \delta)$: $z = 1$. Suppose $z_r = a + bi$ is a complex root on $|z| = 1$ where a and b are real and $|a| < 1$. Note that $a^2 + b^2 = 1$. It is easy to show that

$$|z_r^k[k\rho z_r - (k\rho + 1)]| = |k\rho(a + bi) - (k\rho + 1)| > 1. \tag{140}$$

However, this is impossible if z_r is a root. Therefore, $z = 1$ is the only root on $|z| = 1$, and the denominator must have exactly $(k - 1)$ roots (not necessarily distinct) with magnitude strictly less than 1.

For $H(z)$ to exist for $|z| \leq 1$, the numerator of (136) must have the same $(k - 1)$ roots. All the roots of $(1 - z^k)$ lie on $|z| = 1$. Hence, these roots must be the $(k - 1)$ roots of

$$\sum_{j=0}^{k-1} q_j z^j = 0. \tag{141}$$

We conclude that aside from a constant factor, the denominator of (136) has the form

$$(z - z_0)(z - 1) \sum_{j=0}^{k-1} q_j z^j. \tag{142}$$

Hence, we cancel (!) the factor (141), and

$$H(z) = \frac{c(1 - z^k)}{(z - z_0)(z - 1)}, \tag{143}$$

where c is a constant. Normalizing (143), we solve for c,

$$H(z) = \frac{(z_0 - 1)(1 - z^k)}{k(z_0 - z)(1 - z)}, \tag{144}$$

where $z_0 > 1$ is the only real root of the denominator of (136) in this range. For numerical values of k and $\rho < 1$, the value of z_0 is easily approximated.

Now (144) may be expanded as a power series in z to determine an explicit expression for the elements of $\{q_m\}$ in terms of z_0. Instead, we will find the expected number of phases in the system,

$$L_p = H'(1) = 1/(z_0 - 1) + (k - 1)/2. \tag{145}$$

Let J be the number of phases associated with the next arrival. Since the next arrival box is never empty, $P(J = j) = \lambda/k\lambda = 1/k$, $j = 0, \ldots, k - 1$, and

$$E(J) = (k - 1)/2.$$

Hence,

$$L = [L_p - E(J)]/k = 1/k(z_0 - 1). \tag{146}$$

Results (144) and (146) are valid for $\rho < 1$. For $\rho \geq 1$, it may be shown that the system is unstable.

For $k = 1$, $L = L_p$, $z_0 = 1/\rho$, and we obtain $M/M/1$ results. It can be shown that L is a monotone decreasing function of k.

5-13 WORK AND THE *M/G/1* QUEUE

A wide variety of queueing models can be represented as continuous-time Markov chains. It may come as a surprise that the $M/D/1$ queue, i.e., the single-channel queue with Poisson arrivals and constant service, is not among them. We can approximate the behavior of an $M/D/1$ queue by an $M/E_k/1$ queue with large k. In fact, as $k \to \infty$, correct results for the $M/D/1$ queue can be obtained, e.g., L from (105).

Approximation methods complicate the state space and usually the solution as well. Where methods are known that can handle arbitrary distributions directly, these methods usually provide more insight and require less manipulation to obtain a solution.

In this section, we will obtain results for the single-channel queue with Poisson arrivals and general service, the $M/G/1$ queue. An important concept that will make this task easier is work in system.

In Example 5-2, we observed that when there are separate queues at a multichannel facility, the shortest line is not necessarily the one with the fewest customers. To determine which line to join, we would need to know the service times of all customers in queue and the (remaining) service times of all customers in service. The shortest line is the one where the sum of all the service times of customers in that line is the smallest.

For single and multichannel queues, *work in system* at any epoch $t \geq 0$, $V(t)$, is defined as the sum of the service times of all customers in queue and the remaining service times of all customers in service. It is the sum of the line lengths in the preceding example.

For a single-channel queue, $V(t)$ has a simple representation. At each arrival epoch, $V(t)$ jumps (increases), where the size of the jump is the service time of the arriving customer. Between jumps, $V(t)$ decreases continuously with slope -1, whenever $V(t)$ is positive. A possible realization of $V(t)$ is shown in Figure 5-15. The definition of $V(t)$ here is consistent with that given for the case of deterministic queues in Section 5-4.

For a single-channel queue under a FIFO rule, $V(t)$ would be the delay at epoch t *if* an arrival were to occur there. Because of this, $V(t)$ is sometimes called *virtual delay* in this case. However, $V(t)$ is also well defined for multichannel queues.

The *average* work in system $E(V)$ is defined as the limit

$$E(V) = \lim_{t\to\infty} \int_0^t V(u)\, du/t, \tag{147}$$

when the limit exists. In Chapter 2 notation, $E(V) = \overline{V}$, a time average.

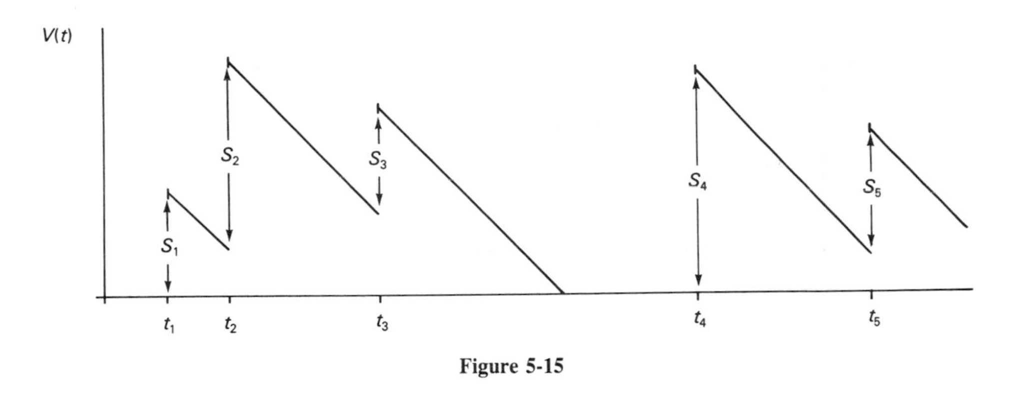

Figure 5-15

Now assume (as we have been doing all along) that whatever the order of service, a customer who enters service is served to completion without interruption. What is the contribution of a customer who has *departed* by t, C_j say, to the integral in (147)? While in queue, C_j contributes S_j to work in system. While in service, C_j contributes the *remainder* of S_j to work in system. Thus, the contribution of C_j to the integral (area) in (147) is the area of the trapezoid in Figure 5-16, $D_jS_j + S_j^2/2$.

Thus, analogous to (9),

$$\int_0^t V(u)\, du = \sum_{j=1}^{\Lambda(t)} (D_jS_j + S_j^2/2) + \text{an error term.} \tag{148}$$

Dividing (148) by t, we obtain an expression analogous to (10). As $t \to \infty$, it can be shown that the error term does not contribute to the limit, and we have

$$E(V) = \lambda[E(SD) + E(S^2)/2], \tag{149}$$

where

$$E(S^2) = \lim_{n\to\infty} \sum_{j=1}^{n} S_j^2/n, \text{ and} \tag{150}$$

$$E(SD) = \lim_{n\to\infty} \sum_{j=1}^{n} S_jD_j/n. \tag{151}$$

Result (149) is analogous to $L = \lambda w$. For the models we have considered, including FIFO and LIFO order of service (but *not* priorities), D_j and S_j are independent. In this case,

$$E(V) = \lambda E(S)d + \lambda E(S^2)/2. \tag{152}$$

As with $L = \lambda w$, the forms of (149) and (152) are independent of the arrival process. These expressions and $L = \lambda w$ are special cases of a result discussed in Section 5-15.

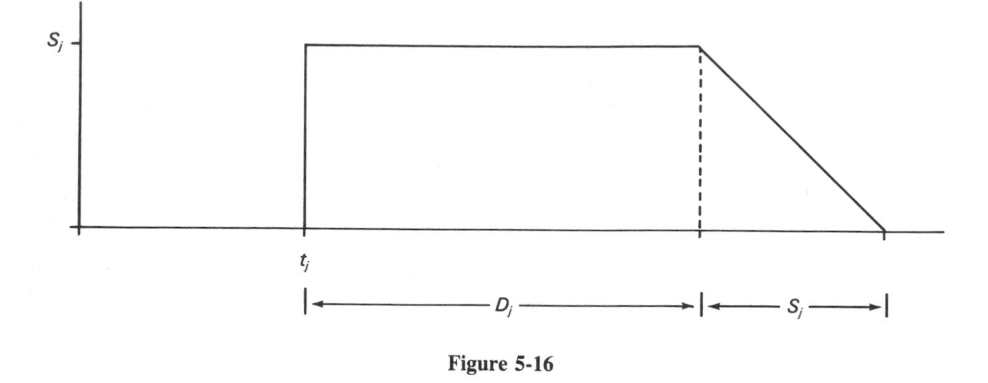

Figure 5-16

For the $M/G/1$ queue, the average work found by a Poisson arrival is the time average work, $E(V)$. For a single-channel queue under FIFO, the average work found by an arrival is the average delay, d. Thus, for an $M/G/1$ queue,

$$d = E(V). \tag{153}$$

By solving (152) and (153) for d, we have

$$d = \lambda E(S^2)/2(1 - \rho), \tag{154}$$

where $\rho = \lambda/\mu$. All the other familiar first moment quantities follow immediately: $w = d + 1/\mu$, $Q = \lambda d$, $L = Q + \rho$.

Notice the form of (154). For fixed λ and μ, d increases with the variance of the service distribution. As $\rho \to 1$, $d \to \infty$.

We could also have derived (154) from $L = \lambda w$, as was done for the $M/M/1$ queue [see (62) and (63)]. The problem here is that the distribution of remaining service is different from that of a full service. Thus, we can write

$$d = Q/\mu + E(S_r), \tag{155}$$

where S_r is the remaining service found by an arrival; $S_r = 0$ if the system is empty. Clearly, $P(S_r > 0) = \rho$, the fraction of time the server is busy. Given $S_r > 0$, S_r is the remaining service as found by a Poisson arrival who finds the server busy. By the results of Example 7 in Section 2-6, the distribution of $(S_r \mid S_r > 0)$ is G_e, the equilibrium distribution corresponding to service distribution G. Thus

$$E(S_r \mid S_r > 0) = E(S^2)/2E(S), \tag{156}$$

and

$$E(S_r) = \rho E(S_r \mid S_r > 0) = \lambda E(S^2)/2, \tag{157}$$

$$d = Q/\mu + \lambda E(S^2)/2. \tag{158}$$

Since $Q = \lambda d$, we now have (154). Also observe that the right-hand sides of (152) and (158) are equivalent.

The concept of work provides more insight and has other applications. For example, suppose for some (non-Poisson) arrival process, it can be shown that

$$d < E(V), \tag{159}$$

i.e., that the average work as found by an arrival is less than the time-average work. From (152), this implies

$$d < \lambda E(S^2)/2(1 - \rho), \tag{160}$$

i.e., the expected delay is less than that for an $M/G/1$ queue. In Problem 5-40, we show (159) is true for an $E_k/G/1$ queue, i.e., expected delay under Erlang arrivals is less than under Poisson. Reversing the inequality in (159) reverses that in (160).

Batch Arrivals

Suppose customers arrive in batches of random size, where the batch sizes, ν_1, $\nu_2, \ldots$, are i.i.d. with distribution $P(\nu_i = j) = \alpha_j, j = 0, 1, \ldots$. If customers are served individually at a single-channel queue, a portion of a customer's delay is caused by other customers within the same batch. This portion of delay for a "typical" customer is the sum of the $J - 1$ preceding service times,

$$\sum_{i=1}^{J-1} S_i, \tag{161}$$

where J is the (random) order of service within a batch.

Since delay is an average over customers, the appropriate distribution of J is

$$P(J = j) = \text{fraction of customers who are } j\text{th in line within their own batch.} \tag{162}$$

In Section 2-5, this was shown to be

$$P(J = j) = \frac{P(\nu \geqq j)}{E(\nu)} = \frac{\sum_{i \geqq j} \alpha_i}{E(\nu)}. \tag{163}$$

The expected number of customers from the same batch *in front of* a typical customer is

$$E(J - 1) = E\{(\nu)(\nu - 1)\}/2E(\nu). \tag{164}$$

Example 5-18: The *M/G/*1 Queue with Batch Poisson Arrivals

Let batches arrive Poisson at rate λ, with batch sizes as given above, where S denotes a customer service time. Thus, the arrival rate of customers is $\lambda E(\nu)$ and the fraction of time the server is busy is $\rho = \lambda E(\nu)/\mu$. Since batches are Poisson, the average work in system found by a batch is the time-average work, from (152) with $\lambda E(\nu)$ replacing λ:

$$E(V) = \rho d + \lambda E(\nu)E(S^2)/2. \tag{165}$$

Now

$$D = V + \sum_{j=1}^{J-1} S_j, \text{ and}$$

$$d = E(V) + E\{(\nu)(\nu - 1)\}/2E(\nu)\mu$$

$$= \frac{\lambda E(\nu)E(S^2)}{2(1 - \rho)} + \frac{E\{(\nu)(\nu - 1)\}}{2E(\nu)(1 - \rho)\mu}. \tag{166}$$

In the preceding example, the presence of batches increases d. For fixed μ and ρ, d increases with $E\{(\nu)(\nu - 1)\}/2E(\nu)$, which will be large if either $E(\nu)$ is large or batch sizes are irregular. We have another example where performance gets worse as the arrival process becomes more irregular.

In this section, only first moment results have been obtained. To find distributions, e.g., (14) and (15), a different approach is required. *Embedded Markov chains* is an important method for doing this. Rather than require that the Markov property hold at all epochs, we require only that it hold at a countable (random) collection of epochs where special events occur. For the $M/G/1$ queue, a chain may be embedded at departure epochs, where X_n is defined as the number of customers in system left behind by the departure of C_n. The sequence $\{X_n\}$ is a discrete-time Markov chain and may be analyzed by the methods in Chapter 3. This is done in Chapter 8.

5-14 THE EFFECT OF ORDER OF SERVICE ON SYSTEM PERFORMANCE; PRIORITIES

In Section 5-4, where LIFO and FIFO were compared, and in Example 5-2, the behavior of $\{N(t)\}$ and the departure process $\{\Omega(t)\}$ were unaffected by the proposed alternatives. Consequently, neither L nor w changed. In these examples, order of service depended on order of arrival and (possibly) extraneous other factors, but did not depend on service time. We consider this case first. Priority queues, where service time can be a factor, follow. Priority queues are treated in more detail in Chapter 10.

Optimality of FIFO for Multichannel Queues

As defined in Section 5-1, the service times in the sequence $\{S_j\}$ are listed in the order of customer arrival. Suppose we rearrange this sequence in the order of customer *departure from queue,* i.e., in the order in which service begins. For single- and multiple-channel facilities fed by a single FIFO queue, customers depart from the queue in the order of their arrival, and no rearranging occurs. In the examples mentioned at the beginning of this section, some rearranging is necessary. The fundamental assumption we make is this:

Assumption. The service times in the *rearranged* sequence are i.i.d. and independent of the arrival process.†

Now consider a realization of the arrival times t_j and service times S_j, $j = 1, 2, \ldots$, where S_j is the jth random variable in the rearranged sequence. These sequences completely determine the number in queue and in system at all epochs, the departure process from the queue $\{\Lambda(t) - N_q(t): t \geq 0\}$, and the corresponding departure epochs.

Let $r_1 \leq r_2 \leq \cdots$ be the ordered departure epochs from the queue, D_{Fj} be the delay in queue of C_j under FIFO, and D_{Aj} be the delay of C_j under any alternative consistent with our assumption, $j = 1, 2, \ldots$. For every j, we have

$$D_{Fj} = r_j - t_j. \tag{167}$$

For some pair j and $j + 1$, suppose

$$D_{Aj} = r_{j+1} - t_j, \qquad D_{A,j+1} = r_j - t_{j+1}. \tag{168}$$

From (167) and (168) the sum of the delays are equal:

$$D_{Fj} + D_{F,j+1} = D_{Aj} + D_{A,j+1}, \tag{169}$$

where D_{Aj} is the *largest* of the four numbers in (169) and $D_{A,j+1}$ is the *smallest*. From this observation, it is easy to see that

$$D_{Fj}^2 + D_{F,j+1}^2 \leq D_{Aj}^2 + D_{A,j+1}^2, \tag{170}$$

with equality holding only if either $t_j = t_{j+1}$ or $r_j = r_{j+1}$.

Now suppose that for some n, the first n arrivals are also the first n departures (not necessarily in order). For any pair of customers among the first n who depart "out of order," i.e., for $j < k$, C_j departs after C_k, we can interchange their departure epochs and reduce the corresponding sum of their squared delays, in accordance with (170). This can be done until all customers depart in order, which is the FIFO case. Since each interchange can only reduce the sum of squares, we have

$$\sum_{j=1}^{n} D_{Fj}^2 \leq \sum_{j=1}^{n} D_{Aj}^2. \tag{171}$$

† In simulation, values of random variables normally are generated only as needed, to minimize storage requirements. Thus, service times usually are generated in the order they are used, where each service time is a random draw from the service distribution, regardless of order of service. Our assumption is implicit in this procedure.

If the first n arrivals are not the first n departures, some of the D_{Aj} will be larger than in (171), and (171) is still true. Thus, (171) is *always* true for *every* $n = 1, 2, \ldots$.

Now divide both sides of (171) by n and let $n \to \infty$. Assuming the limits exist and are the same for all realizations, the limit on the left is the average of the squared delays under FIFO, and the limit on the right is the corresponding average under any alternative. By using expected values to denote these averages, we have

$$E\{D_F^2\} \leq E\{D_A^2\}, \tag{172}$$

where these expected values are also the second moments of the corresponding delay distributions [analogous to (15)]. Since the first moments are equal, $d_F = d_A$, (172) implies that the variances of the delay distributions satisfy the same inequality,

$$V(D_F) \leq V(D_A), \tag{173}$$

and, because delay and service time of the same customer are independent, so do the variances of waiting time in system,

$$V(W_F) \leq V(W_A). \tag{174}$$

These inequalities will be strict if the fraction of customers who depart (strictly) out of order is positive.

Since $d_F = d_A$ and FIFO has smaller variance, we will call FIFO *optimal* under this criterion, among rules for determining order of service that satisfy our assumption. This result can be generalized to convex waiting time costs (see Problem 5-41).

Priorities

There are many reasons for considering alternatives to FIFO. Some customers may have a higher delay cost than others or their presence may tie up more of the service facility. For example, a request for directory assistance placed over the long-distance telephone network has priority over a local request. This is a *priority rule,* i.e., any rule that determines order of service.

When the priority rule (explicitly or implicitly) depends on service time, there is a potential for reducing the average delay as well as delay costs. For example, if there are two customers present, one with a service time of 1 minute and the other with a service time of 1 hour, serving the shorter job first reduces the *average* of their delays by 29.5 minutes. A check-cashing-only line at a bank, which reduces the delays of customers with short service times, also has this effect. (Sometimes check-cashing lines are so long that their potential for reducing average delay is lost.)

Now suppose customers to be served by a single-channel facility fall into k priority *classes*. Let customers in class i have arrival rate λ_i and service distribution G_i, where S_i denotes a draw from G_i and $\rho_i = \lambda_i E(S_i)$, $i = 1, 2, \ldots, k$. Similarly, quantities such as d_i and Q_i denote averages with respect to customers in class i. Let the overall arrival rate be $\lambda = \sum_{i=1}^{k} \lambda_i$, S be a service time drawn from the overall service distribution $G = \sum_{i=1}^{k} \lambda_i G_i/\lambda$, and $\rho = \lambda E(S) = \sum_{i=1}^{k} \rho_i$.

Assume service times are independent of each other and of the arrival process, customers within the same class are served FIFO, and, once service begins, each customer is served to completion without interruption (called the *postponable* case).

Now consider work in system. Notice that realizations of work, e.g., Figure 5-15, are independent of the rule, i.e., work is *conserved*. Hence, $E(V)$ is independent of the rule. Although the way work is represented in terms of delay will change with the rule, Figure 5-16 and (148) still correctly represent each customer's contribution to work. Now $E(V)$ is simply the sum of the time-average work in system associated with each class. When this is taken into account, (152) is replaced by

$$E(V) = \sum_{i=1}^{k} \rho_i d_i + \lambda E(S^2)/2, \tag{175}$$

where the second term on the right is due to the triangular portion of Figure 5-16 and is the sum over i of terms $\lambda_i E(S_i^2)/2$. Note that this term is also independent of the priority rule. [Implicit in the form of (175) is the assumption that order of service may depend on service times only through membership in priority classes with (possibly) different service distributions.]

Example 5-19: The $M/G/1$ Queue with Postponable Priorities

Assume the arrival processes of each class are independent Poisson processes and the specific priority rule: Class 1 has the highest priority, class 2 next highest, etc., where on each service completion, the next customer served is drawn from those in queue having the highest priority.

Since arrivals are Poisson, the average work found by an arrival is also represented by (175). Furthermore, since work is conserved,

$$E(V) = \lambda E(S^2)/2(1 - \rho), \tag{176}$$

the same as d in (154).

For the highest class d_1 is simply the relevant portion of (175), the expected value of the class 1 work found in queue plus the remaining service of whoever may be in service,

$$d_1 = \rho_1 d_1 + \lambda E(S^2)/2; \text{ hence,} \tag{177}$$

$$d_1 = \lambda E(S^2)/2(1 - \rho_1). \tag{178}$$

For the two-class case, $k = 2$, we can now solve for d_2 in (175).

For $k \geq 3$, it is important to observe that for any class i, combining all classes of higher priority than i into a new "super" class 1 will not affect d_i (why?). Thus, as far as class k is concerned, there are only two classes, k and higher than k. From the above results,

$$d_k = \lambda E(S^2)/2(1 - \sum_{j<k} \rho_j)(1 - \rho). \tag{179}$$

With the same argument and (179), we can solve for d_{k-1}, d_{k-2}, and so forth. The general expression is

$$d_i = \lambda E(S^2)/2(1 - \sum_{j<i} \rho_j)(1 - \sum_{j\leq i} \rho_j) \qquad \text{for } i = 1, \ldots, k. \tag{180}$$

As an extension of Example 5-19, suppose customers arrive in identifiable classes a, b, . . . , with known service distributions. How should these classes be assigned priority numbers (e.g., which class should be treated as class 1) so as to minimize overall expected delay $d = \sum_{i=1}^{k} \lambda_i d_i/\lambda$? More generally, if the cost per unit time of delay is class dependent, the expected cost per unit time is minimized under the so-called $c\mu$-*rule*: order the classes by the product of the cost rate and service rate, with the highest being assigned class 1, next highest, class 2, and so forth. See Problem 5-42. The optimality of the $c\mu$-rule is a direct consequence of the form of (180). This result will not hold when interruptions are permitted.

The elementary proof of the optimality of the $c\mu$-rule implicitly assumes that the server is never idle when there is work to be done, so that work is conserved. When we permit the server to be idle, it can be shown that this option should not be used, and the $c\mu$-rule is still optimal. Furthermore, it is optimal even when we permit policies to depend on the exact queue lengths of each class.

When arrivals are not Poisson, however, the past is relevant for predicting when future arrivals will occur, and the $c\mu$-rule will not in general be optimal. For example, if a very costly customer is about to arrive, it may be better to serve a short service time next (ignoring relative costs)—or even to hold the server idle.

If we exclude the idle server option, the $c\mu$-rule is optimal for an arbitrary arrival process when there are $k = 2$ classes, but is not necessarily optimal when $k > 2$. See Section 10-3.

5-15* On $L = \lambda w$ and Generalizations of This Result

In this section, we complete the proof of $L = \lambda w$ that was sketched in Section 5-2. We then generalize this result in a way that includes (149) as a special case. For this purpose, various quantities defined in Section 5-1 are to be regarded as

* Proofs are technical and may be omitted, but this section should be read.

random variables defined on some sample space. For any point ω in this space, these quantities are real numbers.

Our main result as shown in (190) and (199) is that $L = \lambda w$ and its generalizations hold as relations between limits for an arbitrary fixed ω, i.e., it is essentially a deterministic result that holds on sample paths. To simplify notation, we will usually suppress ω, e.g., $t_j \equiv t_j(\omega)$.

For arbitrary ω, we make two assumptions:

$$\lim_{t\to\infty} \Lambda(t)/t = \lambda < \infty, \text{ and} \tag{181}$$

$$\lim_{n\to\infty} \sum_{j=1}^{n} W_j/n = w < \infty, \tag{182}$$

where these are limits in the ordinary sense.

Remark. It is easily shown (see Lemma 2.2 in Stidham [1982]) that (181) is equivalent to

$$\lim_{j\to\infty} t_j/j = 1/\lambda. \tag{183}$$

As observed in Section 5-2, the error term in (9) is negative, and we have

$$\sum_{(j:t_j+W_j\le t)} W_j \le \int_0^t N(u)\,du \le \sum_{j=1}^{\Lambda(t)} W_j \qquad \text{for every } t \ge 0, \tag{184}$$

where the left-hand inequality follows by counting on the far left only those customers who have departed by epoch t. Because the t_j are finite, $\Lambda(t) \to \infty$ as $t \to \infty$, and from (182),

$$\lim_{t\to\infty} \sum_{j=1}^{\Lambda(t)} W_j/\Lambda(t) = w. \tag{185}$$

Dividing (184) by t and writing the right-hand term as in (10),

$$(\Lambda(t)/t) \sum_{j=1}^{\Lambda(t)} W_j/\Lambda(t),$$

we see that this term converges to λw as $t \to \infty$. Hence $L = \lambda w$ would follow if we can show that the left-hand term has the same limit.

Now (181) implies

$$t_j \to \infty \quad \text{as} \quad j \to \infty, \tag{186}$$

and (182) implies

$$W_n/n = \sum_{j=1}^{n} W_j/n - \left[\sum_{j=1}^{n-1} W_j/(n-1)\right](n-1)/n \to 0 \quad \text{as} \quad n \to \infty. \tag{187}$$

From (186) and (187),

$$n/t_n \le \Lambda(t_n)/t_n \to \lambda \quad \text{as} \quad n \to \infty, \text{ and}$$

$$W_n/t_n = (W_n/n)(n/t_n) \to 0 \quad \text{as} \quad n \to \infty. \tag{188}$$

For any $\varepsilon > 0$, (188) implies that for some fixed m, $W_j < t_j\varepsilon$ and $t_j + W_j < t_j(1 + \varepsilon)$ for all $j > m$. Now for any $t > t_m$, a *lower bound* on the left-hand side of (184) is

$$\sum_{(j>m:t_j \le t/(1+\varepsilon))} W_j = \sum_{j=1}^{\Lambda(t/(1+\varepsilon))} W_j - \sum_{j=1}^{m} W_j. \tag{189}$$

Dividing (189) by t and letting $t \to \infty$, this ratio has limit

$$\lambda w/(1 + \varepsilon).$$

Since ε may be arbitrarily small, we have proven (190) in

Theorem 4: $L = \lambda w$. At any sample point ω where limits (181) and (182) exist and are finite, $L(\omega) = \lim_{t\to\infty} \int_0^t N(u)\, du/t$ also exists and is finite, where

$$L(\omega) = \lambda(\omega)w(\omega). \tag{190}$$

If $\lambda(\omega) = \lambda$ and $w(\omega) = w$ with probability 1 (w.p.1), where λ and w are finite constants, then $L(\omega) = L$ w.p.1, where

$$L = \lambda w. \tag{191}$$

If $\lambda(\omega) = \lambda > 0$ w.p.1, and $\{W_j\}$ is a regenerative process with finite mean cycle length and limiting distribution $W(t)$, where $w = \int_0^\infty W^c(t)\, dt = \infty$,

$$L = \lambda w = \infty \quad \text{w.p.1.} \tag{192}$$

Completion of Proof. Equation (191) is immediate because the set $\{\omega\}$ for which $\lambda(\omega) = \lambda$ *and* $w(\omega) = w$ has probability 1. For (192), define regenerative sequence $\{W_j(a)\}$ by $W_j(a) = \min\{W_j, a\}$ for every j and any $a > 0$. For any $t > 2a$, we can write

$$\int_0^t N(u)\, du \ge \sum_{j=1}^{\Lambda(t/2)} W_j(a),$$

and it follows that on a set $\{\omega\}$ with probability 1,

$$\liminf_{t\to\infty} \int_0^t N(u)\, du/t \ge \lambda \int_0^a W^c(t)\, dt/2.$$

By letting $a \to \infty$, we have (192).

Example 5-20

Let $t_j = j$ for each j and

$$W_j = j \qquad \text{if } j \in \{n_i\},$$
$$= 0 \qquad \text{otherwise,}$$

where $n_i = 2^i$, $i = 0, 1, \ldots$. It is easy to see that $\lambda = 1$, and $N(t) = 1$ for $t \geq 2$, so that $L = 1$. However, $\sum_{j=1}^{n} W_j/n$ fluctuates between 1 and 2, and w does not exist. Thus, the existence of λ and L does not imply (190). Notice that limit (187) does not exist here. Without (187), W_j that are both very large and very rare may contribute to w in a way different from L.

If λ and d exist and are finite, then simply defining the queue to be "the system," we have

$$Q = \lambda d. \tag{193}$$

Similarly, if λ and $\overline{S} = \lim_{n\to\infty} \sum_{j=1}^{n} S_j/n$ exist and are finite,

$$L_s = \lambda \overline{S}, \tag{194}$$

where for i.i.d. service times, $\overline{S} = E(S)$, and λ is the arrival rate *of those served.* In fact, (194) holds with $L_s < \infty$, even when $Q = d = \infty$.

To recapitulate, for $I_j(t) = 1$ if $t \in [t_j, t_j + W_j)$, $I_j(t) = 0$ otherwise, $\int_0^\infty I_j(t)\, dt = W_j$ is a portion of the area under $\{N(t)\}$. The result $L = \lambda w$ simply states that the time average of $\{N(t)\}$ is the product of the arrival rate and the average contribution to that area by arriving customers.

We can generalize this result by replacing $I_j(t)$ by some arbitrary nonnegative function $f_j(t)$, $t \geq 0$, associated with customer C_j. Let

$$G_j = \int_0^\infty f_j(t)\, dt \qquad \text{for } j = 1, 2, \ldots, \text{ and}$$

$$H(t) = \sum_{j=1}^{\infty} f_j(t) \qquad \text{for } t \geq 0,$$

where G_j is the contribution of C_j to the area under $\{H(t)\}$. In the notation for time averages introduced in Chapter 2, let

$$\overline{G} = \lim_{n\to\infty} \sum_{j=1}^{n} G_j/n, \text{ and} \tag{195}$$

$$\overline{H} = \lim_{t\to\infty} \int_0^t H(u)\, du/t, \tag{196}$$

when these limits exist.

To prove the next theorem, we make the following

Technical Assumption. At sample point ω, there is an $l_j(\omega) > 0$ for each j such that

$$f_j(t, \omega) = 0 \qquad \text{for every } t \notin [t_j(\omega), t_j(\omega) + l_j(\omega)], \tag{197}$$

where

$$\lim_{j\to\infty} l_j(\omega)/j = 0. \tag{198}$$

Remarks. The purpose of this assumption is to restrict the location of the contribution by C_j to the area under $\{H(t)\}$. The contribution cannot begin before C_j arrives. For $L = \lambda w$, $l_j = W_j$, and (198) is a consequence of $w < \infty$. Thus (198) is a generalization of (187); in Example 5-20, we found that without (187), $L = \lambda w$ may not hold. In applications, the conditions $l_j \le W_j$ and $w < \infty$ are sufficient for (198).

Theorem 5: $\overline{H} = \lambda\overline{G}$. At any sample point ω where limits (181) and (195) exist and are finite, and nonnegative functions $f_j(t)$ satisfy (197) and (198). $\overline{H}(\omega)$ exists, where

$$\overline{H}(\omega) = \lambda(\omega)\overline{G}(\omega). \tag{199}$$

If $\lambda(\omega) = \lambda$ and $\overline{G}(\omega) = \overline{G}$ w.p.1, where λ and $\overline{G}$ are finite constants, $\overline{H}(\omega) = \overline{H}$ w.p.1, where

$$\overline{H} = \lambda\overline{G}. \tag{200}$$

If $\{g_j(t)\}$ is a regenerative process with finite mean cycle length, where $g_j(t) = f_j(t - t_j)$ for all t and j, the $f_j(t)$ satisfy (197), $\lambda(\omega) = \lambda > 0$ w.p.1, and $\overline{G} = \infty$ w.p.1,

$$\overline{H} = \lambda\overline{G} = \infty \quad \text{w.p.1}. \tag{201}$$

Proof. We follow the proof of Theorem 4. Clearly,

$$\sum_{(j:t_j+l_j\le t)} G_j \le \int_0^t H(u)\,du \le \sum_{j=1}^{\Lambda(t)} G_j. \tag{202}$$

From (187) and (198),

$$l_j/t_j = (l_j/j)(j/t_j) \to 0 \qquad \text{as} \quad j \to \infty, \tag{203}$$

which leads to a lower bound on the left-hand term in (202) of form

$$\sum_{(j>m:t_j\le t/(1+\varepsilon))} G_j = \sum_{j=1}^{\Lambda(t/(1+\varepsilon))} G_j - \sum_{j=1}^{m} G_j \qquad \text{for } t > t_m, \tag{204}$$

where for any $\varepsilon > 0$, m is fixed. Dividing (204) by t and letting $t \to \infty$, we have (199), and (200) follows immediately.

Now define regenerative sequence $\{G_j(a)\}$ by $G_j(a) = \int_0^a g_j(t)\,dt$ for all j. In terms of a stationary $g_j(t)$, (discrete) time average $\overline{G}(a) = E\{\int_0^a g_j(t)\,dt\}$, and $\lim_{a\to\infty} \overline{G}(a) = \overline{G} = \infty$. By applying the argument for (192), we have (201).

Remarks. The case $\overline{G} = \infty$ is discussed in Section 5-17. It is easy to extend Theorem 5 to functions f_j that are not sign constrained, when $\lambda(\omega) < \infty$ and both $\overline{G}^+(\omega)$ and $\overline{G}^-(\omega)$ are finite, for corresponding sequences $\{f_j^+\}$ and $\{f_j^-\}$, where for every j and t, $f_j^+(t) = \max\{f_j(t), 0\}$ and $f_j^-(t) = \max\{-f_j(t), 0\}$.

Example 5-21: Time-Average Work

Consider a single- or multiple-channel queue, with an arbitrary arrival process. We make no particular assumption about the order of service except that once service begins on a customer, it is completed without interruption. Any delay occurs before service begins.

Let $f_j(t)$ be the *remaining* service of C_j at epoch t, i.e.,

$$\begin{aligned} f_j(t) &= S_j && \text{for } t \in [t_j, t_j + D_j), \\ &= S_j - (t - t_j - D_j) && \text{for } t \in [t_j + D_j, t_j + W_j), \\ &= 0 && \text{otherwise.} \end{aligned}$$

Hence,

$$G_j = \int_0^\infty f_j(t)\,dt = S_jD_j + S_j^2/2, \qquad j = 1, 2, \ldots,$$

and, as in Section 5-13, let $H(t) = V(t)$, the total *work in system,* where

$$V(t) = \sum_{j=1}^{\infty} f_j(t), \qquad t \geq 0.$$

Under the conditions of Theorem 5, time-average work is

$$E(V) = \lambda\overline{G} = \lambda E(SD) + \lambda E(S^2)/2,$$

which is (149), where $E(SD)$ and $E(S^2)$ are customer averages of sequences $\{S_jD_j\}$ and $\{S_j^2\}$. If the S_j are i.i.d. and S_j and D_j are independent,

$$E(SD) = dE(S).$$

Notice that technical assumption (198) is satisfied if $w < \infty$.

Example 5-22: Limiting Distributions of Excess, Age, and Spread

These distributions were obtained in Section 2-4 for i.i.d. interevent times, and generalized to regenerative interevent times in Problem 2-91. We now obtain a sample-path version of these results, based on Theorem 5, that requires neither independence nor stationarity.

Define $f_j(t, y, \omega)$ for any $y \geq 0$ by

$$f_j(t, y) = 1 \qquad \text{if } t \in [\max(t_j, t_j + l_j - y), t_j + l_j],$$
$$= 0 \qquad \text{otherwise, from which we get}$$
$$G_j(y) = \min(l_j, y).$$

The meaning of the above is that $f_j(t, y)$ is an indicator for the event that measured from t, the remaining length of l_j is $\leq y$, and $G_j(y)$ is the *amount of time* for which this is true.

For every $u \geq 0$, let $I_j(u) = 1$ if $l_j > u$, $I_j(u) = 0$ otherwise, where $\int_0^\infty I_j(u)\, du = l_j$, and we assume that the following limit exists:

$$F^c(u) \equiv \lim_{n\to\infty} \sum_{j=1}^{n} I_j(u)/n, \qquad u \geq 0. \tag{205}$$

We may think of $F(u)$ as the distribution function of a "randomly selected" l_j. Assume that $\{l_j\}$ has customer average $0 < 1/\mu < \infty$, and that $1/\mu$ is the mean of F,

$$\int_0^\infty F^c(u)\, du = 1/\mu.$$

It follows immediately that the fraction of the $G_j(y)$ that exceed u is $F^c(u)$ if $y > u$, and this fraction is equal to 0, otherwise. Furthermore,

$$G_j(y) = \int_0^y I_j(u)\, du \quad \text{and} \quad \sum_{j=1}^{n} G_j(y)/n = \sum \int = \int \sum = \int_0^y \sum_{j=1}^{n} I_j(u)\, du/n.$$

As $n \to \infty$, we interchange $\lim \int = \int \lim$ (the function is bounded), and

$$\overline{G}(y) = \int_0^y F^c(u)\, du.$$

From Theorem 5, when $\lambda > 0$, we have that at sample point ω,

$$\overline{H}(y, \omega) \equiv \overline{H}(y) = (\lambda/\mu)\mu \int_0^y F^c(u)\, du = (\lambda/\mu)F_e(y), \qquad y \geq 0, \tag{206}$$

where F_e is the equilibrium distribution corresponding to F. If $l_j = t_{j+1} - t_j$, $\mu = \lambda$ from (183), and (206) becomes the limiting distribution of excess. For a single-channel queue where $l_j = S_j$, with customer-average distribution G [in the sense of (205)], (206) becomes $(\lambda/\mu)G_e(y)$. The same result holds for multichannel queues.

For spread, let $f_j(t, x) = 1$ if $t \in [t_j, t_j + l_j]$ *and* $l_j \leq x$, $f_j(t, x) = 0$ otherwise. Instead of (206), we get

$$\overline{H}(x) = (\lambda/\mu)\mu[xF(x) - \int_0^x F(u)\, du], \qquad x \geq 0. \tag{207}$$

For $l_j = t_{j+1} - t_j$, $\mu = \lambda$, and (207) becomes $F_s(x)$, the limiting distribution of spread, first found as equation (50) in Chapter 2. The analysis for age is almost identical to that for excess; the result is the same.

5-16 ON POISSON ARRIVALS SEE TIME AVERAGES (PASTA)

For continuous-time Markov chains that are irreducible and positive recurrent, we presented an elementary proof of this result in Section 4-3: For any n, the fraction of Poisson events (arrivals) that find (see) the chain in state n is p_n, the fraction of *time* the chain is in state n, which is a time average. This is the PASTA property. For Markov chains, the arrival rate of the Poisson process is simply a portion of the transition rate out of each state. Because we permit transitions from any state to itself, we can create a Poisson process for any such chain, whenever it is useful to do so.

Unfortunately, our elementary proof does not cover important applications of this result to more general state spaces and stochastic processes. In Section 5-13, for example, we applied PASTA to the work-in-system process $\{V(t)\}$ for an $M/G/1$ queue. While this process is easily seen to be Markov, the state space is uncountable. In this section, we prove PASTA for a general stochastic process.

Let $N \equiv \{N(t): t \geq 0\}$ be a stochastic process called the *system*, which takes on values in some state space, and let B be an arbitrary collection of states. The system interacts in some unspecified way with $\Lambda \equiv \{\Lambda(t): t \geq 0\}$, a Poisson process at rate $\lambda > 0$. For example, if N is the number of customers in system process for a queue, it increases by one customer at every arrival epoch.

We wish to compare the fraction of *time* the system takes on values in B with the fraction of arrivals who see (find on arrival) the system in B. For this purpose, we define the following for every epoch $t \geq 0$:

$$\begin{aligned} U(t) &= 1 \qquad \text{if } N(t) \in B, \\ &= 0 \qquad \text{otherwise}, \\ \overline{U}(t) &= \int_0^t U(s)\, ds/t, \\ A(t) &= \int_0^t U(s)\, d\Lambda(s), \text{ and} \\ \overline{A}(t) &= A(t)/\Lambda(t). \end{aligned}$$

We make the technical assumption that with probability 1, the sample paths of U are left-continuous and have right-hand limits. For the queueing example above, left-continuity means that arrivals are not counted as being in the system until (just) after they arrive. Although not customary for continuous-time processes, left-continuity has no effect on the length of time N spends in any state. With our definitions, $\overline{U}(t)$ is the fraction of time during $[0, t]$ that N spends in B, and $\overline{A}(t)$ is the fraction of arrivals who find N in B.

Typically, we expect the arrivals to affect the system in some way, i.e., Λ and N, and hence Λ and U, will be dependent processes. However, we assume that the system has no *anticipation,* i.e., the future increments of Λ are independent of the past of U. Formally, we make the

Lack of Anticipation Assumption (LAA). For every $t \geq 0$, $\{\Lambda(t+u) - \Lambda(t): u \geq 0\}$ is independent of $\{U(s): 0 \leq s \leq t\}$ and $\{\Lambda(s): 0 \leq s \leq t\}$.

We now state the main result; a proof is given at the end of this section.

Theorem 6: PASTA. Under LAA, $\overline{U}(t) \rightarrow \overline{U}(\infty)$ w.p.1 if and only if $\overline{A}(t) \rightarrow \overline{U}(\infty)$ w.p.1, as $t \rightarrow \infty$.

Remarks.

1. PASTA is true whenever convergence occurs. The only reason for making stronger assumptions, e.g., that the system is regenerative, is to prove convergence.
2. While $\overline{U}(\infty)$ typically will be a constant, it need not be. For example, suppose N is a continuous-time Markov chain with several absorbing states and initial transient state $N(0) = i$. If B is an absorbing state, $\overline{U}(\infty)$ is a random variable, where $\overline{U}(\infty) = 1$ if absorption in B occurs, and $\overline{U}(\infty) = 0$ otherwise.
3. We already know that Λ, a Poisson process, has independent increments. The inclusion of this property in LAA is made for technical reasons. If Λ is a point process that lacks this property, $\{\Lambda(s): 0 \leq s \leq t\}$ may *induce* dependence between $\{U(s): 0 \leq s \leq t\}$ and $\{\Lambda(t+u) - \Lambda(t): u \geq 0\}$.

As Example 5-23 shows, the Poisson process in Theorem 6 need *not* be an arrival process.

Example 5-23: The *GI/M/c* Queue with Arrival Rate λ and Service Rate μ/Server

Let π_n be the fraction of arrivals who find n customers in system, and p_n be the corresponding fraction of time, $n = 0, 1, \ldots$.

We want to show the following well-known result:

$$\mu_n p_n = \lambda \pi_{n-1}, \qquad n = 1, 2, \ldots, \tag{208}$$

where $\mu_n = \mu \min(n, c)$. We shall show (208) by equating transition rates (number of transitions per unit time) across the boundary between states n and $n-1$. Clearly, $\lambda \pi_{n-1}$ is the transition rate from $n-1$ to n.

To find the transition rate in the reverse direction, let $\{\Lambda(t)\}$ be a Poisson "event" process at rate μ_n that generates departures when the number of customers in system is n, i.e., at each of these events, a departure occurs if and only if the number in system is n. (In effect, we define a separate Poisson process for every state.) Let $A(t)$ be the number of transitions from n to $n-1$ in $[0, t]$. We have $\Lambda(t)/t \rightarrow \mu_n$ w.p.1 and, from PASTA, $A(t)/\Lambda(t) \rightarrow p_n$ w.p.1, as $t \rightarrow \infty$. Hence,

$$A(t)/t = [A(t)/\Lambda(t)][\Lambda(t)/t] \rightarrow \mu_n p_n \quad \text{w.p.1},$$

the transition rate from n to $n - 1$, and we have (208). Extension: For $\{\mu_n\}$ an arbitrary function of n, (208) is true.

For Theorem 6, we need the following result, which is of independent interest.

Theorem 7. Under LAA,

$$E\{A(t)\} = \lambda t E\{\overline{U}(t)\} = \lambda E\left[\int_0^t U(s)\, ds\right], \tag{209}$$

i.e., on any finite interval, the *expected* number of arrivals who find the system in state B is equal to the arrival rate times the expected length of time it is there.

Remark. Conditioning on $\Lambda(t)$, the conditional expectations corresponding to (209) typically are not equal. For example, let Λ be the arrival process and N be the number of customers in system process for an $M/M/1$ queue, with $N(0) = 0$. Let B denote that the system is busy. Clearly,

$$E\{A(t) \mid \Lambda(t) = 1\} = 0, \text{ but } E\{\overline{U}(t) \mid \Lambda(t) = 1\} > 0.$$

*Proof of Theorem 7.** The assumption that U is left-continuous and has right-hand limits w.p.1 implies that, w.p.1, U has only a finite number of discontinuities on any finite interval. Hence, for every ω in a set w.p.1, $A(t)$ can be approximated arbitrarily closely by a function of form

$$A_n(t) = \sum_{k=0}^{n-1} U(kt/n)[\Lambda((k + 1)t/n) - \Lambda(kt/n)], \tag{210}$$

for sufficiently large n. From LAA, (210) has expected value

$$E\{A_n(t)\} = \lambda t E\left[\sum_{k=0}^{n-1} U(kt/n)/n\right]. \tag{211}$$

Now $A_n(t) \leq \Lambda(t)$ for all n, and from the dominated convergence theorem,

$$E\{A(t)\} = \lim_{n\to\infty} E\{A_n(t)\} = \lambda t E\{\overline{U}(t)\},$$

which is (209).

Theorem 6 will follow from two lemmas for process R defined as

$$R(t) = A(t) - \lambda t \overline{U}(t), \qquad t \geq 0.$$

See Feller [1971] for the definition of a martingale.

* The remainder of this section is at a higher mathematical level.

Lemma 1. R is a *continuous-time martingale*.

Proof. We need to show that for $t \geq 0$, $h > 0$, the increments of R have the property

$$E\{R(t + h) - R(t) \mid R(s), 0 \leq s \leq t\} = 0. \tag{212}$$

Let $\mathcal{F}_t$ be the σ-field generated by $\{\Lambda(s), U(s); 0 \leq s \leq t\}$. From LAA, we may repeat the argument used to drive (209) for any set in $\mathcal{F}_t$:

$$E\{A(t + h) - A(t) \mid \mathcal{F}_t\} = \lambda E\left[\int_t^{t+h} U(s)\, ds \mid \mathcal{F}_t\right].$$

Because the σ-field generated by $\{R(s); 0 \leq s \leq t\}$ is a subset of $\mathcal{F}_t$, (212) follows immediately.

Lemma 2. $R(t)/t \to 0$ w.p.1 as $t \to \infty$.

Proof. From $A(t) \leq \Lambda(t)$ and $\overline{U}(t) \leq 1$, observe that

$$E\{R^2(t)\} \leq E\{\Lambda^2(t)\} + \lambda^2 t^2 = \lambda t + 2\lambda^2 t^2 \equiv k(t). \tag{213}$$

From Lemma 1, the process $\{R(nh)\}$, with increments

$$X_n = R(nh) - R((n - 1)h), \qquad n = 1, 2, \ldots,$$

is a discrete-time martingale for any $h > 0$.

By the same bounding argument used for (213),

$$E(X_n^2) \leq k(h), \text{ and}$$

$$\sum_{n=1}^{\infty} E(X_n^2)/n^2 < \infty. \tag{214}$$

From (214), by Theorem 3, p. 243 of Feller,

$$R(nh)/n \to 0 \qquad \text{w.p.1} \quad \text{as} \quad n \to \infty. \tag{215}$$

Because $A(t)$ is monotone and changes in $t\overline{U}(t)$ on an interval of length h are bounded by h,

$$R(nh) - \lambda h \leq R(t) \leq R((n + 1)h) + \lambda h, \qquad t \in [nh, (n + 1)h]. \tag{216}$$

The lemma now follows from (215) by dividing (216) by t and letting $t \to \infty$, where $n(t) \equiv [t/h] \to \infty$.

Proof of Theorem 6. Theorem 6 follows immediately from Lemma 2 and $\Lambda(t)/t \to \lambda$ w.p.1, by writing

$$R(t)/t = [A(t)/\Lambda(t)][\Lambda(t)/t] - \lambda\overline{U}(t).$$

Remark. Under LAA, Lemma 2 implies other versions of Theorem 6 under weaker modes of convergence, e.g., in probability measure or in distribution. An expected-value version follows directly from (209).

5-17 CONCLUSIONS AND RELATED LITERATURE

While this chapter is only an introduction, we have emphasized results that are both general and useful for analyzing and understanding the behavior of queues. Relations between time and customer averages play an important role. There are two types: (1) averages exemplified by $L = \lambda w$ and the relationship between work and delay, where the customer average is the average contribution to an integral used to define a time average; and (2) PASTA, where the fraction of all time that a process spends in various states is equal to the fraction of Poisson "arrivals" who find the process in these states. (In applications, the Poisson events need not be arrival epochs.) Results of the first type are essentially free of stochastic assumptions. The second clearly requires that some process be Poisson, but interesting inequalities can sometimes be obtained when the assumption is relaxed in certain ways. A third general result equates rates into and out of a state or a set of states, which, while mathematically trivial, is also very useful.

The result $L = \lambda w$ was taken for granted by early practitioners. The first general proof was by Little [1961], who made certain stationarity assumptions that may be hard to verify. The sample-path proof in Section 5-15 is based on Stidham [1974]; Example 5-20 is similar to one found there. In spite of the intuitive appeal of this result, it is often misunderstood, and conceptually flawed "explanations" have appeared in many places; see Problem 5-20.

The generalization of $L = \lambda w$ in Theorem 5 was first made by Brumelle [1971b]; the sample-path proof given here is based on Heyman and Stidham [1980]. Example 5-22 is from Wolff [1988].

For a particular stochastic model that is either regenerative (with finite mean cycle length) or stationary and ergodic, the limits in Theorems 4 and 5 are constants w.p.1, and our sample-path results hold w.p.1. These results also apply to stationary processes that are not ergodic, where time and customer averages converge to random variables, but some results would have to be restated. In particular, the form of the limiting distributions in Example 5-22 require convergence w.p.1 to constants.

The Case $w = \infty$ and $\overline{G} = \infty$

If we assume that $W_j/j \to 0$ as in (186), $\lambda(\omega) > 0$, and $w(\omega) = \infty$, the proof of $L = \lambda w$ is easily modified to obtain

$$L(\omega) = \lambda(\omega)w(\omega) = \infty.$$

The problem with this approach is that under reasonable stochastic assumptions, $w = \infty$ and $W_j/j \to 0$ are incompatible. For example, if the W_j are i.i.d., $W_j/j \to 0$ w.p.1 if *and only if* $w < \infty$. Weak sufficient conditions for sample-path versions of (finite or infinite) $\overline{H} = \lambda\overline{G}$ are given in Stidham [1982], but they would be very difficult to verify. He also observes that under the technical assumption (197) and (198), Theorem 5 is easily extended to show that when $\lambda(\omega) > 0$ and $\overline{G}(\omega) = \infty$,

$$\overline{H}(\omega) = \lambda(\omega)\overline{G}(\omega) = \infty.$$

Unfortunately, we should not expect (198) to hold when $\overline{G} = \infty$.

For these reasons, we abandoned the intuitively appealing and direct sample-path approach when $\overline{G} = \infty$, and made a regenerative assumption instead. The marked point process approach, e.g., see p. 106 of Franken et al. [1982], also successfully deals with this case, under stationarity assumptions, but only for queueing models of limited complexity.

Multiple Visits

The setup for the proof of $L = \lambda w$ assumed that the W_j are intervals. If customers may return to the system many times, we may wish to exclude their time spent elsewhere from their *total* waiting time. If this is done, counterexamples to $L = \lambda w$ are easily constructed, e.g., see Heyman and Stidham. The reason is that while $w < \infty$, the time average of each customer's time away from the system may be infinite (or may not exist). Stidham [1982] observes that this situation can be modeled in the context of Theorem 5, where W_j is performed within an interval of length l_j, such that (198) holds; under these assumptions $L = \lambda w$ holds. From the preceding discussion, however, this is not a satisfactory formulation for stochastic modeling, but the stronger assumption that $\{l_j\}$ has a finite customer average would be sufficient for $L = \lambda w$ to hold when $\lambda < \infty$ and $w < \infty$.

After $L = \lambda w$, PASTA is the most frequently used result in the queueing literature. It has had a similar history. Many authors, e.g., Cobham [1954], simply used PASTA without comment. While PASTA was generally believed to be true in those days, there was some confusion about what it meant, e.g., does an arrival observe itself? The answer clearly had to be no, for otherwise arrivals would never find the system empty.

Later, when the standards of rigor were higher, PASTA was acknowledged as a property that required proof, and various proofs were constructed under what turn out to be superfluous assumptions.

The treatment of PASTA in Section 5-16 is based on Wolff [1982a], where a discussion of earlier proofs and applications may be found. Watanabe [1964] has shown that the martingale property in Lemma 1 is a characterization of a Poisson process.

Deterministic queues are extensively treated in Newell [1982]. The tandem queue with constant work-force model in Section 5-4 is treated by Oliver and Samuel [1962].

There are different definitions of what we called *phase-type* distributions in Section 5-10. Cox [1955] defined a class of density functions that have rational Laplace transforms. Neuts [1981] defined a class of distributions corresponding to first passage times in finite-state, continuous-time Markov chains. These definitions are not equivalent. Asmussen [1987] reviewed various definitions and showed, among other things, that the class defined in Section 5-10 is a subset of the class defined by Neuts, which in turn is a subset of the class defined by Cox. From our point of view, any of the definitions in Asmussen could be used. The one we have chosen is the smallest class that includes both Erlang and hyperexponential, and has the convergence property in Theorem 2.

These distributions may also be used to approximate moments, *when they are finite*. To approximate a distribution with an infinite second moment, say, we would have to consider phase-type distributions with an infinite number of phases. This means that the actual number of phases a service (say) goes through is an unbounded random variable.

Work in system, which we introduced in Section 5-4, is a very important conceptual and analytic tool. This will be demonstrated many times in subsequent chapters, particularly Chapter 10. Early mathematical treatments of work in system for single-channel queues, where it is called *virtual delay,* are by Beneš [1963] and Takács [1962].

With the help of phase-type distributions, we can, in principle, approximate virtually *any* queue as a continuous-time Markov chain; as a consequence, the literature on Markovian queues is huge. In Chapters 6 and 7, we treat important models of this type. Unfortunately, state spaces become complex, and easy solutions to the balance equations are by no means assured. What appears to be a minor variation of an easy model can greatly complicate the solution; e.g., see Problem 5-25. Numerical solution and approximation methods for complex Markovian queues have been developed, e.g., see Neuts. Queues defined in terms of one or more distributions that need not be of phase-type are dealt with directly in subsequent chapters.

Sometimes the steady-state solution to a complex Markovian queue turns out to be remarkably simple, as we discovered with the product form of equation (99). This and the insensitivity result known as Erlang's loss formula are part of a general theory treated in Chapter 6.

In most queueing models, performance measures *do* depend on the form of distributions. In several places in this chapter, in fact, we have observed that performance measures improve as distributions become more regular. This is usually true and is consistent with our intuition about why queues form at all.

PROBLEMS

5-1. Consider a deterministic queue on interval $[0, T]$ with initial queue length $Q(0) = 0$, arrival rate λ_1 on $[0, t]$ and λ_2 on $(t, T]$, and service rate μ, where $0 < \lambda_2 < \mu < \lambda_1$. Assume T/t is large enough for $Q(T) = 0$ to hold.

(a) Find the average queue length on $[0, T]$.

(b) Find the (average) arrival rate on $[0, T]$.

(c) Under FIFO, find

(i) the delay in queue of an arrival at $x \in [0, T]$,

(ii) the longest delay anyone experiences,

(iii) the fraction of customers who have strictly positive delay, and

(iv) the delay averaged over all customers.

(d) Verify $L = \lambda w$.

(e) Repeat (c) and (d) for LIFO order of service.

5-2. Cumulative input for a pickup truck is $\Lambda(t) = 100\sqrt{t}$ pounds, $0 \le t \le 300$. A dispatch occurs at $T = 300$, and we wish to schedule an intermediate dispatch at some τ, $0 \le \tau \le 300$.

(a) Find the optimal τ, τ^*, that minimizes average delay.

(b) Suppose the pickup truck has a capacity of 900 pounds, i.e., any inventory at τ in excess of 900 pounds will not be picked up until T. What is the optimal τ now?

(c) In (a) and (b), what is the average delay under an optimal dispatch, τ^*?

5-3. *Generalization of Example 5-3 to different processing rates at each station.* Suppose there are two queues in tandem and a total (fixed) work force of m workers, where m is large. A worker at station i can process (serve) r_i customers per unit time, $i = 1, 2$. Thus, if m_i workers are allocated to station i, with $m_1 + m_2 = m$, the service capacity at station i is $m_i r_i$.

For the objective of minimizing overall average delay, what is the optimal allocation of workers to stations? (Treat the m_i as continuous, i.e., ignore the constraint that they are integers.) At which station(s) is the queue permitted to be positive?

5-4. Let $\{\Lambda(t)\}$ be the arrival process of cars at an automotive clinic. The clinic operates as a tandem queue, where, at the first station, repairs are diagnosed. At the second station, repairs are made. Only cars requiring repair go to the second station. Cars not requiring repair leave the system after their diagnosis at the first station is complete. Let α (*not* a function of time) be the fraction of cars that are diagnosed to need repair, $0 < \alpha < 1$, and r_i be the number of cars per hour that can be processed by one worker at station i, $i = 1, 2$.

Analyze as a deterministic queue. For a constant work force (total number of workers) show that (ignoring integer constraints) if

$$1/r_1 + \alpha/r_2 > 1/r_2 \quad \text{or, equivalently,} \quad r_2 > (1 - \alpha)r_1,$$

average overall delay is minimized by allocating workers efficiently in such a way that a queue is allowed to form only at the first station.

If the preceding inequality is reversed, the policy just described is *not* optimal. What should be done in this case?

5-5. For an $M/M/1$ queue with $\lambda < \mu$, what fraction of time is the server busy? Verify your result from $L = \lambda w$.

5-6. What is the expected waiting time in system for an $M/M/1$ queue where order of service is LIFO, i.e., on service completion, the next customer served is the most recent arrived? Supppose we don't even know the order of service. (Think of a busy retail shop that does not have a "take a number" system.)

5-7. For an $M/M/1$ queue, $\{N_q(t): t \geq 0\}$ is not a Markov process. Explain why not.

5-8. Some customers may be dissatisfied with the service they received and must be served again. Suppose that on completion of service, a customer is dissatisfied with probability $1 - \alpha$, for some $\alpha \in (0, 1)$, independent of whether that customer had been dissatisfied (one or more times) before. Otherwise, we have a standard $M/M/1$ queue. Subsequent service times on the same customer, if any, are also independent and $\sim$ exp (μ).

(a) Let S_T denote the total time spent in service of a customer, *until satisfied*. Show that $S_T \sim$ exp $(\alpha\mu)$. [*Hint:* Represent S_T as a sum and find $E(e^{-sS_T})$.]

(b) Suppose dissatisfied customers are served again immediately, until satisfied, and that the order of service is FIFO. What are the conditions for the queue to be stable? Under these conditions, find the expected *total* waiting time of a customer in system.

(c) Suppose customers are served in the order in which they join the queue, and dissatisfied customers join the *end* of the queue. What are your answers to (b) now?

5-9. For an $M/M/1$ queue with queue limit l, $w = L/\lambda$ is an average waiting time over what stream of customers? Show that your answer is consistent with (68).

5-10. The manager of a small market must decide which of two potential checkers to hire: Bill, who works slowly, can be hired for $C_1 = \$6.00$ per hour, or Alice, who works faster, demands a higher rate $\$C_2$/hour. Both give exponential service at rates $\mu_1 =$ 20 customers/hour and $\mu_2 = 30$ customers/hour, respectively.

The arrival of customers at the checkout counter is Poisson at rate $\lambda =$ 10/hour. The manager estimates that on average, each customer's time is worth 5 cents/minute and should be accounted for in the model.

(a) Set up the expected cost/hour incurred by hiring either Bill or Alice.

(b) With the above data, how much would you be willing to pay Alice?

(c) If you didn't know Alice's service rate, find an upper bound on the amount you would pay her.

5-11. *Recursive formula for loss probabilities*. For an $M/M/c$ *loss* system with arrival rate λ and service rate μ, define $B(c, a) = p_c$, the fraction of customers lost, where $a = \lambda/\mu$ is the offered load. We will obtain a recursive formula (in c) for $B(c, a)$ for any fixed a. To do this, we compare having c channels with having only $(c - 1)$.

(a) Without writing out explicit formulas for the p_n, show that

$$p_{c-1} = B(c - 1, a)(1 - B(c, a)) \qquad \text{for } c \geq 1,$$

where p_{c-1} is a probability for the c-channel loss system. (*Hint:* How does having channel c affect the ratio p_n/p_{n-1}, $n = 0, 1, \ldots, c-1$?)

(b) From (a), show that for $c \geq 1$,

$$B(c, a) = \frac{aB(c-1, a)}{aB(c-1, a) + c}, \tag{P.1}$$

where $B(0, a) = 1$. [For brevity, $B(c, a) = B_c$ in Chapter 7.]

5-12. *Continuation. Recursive formula for delay probabilities.* For a stable $M/M/c$ queue (with no queue limit) with state probabilities $\{p_n\}$, let $C(c, a) = P(D > 0) = \sum_{n=c}^{\infty} p_n$. Show that

(a) $C(c, a) = cp_c/(c - a)$,

(b) $p_{c-1} = B(c-1, a)(1 - C(c, a))$,

(c) $$C(c, a) = \frac{aB(c-1, a)}{aB(c-1, a) + c - a}, \quad c \geq 1. \tag{P.2}$$

5-13. The arrival process of customers at a taxi stand is Poisson at rate λ, and the arrival process of taxis at the stand is Poisson at rate μ. Arriving customers who find taxis waiting (being delayed) leave immediately in a taxi. Arriving taxis who find customers waiting leave immediately with *one* customer each. Otherwise, customers will queue up to a limit of c customers, i.e., arriving customers who find c other customers waiting leave immediately *without* a taxi. Similarly, taxis queue up to a limit of t taxis.

(a) Define a state space and find the corresponding state probabilities. In terms of the probabilities in (a) find:

(b) The average number of customers and the average number of taxis in the queue.

(c) The average delay of those customers who leave in a taxi.

(d) The average delay of those taxis who leave with a customer.

5-14. *Continuation.* What are the conditions for the queue to be stable if:

(a) there is a queue limit on taxis but not on customers,

(b) there is a queue limit on customers but not on taxis, or

(c) there is no queue limit on either customers or taxis?

5-15. A fleet of m planes is said to be in a "state of readiness" if at least r of these planes are operational, $0 < r < m$. The time until failure of a plane in use is $\sim$ exp (λ). Planes are repaired at a k channel facility, $0 < k < m$; repair times are exp (μ). When the number of operational planes exceeds r, the excess planes are treated as spares. Spares do not fail when they are not in use. When a failure causes the number of planes in use to fall to $r - 1$, a spare is brought into use instantaneously.

(a) Define a state space and find a recursion formula for the corresponding state probabilities.

(b) In terms of the state probabilities in (a), what fraction of time is the fleet in a state of readiness?

5-16. A small gas station has space for two cars. Potential customers are Poisson at unknown rate but will not stop if space is filled (they are lost). (These were the good old days.) The mean time between the arrival of actual customers (those who stop and are served) is estimated to be 6 minutes. There is one attendant, whose service time is exponential with mean 5 minutes.

(a) What fraction of potential customers are lost?

(b) Find the expected waiting time, w, of those who are served.

5-17. Consider an $M/M/1$ queue with arrival rate λ and service rate μ, where customers in queue (but not the one in service) may get discouraged and leave without receiving service. Each customer who joins the queue will leave after a time distributed exp (γ), if the customer has not entered service before that time.

(a) Represent this model as a birth and death model. Write the recursion formula for the state probabilities. Solve. (The solution won't be closed form, but it is easy to obtain numerical results.) Under what conditions is this system stable?

(b) What fraction of arrivals are served?

(c) Suppose an arrival finds 1 customer in system. Under a FIFO rule, find

(i) the probability that this customer is served, and

(ii) the expected delay in queue, given this arrival is served

5-18. Derive (84).

5-19. Consider an $M/M/2$ queueing model with no queue allowed (i.e., arrivals when both channels are busy do not receive service and are lost).

Each channel is operated by one man. Each man records the number of hours required to service each customer and the number of customers he serves. The following data were collected: During 10,000 hours of operation, 40,000 customers received service and 8,000 man-hours of service were recorded (total for both channels). (Assume that the sample is large enough so that statistical fluctuations in these numbers are negligible.)

(a) During the 10,000 hours above, how many customers were lost?

(b) Suppose lost-revenue/lost-customer is \$5.00, and the cost of operating a channel (busy or not) is \$4.00/hour. Would it be desirable to add another channel? (Assume no queue is allowed.)

5-20. The following intuitive argument for $L = \lambda w$ has been proposed: "The average number of customers in system *found* by an arriving customer is equal to the average number of customers in system *left behind* by a departing customer. The second quantity is the arrival rate times the average waiting time in system." Note that there are two steps in this argument.

(a) Is the argument complete, i.e., if each step is true, does this prove $L = \lambda w$?

(b) Give an example where at least one of the steps is false but $L = \lambda w$ is true.

(c) Give an example where the second step is true.

(d) How general is the first step?

5-21. Verify (99).

5-22. Consider an $M/M/c$ loss system (called *station 1*) with arrival rate λ, where customers who find all channels busy (and are lost) are served at a secondary *infinite* channel facility (called *station 2*). (Customers who are served at station 1 leave the system. They do *not* go to station 2.) Service times at both stations are distributed exp (μ).

Let N_i be the number of customers at station i in steady state, $i = 1, 2$, and denote their joint distribution by

$$P(N_1 = m, N_2 = n) = p_{mn};\ m = 0, 1, \ldots, c;\ n = 0, 1, \ldots.$$

(a) Write out balance equations for the p_{mn}. Under what conditions would you expect this two-station system to be stable?
(b) What is the marginal distribution $P(N_1 = m) = p_m$, say, $m = 0, 1, \ldots, c$? In terms of this distribution, represent $E(N_1)$.
(c) What is the distribution of $N_1 + N_2$?
(d) Use (b) and (c) to find an expression for $E(N_2)$.
(e) What is the arrival rate of customers at station 2?
(f) Use (e) to find another expression for $E(N_2)$.

5-23. An item inventory at a warehouse is stocked up to a maximum of S units. Demand for the item is Poisson at rate λ units/day. When the inventory level drops to s, $0 < s < S/2$, a reorder of size $S - s$ units is placed with the factory. Factory delivery time (of the entire $S - s$ units) is exponential with mean $1/\mu$ days. The inventory level will continue to drop with demand while the factory reorder is outstanding. When the inventory level is zero, demand is lost.
(a) Define an appropriate state space for this model.
(b) This problem cannot be formulated as a birth and death model. Why not?
(c) Set up balance equations for the state probabilities corresponding to the state space in (a).
(d) In terms of the state probabilities in (c), what is the average inventory level?
(e) In terms of the state probabilities in (c), at what rate (number of units/day) is demand lost?
(f) Suppose the cost of lost sales is $\$R$/unit, the cost of placing a factory order is $\$C$/order, and the cost of holding inventory is $\$h$/unit/day, based on the *average* inventory held at the warehouse. Find an expression for the total cost/unit time in terms of the state possibilities in (c).

5-24. *Continuation.* Formulate Problem 5-23 in terms of renewal-rewards, and find an expression for the (total cost)/(unit time) as a ratio of two expectations. Verify that your expression agrees with that found in Problem 5-23(f).

5-25. Consider an $M/M/2$ queue with arrival rate λ and service rate μ/server, where each channel has its own queue. Arrivals join the shortest line (number in queue plus service), but cannot switch later (no jockeying).
(a) Formulate two ways; for one of these, you will have to make an additional assumption. Write down balance equations for one formulation; do *not* solve. (This problem is notorious; see remark below.)
(b) Would you expect w for this model to be greater, less, or the same as that for the standard (single queue) $M/M/2$ model?

Remark. Textbook problems usually have "closed form" solutions, but this one does not (an incorrect closed form solution has been published). Of course, a unique solution to the balance equations exists when $\lambda < 2\mu$. Using generating functions, Kingman [1961] finds properties of this solution.

5-26. For a finite source model (the arrival process in Example 5-6), show that the results for L and Q in Example 5-7 are still valid. Also compare w.

5-27. Poisson arrivals at rate λ stop at a single-channel gas station for gas, oil, or both. Service times for gas are $\sim$ exp (μ_g), for oil are $\sim$ exp (μ_o), and for both are $\sim$ as the convolution of these distributions. Customer service requirements are i.i.d. With probabilities α_g, α_o, and α_b, a customer will want gas, oil, or both, respectively, where $\alpha_g + \alpha_o + \alpha_b = 1$.

Define an appropriate state space and write out the set of balance equations for the corresponding steady-state probabilities.

5-28. For an $E_2/M/1$ *loss system* (no queue allowed) with arrival rate λ, service rate μ, and $\rho = \lambda/\mu$,

(a) Define a state space and set up balance equations for the corresponding state probabilities.

(b) Find an expression for the fraction of arrivals lost (it will be *different* from the fraction of time the server is busy), and show that this fraction is *less* than $\rho/(1 + \rho)$, the fraction lost for the corresponding $M/M/1$ system. (*Hint:* Use $L = \lambda w$ to find the fraction of arrivals *served*.)

(c) Repeat for an $H/M/1$ loss system and show that the fraction lost is *greater* than $\rho/(1 + \rho)$.

5-29. For an $M/G/1$ loss system with arrival rate λ and mean service time $1/\mu$ (service rate μ), use renewal theory to show that the fraction of arrivals lost is $\rho/(1 + \rho)$, independent of the form of G.

5-30. For an $M/E_2/c$ loss system with arrival rate λ, service rate μ, and $\rho = \lambda/\mu$, let N denote the number of customers in system and J be the number of these customers who are undergoing their first phase of service. Let $P(N = n, J = j) = p_{nj}$, $0 \leq j \leq n \leq c$.

(a) Write out balance equations for the state probabilities for this state space.

(b) Verify (plug in) that the balance equations have the solution

$$p_{nj} = [\rho^n p_0/n!]\left[\binom{n}{j}(1/2)^n\right], \qquad 0 \leq j \leq n \leq c,$$

where $p_0 = \left[\sum_{n=0}^{c} \rho^n/n!\right]^{-1}$.

(c) From (b), find the marginal distribution of N and the conditional distribution of J (given $N = n$). Verify that N has the same distribution here as it does for the corresponding $M/M/c$ loss system.

(d) Show that the equilibrium distribution of E_k can be written as a mixture of convolutions of an exponential. (*Hint:* Review how the equilibrium distribution arises as a limit for the excess random variable in a renewal process and adapt the argument.)

(e) From (d), interpret the conditional distribution of J in (c).

5-31. *Continuation.* Suppose the arrival rate λ_n is a function of the number of customers in system (as in a birth and death process), but the rest of the model is unchanged. For the same state space, show that

$$p_{nj} = \binom{n}{j}(1/2)^n q_n \qquad \text{for } 0 \leq j \leq n \leq c,$$

where (q_n) is the unique solution to

$$\lambda_n q_n = (n + 1)\mu q_{n+1}, \qquad n = 0, 1, \ldots, c - 1; \qquad \sum_{n=0}^{c} q_n = 1.$$

5-32. Verify (111).

5-33. In Problem 5-22, suppose the service times at stations 1 and 2 have some arbitrary distribution with service rate μ. Now find $E(N_2)$.

5-34. For Erlang's loss formula, apply $L = \lambda w$ to find the fraction of customers lost, f_l, and verify that $f_l = p_c$.

5-35. Consider an $M/P_2/2$ loss system where P_2 denotes a phase-type distribution (116) with $k = 2$ phases.

(a) Define a state space and write out balance equations for the corresponding steady-state probabilities.

(b) What is the equilibrium distribution of P_2, P_{2e}?

(c) From what you know about $M/G/c$ loss systems, deduce the solution to the balance equations in (a).

5-36. Show that the hyper-exponential distribution,

$$H = \alpha \exp(\mu_1) + (1 - \alpha) \exp(\mu_2), \qquad \alpha \in (0, 1),$$

is of phase-type, i.e., H may be written in form

$$H = (1 - \beta) \exp(\gamma_1) + \beta \exp(\gamma_1) * \exp(\gamma_2).$$

[*Hint:* Equate Laplace transforms of the density functions and determine the exact correspondence between (α, μ_1, μ_2) and $(\beta, \gamma_1, \gamma_2)$.]

5-37. For a $BM/M/1$ queue with batch arrival rate λ, batch size distribution $P(\nu_j = n) = \alpha_n$, $n = 0, 1, \ldots$, find the generating function $P(z) = \sum_{n=0}^{\infty} p_n z^n$:

$$P(z) = \frac{\mu(1 - z)(1 - \rho)}{\lambda z A(z) - (\lambda + \mu)z + \mu},$$

where $\rho = \lambda E(\nu)/\mu$, $A(z) = \sum_{n=0}^{\infty} \alpha_n z^n$, and $\rho < 1$. Use this result to find

$$L = \frac{\rho}{1 - \rho} \cdot \frac{E\{\nu(\nu + 1)\}}{2E(\nu)}. \tag{P.3}$$

5-38. For a $BM/M/2$ queue with batch arrival rate λ, constant batch size $\nu_j = 4$, exponential service at rate μ, and server utilization $\rho = 2\lambda/\mu < 1$, find the generating function $P(z) = \sum_{n=0}^{\infty} p_n z^n$ for the distribution of the number of customers in system:

$$P(z) = \frac{(1 - z)(1 - \rho)(16 + 4\rho z)}{[\rho z^5 - (\rho + 4)z + 4](4 + \rho)}, \text{ and} \tag{P.4}$$

$$L = P'(1) = \rho/(4 + \rho) + 5\rho/[2(1 - \rho)]. \tag{P.5}$$

5-39. *Continuation.* Consider a $BM/D/2$ queue with the same arrival process and service rate. (The comparison of this model with the one above is from Wolff [1977b].)

(a) Argue that this model is equivalent to *two* $BM/D/1$ queues, each with constant batch size $\nu_j = 2$.

(b) From (a), find

$$L = 2\rho + \rho(\rho + 1)/(1 - \rho). \tag{P.6}$$

(c) Compare (P.5) and (P.6). In particular, show that for $\rho_0 \approx 0.35$,

$$L_{BM/M/2} < L_{BM/D/2} \quad \text{for } \rho < \rho_0 \text{, and}$$

$$L_{BM/M/2} > L_{BM/D/2} \quad \text{for } \rho > \rho_0.$$

Important Remarks. For multichannel queues, we have shown that performance (L here) can become worse as the service time distribution becomes more regular, at least under light traffic (low server utilization). This is an example of what we call *contrary* behavior. This does not occur for standard single channel queues (see Chapter 11), but it can occur in other contexts, e.g., see Sections 7-6 and 7-7. Contrary behavior with respect to service can occur when *arrivals* are sufficiently irregular in some sense.

5-40. For an $E_2/G/1$ queue, suppose the Erlang arrival process is generated as follows: For a Poisson process at rate 2λ, the *even*-numbered arrivals stop and are served (they are the actual customers). The odd-numbered Poisson arrivals, called *observers,* merely observe the system; they are not served, and they contribute no work to the work-in-system process. Let $V(t_n^-)$ be the work found by the nth Poisson arrival, and $E(V_a)$ and $E(V_o)$ be the arrival average work found by an arriving customer and an observer, respectively.

(a) Show that

$$V(t_{2n-1}^-) \geq V(t_{2n}^-) \quad \text{for all } n,$$

with equality holding if and only if $V(t_{2n-1}^-) = 0$.

(b) From (a), show that $E(V_a) < E(V_o)$.

(c) Argue that $E(V) = [E(V_a) + E(V_o)]/2$, and therefore that $E(V_a) < E(V)$.

(d) From (c), show that $d_{E_2/G/1} < d_{M/G/1} = \lambda E(S^2)/2(1 - \rho)$.

(e) Extend to an $E_k/G/1$ queue.

5-41. *Optimality of FIFO.* Let $f(t)$ be a nondecreasing *convex* function, where convex means that for every real a and b, and $\beta \in [0, 1]$,

$$f[\beta a + (1 - \beta)b] \leq \beta f(a) + (1 - \beta)f(b).$$

(a) Draw a picture showing the geometric property of a convex function: On any $[a, b]$, $f(t)$ lies below the straight line connecting $f(a)$ and $f(b)$.

(b) For $d_1 \leq d_2 \leq d_3 \leq d_4$ such that $d_1 + d_4 = d_2 + d_3$, show that

$$f(d_2) + f(d_3) \leq f(d_1) + f(d_4).$$

(c) Extend the argument in Section 5-14 to show that when customer delay cost is a nondecreasing convex function of delay, FIFO is optimal.

5-42. *The $c\mu$-rule.* Consider the postponable priority queue in Example 5-19. Show that for the linear cost model discussed there, the $c\mu$-rule is optimal. Do this by showing that if classes i and $i + 1$ are out of their $c\mu$-order, cost is reduced by interchanging these classes. Note that from (180), the other classes are not affected by this change.

6

REVERSIBILITY AND QUEUEING NETWORKS

We now treat certain queueing networks and related models that turn out to have remarkably simple stationary behavior. These are generalizations of the tandem queue in Section 5-10 and the $M/G/c$ loss system in Section 5-11. The models in this chapter are continuous-time Markov chains with special structure. We assume throughout that these chains are irreducible and regular, where the latter term means that the number of jumps a chain has in any finite interval is *finite with probability 1*.

Under these assumptions, chain $\{X(t)\}$ is completely determined by the distribution of $X(0)$ and a matrix of conditional transition rates $\mathbf{A} = (a_{ij})$ defined in equations (12) and (13) of Chapter 4. Furthermore, $\{X(t)\}$ is *positive* (recurrent) if and only if there is a unique solution $\{p_j\}$ to balance equations (23) in Chapter 4, where $\sum_j p_j = 1$, and for each j, $p_j > 0$ is the *fraction of time* $\{X(t)\}$ spends in state j. Finally, if $X(0) \sim \{p_j\}$, $\{X(t)\}$ is a stationary stochastic process, where in particular, $X(t) \sim \{p_j\}$ for all t, and we will call $\{p_j\}$ a *stationary* (or *steady-state*) distribution.

In this chapter, we will consider stochastic processes $\{X(t)\}$ where $t \in (-\infty, \infty)$. For arbitrary real τ, we define a *reversed* process $\{Y(t)\}$ by $Y(t) = X(\tau - t)$ for all t.

6-1 REVERSIBLE CHAINS

As defined in Section 3-10, a stochastic process $\{X(t): -\infty < t < \infty\}$ is called *reversible* if for every integer $n \geq 1$, $t_1, \ldots, t_n$, and τ, $(X(t_1 + \tau), \ldots, X(t_n + \tau))$ has the same distribution as $(X(-t_1), \ldots, X(-t_n))$.

The original (*forward*) and reversed processes are said to be *stochastically equivalent*, and, because τ is arbitrary, a reversible process is stationary.

Now suppose $\{X(t)\}$ is a stationary continuous-time chain with stationary distribution $\{p_j\}$. From the Markov property, the reversed process is a Markov chain. For $i \neq j$, let

$$\begin{aligned} Q_{ij}(h) &= P(X(t) = j \mid X(t+h) = i) \\ &= P(X(t) = j, X(t+h) = i)/p_i \\ &= P(X(t+h) = i \mid X(t) = j)p_j/p_i, \end{aligned} \tag{1}$$

from which we have

$$b_{ij} \equiv \lim_{h \to 0} Q_{ij}(h)/h = a_{ji}p_j/p_i,$$

which we write as

$$p_i b_{ij} = p_j a_{ji} \qquad \text{for all } i \text{ and } j. \tag{2}$$

A $j \to i$ transition in the original chain will be seen as an $i \to j$ transition in the reversed chain, and (2) simply states the obvious fact that the corresponding rates must be equal.

Thus the reversed chain has stationary transition probabilities with conditional transition rate matrix $\mathbf{B} = (b_{ij})$. It is easy to see that the reversed chain is irreducible, positive, and also has stationary distribution $\{p_j\}$ (the p_j are time averages for the reversed chain too!). It is also obviously true (and can be algebraically verified) that for all i,

$$b_i \equiv \sum_{j \neq i} b_{ij} = a_i, \tag{3}$$

i.e., the time the reversed chain spends in state i has the same exponential distribution!

We now state a result that turns out to be very useful later in this chapter.

Theorem 1. Let $\{X(t)\}$ be an irreducible Markov chain with state space $\mathcal{S}$ and transition rates (a_{ij}). If we can find a probability distribution $\{p_j\}$ over $\mathcal{S}$ and a collection of numbers (b_{ij}) for $i, j \in \mathcal{S}$ that satisfy (2) and (3), then $\{X(t)\}$ is positive recurrent with stationary distribution $\{p_j\}$, and the corresponding reversed chain has transition rates (b_{ij}).

Proof. Summing (2) on j, and using (3), we have a solution to balance equations (23) in Chapter 4, and the theorem follows.

Remarks. Theorem 1 holds when we permit $i \to i$ transitions. To apply this result, we need to "guess" $\{p_j\}$. If we also understand the reversed chain well enough to guess (b_{ij}), verifying (2) often turns out to be much easier than verifying the full balance equations about every state.

We now return to reversible chains. Because they are necessarily stationary, we will call a Markov chain reversible when the *stationary version* of the chain has this property.

Now **B** uniquely determines joint distributions for the reversed chain, and it follows that a chain is reversible if and only if **B** = **A**. We now state

Theorem 2. A Markov chain is reversible if and only if there exists probability distribution $\{p_j\}$ over the state space such that

$$p_i a_{ij} = p_j a_{ji} \qquad \text{for all pairs } i \neq j; \tag{4}$$

$\{p_j\}$ is the unique stationary distribution for the chain.

Proof. The chain is reversible if and only if **B** = **A**, which is equivalent to (4), where $\{p_j\}$ is the unique stationary distribution. Now suppose (4) holds for some $\{p_j\}$. Summing (4) on j, $\{p_j\}$ satisfies balance equations (23) in Chapter 4, $\{p_j\}$ is stationary, and **B** = **A**.

Thus "guessing" a $\{p_j\}$ and verifying (4) determines reversibility.

Equation (4) means that the transition rates between *every pair of states* are equal. We call (4) *detailed* balance equations, and (23) in Chapter 4 *full* balance equations. Unlike (23) in Chapter 4, which for each state simply equates transition rates across a boundary, there is no reason to expect detailed balance to hold in general. Nevertheless, important special-structure chains turn out to be reversible, and this will have interesting and useful consequences.

Remark. A bothersome technical point is our convention of right-continuity, i.e., if a jump occurs at t, $X(t)$ is the state entered at (just after) the jump. With this convention, the reversed process is *left*-continuous. However, the probability of a jump at any epoch is 0, and transition probabilities $P_{ij}(t)$ are continuous functions of time. It follows that distributions are not affected by the convention of right- vs. left-continuity.

Example 6-1: The Birth and Death Process

We introduced this important process in Section 4-4 and used it as the basis for several elementary queueing models in Chapter 5. Its salient feature is that all jumps are between adjacent states, which are ordered 0, 1, ... We defined positive transition rates $\lambda_j = a_{j,j+1}$ for $j \geq 0$, and $\mu_j = a_{j,j-1}$ for $j \geq 1$. Equating transition rates between sets $\{0, \ldots, j\}$ and $\{j + 1, \ldots\}$, we have

$$\lambda_j p_j = \mu_{j+1} p_{j+1}, \qquad j = 0, 1, \ldots,$$

which are the only positive terms in (4). Hence, a *positive* birth and death process is reversible.

Example 6-2: Trees

This is a generalization of the birth and death process to chains where $i \to j$ jumps must alternate with $j \to i$ jumps for all i and j. The discrete-time version of this structure is Example 4-14, and it follows that a positive continuous-time chain with tree structure is reversible.

Let $\{X(t)\}$ have embedded chain $\{X_n\}$ (the EMC defined in Section 4-1), where $\{X_n\}$ is recurrent, and we assume that it also is positive. From equation (115) in Chapter 3 and the transition rate interpretation of (4), we have

Theorem 3. Continuous-time chain $\{X(t)\}$ is reversible if and only if the corresponding EMC $\{X_n\}$ is reversible.

One application of Theorem 3 is to prove the continuous-time version of Kolmogorov's criterion, Theorem 18 in Chapter 3. Define a *loop* to be any finite sequence of transitions with the same initial and terminal state. From $a_{ij} = a_i p_{ij}$ for all i and j, it is clear for any loop, e.g., $i \to j \to k \to i$, that

$$p_{ij} p_{jk} p_{ki} = p_{ik} p_{kj} p_{ji} \qquad \text{if and only if } a_{ij} a_{jk} a_{ki} = a_{ik} a_{kj} a_{ji},$$

and we have

Theorem 4: Kolmogorov's Criterion. A positive continuous-time chain is reversible if and only if the product of the transition rates corresponding to any loop is equal to the product of the transition rates for the same loop traversed in the reverse direction.

Remark. The above proof assumes that the EMC is positive, but this is not required for the validity of the theorem. In fact, in Problems 3-48 and 3-49, we show how to extend reversibility and Kolmogorov's criterion to the null-recurrent case. Once this is done, the above proof still works.

The advantage of Kolmogorov's criterion is that reversibility can be determined directly from $\mathbf{A}$, without explicit use of $\{p_j\}$. In applications, verification for a few well-chosen loops will often imply that the criterion holds. In particular, loops that involve multiple returns to the same state can be decomposed into "simple" loops. For the birth and death process and for trees, the criterion holds trivially because for any loop, the reversed loop is the same sequence of transitions.

Example 6-3: The *M/M/2* Queue with Different Service Rates

Consider an *M/M/2* queue where μ_i is the service rate of channel i, $i = 1, 2$. When there is one customer in system, we need to keep track of which channel is busy; denote these states by $(1, i)$. When both channels are empty, assume that an arrival enters channel 1 with probability α. We have a Markov chain with state space and transition rates indicated in Figure 6-1. It should be clear that we need only check loop $0 \to (1, 1) \to 2 \to (1, 2) \to 0$, for which the product of the forward and reversed transition rates are

$$\alpha\lambda \cdot \lambda \cdot \mu_1 \cdot \mu_2 \quad \text{and} \quad (1 - \alpha)\lambda \cdot \lambda \cdot \mu_2 \cdot \mu_1,$$

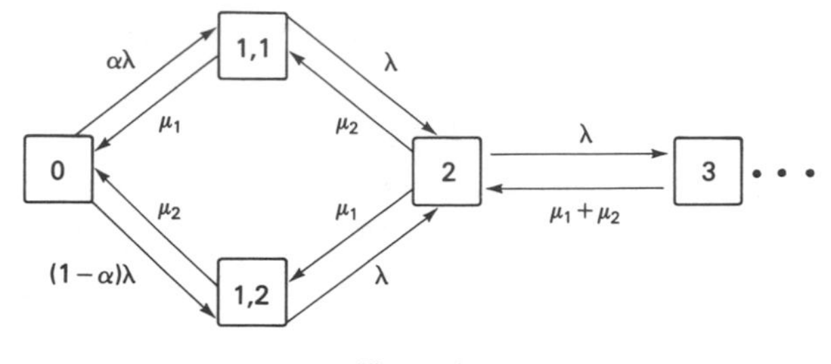

Figure 6-1

respectively. When this chain is positive ($\lambda < \mu_1 + \mu_2$), it is reversible if and only if $\alpha = 1/2$.

It is interesting to observe that for $\alpha \neq 1/2$ and $\mu_1 = \mu_2$, the above chain is not reversible, even though the birth and death process we get by combining states (1, 1) and (1, 2) into state 1 is reversible.

A chain that is not necessarily reversible is said to be *dynamically reversible* if for some relabeling of the states, the reversed chain is stochastically equivalent to the original chain. Example: Consider a closed system consisting of three single-channel stations with exponential service at rate μ. A single customer visits the stations in cyclic order, $1 \to 2 \to 3 \to 1$, etc., where state j means the customer is at station j, and j' is a relabeled state. The chain is dynamically reversible, e.g., for $\{1' = 1, 2' = 3, 3' = 2\}$.

For the next result, imagine that for each t, $X(t)$ may be a vector of random variables. The definition of reversible processes in Section 6-1 may easily be applied to vector-valued processes. An immediate consequence of this definition, which is useful for creating more complex reversible processes out of simple ones, is

Theorem 5. If $\{X_1(t)\}$ and $\{X_2(t)\}$ are independent stochastic processes that are each reversible, the vector-valued process $\{X(t)\}$, defined by $X(t) = (X_1(t), X_2(t))$ for all t, is also reversible.

For Markov chain $\{X(t)\}$ with state space $\mathcal{S}$ and transition matrix $\mathbf{A}$, suppose we create a new Markov chain with state space E, a proper subset of $\mathcal{S}$, by *truncation* of the state space $\mathcal{S}$ as follows: For any $i \in E$, let the transition rate from i to j be a_{ij} if $j \in E$, and 0 otherwise, where E is chosen so that the new chain is irreducible. Now suppose $\{X(t)\}$ is reversible with stationary distribution $\{p_j\}$. Thus we have

$$p_i a_{ij} = p_j a_{ji} \qquad \text{for all } i, j \in E, \tag{5}$$

even after truncation, but $P(E) \equiv \sum_{j \in E} p_j < 1$, i.e., $\{p_j\}$ restricted to E is not a probability distribution. However, by renormalizing $\{p_j\}$, we have

Theorem 6: Truncation. From a reversible Markov chain with stationary distribution $\{p_j\}$, create a new chain by truncation of the state space to E, for any E such that the new chain is irreducible. The new chain is reversible and has stationary distribution $\{p_j/P(E)\}$.

Remark. Other alterations of a reversible chain also preserve reversibility. For example, suppose we set $a_{ij} = a_{ji} = 0$ for one or more pairs (i, j), without truncation. If the altered chain is irreducible, it is also reversible.

Example 6-4: Two Queues with a Common Queue Limit Constraint

Suppose a doctor's office is an $M/M/1$ queue with arrival rate λ_1 and service rate μ_1, and a dentist's office is an independent $M/M/1$ queue with arrival rate λ_2 and service rate μ_2. Let $\rho_i = \lambda_i/\mu_i$, $i = 1, 2$. Consider the vector-valued Markov chain, where p_{mn} is the steady-state probability of m patients at the doctor's office and n patients at the dentist's office. For $\rho_i < 1$, we have

$$p_{mn} = (1 - \rho_1)\rho_1^m(1 - \rho_2)\rho_2^n \qquad \text{for } m, n \geq 0,$$

and, from Theorem 5, this chain is reversible.

Now suppose the offices share a waiting room that can hold a maximum of $l \geq 1$ patients, which constrains the state space to set

$$E = \{(m, n): m, n \geq 0, \text{ and } (m - 1)^+ + (n - 1)^+ \leq l\}.$$

Arrivals who find both the waiting room full and the server (doctor or dentist, as appropriate) busy are lost. From Theorem 6, the chain with this constrained state space is reversible and has steady-state probabilities

$$p_{mn}/P(E) \qquad \text{for all } (m, n) \in E.$$

Remark. When one or both of the original queues is unstable ($\rho_i \geq 1$), the constrained system is not, and it is easily shown that the constrained system is still reversible with solution of essentially the same form, i.e., the steady-state probability of state (m, n) is *proportional* to $\rho_1^m\rho_2^n$.

6-2 TANDEM QUEUES

One of the most interesting applications of reversible chains is to tandem queues. We begin with positive birth and death process $\{X(t)\}$ with stationary distribution $\{p_j\}$. We call points of increase *arrival epochs* and points of decrease *departure epochs*; the corresponding point processes are arrival and departure *processes*. The birth and death process *determines* these processes, e.g., the joint distribution of the number of departures in disjoint intervals. As a slight generalization of birth and death, we sometimes permit transitions of type $j \to j$, which are counted as arrivals *and* departures. A separate Poisson process may be defined to generate these transitions for each state that permits them.

Now consider the special case $\lambda_j = \lambda$ for all j, i.e., the arrival process is Poisson. Denote this special case of a birth and death process by a (λ, μ_j)-*process*.

Stable $M/M/1$ and $M/M/c$ queues are examples of this case, as are corresponding queue limit models, provided that every "lost customer" is counted as an arrival *and* a departure. The Poisson arrival process has the conventional *lack of anticipation* (LAA) property (see Section 5-16), meaning that for each epoch t, the future (increments) of the arrival process are independent of $\{X(s): s \le t\}$. It also has the PASTA property: If t is an arrival epoch, $X(t^-) \sim \{p_j\}$, and it is independent of the arrival process after t. From Example 6-1, a positive (λ, μ_j)-process is reversible, and hence the reversed process also has Poisson arrivals at rate λ with the same (LAA and PASTA) properties. Arrivals for the reversed process are departures for the original process, and the "future" for the reversed process is the "past" for the original process. We have shown

Theorem 7. The departure process for stationary (λ, μ_j)-process $\{X(t)\}$ is Poisson at rate λ. For every epoch t, the departure process prior to t is independent of $\{X(s): s \ge t\}$. If t is a departure epoch, $X(t^+) \sim \{p_j\}$, and is independent of the departure process prior to t.

Consider a system of $J \ge 2$ single-channel queues in tandem, as shown in Figure 6-2, which pictures customers in queue and in service at various stations. Arrivals at station 1 are Poisson at rate λ. Service times are independent exponential random variables, with service rate μ_i, and $\rho_i \equiv \lambda/\mu_i$ at station i, $i = 1, \ldots, J$. Assume $\rho_i < 1$ for all i.

We want to find the joint stationary distribution of $N_1, \ldots, N_J$, where N_i is the number of customers "in system" at station i. Now station 1 is an $M/M/1$ queue, and if it is stationary, we know from Theorem 7 that station 2 is also $M/M/1$. If stations 1 and 2 are jointly stationary, station 3 is $M/M/1$, and so on, which determines the marginal distributions of the N_i,

$$P(N_i = j) = (1 - \rho_i)\rho_i^j \qquad \text{for } j = 0, 1, \ldots \quad \text{and} \quad i = 1, \ldots, J.$$

To obtain the corresponding joint distribution, we need to use more of Theorem 7. Suppose station 1 is stationary, so that $N_1(t) \sim N_1$ for all t, and consider the joint distribution of the number of customers at the first two stations at arbitrary epoch t. $N_1(t)$ is independent of the departure process from station 1 prior to t, and it follows that for all t,

$$P(N_1(t) = m, N_2(t) = n) = (1 - \rho_1)\rho_1^m P(N_2(t) = n). \tag{6}$$

Now let $t \to \infty$, and we have that the joint stationary distribution of N_1 and N_2 is the product of their marginals. For arbitrary J and stationary first station, the

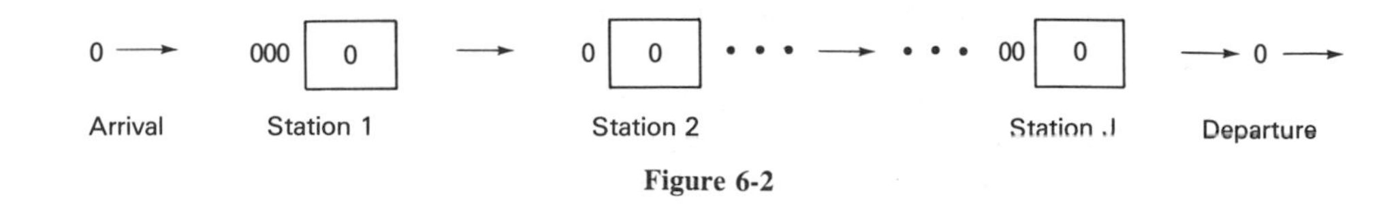

Figure 6-2

remaining system has Poisson arrivals and $J - 1$ stations. We have

Theorem 8. The joint stationary distribution for the number of customers in system at each of the J single-channel queues in tandem is

$$P(N_i = n(i), i = 1, \ldots, J) = \prod_{i=1}^{J} (1 - \rho_i)\rho_i^{n(i)} \quad \text{for } n(i) \geq 0, \quad i = 1, \ldots, J. \tag{7}$$

Important Remarks. Theorem 7 may be used to generalize Theorem 8 to the case where the service rate at each station i is a function of the number of customers at that station. The joint stationary distribution is the product of the marginals, where each station is a (λ, μ_j)-process. Only the form of the marginal distributions would change. The factorization of the joint distribution as in (7) is what we will call *product form*. Product form holds even if there is an upper bound on the number of customers at one or more stations (queue-limit models), provided that customers who find a station full immediately go to the next station. The important point is that at a queue-limit station, neither the departure process of served customers nor the departure process of lost customers is Poisson, but the composite of these processes is Poisson. If only the served customers go to the next station (or only the lost customers as in the overflow models in Chapter 7), the solution becomes much more complicated, and is *not* of product form.

Now (7) means that if we observe the tandem queue at some random moment, the number of customers at each of the stations are independent random variables. However, the *stochastic processes* $\{N_i(t)\}$ clearly (?) are dependent. Why is this so?

Theorem 8 can be generalized in other ways, e.g., a departure from one station goes to the next station with probability q, but we defer generalizations of this kind until the next section.

For the rest of this section, we consider FIFO waiting times for single or multiple channel queues in tandem. Let W_i be the (stationary) waiting time of a customer at station i, $i = 1, \ldots, J$. Because each station has Poisson arrivals, the marginal distributions of the W_i are those found for $M/M/1$ and $M/M/c$ queues in Chapter 5. We now determine some properties of the joint distribution of the waiting times of the same customer at each station. Our first result is

Theorem 9. The stationary waiting time of a customer in a FIFO $M/M/c$ queue is independent of the departure process prior to that customer's departure epoch.

Sketch of Proof. The idea is to use reversibility to assert that because the waiting time of a customer is independent of the arrival process after that customer's arrival epoch, it is also independent of the departure process prior to that customer's departure epoch. This will be true if we can show that the waiting time for the forward and reversed processes is the same random variable. This is easy to do for $c = 1$; consider the busy period in Figure 6-3, where it is easy to see that the waiting times for the forward and reversed chains are the same random variables, in reverse order. For $c > 1$, customers need not depart in the order that they arrive. For example, if Figure 6-3 represents this case as well for some fixed c, the first departure is the same customer as the first arrival only with probability 1/2. In the reversed chain, the *fourth* arrival is the *fourth* departure with the *same* probability, and if both events occur, $W^{(1)}(\text{forward}) = W^{(4)}(\text{reversed})$. It is possible to do this with an arbitrary sample path of jumps, and, with indicator variables, to represent the forward and reversed waiting times as the same random variables.

Now suppose we have two FIFO queues in tandem, where the first station is $M/M/c$. The second station may also be multichannel, and can have a *general* service distribution. The waiting time of an arrival at the second station depends only on the "status" of that station, which in turn depends on prior arrivals and their service times at that station. By Theorem 9, we have

Theorem 10. For two FIFO queues in tandem, where the first station is $M/M/c$, the stationary waiting times of the same customer at each station are independent random variables.

The extension of Theorem 10 to J queues in tandem is immediate, provided that stations $2, \ldots, J - 1$ are *single channel* and exponential. We state a special case of this result as

Theorem 11. For $J \geq 3$ FIFO single-channel queues in tandem with Poisson arrivals and exponential service, the stationary waiting times of the same customer at each of the stations are independent random variables.

Remark. The single-channel requirement in Theorem 11 is necessary to prevent customers from "passing," e.g., the second departure from station 1 cannot arrive at station 3 before the first departure from station 1. If passing can occur, future departures can cause delay at later stations. In fact, if at some station i, $2 \leq i \leq J - 1$, $c_i > 1$, waiting times at later stations are dependent on waiting times at earlier ones. See Problem 6-6.

In the remainder of this chapter, we will greatly expand the models for which product-form results (e.g., Theorem 8) hold. Waiting times, however, will not be independent. Even for models where waiting times are independent, it is easy to show that the corresponding delays are not. For example, for two single-channel

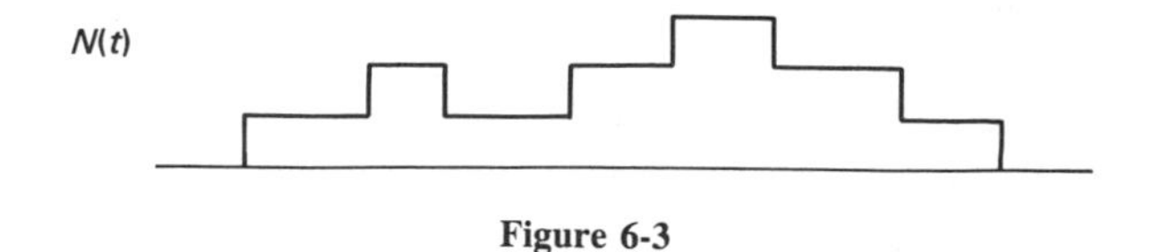

Figure 6-3

exponential queues in tandem where D_i is the stationary delay in queue at station i, it is immediate that

$$P(D_1 = 0, D_2 = 0) > P(\text{arrival finds entire system empty})$$
$$> (1 - \rho_1)(1 - \rho_2) = P(D_1 = 0)P(D_2 = 0). \quad (8)$$

6-3 JACKSON NETWORKS

The analysis in Section 6-2 breaks down if customers are permitted to return to stations visited before, and it turns out that the arrival process of (new and returning) customers at each station is no longer Poisson. Nevertheless, for the more general models considered in this section, we still have product-form solutions to the full balance equations.

We now consider queueing networks consisting of $J \geq 1$ stations, where customers may move from station to station in arbitrary order. A network is called *open* if external arrivals are permitted at one or more stations and departures from the system are permitted from one or more of the stations. If neither is permitted, a network is called *closed.*

For a closed network, the number of customers in the entire system is fixed and will be denoted by $M \geq 1$. For example, each of M trucks may be at a repair facility when in need of repair, and in the field when working properly. We have two stations, the repair facility, and "the field." If M is very large and breakdowns are rare, we might ignore the exact value of M and model the arrival of trucks at the repair facility as (say) Poisson. This model is an open network with $J = 1$ station.

Thus we may choose to use an open (closed) network model even when the physical system is closed (open). We will find that open networks are easier to solve numerically; this is a serious issue when both the number of stations and customers are large. When we model a portion (subset of the stations) of a large network as a smaller open network, the computational and conceptual advantages can be enormous. Balanced against these advantages is the hazard that the simpler open model will fail to capture important aspects of reality. Deciding what to leave in and what to leave out is the essence of modeling.

At every epoch $t \geq 0$, let $N_i(t)$ be the number of customers in system at station i, $i = 1, \ldots, J$, and $\mathbf{N}(t) = (N_1(t), \ldots, N_J(t))$. We model $\{\mathbf{N}(t)\}$ as a continuous-time Markov chain with special structure, where state $\mathbf{n} = (n_1, \ldots, n_J)$ means that there are n_i customers at station i, $i = 1, \ldots, J$.

The state spaces for open and closed networks, respectively, are

$$\mathcal{S} = \{\mathbf{n}: n_i \geq 0, i = 1, \ldots, J\} \text{ and}$$

$$\mathcal{S} = \left\{\mathbf{n}: n_i \geq 0, i = 1, \ldots, J, \text{ and } \sum_{i=1}^{J} n_i = M\right\}.$$

The assumptions we make in this section will ensure that $\{\mathbf{N}(t)\}$ is irreducible.

Let $a(\mathbf{n}, \mathbf{m})$ be the (conditional) transition rate from state $\mathbf{n}$ to state $\mathbf{m}$, $p(\mathbf{n})$ be the steady-state (stationary) probability that the chain is in state $\mathbf{n}$, and $p_i(n_i)$ be the corresponding probability that there are n_i customers at station i.

Transitions out of state $\mathbf{n}$ are restricted to states $\mathbf{m}$ that can be reached from $\mathbf{n}$ by the movement of one customer, either from one station to another, or by being either an arrival or a departure. To be precise, for every station i, let $\varepsilon_i = (e_1, \ldots, e_J)$ be the vector with components $e_i = 1$, and $e_j = 0$ for $j \neq i$. We permit an $\mathbf{n} \to \mathbf{m}$ transition to occur only for $\mathbf{m}$ that is one of three functions of $\mathbf{n}$. For every i and $j \in \{1, \ldots, J\}$, denote these functions and the corresponding transition rates as follows:

$$\begin{aligned} &\textit{shift} \text{ from } i \text{ to } j\text{:} \quad a(\mathbf{n}, S_{ij}\mathbf{n}) = \mu_{ij}(\mathbf{n}), && \text{where } S_{ij}\mathbf{n} = \mathbf{n} - \varepsilon_i + \varepsilon_j, \\ &\textit{arrival} \text{ at } i\text{:} \quad a(\mathbf{n}, A_i\mathbf{n}) = \lambda_i(\mathbf{n}), && \text{where } A_i\mathbf{n} = \mathbf{n} + \varepsilon_i, \text{ and} \\ &\textit{departure} \text{ from } i\text{:} \quad a(\mathbf{n}, D_i\mathbf{n}) = \mu_{i,J+1}(\mathbf{n}), && \text{where } D_i\mathbf{n} = \mathbf{n} - \varepsilon_i, \end{aligned}$$

and $J + 1$ is simply a way of accounting for departures from station i who leave the system. Let the service rate at station i be

$$\mu_i(\mathbf{n}) = \sum_{j=1}^{J+1} \mu_{ij}(\mathbf{n}) \qquad \text{for } i = 1, \ldots, J,$$

where $\mu_i(\mathbf{n}) > 0$ if and only if $n_i > 0$.

To be a *Jackson* network, we impose the following additional conditions:

$$\lambda_i(\mathbf{n}) = \lambda_i, \tag{9}$$

$$\mu_i(\mathbf{n}) = \mu_i(n_i), \text{ and} \tag{10}$$

$$\mu_{ij}(\mathbf{n}) = \mu_i(n_i)p_{ij} \qquad \text{for } i \leq J, \quad j \leq J + 1, \quad \text{and} \quad \mathbf{n} \in \mathcal{S}, \tag{11}$$

where $\sum_{j=1}^{J+1} p_{ij} = 1$ for all i. Equation (9) means that the arrival rate at each station is independent of the state of the system, i.e., the arrival processes at each of the stations are *independent Poisson processes*. Equation (10) means that the service rate at each station may depend on the number of customers at that station, but is otherwise independent of $\mathbf{n}$. For example, a station may be multichannel with exponential service. With (10), (11) means that the station visited next by a departure from station i may depend on i, but is otherwise independent of $\mathbf{n}$. Because $\{\mathbf{N}(t)\}$ is a Markov process, (11) is equivalent to assuming that given the station a customer is departing from, and the past history of stations visited by that customer, the next station visited is independent of this past history.

Thus the meaning of (11) is that the sequence of states visited by every customer is a Markov chain with transition probability matrix $\mathcal{P} = (p_{ij})$. When we make this assumption, we say that customers have *Markovian paths*.

Unfortunately, the matrix $\mathscr{P}$ has J rows and $J + 1$ columns, i.e., it is not square. To complete our definition of a Jackson network, we need to distinguish the open and closed cases.

For a *closed* network, $p_{i,J+1} = 0 = \lambda_i$, for $i = 1, \ldots, J$, and for some fixed $M \geq 1$, the same M customers circulate about the network. In this case, delete column $J + 1$ from matrix $\mathscr{P}$, which is now of size $J \times J$.

For an *open* network, $\lambda_i > 0$ for some i, and $p_{j,J+1} > 0$ for some j. Thinking of $J + 1$ as the "outside world," define

$$p_{J+1,i} = \lambda_i / \sum_{i=1}^{J} \lambda_i \qquad \text{for } i = 1, \ldots, J \quad \text{and} \quad p_{J+1,J+1} = 0,$$

and include these probabilities as row $J + 1$ in $\mathscr{P}$, which is now of size $(J + 1) \times (J + 1)$.

We are now ready for the following

Definition. A *Jackson network* is the continuous-time Markov chain $\{\mathbf{N}(t)\}$ previously described where the nonzero transition rates are of form (9) through (11), and $\mathscr{P}$ is a transition probability matrix of an *irreducible* discrete time chain.

Remarks. The irreducibility of $\mathscr{P}$ is necessary and sufficient for the irreducibility of $\{\mathbf{N}(t)\}$. We permit $p_{ii} > 0$; when this is the case, there will be transitions of type $\mathbf{n} \to \mathbf{n}$ that result from shifts where $i = j$.

Open Jackson Networks

Let Λ_i be the composite (external and internal) arrival rate (departure rate) at station i, $i = 1, \ldots, J$. For a stable network, all arrivals are eventually served, and we can compute the Λ_i in advance by equating departure rates (on the left in (12)) and arrival rates at every station:

$$\Lambda_i = \lambda_i + \sum_{j=1}^{J} \Lambda_j p_{ji}, \qquad i = 1, \ldots, J. \tag{12}$$

Because $\mathscr{P}$ is a $(J + 1) \times (J + 1)$ irreducible matrix, we know from the theory in Section 3-7 that (12) has a unique solution.

Now define what later will be normalizing constants $k_1, \ldots, k_J$ by

$$k_i^{-1} = \sum_{n=0}^{\infty} \Lambda_i^n / [\mu_i(1) \cdot \ldots \cdot \mu_i(n)], \tag{13}$$

where we set the first ($n = 0$) term in the sum equal to 1. We now state

Theorem 12. An open Jackson network is stable (positive recurrent) if and only if (13) converges for all i. In this case, the joint stationary distribution is of form

$$p(\mathbf{n}) = p_1(n_1) \cdot \ldots \cdot p_J(n_J), \text{ where} \tag{14}$$

$$p_i(n_i) = k_i \Lambda_i^{n_i} / [\mu_i(1) \cdot \ldots \cdot \mu_i(n_i)] \qquad \text{for } n_i \geq 0, \quad i = 1, \ldots, J. \tag{15}$$

Proof. The full balance equations are

$$\begin{aligned} p(\mathbf{n})\Big[\sum_j a(\mathbf{n}, A_j\mathbf{n}) + \sum_i a(\mathbf{n}, D_i\mathbf{n}) + \sum_i \sum_j a(\mathbf{n}, S_{ij}\mathbf{n})\Big] \\ = \sum_j p(A_j\mathbf{n})a(A_j\mathbf{n}, \mathbf{n}) + \sum_i p(D_i\mathbf{n})a(D_i\mathbf{n}, \mathbf{n}) + \sum_i \sum_j p(S_{ij}\mathbf{n})a(S_{ij}\mathbf{n}, \mathbf{n}), \end{aligned} \tag{16}$$

for every state $\mathbf{n}$, where for any state outside the state space (e.g., $D_i\mathbf{n}$ when $n_i = 0$), set the corresponding state probability equal to zero. All sums are over all stations. It is easy to see that the full balance equations will be satisfied if $\{p(\mathbf{n})\}$ satisfies what are called *partial* balance equations

$$\begin{aligned} p(\mathbf{n})\Big[a(\mathbf{n}, D_i\mathbf{n}) + \sum_j a(\mathbf{n}, S_{ij}\mathbf{n})\Big] \\ = p(D_i\mathbf{n})a(D_i\mathbf{n}, \mathbf{n}) + \sum_j p(S_{ij}\mathbf{n})a(S_{ij}\mathbf{n}, \mathbf{n}), \qquad i = 1, \ldots, J, \text{ and} \\ p(\mathbf{n}) \sum_j a(\mathbf{n}, A_j\mathbf{n}) = \sum_j p(A_j\mathbf{n})a(A_j\mathbf{n}, \mathbf{n}). \end{aligned} \tag{17}$$

Substituting (14) and (15) into (17), (17) reduces to (12) after some cancellation, which means that a solution of form (15) with positive k_i always satisfies the balance equations, but may not be a probability distribution. The convergence of (13) for all i determines the k_i, and we have the positive case. If (13) diverges for some i, we have found a *stationary measure* $\{p(\mathbf{n})\}$ where $p(\mathbf{n}) > 0$ for all $\mathbf{n}$, and $\sum_{\mathbf{n}} p(\mathbf{n}) = \infty$, which implies (by Theorem 6 in Chapter 4) that the chain is either transient or null.

The meaning of Theorem 12 is that the *stationary* number of customers in system at station i, $i = 1, \ldots, J$, are independent random variables with marginal distributions (15); we again have product form. Of course, the corresponding *stochastic processes* $\{N_i(t)\}$ are not independent.

If station i were operating in isolation with Poisson arrivals at rate Λ_i, it would be a (λ, μ_n)-type birth and death process, and it would have stationary distribution (15). However, except in special cases such as tandem queues where a customer cannot visit the same station more than once, the composite arrival process at station i is *not* a Poisson process. See Example 6-5.

Nevertheless, the composite process does have the PASTA property: Consider station i in isolation and let $\Lambda_{in} p_i(n)$ be the composite arrival rate at station i when there are n customers there. (Note that Λ_{in} is not a Markovian conditional rate, but it is well defined as a limit as $t \to \infty$) By equating transition rates between having n and $n + 1$ customers at station i, we have

$$\Lambda_{in} p_i(n) = \mu_i(n + 1) p_i(n + 1), \qquad n = 0, 1, \ldots, \tag{18}$$

which determine $\{p_i(n)\}$. From the form of (15), it follows that

$$\Lambda_{in} = \Lambda_i \qquad \text{for all } n, \tag{19}$$

which implies that *for every n, the fraction of composite arrivals at station i who find n customers there is equal to the corresponding fraction of time.* Note that when $p_{ii} > 0$, both sides of (18) are inflated by transitions that do not change the state. However, from (11), they are inflated by the same factor $(1 - p_{ii})^{-1}$, and (18) is still valid.

Partial balance equations (17) have interesting interpretations in terms of transition rates [see the interpretation of (22)], and are intermediate in form between full balance equations (16) and the detailed balance equations (3) that hold for reversible chains. Unlike (16), there is no conservation principle that requires (17) to hold. Jackson networks will not in general be reversible, but they do have this useful property:

Theorem 13. The reversed chain of a stationary open Jackson network is also a stationary open Jackson network.

Sketch of Proof. The reversed chain is irreducible and positive, and it is straightforward to verify that the transition rates (2) for the reversed chain are of form (9) through (11).

An important consequence of Theorem 13 is

Theorem 14. For a stationary open Jackson network, the departure process from the system at every station i is Poisson at rate $\Lambda_i p_{i,J+1}$. The departure processes are independent of each other, and for every epoch t, the departure processes prior to t are independent of the state of the system $N(t)$.

Sketch of Proof. From the meaning of the Λ_i, the departure *rates* are immediate. The departure processes for a stationary Jackson network are the arrival processes for the reversed Jackson network, and therefore are independent and Poisson.

Example 6-5

Consider an $M/M/1$ queue with arrival rate λ and service rate μ, where with probability α, a customer leaves the system on service completion, and with probability $1 - \alpha$, this customer rejoins the queue. Let S_T be the *total* time a customer spends in service. It is easily shown (Problem 5-8) that $S_T \sim \exp(\alpha\mu)$. Assume that $\lambda/\alpha\mu < 1$.

If the server continues to serve the same customer each time that customer rejoins the queue, until that customer eventually departs, we have a conventional

$M/M/1$ queue with arrival rate λ, service rate $\alpha\mu$, and stationary distribution of number of customers in system

$$p_n = (1 - \rho)\rho^n, \qquad n = 0, 1, \ldots,$$

where $\rho = \lambda/\alpha\mu$. This system has Poisson departures.

Alternatively, customers could rejoin the *end* of the queue. The number-in-system process is a Markov chain with the same transition rates, so that $\{p_n\}$ and the departure process are unchanged. It is easy to see that this chain is reversible.

This alternative is also an open Jackson network with $J = 1$ station and $\alpha = p_{12}$. The composite arrival rate is [from (12) or by direct argument] $\Lambda = \lambda/\alpha$. From Theorem 12, the stationary distribution is the same $\{p_n\}$, but the *interpretation* of ρ is different,

$$\rho = (\lambda/\alpha)/\mu = \Lambda/\mu.$$

It is easy to see that the composite arrival process does not have independent increments, and hence cannot be Poisson: If the chain is stationary, the probability of a composite arrival in $(t, t + h]$ is

$$\lambda h + \rho\mu h(1 - \alpha) + o(h) = \Lambda h + o(h).$$

On the other hand, given that a composite arrival occurred in $(t - h, t]$, the probability of a composite arrival in $(t, t + h]$ is

$$\lambda h + \mu h(1 - \alpha) + o(h) > \Lambda h + o(h).$$

From PASTA, the fraction of external arrivals who find n customers in system is p_n, the corresponding fraction of time, which is also the fraction of departures *from the system* who leave n customers behind. The event that a customer rejoins the queue is independent of the number of other customers present. Thus, as was shown in general as (19), the fraction of *composite* arrivals who find n customers in system is also p_n, even though the composite process is not Poisson.

Closed Jackson Networks

Because the state space is finite, a closed Jackson network is always stable. Unlike open networks, however, the composite arrival rates cannot be computed in advance because they depend on the length of time required for customers to circulate through the system. *Relative* arrival rates are determined by transition probability matrix $\mathscr{P}$.

Every transition is both a departure from some station and an arrival at some (possibly the same) station. For every i, let Λ_i be the *fraction* of transitions that are arrivals at (departures from) station i. From the irreducibility of $\mathscr{P}$, $\{\Lambda_i\}$ is the *unique* solution (that sums to 1) to

$$\Lambda_i = \sum_{j=1}^{J} \Lambda_j p_{ji}, \qquad i = 1, \ldots, J. \tag{20}$$

We are now ready to state

Theorem 15. The joint stationary distribution for a closed Jackson network is

$$p(\mathbf{n}) = K_M \prod_{i=1}^{J} \Lambda_i^{n_i}/[\mu_i(1) \cdot \ldots \cdot \mu_i(n_i)] \qquad \text{for} \left\{ \mathbf{n}: n_i \geq 0 \text{ and } \sum_{i=1}^{J} n_i = M \right\}, \tag{21}$$

where K_M is a normalizing constant that is a function of M, the fixed number of customers in the system.

Proof. The full balance equations are (16) with terms involving either A_j or D_i deleted. Similarly, the partial balance equations (17) simplify to

$$p(\mathbf{n}) \sum_j a(\mathbf{n}, S_{ij}\mathbf{n}) = \sum_j p(S_{ij}\mathbf{n}) a(S_{ij}\mathbf{n}, \mathbf{n}), \qquad i = 1, \ldots, J. \tag{22}$$

Substituting (21) into (22), (22) reduces to (20) after some cancellation; i.e., (21) solves the balance equations. Normalization determines K_M.

Remarks. Solution (21) is still called *product form*, even though the number of customers at each of the stations are now dependent random variables (their sum is the constant M). Formally, the dependence arises in the determination of K_M, because the ranges of summation over different stations are interdependent. The number of states grows rapidly with J and M, and computing K_M as the reciprocal of a direct sum is often impractical. The form of (21) appears to be different from equation (2.3) in Kelly [1979], but it can be shown that these expressions are equivalent.

Equation (22) has the following interpretation: {The departure rate out of station i when the chain is in state $\mathbf{n}$} = {the arrival rate to station i for those arrivals that induce transitions into state $\mathbf{n}$}.

Example 6-6

We have M machines, where the time until breakdown of a working machine is $\sim$ exp (λ). Machines are repaired at a single-channel repair facility, where repair times are $\sim$ exp (μ). (This was Example 5-6.) This is a closed Jackson network with $J = 2$ stations. Let the repair facility be station 1, the location of the working machines be station 2, and p_n denote the probability of state $(n, M - n)$, $n = 0, \ldots, M$.

We have $p_{12} = p_{21} = 1$, and $\Lambda_1 = \Lambda_2 = 1/2$, i.e., each machine alternates between working and not working. It is easy to see that for $n > 0$, $\mu_1(n) = \mu$, and $\mu_2(n) = n\lambda$. From (21), p_n is proportional to

$$[\mu^n \lambda^{(M-n)} (M - n)!]^{-1},$$

which is equivalent to the birth and death result

$$\lambda(M - n)p_n = \mu p_{n+1}, \qquad n = 0, \ldots, M - 1.$$

The closed version of Theorem 13 is straightforward to verify:

Theorem 16. The reversed chain of a stationary closed Jackson network is also a closed Jackson network.

Arrivals at a station in a closed Jackson network do not have the PASTA property, but we will find that the distribution of the number of customers they find has a very interesting time-average interpretation.

First we need to relate a closed network to a corresponding open network that has the same service rate functions $\{\mu_i(n)\}$, and composite arrival rates (12) that are proportional to relative rates (20). The Fixed-Population-Mean Method in Section 6-5 presents one way to define an open network with these properties. Let N_i^o and N_i^c be the stationary number of customers at station i in the corresponding open and closed networks, $N^o = \sum N_i^o$, and Λ_i^o be the composite arrival rate at station i in the open network.

Comparing (14) and (15) with (21), notice that (21) is simply a re-normalization of (14) under the constraint that $N^o = M$, i.e., for $\mathbf{n} \in \mathcal{S}$,

$$P(N_i^c = n_i, i = 1, \ldots, J) = P(N_i^o = n_i, i = 1, \ldots, J \mid N^o = M). \tag{23}$$

Note that (23) is independent of the magnitude of the composite arrival rates Λ_i^o in the open network, and we will always be able to scale them to ensure that the open network is stable. Of course, their ratios Λ_i^o/Λ_j^o are determined by the relative rates for the closed network.

For the closed network, let f_{in} be the fraction of departures from station i who leave n customers there, and a_{in} be the fraction of arrivals at station i who find n customers there, $n = 0, \ldots, M - 1$. Now (18) is valid for closed networks as well, where the transition rate on the right corresponds to f_{in}, and the rate on the left to a_{in}. The arrival and departure rates at station i are equal, and normalizing with this rate, it follows that

$$f_{in} = a_{in}, \qquad n = 0, \ldots, M - 1, \text{ where} \tag{24}$$

$$f_{in} \propto \mu_i(n + 1)P(N_i^c = n + 1) = \mu_i(n + 1)P(N_i^o = n + 1 \mid N^o = M). \tag{25}$$

Now write

$$\begin{aligned}
\mu_i(n + 1)&P(N_i^o = n + 1 \mid N^o = M) \\
&= \mu_i(n + 1)P(N_i^o = n + 1)P(N^o - N_i^o = M - 1 - n)/P(N^o = M) \\
&= \Lambda_i^o P(N_i^o = n)P(N^o - N_i^o = M - 1 - n)/P(N^o = M) \\
&= \Lambda_i^o P(N_i^o = n \mid \mathrm{N}^o = \mathrm{M} - 1)P(N^o = M - 1)/P(N^o = M).
\end{aligned}$$

Combining with (25), we have

$$f_{in} \propto P(N_i^o = n \mid N^o = M - 1), \tag{26}$$

and, because both sides of (26) are distributions, it follows that

$$f_{in} = P(N_i^o = n \mid N^o = M - 1), \qquad n = 0, \ldots, M - 1. \tag{27}$$

Combining (23), (24), and (27), we have this remarkable result:

Theorem 17. For every station i in a closed Jackson network with M customers, the fraction of arrivals at station i who find n customers there is equal to the fraction of time there are n customers at station i in a closed network with the same structure, but with *one less customer*, $n = 0, \ldots, M - 1$.

In electrical networks there is a result known as *Norton's Theorem*, which states that for certain purposes, a network with J nodes (stations) can be reduced to an equivalent network with just two nodes. We now derive an analogous result for closed Jackson networks that goes by the same name.

Fix arbitrary station i in a closed Jackson network, and let Λ_n^c be the composite arrival rate of customers to station i when $N_i^c = n$. Now Λ_n^c is not a Markovian rate, but is a sum of relevant rates over all possible states when $N_i^c = n$. The product $\Lambda_n^c P(N_i^c = n)$ is the long-run rate of arrivals at station i who find n customers there. We are now ready to state

Theorem 18. For every station i in a closed Jackson network, the joint distribution of $\{N_j^c, j \neq i \mid N_i^c = n\}$, for $n = 0, \ldots, M$, is independent of service rate function $\{\mu_i(n)\}$ at that station. If $p_{ii} = 0$, $\{\Lambda_n^c\}$ is also independent of $\{\mu_i(n)\}$.

Proof. We use (23) and the independence of the N_j^o to write

$$\begin{aligned} P(N_j^c = n_j, j \neq i \mid N_i^c = n) &= P(N_j^c = n_j, j \neq i,\ N_i^c = n)/P(N_i^c = n) \\ &= \frac{P(N_j^o = n_j, j \neq i,\ N_i^o = n \mid N^o = M)}{P(N_i^o = n \mid N^o = M)} \\ &= P(N_j^o = n_j, j \neq i,\ N_i^o = n)/P(N_i^o = n,\ N^o = M) \\ &= P(N_j^o = n_j, j \neq i)/P(N^o - N_i^o = M - n). \end{aligned}$$

The last expression holds for all n_j with $\sum_{j \neq i} n_j = M - n$, and is independent of N_i^o and hence of $\{\mu_i(n)\}$. Now suppose $p_{ii} = 0$, and let $\Lambda^c(\mathbf{n})$ be the conditional arrival rate at station i when the network is in state $\mathbf{n}$. Summing the relevant rates over all possible states, we have

$$\Lambda_n^c P(N_i^c = n) = \sum_{\{\mathbf{n}: N_i^c = n\}} \Lambda^c(\mathbf{n}) p(\mathbf{n}) = \sum_{\{\mathbf{n}: N_i^c = n\}} \Lambda^c(\mathbf{n}) p(\mathbf{n} \mid N_i^c = n) P(N_i^c = n),$$

which determines $\{\Lambda_n^c\}$ and shows that it is also independent of $\{\mu_i(n)\}$.

Suppose we were thinking of changing the operation of station i in some way, e.g., by adding channels, and would like to determine the effect of this change on performance at station i and on overall throughput. Once we have found $\{\Lambda_n^c\}$ for some specified rate function $\{\mu_i(n)\}$, we can change $\{\mu_i(n)\}$ to some proposed function $\{\mu_i'(n)\}$, and find the new distribution of N_i^c, $\{P(N_i^c = n) = p_n^{c'}\}$, from the recursion

$$\Lambda_n^c p_n^{c'} = \mu_i'(n+1) p_{n+1}^{c'}, \qquad n = 0, \ldots, M-1. \tag{28}$$

The new throughput at station i is $\sum \Lambda_n^c p_n^{c'}$. Equation (28) is equivalent to the next theorem.

Theorem 19: Norton's Theorem for Closed Jackson Networks. Suppose $p_{ii} = 0$. When changing service rate function $\{\mu_i(n)\}$ at station i, we can determine the effects of this change on station i and on throughput by replacing the J-station network by an equivalent two-station network, where one is station i, and the other is new station 0 with $p_{oi} = 1$, and service rate function defined by $\mu_0(n) = \Lambda_{M-n}^c$, $n = 0, \ldots, M$.

Remarks. This assumes $p_{ii} = 0$, but the recursion formula (28) is easily modified when $p_{ii} > 0$. Another use of these results is this: Suppose we want to calculate throughput. We can change $\{\mu_i(n)\}$ at station i of our choice, in any way that makes finding $\{\Lambda_n^c\}$ easier. The throughput is computed using the correct $\{\mu_i(n)\}$.

Modeling and Performance Measures

As mentioned early in this section, the same application can sometimes be modeled either as an open or a closed network.

Open networks are appropriate when we want to decouple a collection of queues from some large population of potential customers. These potential customers generate arrival processes that we assume are independent of congestion at the stations. When an open network is stable, the arrival and departure rates at every station (*throughput*) are determined by external demand, and can easily be computed. Throughput is fixed in the sense that it does not depend on the level of service [the functions $\mu_i(n_i)$ in our model]. Performance measures such as average delay or queue length, possibly at each station, are of primary interest.

For a closed network, overall $L = M$ is fixed; throughput, which now does depend on level of service, is usually the primary performance measure. Under the assumption of Markovian paths, relative throughputs (20) are fixed; the throughput at one station determines the throughput at the others.

In later sections, we will consider more general customer paths. Among other things, this will permit "mixed" networks that are closed for some customers and open for others, but we will not treat these generalizations.

6-4 COMPUTING THE THROUGHPUT OF CLOSED JACKSON NETWORKS

Throughput is particularly easy to compute for open Jackson networks. This is not the case for closed networks because K_M must be found. Computing K_M as a direct sum is impractical for large J and M, and alternatives have been devised that are more efficient. We now describe a method for computing throughput and the usual first-moment performance measures for a closed network of *single-channel* stations that is based on Theorem 17.

First we consider the symmetric case, where the Λ_i in (20) and $\mu_i(n)$ are the same for all i, where $\mu_i(n) = 1$ for $n \geq 1$. One way to produce equal Λ_i is to put the stations in a *cyclic arrangement*, where every customer visits each station in the same fixed order. However, it is important to observe that the result we get does not require that the stations actually be arranged in this manner, but only that the Λ_i be the same for all i. We also assume FIFO order of service, but our results hold for (say) LIFO as well.

Example 6-7: Symmetric Single-Channel Cyclic Queues

Suppose we have J single-channel stations, M customers who each visit the stations in the same cyclic order, and $\mu_i(n) = 1$ for $n \geq 1$. From symmetry, the expected number of customers at station i is $L_i(J, M) = M/J$. Denote the expected delay of an arrival at station i by $d_i(J, M)$. The expected "round-trip time" of a customer between visits to the same station is

$$\varkappa(J, M) = Jd_i(J, M) + J.$$

From Theorem 17 and exponential service, we have

$$d_i(J, M) = L_i(J, M - 1) = (M - 1)/J,$$

and hence

$$\varkappa(J, M) = J + M - 1.$$

The rate that a particular customer visits station i is $1/\varkappa(J, M)$, and the throughput is

$$\Lambda^c(J, M) = M/\varkappa(J, M) = M/(J + M - 1). \tag{29}$$

In the asymmetric case, we can find the throughput recursively. Let $\mu_i(n) = \mu_i$ for $n \geq 1$. Suppose we decide to find the throughput at some fixed but arbitrary station i, $\Lambda_i^c(J, M)$. Set relative throughput $\Lambda_i = 1$, and for $j \neq i$, let Λ_j be the number of visits to station j per visit to station i. We use the same notation introduced in Example 6-7, except that now $\varkappa_i(J, M)$, the expected round-trip time to station i, depends on i. Also let $w_j(J, M) = d_j(J, M) + 1/\mu_j$, the mean waiting time at station j.

It can be shown that

$$\varkappa_i(J, M) = \sum_{j=1}^{J} \Lambda_j w_j(J, M), \tag{30}$$

and from Theorem 17, we also have

$$d_j(J, M) = L_j(J, M - 1)/\mu_j, \qquad j = 1, \ldots, J. \tag{31}$$

The throughput at station i is

$$\Lambda_i^c(J, M) = M/\varkappa_i(J, M), \tag{32}$$

and of course,

$$L_j(J, M) = \Lambda_i^c(J, M)\Lambda_j w_j(J, M), \qquad j = 1, \ldots, J. \tag{33}$$

Starting with $M = 1$, we compute $\varkappa_i(J, 1) = \sum \Lambda_j/\mu_j$, and then (32) and (33). We then compute (31) and (30) for $M = 2$, and repeat the process. The amount of computation involved is linear in both J and M, and therefore is of order JM. Kelly [1979] deals with computational methods for computing K_M (see pp. 47–48, and chapter notes) that are of order KM^2 for general closed Jackson networks, and become of order KM for the single-channel case.

6-5 APPROXIMATING CLOSED NETWORKS; THE FIXED-POPULATION-MEAN (FPM) METHOD

An alternative to computing either throughput or K_M for a closed Jackson network is to bound throughput for a closed network by throughput for a corresponding open network. We now describe and analyze such a method.

For a closed network, suppose arrivals to station 1 leave the system instead. They are replaced by Poisson arrivals at station 1 at some rate $\lambda_1 = \lambda$ that we may choose. Formally, replace $J \times J$ matrix $\mathscr{P}$ by $(J + 1) \times (J + 1)$ matrix $\mathscr{P}'$ where $p'_{i,J+1} = p_{i1}$, $p'_{i1} = 0$ for $i = 1, \ldots, J$, $p'_{J+1,1} = 1$, and $a(\mathbf{n}, A_1\mathbf{n}) = \lambda$. The approximation will not depend on which station we select.

We now have an open network that is irreducible, where the Λ_i in (12) are increasing functions of λ. Let $N(\lambda) = N_1 + \cdots + N_J$ be the stationary number of customers in system, with expectation $E\{N(\lambda)\}$, a strictly increasing continuous function of λ (where stable), with the properties $E\{N(0)\} = 0$ and $\lim_{\lambda\to\infty} E\{N(\lambda)\} = \infty$. Now suppose we choose λ such that

$$E\{N(\lambda)\} = M, \tag{34}$$

the fixed number of customers in system for our original closed network. This way of selecting an open network is called the *Fixed-Population-Mean (FPM)* method.

We will show (Theorem 22) that for most cases of practical importance, the throughput of an FPM open network is a *lower bound* on the true throughput of the corresponding closed network. Actually, we first obtain a generalization of this result for a single station of a closed network considered in isolation.

Let $N_i^c \sim \{p_n^c\}$ be the stationary number of customers in system at some station i in a closed network, $N_i^o \sim \{p_n^o\}$ be the stationary number of customers in system at the same station in the corresponding FPM open network, where $N^c = M$ and N^o are the total number of customers in system for each case.

These distributions satisfy recursion relations

$$\Lambda_n^c p_n^c = \mu_{n+1} p_{n+1}^c \quad \text{and} \quad \Lambda^o p_n^o = \mu_{n+1} p_{n+1}^o, \qquad n = 0, 1, \ldots, \tag{35}$$

where Λ^o is the composite arrival rate (throughput) at this station in the open network, and Λ_n^c is the composite arrival rate at this station in the closed network when there are n customers there. The throughput at this station in the closed network is

$$\Lambda^c = \sum_n \Lambda_n^c p_n^c = \sum_n \mu_n p_n^c, \tag{36}$$

which is an actual rate, in contrast to the relative rates in (20).

It is readily verified from the form of the open network recursion (35) that $\{p_n^o\}$ has the property

$$(p_n^o)^2 \geq p_{n+1}^o p_{n-1}^o \qquad \text{for } n \geq 1, \tag{37}$$

if and only if $\{\mu_n\}$ is a *nondecreasing* function of n. Distributions with property (37) are said to be *log-concave*. It is easy to see that (37) is equivalent to this: p_{n+1}^o/p_n^o is *nonincreasing in* n.

We state without proof (see Theorem 1.2 on p. 394 of Karlin [1968]) that

$$\textit{convolutions} \text{ of log-concave distributions are log-concave.} \tag{38}$$

Thus, for an open Jackson network where the $\mu_i(n_i)$ are nondecreasing functions of n_i for every i, *distributions of sums such as* $N_i^o + N_j^o$ *are also log-concave.*

From (23),

$$p_n^c = P(N_i^o = n \mid N^o = M) = P(N_i^o = n)P(N^o - N_i^o = M - n)/P(N^o = M),$$

and

$$p_{n+1}^c/p_n^c = \frac{P(N_i^o = n+1)P(N^o - N_i^o = M - n - 1)}{P(N_i^o = n)P(N^o - N_i^o = M - n)}, \qquad n = 0, 1, \ldots. \tag{39}$$

We are now ready for

Theorem 20. For a closed Jackson network where $\mu_i(n)$ is a nondecreasing function of n, $i = 1, \ldots, J$, the arrival rate at each station is a decreasing function of the number of customers there; in the preceding notation,

$$\{\Lambda_n^c\} \text{ is a decreasing function of } n; \ \Lambda_M^c = 0. \tag{40}$$

Proof. We already know that $\Lambda_M^c = 0$. Combining (35) and (39), we have

$$\Lambda_n^c = \Lambda^o P(N^o - N_i^o = M - n - 1)/P(N^o - N_i^o = M - n).$$

When the $\{\mu_i(n_i)\}$ are nondecreasing, N_i^o is $\sim$ log-concave for every i. From (38), $N^o - N_i^o$ is also $\sim$ log-concave, and we have (40).

Remarks. For $J \geq 3$, (40) will be strictly decreasing, but for $J = 2$ single-channel stations, Λ_n^c is a positive constant, except for $n = M$. $\{p_n^c\}$ is also log-concave, and from (40), "even more so" than is $\{p_n^o\}$. If the $\{\mu_i(n)\}$ are not monotone, it is easy to construct examples where (40) does not hold. When {40} holds, it is easy to see that

$$N_{ia}^c \overset{\text{st}}{\leq} N_i^c, \tag{41}$$

where N_{ia}^c is stationary number of customers at station i found by an arrival there.

Now assume $L_i^o \equiv E(N_i^o) \leq L_i^c \equiv E(N_i^c)$, and that (40) holds. This implies $\Lambda^o < \Lambda_0^c$ (otherwise we would have $N_i^o \overset{\text{st}}{\geq} N_i^c$), and that for some $m \geq 1$,

$$\Lambda_n^c \leq \Lambda^o \quad \text{for } n \geq m \quad \text{and} \quad \Lambda_n^c > \Lambda^o \quad \text{for } n < m.$$

From this and (35), we can solve for p_n in terms of p_m, writing

$$p_n^c = b_{m,n}^c p_m^c \quad \text{and} \quad p_n^o = b_{m,n}^o p_m^o \qquad \text{for all } n, \tag{42}$$

where for $r_{m,n} \equiv b_{m,n}^c / b_{m,n}^o$,

$$0 \leq r_{m,n} \leq 1.$$

With m fixed, $r_{m,n}$ decreases in both $(n - m)^+$ and $(n - m)^-$, and for some (m, n), e.g., at $n = M + 1$, $r_{m,n} < 1$.

Normalizing (42), we have $p_m^c > p_m^o$, and that

$$\{p_n^c / p_n^o\} \text{ is a } \textit{unimodal} \text{ function of } n, \tag{43}$$

with maximum at $n = m$ (and at $n = m + 1$ when $\Lambda_m^c = \Lambda^o$).

Now (43) and $p_0^c \geq p_0^o$ would imply that N_i^o is (strictly) stochastically larger than N^c, a contradiction of $L_i^o \leq L_i^c$. Consequently, $p_0^c < p_0^o$. Thus, $\{p_n^c\}$ has more mass at states "in the middle" and less mass in the tails than $\{p_n^o\}$, and the corresponding tail distributions *cross once* in the manner shown in Figure 6-4, where for visual purposes, we have chosen to illustrate this property for continuous distributions.

From

$$L_i^o = \sum_{j=0}^{\infty} P(N_i^o > j) \leq \sum_{j=0}^{\infty} P(N_i^c > j) = L_i^c,$$

and the crossing property in Figure 6-4,

$$\sum_{j=0}^{n} P(N_i^o > j) \leq \sum_{j=0}^{n} P(N_i^c > j) \qquad \text{for } n = 0, 1, \ldots. \tag{44}$$

Now (44) is equivalent to

$$E\{(n - N_i^o)^+\} \geq E\{(n - N_i^c)^+\}, \qquad n = 1, 2, \ldots, \tag{45}$$

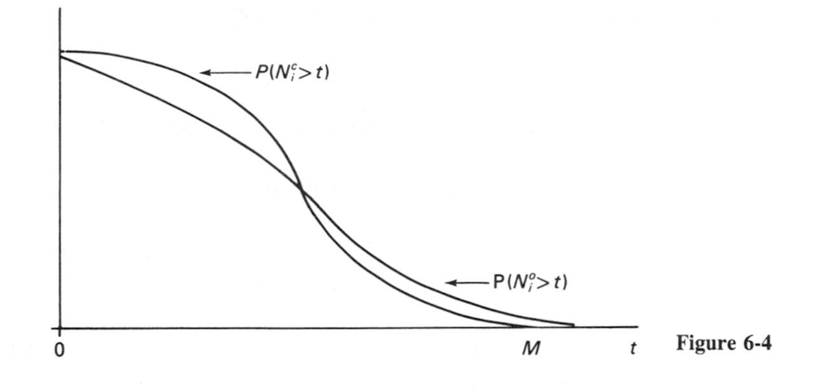

Figure 6-4

which is the discrete version of what Stoyan [1983] calls a *concave* ordering (cv), denoted by $N_i^o \leq_{\text{cv}} N_i^c$, between pairs of random variables, and their distributions. In fact, [see (49)], (45) is equivalent to

$$E\{g(N_i^o)\} \leq E\{g(N_i^c)\}, \tag{46}$$

for every function $g(\cdot)$ that is *nondecreasing and concave*, which is the reason for the name ($g(\cdot)$ is *concave* if $-g(\cdot)$ is convex).

If $\{\mu_n\}$ is a nondecreasing concave function of n, i.e., $g(\cdot) = \mu_\cdot$, we immediately have (47) in the next theorem.

Theorem 21. Let some station i in a closed network, with throughput Λ^c and property (40), be approximated by the same station in a corresponding open network with throughput Λ^o. We have

$$L_i^o \leq L_i^c \quad \text{and} \quad \{\mu_n\} \text{ nondecreasing and concave} \Rightarrow \Lambda^o \leq \Lambda^c; \tag{47}$$

$$L_i^o \geq L_i^c \quad \text{and} \quad \{\mu_n\} \text{ nondecreasing and convex} \Rightarrow \Lambda^o \geq \Lambda^c. \tag{48}$$

Completion of Proof. Suppose $L_i^o \geq L_i^c$. If $\Lambda^o \geq \Lambda_0^c$, $N_i^o \overset{\text{st}}{\geq} N_i^c$, and (48) holds. Now suppose $\Lambda^o < \Lambda_0^c$. We have (40) and (43). If $p_0^c \geq p_0^o$, we again have $N_i^o \overset{\text{st}}{\geq} N_i^c$. Otherwise, the corresponding distributions have the crossing property in Figure 6-4. This property and $L^o \geq L^c$ imply

$$\sum_{j=n}^{\infty} P(N_i^o > j) \geq \sum_{j=n}^{\infty} P(N_i^c > j) \qquad \text{for } n = 0, 1, \ldots,$$

which is equivalent to

$$E\{(N_i^o - n)^+\} \geq E\{(N_i^c - n)^+\}, \qquad n = 0, 1, \ldots, \tag{49}$$

and is called a *convex* ordering (c), denoted by $N_i^c \leq_{\text{c}} N_i^o$. In Section 11-5, we find (Stoyan and also Problem 11-13) that (49) is equivalent to

$$E\{g(N_i^c)\} \le E\{g(N_i^o)\},$$

for every *nondecreasing convex* function g, and (48) follows immediately.

We are finally ready to state the main result in this section.

Theorem 22. Let a closed Jackson network be approximated by a corresponding FPM open network. (a) If the $\{\mu_i(n)\}$ are nondecreasing and *concave* in n for every i, the throughput of the open network is a *lower bound* on the throughput of the closed network. (b) If the $\{\mu_i(n)\}$ are nondecreasing and *convex* in n for every i, the throughput of the open network is an *upper bound* on the throughput of the closed network.

Proof. Since $L^o = \sum L_i^o = \sum L_i^c = M$, we must have $L_i^o \le L_i^c$ for some i. Equation (40) holds for every station, and hence (47) holds for station i. The relative throughputs are governed by (20), and we have case (a). Similarly, $L_i^o \ge L_i^c$ for some i, and we have case (b).

Remarks. Concave case (a) is of primary importance because it includes the case where all stations operate as conventional single or multichannel queues. When all stations operate as infinite-channel queues, the $\mu_i(n)$ are *linear functions*, both cases apply, and the open and closed throughputs must be equal. Independent of this analysis, this conclusion is an easy consequence of $L = \lambda w$. Otherwise, it can be shown that the bounds on the throughput are strict.

To apply FPM, we need to solve (34) for λ. Because $E\{N(\lambda)\}$ is monotone in λ, this is easy to do numerically.

We now briefly explore the quality of the lower bound in case (a) of Theorem 22 as an approximation for closed network throughput. For ease of computation, return to the symmetric single-channel case in Example 6-7, where for all i, $\mu_i(n) = 1$ for $n \ge 1$, and we found a simple exact expression for Λ^c. The throughput is the fraction of time each station is busy. For the open network, $\Lambda^o = \rho$, where $L_i^o = M/J = \rho/(1 - \rho)$, and from this and (29), we have

$$\Lambda^o = M/(J + M) \quad \text{and} \quad \Lambda^c = M/(J + M - 1).$$

The percentage error in the lower bound Λ^o is low ($<5\%$) for $J + M > 20$, and the FPM method works very well for symmetric single-channel networks of moderate size.

Of greater importance, of course, is the performance of FPM with asymmetric and/or multichannel stations. From Theorem 22, and the role of ∞-channel queues, we expect FPM to perform better for networks of multichannel stations, and in general to perform well when both M and J are large, and there is no clear "bottleneck" station.

6-6 OPEN NETWORKS WITH GENERAL CUSTOMER ROUTES

One drawback to the applicability of Jackson networks is the assumption of Markovian paths or routes. In a machine shop, for example, a particular type of job would visit a sequence of machines in some predetermined order. Knowledge about where a job had been previously would affect the probability that it visits a particular machine next.

We now show that this assumption is not required for product-form results to hold. In fact, routes can be essentially arbitrary. We will include general routes in our model in a way that preserves the Markov property. This will be done by enlarging the state space to include information about the route that each customer is on.

Now suppose we have an open network with R different customer *routes*, where R is finite or countable. Arrivals who take route r visit the stations in the network in some deterministic sequence,

$$v(r, 1), v(r, 2), \ldots, v(r, e(r)),$$

before they leave the system. If $v(r, s) = i$, we say a route-r customer visits station i at *stage* s, $s = 1, \ldots, e(r)$. A route-r customer may visit the same station more than once. Because of this, it will be necessary to keep track of the stage as well as the route of every customer. Let $\#(r, i)$ be the number of times a route-r customer visits station i.

Clearly, Markovian paths can be represented in this manner, but doing so would make the analysis of Jackson networks needlessly complicated. On the other hand, consider a network with five stations and only two routes,

$$1,3,5 \quad \text{and} \quad 2,3,4.$$

This network cannot be formulated in terms of Markovian paths. (An alternative to deterministic routes is to let each route be a Markov chain. This alternative may be advantageous in applications where a small number of chains would suffice.)

We assume that the R arrival streams of r-route customers are independent Poisson processes at rate λ_r, with $\sum \lambda_r < \infty$. The composite arrival rate of customers at station i is

$$\Lambda_i = \sum_{r \in R} \#(r, i)\lambda_r, \qquad i = 1, \ldots, J. \tag{50}$$

The description of the operation of the stations is more complicated than for Jackson networks, where all customers are treated alike, and there is no need to make any particular assumption about order of service. (We did assume FIFO in Section 6-4, but that was to simplify the analysis; the results obtained do not depend on FIFO.) Not only do we now have to keep track of the route and stage of every customer, we also have to specify how the service capacity at each station is allocated to the customers there. We call this *queue management*.

It will be convenient to treat service times in a manner different from Section 6-3: For every customer at every station, we assume that

$$\text{service requirements are} \sim \exp(1).$$

The line-length-dependent service-rate functions $\mu_i(n)$ are now defined to be *service capacity* scale factors on these requirements. Service requirements, even of the same customer at the same or different stations, are assumed to be independent of each other, and of all else.

Consider (arbitrary) station i. When there are n customers there, we impose an ordering on the positions of these customers, $1, \ldots, n$. We make the following additional *queue management* assumptions:

(a) A proportion $\gamma_i(l, n)$ of service capacity $\mu_i(n)$ is allocated to the customer in position l, $l = 1, \ldots, n$; $\sum_l \gamma_i(l, n) = 1$. When a customer in position l departs from station i, customers in positions $l + 1, \ldots, n$ move to positions $l, \ldots, n - 1$, respectively.

(b) An arrival at station i who finds n customers there moves into position l with probability $\delta_i(l, n + 1)$, $l = 1, \ldots, n + 1$; $\sum_l \delta_i(l, n) = 1$. Customers found in positions $l, \ldots, n$ move to positions $l + 1, \ldots, n + 1$, respectively.

Thus a customer in position l at station i completes service at (conditional) rate $\gamma_i(l, n)\mu_i(n)$ when there are n customers there. It is important to observe that this rate is *independent of the route and the stage* of the customer. This rules out priority queues of the kind considered in Chapter 10.

Consider these examples of queue management:

$$\text{Single-channel FIFO:} \quad \gamma_i(1, n) = 1 \quad \text{and} \quad \delta_i(n, n) = 1. \tag{51}$$

$$\text{Single-channel LIFO:} \quad \gamma_i(1, n) = 1 \quad \text{and} \quad \text{for } n \geq 2, \quad \delta_i(2, n) = 1. \tag{52}$$

$$\gamma_i(1, n) = 1 \quad \text{and} \quad \delta_i(1, n) = 1, \tag{53}$$

which is called *preemptive LIFO*, and means that the entire service capacity is allocated to the most recent arrival.

Given the queue management, we need to keep track of the route and stage of every customer in each position at each station. Let $r_i(l)$ and $s_i(l)$ be the respective route and stage of the customer in position l at station i, and $c_i(l) = (r_i(l), s_i(l))$, which we call the *class* of this customer. The *state* of the system is a hierarchical representation of this information: The state of station i is $\mathbf{c}_i = (c_i(1), \ldots, c_i(n_i))$, and the state of the system is

$$\mathbf{C} = (\mathbf{c}_1, \ldots, \mathbf{c}_J),$$

where $\{C(t)\}$ is a continuous-time Markov chain, with states of form $\mathbf{C}$ and transition rates determined by the preceding assumptions. This chain is irreducible, provided we restrict the $\mathbf{c}_i$ to "possible" states in terms of the route and stage in each position. Also let N_i be the stationary number of customers at station i, for every i.

Let $p(\mathbf{C})$ and $p_i(\mathbf{c}_i)$ be the corresponding stationary distributions for this chain, when they exist, and define the function

$$f_i(r, s) = \lambda_r \qquad \text{if } v(r, s) = i,$$
$$= 0 \qquad \text{otherwise, and also define}$$

$$k_i^{-1} = \sum_{n=0}^{\infty} \Lambda_i^n/[\mu_i(1) \cdot \ldots \cdot \mu_i(n)], \qquad i = 1, \ldots, J. \tag{54}$$

Stability conditions and other results will closely parallel those for an open Jackson network. We are now ready to state

Theorem 23. $\{C(t)\}$ is stable (positive recurrent) if and only if (54) converges for all i. When stable, it has stationary distribution of form

$$p(\mathbf{C}) = p_1(\mathbf{c}_1) \cdot \ldots \cdot p_J(\mathbf{c}_J), \tag{55}$$

where

$$p_i(\mathbf{c}_i) = P(N_i = n_i) \prod_{l=1}^{n_i} [f_i(r_i(l), s_i(l))/\Lambda_i], \qquad i = 1, \ldots, J, \text{ and} \tag{56}$$

$$P(N_i = n_i) = k_i \Lambda_i^{n_i}/[\mu_i(1) \cdot \ldots \cdot \mu_i(n_i)],\ n_i \geq 0, \qquad i = 1, \ldots, J. \tag{57}$$

Sketch of Proof. We refer to p. 61 of Kelly [1979] for details, but the idea is to "guess" the form of the transition rates for the reversed chain, and then use Theorem 1. When doing this, plug in (56) and (57) without the normalizing constants, and we will get a solution to the balance equations whether or not normalization is possible, i.e., we have a *stationary measure*. When (54) converges for all i, we have the positive case. Otherwise, all states are either transient or null.

Remark. Both here and for open Jackson networks, it is possible for one or more stations to be stable, even when the entire system is not.

Important Remarks. Equation (55) is again of product-form, meaning that the collections of random variables that represent the status of each station are independent. In particular, the N_i are independent. From (57) we see that the distribution of each N_i has the same form (15) that we found for an open Jackson network, except that the Λ_i now are found from (50) rather than (12). From (56), the route and stage of the customers at station i are independent from position to position. Given $N_i = n$, the probability that a route-r, stage-s customer is in position $l \leq n$ at station i is λ_r/Λ_i, provided that $v(r, s) = i$. These results are independent of the particular way that the queues are managed.

Let $N_i(r, s)$ be the number of route-r, stage-s customers at station i. It is immediate from (56) that $(N_i(r, s) \mid N_i)$ has a binomial distribution. When N_i is geometric (the single-channel case), it is easily shown that $N_i(r, s)$ will also have a geometric distribution. For distinct (r, s) pairs and the same i, the $N_i(r, s)$ are dependent random variables.

Example 6-8

Consider a two-station network with two routes: 1,2,1 with arrival rate λ_1, and 2,1,2 with arrival rate λ_2. Each station is a single-channel FIFO queue with service rate μ_i at station i. The composite arrival rates are $\Lambda_1 = 2\lambda_1 + \lambda_2$, and $\Lambda_2 = \lambda_1 + 2\lambda_2$. Let $\rho_i = \Lambda_i/\mu_i$. From Theorem 23, the N_i are independent with marginal distributions

$$P(N_i = n) = (1 - \rho_i)\rho_i^n, \qquad n \geq 0.$$

Given $N_1 > 0$, the probability that a route-1, stage-1 customer is in service at station 1 is λ_1/Λ_1.

It is easily shown that the reversed process $\{C(-t)\}$ is also an open network with general customer routes, with routes that are the reverse of the original ones, and the analog of Theorem 14 follows.

6-7 NETWORKS OF QUASI-REVERSIBLE STATIONS

From (57), observe that the distribution of $(N_1, \ldots, N_J)$ depends on customer routes only through their contribution to the composite rates Λ_i. This distribution is *insensitive* to the detailed route structure. We treat cases of service distribution insensitivity, e.g., as in Erlang's loss formula, later in this chapter.

For this purpose, and also to investigate other generalizations of what we have done, we now identify a property called *quasi-reversibility*. It is a property that individual stations in this chapter have when they are taken in isolation, i.e., when they are *not* part of a network. It will also be a property of the entire network when the stations are put together by certain rules to form the network. The simple form of the results we have obtained is a consequence of the fact that all the stations in the network have this property.

We now model a single station in isolation as an irreducible continuous-time Markov chain with states x belonging to countable state space $\mathcal{S}$. Let the (conditional) transition rates be denoted by $\{a(x, y), x, y \in \mathcal{S}\}$. To avoid technicalities, assume that $a(x, x) = 0$ for all x, so that the transition rate out of every x is unambiguously given by

$$a(x) = \sum_{y \neq x} a(\mathrm{x}, \mathrm{y}).$$

Our results will hold when we permit $a(x, x) > 0$.

Suppose this chain has stationary distribution $\{p(x), x \in \mathcal{S}\}$, and we use

$\{p(x)\}$ to define transition rates $\{b(x, y)\}$ for the reversed chain:

$$p(x)b(x, y) = p(y)a(y, x) \qquad \text{for all } x, y \in \mathcal{S}. \tag{58}$$

Suppose there are well-defined arrivals and departures of customers of class $c \in C$ (*c-customers*), where C is finite or countable (e.g., the classes defined in Section 6-6). When the chain is in state x and a c-customer arrives, there is a transition to some new state, and let $\mathcal{S}^a(c, x)$ be the set of possible states visited when this occurs. Similarly, let $\mathcal{S}^d(c, x)$ be the set of possible states visited when the chain is in state x and a c-customer departs. (We assume that only the arrivals and departures of c-customers can induce these transitions.)

Definition. A station is called *quasi-reversible* if for every state x, and each class c, there is a positive $\lambda(c)$ such that

$$\sum_{y \in \mathcal{S}^a(c,x)} a(x, y) = \lambda(c), \text{ and} \tag{59}$$

$$\sum_{y \in \mathcal{S}^b(c,x)} b(x, y) = \lambda(c). \tag{60}$$

Thus the arrival rates in the forward and reversed chains are independent of the state of the chain. Looking first at the forward and then the reversed chain, this implies

Theorem 24. The arrival processes of c-customers, $c \in C$, are independent Poisson processes that have the lack of anticipation property with respect to the chain. The departure processes of c-customers, $c \in C$, are independent Poisson processes, with lack of anticipation for the reversed chain.

Remark. We usually start with (59) as an *assumption*. If the routes in Section 6-6 are taken as the classes, we showed there that the entire network is quasi-reversible.

The easiest way to construct a network out of quasi-reversible stations is to assume the classes and their arrival rates at each station are generated by the route structure in Section 6-6. A customer's class is a route *and* stage; it remains the same until the customer moves on to the next stage. We refer to Kelly [1979], p. 68, for details; he generates networks in a more general way in [1982]. In the next theorem, it is assumed that with the generated arrival rates, every station in isolation is stable. In applications, this means that appropriate normalizing constants must be finite.

Theorem 25. A stationary open network of quasi-reversible stations has these properties:

(i) The state variables for each station are independent of those for other stations.

(ii) The arrival process of every class of customers at each station has the PASTA property.

(iii) The *network* is quasi-reversible.

(iv) The reversed process corresponding to the network is another quasi-reversible network.

Important Remark. The arrival processes in (ii) usually will not be Poisson, and when they are not, the individual stations, once inserted into the network, no longer are quasi-reversible.

6-8 SYMMETRIC QUEUES

Consider again a single station in isolation. We present a summary of a general approach to Erlang's loss formula (Section 5-11) and related insensitivity results under other modes of operation.

We begin by recalling that an $M/G/1$ loss system can be analyzed as an alternating renewal process, and it is immediate that the stationary distribution of the number of customers in system is insensitive to G (depending only on the mean). To make a Markov process out of the continuous-time process $\{N(t)\}$, we need more information, such as the remaining service S_r of the customer in service when $N(t) = 1$, where S_r is called a *supplementary variable*. (We can use *attained* service instead of S_r.) Given the system is busy at some random moment, $S_r \sim G_e$.

For the $M/G/2$ loss system and more complicated models, this elementary approach doesn't work, but we can introduce as many supplementary variables as needed so that with the inclusion of these variables, we have a Markov process. We also have a choice about whether to analyze the continuous-time process directly, or to analyze the corresponding discrete-time process at arrival epochs, say (or *both* arrival and departure epochs; e.g., see Brumelle [1978]).

As in Section 6-6, we have service capacity scale factors $\mu(n)$, a function of n, the number of customers at the station. The number of customers at the station will either be unbounded, or in the queue limit case, will have upper limit denoted by m.

We permit different classes of customers, where c-customer arrivals are Poisson at rate λ_c, and c-customer service requirements are i.i.d., with some general distribution G_c, and mean r_c. Let $\lambda = \sum \lambda_c < \infty$, and overall $G = \sum \lambda_c G_c/\lambda$, with mean r.

To obtain the following results, it is necessary to restrict the queue management rules in Section 6-6 to the following

Definition. A single station is called a *symmetric queue* if for every n,

$$\gamma_i(l, n) = \delta_i(l, n), \qquad i = 1, \ldots, n. \tag{61}$$

Thus, Preemptive LIFO (PL) is symmetric, but FIFO and standard LIFO are not. The other important symmetric rule is

$$\gamma_i(l, n) = 1/n. \tag{62}$$

When $\mu(n) = 1$, for $n \geq 1$, (62) is called *processor sharing* (PS). When $\mu(n) = n$, $n \geq 1$, we have an *infinite server* queue, and when there is a single customer class with $\mu(n) = n$, $n \leq m$, we have the classical *Erlang's loss model*. With (62) and $\mu(n) = \min(n, c)$, $c \geq 2$, we have a processor-shared *multichannel* queue.

Let N be the stationary number of customers in system. We now state without proof

Theorem 26. A symmetric queue is stable with $N \sim \{p_n\}$, where $\{p_n\}$ satisfies

$$\lambda r p_n = \mu(n + 1)p_{n+1}, \qquad n \geq 0, \tag{63}$$

if and only if these equations can be normalized. When stable, and when $N = n$ for any n, the status of each of these customers is independent of the others, where the probability that customer i is a c-customer is

$$\lambda_c r_c / \lambda r,$$

and given customer i is a c-customer, it has remaining service time

$$S_{ir} \sim G_{ce},$$

the corresponding equilibrium distribution.

There are direct ways to prove Theorem 26 by showing that the proposed solution satisfies certain stationarity conditions, e.g., as Brumelle does in the single-class case. With Brumelle's approach, the arrival rate can depend on n, in which case λ_n replaces λ in (63). Other authors have more general formulations that include the class structure.

Kelly's approach [1979] is to first suppose that the service requirements distributions G_c are of *phase-type*. He obtains the distribution of the number of completed phases, which is equivalent to the equilibrium distribution. He also shows that the symmetric queues are quasi-reversible, and hence can serve as stations in a *quasi-reversible* network. Doing this directly in terms of general distributions would be extremely difficult. He then approaches general distributions by considering sequences of phase-type distributions that converge to general distributions. Barbour [1976] shows that this limit argument is valid for obtaining the stationary distribution when the service requirements distributions of all classes in the entire network are arbitrary.

Kelly also shows that the process $\{N(t)\}$ for a symmetric queue in isolation is *reversible*, even though it is not Markov.

6-9 LITERATURE NOTES

R. R. P. Jackson [1954] was the first to obtain the product-form solution for two single-channel exponential queues in tandem. This topic received a lot of attention from many authors, leading up to J. R. Jackson [1963] and what has come to be called the Jackson networks in Section 6-3.

In parallel with this, Burke [1956] showed Poisson departures for the $M/M/c$ queue, by an argument somewhat similar to Problem 6-5. This was quickly followed by Reich [1957], who used reversibility. While these results helped explain R. R. P. Jackson's result, those of J. R. Jackson seemed all the more surprising, when it was realized that the internal flows in a network are not Poisson.

Reich also showed that the waiting times of the same customer at a series of single-channel exponential queues are independent random variables. Burke [1969] showed that when a multichannel queue is inserted between single channel queues (Problem 6-6), waiting times are dependent. He also proved Theorems 9 and 10. It has been shown (Walrand and Varaiya [1980]) that in fairly general networks without "overtaking" (which allows passing in Problem 6-6), waiting times are independent.

The earliest proof of Erlang's loss formula seems to be by Sevast'yanov [1957], and many generalizations followed in what was for a time a separate literature. Since about the mid-1970s these literatures have merged and exploded. See Kelly [1979, 1982], Franken et al. [1982], and Sauer and Chandy [1981]. Quasi-reversibility is not the only unifying idea, but we will not attempt to explain alternatives here.

There is also an active literature on networks of *reversible* stations, e.g., Pollett [1986].

There are a number of proofs of Theorems 17 through 19, and they have been extended to multiple classes of customers; see Lavenberg and Reiser [1980], Chandy et al. [1975], and Walrand [1983].

See Kelly [1979] for a treatment of closed networks with multiple customer classes, and also for conditions under which the arrival rate can depend on the state of the system. The approximation in Section 6-5 can be applied to multiclass queues.

Of all the generalizations of Jackson networks, general customer routes is probably of greatest practical importance. Note, however, that these results do not hold under congestion-dependent routing. For example, if a station operates as a loss system, lost customers of the same class continue on the same route as served customers.

Kelly observes that through class-dependent service distributions, we can construct networks where the service times of the same customer at different stations are dependent random variables. However, the network will not be quasi-reversible unless the stations operate as symmetric queues. See the discussion at the end of Section 11-9.

Rules such as PL and PS can be used to understand, analyze, and approximate conventional FIFO queues; see Sections 8-2, 9-11, and 11-8.

PROBLEMS

6-1. Give a formal argument for Theorem 5.

6-2. Partition the states of a reversible Markov chain into A and A^c. Suppose that for all $i \in A$ and $j \in A^c$, transition rates a_{ij} are changed to ka_{ij}, where $k > 0$ is an arbitrary constant. Show that the altered chain is also reversible. Find the new stationary distribution in terms of the old one.

6-3. In Example 6-4, suppose arrivals who find the waiting room full [i.e., $(m - 1)^+ + (n - 1)^+ = l$] are lost, even if the person they wish to see is free. Is this model reversible? Does Theorem 6 apply?

6-4. Following Example 6-4, it was remarked that even when the original queue is unstable, the truncated model is reversible with state probabilities proportional to $\rho_1^m\rho_2^n$.

(a) Verify from (4).

(b) Verify by first truncating the separate unconstrained doctor and dentist queues; then apply Theorem 6.

6-5. *An alternative derivation of Poisson departures.* Consider a positive (λ, μ_j)-process (queue) $\{N(t)\}$, with state probabilities uniquely determined by

$$\lambda p_j = \mu_{j+1}p_{j+1}, \qquad j = 0, 1, \ldots .$$

Assume $N(0) \sim \{p_j\}$, so that the process is stationary. Let τ be the first departure epoch, $Y = N(\tau^+)$, which is the number of customers left behind by the first departure, and $\{p_j\}$ have generating function $G(z)$. Define conditional joint transform

$$r_j = E\{e^{-s\tau}z^Y \mid N(0) = j\}, \qquad j \geq 0.$$

(a) By conditioning on $N(0)$ and whether the *first transition* is either an arrival or a departure, show that the r_j satisfy

$$(\lambda + \mu_j + s)r_j = \mu_j z^{j-1} + \lambda r_{j+1}, \qquad j = 0, 1, \ldots .$$

(b) Multiply the expression in (a) by p_j, sum over j, and, after some cancellation, obtain

$$(\lambda + s)E\{e^{-s\tau}z^Y\} = \sum_{j=0}^{\infty} \lambda p_j z^j, \text{ or}$$

$$E\{e^{-s\tau}z^Y\} = [\lambda/(\lambda + s)]G(z).$$

(c) From (b), what are the marginal distributions of τ and Y? What is their joint distribution?

(d) Now consider doing this for the sequence of interdeparture times $\{\tau_n\}$ in terms of $\{Y_n\}$, where for each n, τ_n depends on Y_{n-1} $[Y_0 \equiv N(0)]$. Argue that the departure process is Poisson.

(e) What other parts of Theorem 7 follow from this argument?

6-6. *Dependent waiting times.* Consider a stationary tandem queue with $J = 3$ stations, Poisson arrivals at rate λ, exponential service at rate μ/channel, and c_i channels at station i, where $c_1 = c_3 = 1$, and $c_2 = 2$:

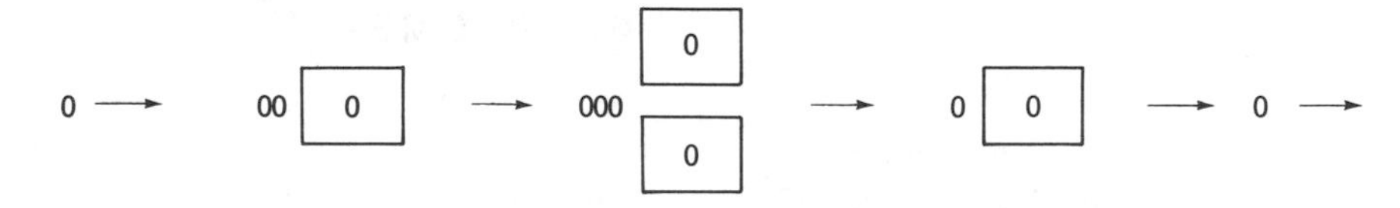

Tag a customer on arrival, and let W_i be the stationary FIFO waiting time of this customer at station i. We know from Theorem 9 that W_1 and W_2 are independent, as are W_2 and W_3. Let N_i be the number of customers at station i just after the tagged customer departs from station 1. We also know that W_1, N_2, and N_3 are independent. However, N_1 is the number of Poisson arrivals during W_1, and is stochastically increasing as W_1 increases!

(a) If $N_2 = 0$, later departures from station 1 may pass the tagged customer at station 2, and arrive at station 3 before the tagged customer does. How will this affect W_3?

(b) Argue that W_3 stochastically increases as W_1 increases.

6-7. Consider an open Jackson network consisting of J *single-channel* queues, where μ_j is the service rate at station j, $j = 1, \ldots, J$.

(a) What is L_j, the expected number of customers at station j?

(b) Suppose we could allocate service capacity μ_j, under the constraint

$$\sum_{i=1}^{J} \mu_i = \mu, \qquad \text{a constant.}$$

Determine the optimal allocation in the sense that $L = L_1 + \cdots + L_J$ is minimized. (At least formulate the problem; the easiest method of solution uses a Lagrange multiplier.)

Remark. It is easy to generalize to weighted linear functions when appropriate, i.e., minimize $\sum \alpha_i L_i$ subject to $\sum \beta_i \mu_i$ = a constant.

6-8. Consider an open Jackson network with $J = 2$ stations, where station j operates as a c_j-channel facility, where each channel is a single worker with exponential service rate μ. Suppose $c_1 + c_2 = c$ workers, a constant. Describe how you would determine the optimal allocation of workers to stations (to minimize overall L, say).

6-9. Even though the waiting times of the same customer at different stations in "typical" open Jackson networks are dependent random variables, and composite arrivals at each station are not Poisson, the delay *distributions* at single-channel or multichannel FIFO stations are easy to obtain. Explain how you would do it.

6-10. Why are the stochastic processes $\{N_i(t)\}$ in Theorems 8 and 12 *not* independent?

6-11. The arrival of trucks carrying tomatoes to a cannery is Poisson at rate λ. Arriving trucks are weighed at scales #1, unloaded into a bin, weighed at scales #2, and then depart. A load of tomatoes arrives at the bin when the truck they were on completes unloading. Unloaded tomatoes are fed immediately to a canning line:

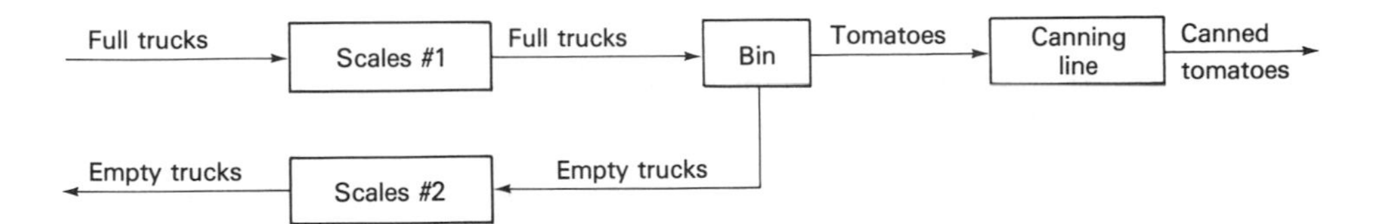

Weighing, unloading, and canning times (of a load) are independent, where weighing times are $\sim$ exp (μ), unloading times are $\sim$ exp (α), and canning times are $\sim$ exp (γ).

(a) Under what conditions is the entire system stable? Assume these conditions hold below.

(b) Notice that the flow of trucks is not affected by unloaded tomatoes. Use this observation to write down the product-form solution for the distribution of the number of trucks at each of the scales and the bin.

(c) Notice that empty trucks have no effect on the flow of tomatoes. Use this observation to write down the product-form solution for the number of trucks at the first scales and the bin, and the number of loads of tomatoes at the canning line.

(d) Would you expect the entire four-station system to have a product-form solution?

(e) If trucks are all the same size and are full on arrival, we would expect that the time to can a load of tomatoes would be a constant. Analyze the canning line under this assumption.

(f) It has been proposed to combine the two single-channel scales into one two-channel station that would serve both empty and full trucks *in their order of arrival*. What is the product-form solution corresponding to (b)? Compare this proposal with the original mode of operation, from the point of view of trucks. [*Hint:* We are *pooling*. (Why is the truck process no longer a Jackson network? How come we can still solve it?)]

(g) How would the proposal in (f) complicate the *analysis* of the canning line?

6-12. In a closed Jackson network with M customers and J stations, argue that the number of states in the Markov chain is

$$\binom{J + M - 1}{M}.$$

6-13. *Continuation.* Suppose we have a closed Jackson network with symmetric single-channel stations, as in Example 6-7.

(a) Show that all states are equally likely.

(b) How many states correspond to the event that station i is empty?

(c) From (a), (b), and the result above, find (29).

6-14. Describe (in principle) how you would find the delay distribution at a single-channel FIFO station in a closed Jackson network.

6-15. A machine shop has three machines, A, B, and C, where each machine operates as a single-channel FIFO queue. Service times are exponential at respective rates μ_a,

μ_b, and μ_c. The shop gets four types of jobs, numbered 1 through 4, where each type requires service on machines in a particular sequence:

1: *ABCA* 2: *CAB* 3: *ACBC* 4: *BCAB*.

The arrival process of type i jobs is Poisson at rate λ_i.

(a) Under what conditions is this system stable?

(b) When stable, what is the joint stationary distribution of the number of jobs at each machine?

(c) For a type 3 job, find the delay distribution until service begins on machine A.

6-16. (Wolff and Wrightson [1976]) A city has two fire stations with one fire engine at each. The city is partitioned into two regions, where a fire in region i will be served by a fire engine from station i, if free, $i = 1, 2$. Otherwise, the other fire engine is sent, if it is free. If both are busy, the fire is "lost." The occurrence of fires in region i is Poisson at rate λ_i, and the times to put them out (including fire engine travel time) are independent with some general distribution G_i with finite mean $1/\mu_i$, *independent* of which engine is sent. Let $\rho_i = \lambda_i/\mu_i$.

(a) Argue that from the point of view of fires, Erlang's loss formula applies. Define an appropriate service distribution G and service rate μ for this purpose, and let $\rho = (\lambda_1 + \lambda_2)/\mu$.

(b) In terms of the notation in (a), what fraction of fires are lost? Let $\{p_j\}$ be the now known distribution of the number of busy engines below.

(c) Use PASTA and $L = \lambda w$ to find p_{bi}, the fraction of time engine i is busy,

$$p_{bi} = (\rho_j + \rho^2/2)p_0/(1 + \rho) + p_2, \qquad i \neq j.$$

Remarks. Not only are the p_j insensitive to G, but the p_{bi} are insensitive to the G_i. However, it can be shown through example that the p_{bi} are *not* insensitive under any of the following:

(i) three or more servers, with two or more regions with preferences, or

(ii) server-dependent service distributions, or

(iii) preference-dependent service distributions (e.g., region-engine pairwise-dependent travel times)

7

OVERFLOWS AND RETRIALS

In Chapter 5, a standard c-channel queue with no queue allowed was called a *loss* system. Arrivals finding all channels busy depart immediately without being served, and were *said* to be lost.

In practice, these customers are not necessarily lost; they may either be served somewhere else, or they may return later for service at the original system. We now model these possibilities. Arrivals who find all channels busy are said to be *blocked*, and the departure process of these customers is called an *overflow* process. (In tandem queues, blocking has a different meaning; see Section 5-3.) The original c-channel system is called the *primary* station, and overflows who are sent elsewhere for service go to a *secondary* station. A *two-stage* model is a system with one or more primary stations and one secondary station. The *blocking probability* at a station is the fraction of arrivals at that station who overflow from that station. The *offered load* at each station is the arrival rate divided by the service rate.

In telephone systems, for example, groups of customers may have exclusive access to some trunk lines, whereas other lines are shared. Each group's lines form a primary station. The shared lines form the secondary station.

The theory for one primary station is developed in Sections 7-1 through 7-4. An exact analysis is intractable for more primary stations. In Sections 7-5 through 7-8, we present and evaluate various approximation methods for this case, where some of these methods are based on the preceding theory.

Overflows who return (later) to the primary station are called *retrials*, and are said to be *in orbit* between trials. (Everyday examples are reservation systems.) This model is treated in Sections 7-9 through 7-13.

7-1 THE TWO-STAGE MODEL

Consider a two-stage model, where arrivals go to a primary station that is a c-channel loss system, $c \geq 1$. Overflows from the primary station go to a secondary station that is either ∞-channel or a finite channel loss system. Service times are i.i.d., $\sim$ exp (μ), regardless of where service takes place. The arrival process at the primary station is a renewal process at rate λ, $0 < \lambda < \infty$, with inter-arrival distribution A. (The important special case of Poisson arrivals is treated in Section 7-4.) See Figure 7-1.

An overflow occurs at the primary station when all c channels there are busy. Because we have renewal arrivals and exponential (memoryless) service, overflow epochs at the primary station are regeneration points for that station, and this overflow process is a renewal process. We denote the corresponding inter-overflow distribution by A_c, and the overflow rate by λ_c. In this notation, the blocking probability at the primary station is

$$B_c \equiv \lambda_c/\lambda, \tag{1}$$

and the offered load to the secondary station is

$$m_c \equiv \lambda_c/\mu. \tag{2}$$

Now c is arbitrary, and a standard device for the analysis of this model is to create not two stations, but an infinite sequence of one-channel stations, called the *sequential* model.

Suppose we have ∞ channels, numbered arbitrarily 1, 2, . . . , where each arrival is served by the lowest numbered channel that is free. An arrival at channel 1 is counted as an arrival at channel c if channels 1, . . . , $c - 1$ are all busy, and is counted as an overflow at channel c if it also is busy. For any fixed c, the overflow process from channel c is the same as that from a standard c-channel loss system (the primary station), but now we have defined overflow processes for every c, with distribution A_c and rate λ_c, $c = 1, 2, \ldots$. The sequential model is represented in Figure 7-2.

Suppose the secondary station in our original two-stage model has ℓ channels. The arrival rate at the secondary station is λ_c. Overflows at the secondary station occur only when an arrival at the primary station finds channels 1, . . . , $c + \ell$ busy (in our sequential model). The secondary station has overflow rate

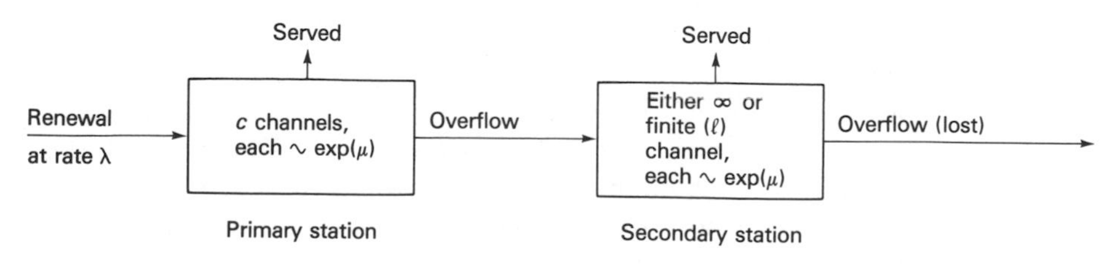

Figure 7-1

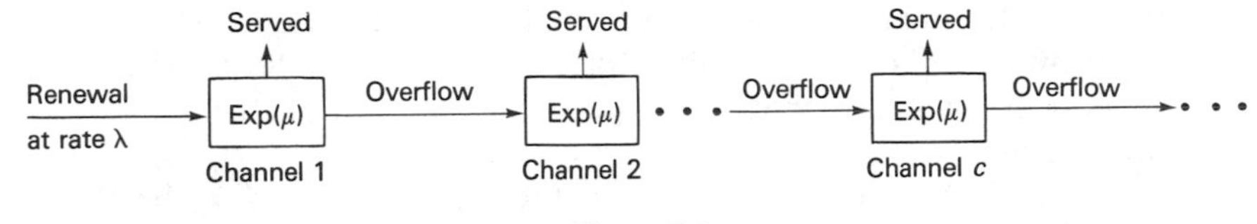

Figure 7-2

$\lambda_{c+\ell}$, and blocking probability

$$\lambda_{c+\ell}/\lambda_c = m_{c+\ell}/m_c. \tag{3}$$

Finding methods to compute and/or approximate performance measures such as (3) is the objective of this chapter. We begin this quest by using the sequential model to find an important equation that relates transforms of inter-overflow distributions.

Note that overflows at channel c are also overflows at channels $0, \ldots, c-1$, and would be lost customers if we had a $GI/M/c$ loss system. Overflows at channel 0 are the arrivals, where in expressions below, $A_0 = A$, and $m_0 = \lambda/\mu \equiv a$, the offered load at channel 1.

At epoch 0, suppose an overflow occurs at channel $c+1$ (and c), and let $Y_{c+1} \sim A_{c+1}$ and $Y_c \sim A_c$ be the times until the next overflow at each of these channels, where either $Y_{c+1} = Y_c$ or $Y_{c+1} > Y_c$. We now relate their transforms, $\tilde{A}_c(s) = E(e^{-sY_c})$, and their means, $E(Y_c) = 1/\lambda_c$, $c = 1, 2, \ldots$.

At epoch 0, both channels c and $c+1$ are busy. The next c-overflow will also be an $(c+1)$-overflow if and only if the remaining service at channel $c+1$ exceeds Y_c. Otherwise, $(Y_{c+1} - Y_c \mid Y_{c+1} > Y_c) \sim A_{c+1}$, *independent* of Y_c. Hence,

$$E(e^{-sY_{c+1}} \mid Y_c) = e^{-sY_c}[e^{-\mu Y_c} + \tilde{A}_{c+1}(s)(1 - e^{-\mu Y_c})],$$

$$\tilde{A}_{c+1}(s) = \tilde{A}_c(s+\mu) + \tilde{A}_{c+1}(s)[\tilde{A}_c(s) - \tilde{A}_c(s+\mu)], \quad \text{and}$$

$$\tilde{A}_{c+1}(s) = \tilde{A}_c(s+\mu)/[1 - \tilde{A}_c(s) + \tilde{A}_c(s+\mu)], \qquad c = 0, 1, \ldots. \tag{4}$$

By the same argument, it is easily shown that

$$E(Y_{c+1}) = E(Y_c)/\tilde{A}_c(\mu), \quad \text{and hence,} \tag{5}$$

$$m_{c+1} = m_c\tilde{A}_c(\mu), \qquad c = 0, 1, \ldots. \tag{6}$$

7-2 THE $GI/M/\infty$ QUEUE

We now analyze the secondary station in the two-stage model, where the primary station has c channels, and the secondary station has ∞ channels. Thus, the secondary station is a $GI/M/\infty$ queue with inter-arrival distribution A_c and arrival rate λ_c. Our analysis is general in that it does not depend on A_c being an inter-overflow distribution. For example, it holds for deterministic arrivals. In this section, *arrivals* are arrivals at the secondary station, which for brevity we call the *system*.

Let S_n be the service time of the nth arrival and T_n be the time between arrivals n and $n + 1$, $n = 1, 2, \ldots$, where $\{S_n\}$ and $\{T_n\}$ are i.i.d. sequences, with $S_n \sim \exp(\mu)$, $T_n \sim A_c$, $E(e^{-sT_n}) = \tilde{A}_c(s)$, and $E(T_n) = 1/\lambda_c$.

For $j = 0, 1, \ldots$, let p_j be the proportion of *time* there are j customers in system, and π_j be the proportion of *arrivals* who find j customers in system. Applying PASTA to the exponential channels, the transition rate from j to $j - 1$ is $j\mu p_j$. Equating that rate to the rate from $j - 1$ to j, we have

$$j\mu p_j = \lambda_c \pi_{j-1}, \qquad j = 1, 2, \ldots. \tag{7}$$

From (7) (or $L = \lambda w$),

$$\sum_{j=1}^{\infty} j p_j = \lambda_c/\mu = m_c,$$

which of course is the offered load at the secondary station. More generally, the kth factorial moment of $\{p_j\}$ is proportional to the $(k - 1)$st factorial moment of $\{\pi_j\}$: For $k = 1, 2, \ldots$,

$$\begin{aligned} M_k &\equiv \sum_{j=k}^{\infty} j(j-1)\cdots(j-k+1)p_j \\ &= m_c \sum_{j=k}^{\infty} (j-1)\cdots(j-k+1)\pi_{j-1} \equiv m_c \Pi_{k-1}. \end{aligned} \tag{8}$$

In particular, $M_2 = m_c\Pi_1$, where Π_1 can be found by a conditional argument: Let X_n and X_{n+1} be the respective number of customers found by arrivals n and $n + 1$, where $\{X_n\}$ is a Markov chain and $X_n \sim \{\pi_j\}$, the stationary distribution. Given X_n and T_n, X_{n+1} has a *binomial* distribution, where each of the $X_n + 1$ customers has probability $P(S > T_n) = e^{-\mu T_n}$ of being present when the next arrival occurs. Hence,

$$E(X_{n+1} \mid X_n, T_n) = (X_n + 1)e^{-\mu T_n},$$

where X_n and T_n are independent, and

$$E(X_{n+1}) = \Pi_1 = (\Pi_1 + 1)\tilde{A}_c(\mu), \quad \text{or}$$

$$\Pi_1 = \tilde{A}_c(\mu)/[1 - \tilde{A}_c(\mu)] \quad \text{and} \quad M_2 = m_c\tilde{A}_c(\mu)/[1 - \tilde{A}_c(\mu)]. \tag{9}$$

Higher factorial moments may also be obtained:†

† This is the form of the factorial moments of a binomial distribution. It may be obtained by differentiating the binomial generating function.

$$E\{X_{n+1}(X_{n+1}-1)\cdots(X_{n+1}-k+1)\mid X_n, T_n\}$$
$$=(X_n+1)(X_n)\cdots(X_n-k+2)e^{-k\mu T_n}$$
$$=[X_n(X_n-1)\cdots(X_n-k+1)+kX_n(X_n-1)\cdots(X_n-k+2)]e^{-k\mu T_n},$$

from which we get factorial moment ratios

$$M_{k+1}/M_k=\Pi_k/\Pi_{k-1}=k\tilde{A}_c(k\mu)/[1-\tilde{A}_c(k\mu)], \qquad k=1,2,\ldots, \tag{10}$$

where we define $\Pi_0=1$.

Let $\mathscr{V}_c$ be the *variance* of $\{p_j\}$, where

$$\mathscr{V}_c=M_2+m_c-m_c^2, \quad \text{and, from (9),}$$
$$z_c\equiv\mathscr{V}_c/m_c=M_2/m_c+1-m_c=[1-\tilde{A}_c(\mu)]^{-1}-m_c, \tag{11}$$

where z_c is called *peakedness*. If the arrivals at the secondary station were Poisson, we would have $z=1$. Thus peakedness is a measure of the deviation from this idealized case.

7-3 THE $GI/M/\ell$ LOSS SYSTEM

Consider the same model as in Section 7-2, except that the secondary station now has a finite number of channels, $\ell=1,2,\ldots$. Arrivals who find ℓ customers in system (ℓ busy channels) are lost, i.e., we have a $GI/M/\ell$ loss system. We again define a Markov chain at arrival epochs. Notice that for $i<\ell$, the chain for this loss system has the same transition probabilities as the chain in Section 7-2, because the conditional distribution of $(X_{n+1}\mid X_n=i<\ell)$ does not depend on having upper limit ℓ. Row ℓ of the matrix is now identical with row $\ell-1$. The matrix has this form:

$$\begin{pmatrix} a_{00} & a_{01} & 0 & & \cdots & 0 \\ a_{10} & a_{11} & a_{12} & 0 & \cdots & \vdots \\ & & & & & 0 \\ \vdots & & & & \cdots & a_{\ell-1,\ell} \\ a_{\ell-1,0} & a_{\ell-1,1} & & & \cdots & a_{\ell-1,\ell} \end{pmatrix},$$

where $a_{i,i+1}=\tilde{A}_c((i+1)\mu)$, $i=0,\ldots,\ell-1$. The other a_{ij} are easy to find, but we shall not need them explicitly.

Now $\{p_j\}$ and $\{\pi_j\}$ obviously have changed, but they still satisfy (7) on the range $j=1,\ldots,\ell$. In particular,

$$\ell p_\ell=m_c\pi_{\ell-1}. \tag{12}$$

To prevent confusion with the notation in Section 7-2, we should indicate the dependence of these distributions on ℓ, by $\{p_j(\ell)\}$ say. However, we omit this extra notation until equation (14) below.

From the transition probability matrix, we have

$$\pi_\ell = (\pi_\ell + \pi_{\ell-1})\tilde{A}_c(\ell\mu), \quad \text{or}$$

$$\pi_\ell = \pi_{\ell-1}\tilde{A}_c(\ell\mu)/[1 - \tilde{A}_c(\ell\mu)]. \tag{13}$$

From (12) and (13) we have an expression for the ratio

$$\varkappa(c, \ell) \equiv \pi_\ell(\ell)/p_\ell(\ell) = \ell\tilde{A}_c(\ell\mu)\,[1 - \tilde{A}_c(\ell\mu)]^{-1}/m_c. \tag{14}$$

Note the similarity between (14) and (10).

Clearly,

$$\pi_\ell(\ell) = m_{c+\ell}/m_c\,, \tag{15}$$

but there is no corresponding expression for $p_\ell(\ell)$, except for $\ell = 1$. Since $p_1(1)$ is the fraction of time channel $c + 1$ is busy, it is immediate that

$$p_1(1) = m_c - m_{c+1}. \tag{16}$$

Remark. Typically, we will find that time and arrival averages are very different. We now show that *only* in the Poisson case are they equal: Suppose they are equal. From (6), (15), and (16), it easily follows that

$$\pi_1(1) = p_1(1) \Leftrightarrow \tilde{A}_c(\mu) = \lambda_c/(\lambda_c + \mu). \tag{17}$$

The right side of (17) holds for all $\mu > 0$ if and only if A_c is exponential, i.e., the overflow process is Poisson! Except for $c = 0$, this is *never* true.

7-4 THE TWO-STAGE MODEL WITH POISSON ARRIVALS

For any inter-arrival distribution A at the primary station, (4) may be applied recursively to (in principle) determine $\tilde{A}_c(s)$ for any c, which in turn determines measures of performance at the secondary station. Numerically, computing an expression such as (14) can be done by building up a triangular array of transform values at multiples of μ.

We now assume Poisson arrivals at the primary station. Not only is this the case of greatest practical interest, it is also the easiest to work out.

Let q_i be the probability (proportion of time) that there are i customers at the primary station, where

$$\lambda q_{i-1} = i\mu q_i, \qquad i = 1, 2, \ldots, c, \tag{18}$$

i.e., $\{q_i\}$ is a truncated Poisson distribution. Because arrivals are Poisson, q_c is also the fraction of customers lost by (who overflow from) the primary station, which is the *blocking probability* B_c in (1).

Let $a = \lambda/\mu$, the offered load at the primary station. For fixed a, there is an easily derived recursion formula for B_c as a function of c:

$$B_{c+1} = aB_c/(aB_c + c + 1), \qquad c = 0, 1, \ldots, \tag{19}$$

where $B_0 = 1$. [The derivation uses the ratio of terms in (18). See Problem 5-11.] From (1) and (2), $\mathscr{m}_c = aB_c$, and from (19),

$$\mathscr{m}_{c+1} = a\mathscr{m}_c/(\mathscr{m}_c + c + 1), \qquad c = 0, 1, \ldots, \tag{20}$$

where $\mathscr{m}_0 = a$.

Substituting (6) into (20), we have

$$\tilde{A}_c(\mu) = a/(\mathscr{m}_c + c + 1). \tag{21}$$

If the secondary station has ∞ channels, the distribution $\{p_j\}$ has mean $\mathscr{m}_c$, and, substituting (21) into (11), peakedness

$$\mathscr{z}_c = 1 - \mathscr{m}_c + a/(\mathscr{m}_c + c + 1 - a), \tag{22}$$

and variance

$$\mathscr{V}_c = [1 - \mathscr{m}_c + a/(\mathscr{m}_c + c + 1 - a)]\mathscr{m}_c. \tag{23}$$

The quantities in (19) through (23) are functions of a and c. The notation emphasizes c because for any fixed a, these expressions may be numerically evaluated through recursive use of either (19) or (20). We now discuss some properties of these functions.

Obviously, $\mathscr{m}_c$ and B_c are strictly decreasing functions of c. From (20), it can be shown that they are *convex* functions. Thus, as c increases, the marginal reduction in the overflow rate from the primary station decreases.

The behavior of the peakedness is of particular interest. For either $a = 0$ or $c = 0$, $\mathscr{z}_c = 1$. Otherwise, for any $a > 0$ and $c \geq 1$, it can be shown that

$$\mathscr{z}_c > 1. \tag{24}$$

If the overflow process from the primary station were Poisson, the secondary station would be an $M/M/\infty$ queue, $\{p_j\}$ would be Poisson with mean and variance equal to $\mathscr{m}_c$, and peakedness equal to 1. As we have seen, overflow processes are *not* Poisson. Overflows from the primary station occur only when it is full, and the arrival process to the secondary station is Poisson during intervals for which this is true. Thus, an overflow process is an arrival process that is generated by turning a Poisson process "on" and "off" for intervals of random duration, and is an example of what is called an intermittent or *stuttering* Poisson process.

The effect of turning a Poisson process on and off is to make an overflow process *less regular* than a Poisson process. For example, the coefficient of variation of the inter-overflow distribution can be shown to exceed 1. The effect on the secondary station is to degrade performance; e.g., (24), and for loss systems, to increase $\pi_\ell(\ell)$.

For large a and c, peakedness can be quite large. For example, for $a = 500$ and $c = 510$, $\mathscr{z}_c = 11.3$. For arbitrary fixed a, however, it easily follows from (22) that $\mathscr{z}_c \to 1$ as $c \to \infty$.

In more complex situations, peakedness has been used to characterize the arrival process at a loss system for the purpose of approximating loss rates. A famous example, called the equivalent random method, is treated in Section 7-5.

To derive recursion relations for calculating expressions such as (14), we begin by rearranging (4) for the special case where s is a multiple of μ: For any integers $j \geq 1$ and $c \geq 0$.

$$\tilde{A}_c((j+1)\mu) = \tilde{A}_{c+1}(j\mu)[1 - \tilde{A}_c(j\mu)][1 - \tilde{A}_{c+1}(j\mu)]^{-1}. \tag{25}$$

So as not to be lost in algebra, we state our result as

Theorem 1. For any $j \geq 1$,

$$\tilde{A}_{c+1}(j\mu) = a[c + j + 1 + a - (c+1)\tilde{A}_c(j\mu)]^{-1}, \qquad c = 0, 1, \ldots, \tag{26}$$

where $\tilde{A}_0(j\mu) = a/(a+j)$, and, for any $c \geq 0$,

$$\tilde{A}_c((j+1)\mu) = a[1 - \tilde{A}_c(j\mu)][c + j + 1 - (c+1)\tilde{A}_c(j\mu)]^{-1}, \qquad j = 1, 2, \ldots, \tag{27}$$

where $\tilde{A}_c(\mu)$ is given by (21), and for any $c \geq 0$,

$$m_c\varkappa(c, j+1) = a(j+1)\,[c + j + 1 - a + m_c\varkappa(c, j)]^{-1}, \qquad j = 0, 1, \ldots,$$

where $\varkappa(c, 0) = 1$. (28)

Proof. From (25), it is straightforward algebra that for any fixed pair (c, j), (26) and (27) are equivalent. From (21),

$$\begin{aligned} m_c &= a/\tilde{A}_c(\mu) - c - 1, \quad \text{and} \\ \tilde{A}_{c+1}(\mu) &= a/(m_{c+1} + c + 2) \\ &= a[m_c\tilde{A}_c(\mu) + c + 1]^{-1} \\ &= a[c + 2 + a - (c+1)\tilde{A}_c(\mu)]^{-1}, \end{aligned}$$

which is (26) for $j = 1$ and any c. To complete an inductive proof, assume (26) is true for arbitrary fixed j and any c. From (27),

$$\tilde{A}_{c+1}((j+1)\mu) = a[1 - \tilde{A}_{c+1}(j\mu)][c + j + 2 - (c+2)\tilde{A}_{c+1}(j\mu)]^{-1}. \tag{29}$$

By substituting (26) for $\tilde{A}_{c+1}(j\mu)$ in (29), we have, after considerable rearrangement,

$$\tilde{A}_{c+1}((j+1)\mu) = a/(c + j + 2 + a - x), \quad \text{where} \tag{30}$$

$$\begin{aligned} x &= a(c+1)\,[1 - \tilde{A}_c(j\mu)][c + j + 1 - (c+1)\tilde{A}_c(j\mu)]^{-1} \\ &= (c+1)\tilde{A}_c((j+1)\mu). \end{aligned}$$

Thus, (30) is (26) for $j + 1$ and any c. From (14), (28) is just a rearrangement of

(27) for any $j \geq 1$, and for $j = 0$, (28) is equivalent to (21). This completes the proof.

Computing (14) for specified a, c, and ℓ can be done in several ways. For $c < \ell$, use (26), (20), and (14). For $c > \ell$, use (27) or (28) instead of (26). To illustrate, for $a = c = 500$, we compute $\mathscr{m}_c = 17.42$, and list $\varkappa(c, \ell)$ for several values of ℓ:

ℓ	1	3	5	10	18
$\varkappa(c, \ell)$	1.56	2.31	2.86	3.86	5.01

This list illustrates two properties that we state without proof: For $\ell \geq 1$, $\varkappa(c, \ell) > 1$, and for any fixed c, $\varkappa(c, \ell)$ is increasing in ℓ.

Now $\varkappa(c, \ell)$ is the ratio of an arrival average to a time average, which again illustrates the irregularity of overflow processes. We might expect $\ell = 18$ channels to serve "most" of an offered load of 17.42. In fact, in this example, 46 percent of the arrivals at the secondary channel overflow. If the same offered load were Poisson, only 15 percent would overflow. [Do not confuse this value with $p_{\mathscr{A}}(\ell) = .09$, the actual time average in this example.]

7-5 THE EQUIVALENT RANDOM METHOD AND HAYWARD'S APPROXIMATION

Suppose g independent primary stations of the type in Section 7-4 (same μ, Poisson arrivals, but λ and c may depend on the station) overflow into one secondary station of ℓ channels. (Although overflows at each of the primary stations are renewal processes, their superposition is not.) We wish to approximate the fraction of lost customers at the secondary station.

The *equivalent random method* (*EQRM*) does this by finding an *equivalent* single primary station in the sense that an infinite-channel secondary station has the same $\mathscr{m}$ and $\mathscr{V}$ (and $\mathscr{z}$) for this single station as it had for the original collection of g primary stations.

For an infinite-channel secondary station, the (stationary) number of busy channels is the sum of the corresponding quantities generated by each primary station, where these quantities are independent random variables.

Let $\mathscr{m}_i$ and $\mathscr{V}_i$ be the mean and variance at the secondary station generated by group i, $i = 1, 2, \ldots, g$, and $\mathscr{m}$ and $\mathscr{V}$ be the corresponding quantities generated by the collection of g primary stations. It follows that

$$\mathscr{m} = \sum_{i=1}^{g} \mathscr{m}_i \quad \text{and} \quad \mathscr{V} = \sum_{i=1}^{g} \mathscr{V}_i. \tag{31}$$

The equivalent primary station is the pair (c, a) such that $\mathscr{m}$ and $\mathscr{V}$ in (31) are equal to $\mathscr{m}_c$ in (20) and $\mathscr{V}_c$ in (23), respectively. Formally, this means that we must invert a function, $(\mathscr{m}, \mathscr{V}) = f(c, a)$ say.

Once (c, a) is determined, we use $\mathscr{m}_{c+\ell}/\mathscr{m}_c$ to *approximate* the blocking probability at the secondary station.

In practice, the inversion is performed from a set of graphs or tables. Provided $\mathscr{z} > 1$, it is known that there will be a unique inverse, but c may not be an integer. (For noninteger c, we can interpolate from integer values.)

Rapp [1964] proposed an easy approximation for (c, a): For a, let

$$\hat{a} = \mathscr{m}\mathscr{z} + 3\mathscr{z}(\mathscr{z} - 1), \tag{32}$$

and, solving (22) for c, we have

$$\hat{c} = \hat{a}(\mathscr{m} + \mathscr{z})/(\mathscr{m} + \mathscr{z} - 1) - \mathscr{m} - 1. \tag{33}$$

For $c \leq a$, the approximation is quite accurate, but tends to be high. In this case, we can round $\hat{c}$ down to the nearest integer, $[\hat{c}]$, and lower $\hat{a}$ so that (33) still holds. When $(c - a)/\sqrt{a} > 1$, however, the approximation is too low, and can be very poor, as we illustrate in Table 7-1.

For the worst case, $\hat{a} = 39.7$, and, rounding up slightly, $\hat{c} = 51$. Using these values for a and c, we compute $\mathscr{m} = 0.525$ and $\mathscr{z} = 3.57$. Remarkably, large errors in estimates of a and c have much less effect on blocking probability estimates at the secondary station, at least in this example.

Suppose a single primary station with $a = 100$ and $c = 130$ is followed by a secondary station with ℓ channels. The EQRM estimate (by assumption, exact) of the blocking probability at the secondary station is $\mathscr{m}_{c+\ell}/\mathscr{m}_c$. For the Rapp approximation, we compute the corresponding ratio based on the values of $\hat{a}$ and $\hat{c}$ found previously. See Table 7-2. As might have been anticipated, the relative difference increases as the blocking probability decreases. (Fortunately, blocking probabilities are ratios of offered loads, rather than the loads themselves.)

As a precaution, we should not use the Rapp approximation when $\mathscr{m} < \mathscr{z}$.

We now describe a simple alternative to the EQRM, known as *Hayward's approximation* (*HA*): Let the secondary station have ℓ channels, offered load $\mathscr{m}$,

TABLE 7-1

a	c	$\mathscr{m}$	$\mathscr{z}$	$\hat{a}$
10	5	5.64	1.46	10.25
	10	2.15	2.03	10.64
	13	0.84	2.22	9.99
	16	0.24	2.15	7.93
100	20	80.2	1.24	100.4
	50	50.9	1.87	100.1
	100	7.57	5.10	101.3
	110	2.75	5.53	90.4
	120	0.57	5.07	64.8
	130	0.058	4.16	39.7

TABLE 7-2

ℓ	EQRM blocking probability	Rapp blocking probability
1	.764	.756
3	.434	.411
5	.240	.209
7	.129	.099
9	.067	.044

peakedness $z > 1$, and blocking probability denoted by $B(\ell, m, z)$, which we approximate by

$$B(\ell, m, z) \approx B(\ell/z, m/z, 1), \tag{34}$$

where the right-hand term in (34) is the blocking probability for a station with ℓ and m reduced by factor z, and Poisson arrivals.

Let $i = \ell/z$. Provided i is an integer, the right-hand term in (34) is easily computed as B_i in (19), where $a = m/z \equiv m_0$. Otherwise, we must interpolate between integer values. Interpolation is more delicate here because we have much smaller numbers [by a factor of about 100 when $z = 10$, e.g., compare m_0 with (32)]. Our interpolation is based on the (approximate) geometric decay of m_c in (20): For any $i > 0$, we define

$$B_i = B_{[i]}(B_{[i+1]}/B_{[i]})^{(i-[i])}, \tag{35}$$

where $[i]$ is the integer part of i, and $B_0 = 1$.

We also present an elementary lower bound on blocking probabilities: For any loss system (or, more generally, any multichannel queue that also has losses), the arrival rate is the sum of the rates of those lost and of those served. A (strict in our models) upper bound on the rate served is the service rate times the number of channels. It follows immediately that a lower bound on the blocking probability is

$$B(\ell, m, z) > (m - \ell)/m \equiv \text{LB}. \tag{36}$$

This bound is worthless for $\ell > m$, but quite close in "heavy traffic"; see Section 7-8.

In Table 7-3, we compare HA with EQRM, and, where relevant, with the LB. In each case, (m, z) at the secondary station is computed for specified (c, a) at the primary station. The blocking probabilities for each method are listed as functions of ℓ, the number of channels at the secondary station. We have chosen (a, c) so that z is very close to an integer, and rounded to that integer. This choice facilitates comparing HA for (near) integer ℓ/z with interpolations using (35). The rounding has a negligible effect on HA.

These examples illustrate that HA blocking probability estimates are often

TABLE 7-3

Case	ℓ	EQRM	$\ell/2$	HA	LB
	1	.893	0.5	.887	.865
	2	.791	1	.787	.729
	3	.692	1.5	.683	.594
$a = 20$	7	.356	3.5	.344	.053
$c = 14$	8	.289	4	.280	—
$m = 7.388$	9	.230	4.5	.219	
$z = 1.987$	12	.101	6	.096	
≈ 2	13	.073	6.5	.068	
	14	.051	7	.048	
	15	.0346	7.5	.0323	
	21	.00186	11.5	.00188	
	22	.00103	12	.00109	

Case	ℓ	EQRM	$\ell/5$	HA	LB
	2	.9786	0.4	.9782	.9744
	5	.9466	1	.9465	.9434
	8	.9147	1.6	.9142	.9095
$a = 500$	40	.5841	8	.5834	.5476
$c = 416$	42	.5643	8.4	.5631	.5249
$m = 88.411$	80	.2262	16	.2263	.0951
$z = 4.9974$	82	.2114	16.4	.2113	.0725
≈ 5	100	.1000	20	.1016	—
	102	.0903	20.4	.0918	
	120	.0290	24	.0313	
	122	.0249	24.4	.0270	

Case	ℓ	EQRM	$\ell/10$	HA	LB
	5	.850	0.5	.837	.787
	10	.718	1	.701	.573
$a = 500$	15	.592	1.5	.562	.359
$c = 491$	20	.478	2	.451	.146
$m = 23.416$	25	.378	2.5	.343	—
$z = 10.019$	30	.290	3	.260	
≈ 10	35	.217	3.5	.186	
	40	.157	4	.132	
	45	.110	4.5	.088	
	50	.074	5	.058	

remarkably close to those given by EQRM, particularly when m/z is of moderate size, say 4 or larger. The HA estimates are usually lower, but may be higher, particularly when blocking probabilities, by either estimate, are very low. In these examples, interpolation using (35) has little effect on these comparisons. Interpolation may have more effect if both m/z and ℓ/z are small, and there is a substantial gap between these estimates and the lower bound.

Over the years, the EQRM has become revered; often forgotten is that it too is an approximation. It is not the standard of comparison for judging the accuracy of other methods. The EQRM was invented to treat the difficult problem of determining blocking probabilities when there are several (or many) primary stations. While the accuracy of this method was investigated years ago for systems of modest size, it is often applied in circumstances where its accuracy is unknown. Of course, it does provide "answers," but then, so does the more elementary HA. For further discussion, see Section 7-8.

The blocking probability estimated by the EQRM, HA, or any other method is an overall measure for the combined overflows from all the primary stations. We now turn our attention to the blocking experienced by individual overflow streams.

Let B be the overall blocking probability and let B_i be the blocking probability (fraction of overflows blocked) for the overflow process from station i, where that process has peakedness $\mathscr{z}_i = \mathscr{V}_i/\mathscr{m}_i$, $i = 1, \ldots, g$. Let B_T be the fraction of *time* that the secondary station is blocked. It is reasonable to expect B_i to increase with $\mathscr{z}_i$. As a first approximation, suppose B_i is linear, i.e.,

$$B_i \approx x\mathscr{z}_i + y,$$

where, to be consistent, we must have

$$\sum_{i=1}^{g} \mathscr{m}_i B_i = \mathscr{m}B, \quad \sum_{i=1}^{g} \mathscr{m}_i \mathscr{z}_i = \mathscr{m}\mathscr{z}, \quad \sum_{i=1}^{g} \mathscr{m}_i = \mathscr{m},$$

and from above,

$$B = x\mathscr{z} + y.$$

Furthermore, if the overflow process from primary station j, say, were Poisson, we would have (from PASTA) that $B_j = B_T$, and $\mathscr{z}_j = 1$, which imply that

$$B_T = x + y.$$

(For example, the secondary station could also serve as a primary station for some customers.) By combining these expressions, we can eliminate x and y,

$$B_i \approx B_T + (\mathscr{z}_i - 1)(B - B_T)/(\mathscr{z} - 1), \tag{37}$$

which we call the *Katz approximation* [1967]. If there are two primary stations, where the overflow process from one is Poisson, the consistency requirements we imposed will make (37) exact. (One can always pass a linear function through two points.) As we show in Section 7-7, however, it is possible for the true relation to be highly nonlinear.

To use (37), we need B_T. If we use the EQRM to find an equivalent primary station (c, a), $B \approx \pi_\ell(\ell) = \mathscr{m}_{c+\ell}/\mathscr{m}_c$, and

$$B_T \approx p_\ell(\ell) = \varkappa(c, \ell)\pi_\ell(\ell),$$

where $\varkappa(c, \ell)$ may be found by the recursion methods in Section 7-4.

7-6 THE *BM/G/∞* QUEUE

As we found in Section 7-5, the superposition of overflow processes can be a very irregular arrival process. A batch Poisson process, denoted in earlier chapters by *BM*, is also irregular, and could be used to model arrivals at a secondary station.

For that reason, we now consider an ∞-channel queue with *BM* arrivals. We will also explore the effect of the service distribution, denoted by G here, with service rate μ, $0 < \mu < \infty$, where in earlier sections of this chapter we assumed that G was exponential. Let the arrival of batches be Poisson at rate λ, where batch sizes are i.i.d. Let ν_i be the size of batch i, where $E(\nu_i) = E(\nu) = v$, and $V(\nu_i) < \infty$.

For any $t \geq 0$, let $\Lambda(t)$ be the number of batches and $C(t)$ be the number of customers that have arrived by epoch t, where

$$C(t) = \sum_{i=1}^{\Lambda(t)} \nu_i,$$

where, by conditioning on $\Lambda(t)$, it is easily shown that

$$\iota \equiv V\{C(t)\}/E\{C(t)\} = E(\nu^2)/v, \tag{38}$$

where $\iota \geq 1$, and $\iota = 1$ only if $P(\nu \in \{0, 1\}) = 1$, i.e., we have a standard Poisson process. The larger ι, the more *irregular* the arrival process.

We want to relate ι and G to the peakedness of N, the (stationary) number of customers in system. For an initially empty system, let R_i be the number of customers from batch i that remain in the system at t, and $N(t)$ be the number of customers in system at t, where

$$N(t) = \sum_{i=1}^{\Lambda(t)} R_i.$$

Given $\Lambda(t)$, the (unordered) batch arrival epochs are i.i.d., uniformly distributed on $(0, t)$. It follows that the R_i are i.i.d., and their distribution is independent of $\Lambda(t)$. As we found (38),

$$E\{N(t)\} = \lambda t E(R_i), \quad \text{and} \tag{39}$$

$$V\{N(t)\} = \lambda t E(R_i^2). \tag{40}$$

To find the moments of R_i, it is convenient to think of R_i as a sum of indicators,

$$R_i = \sum_{j=1}^{\nu_i} I_j, \tag{41}$$

where $I_j = 1$ if the jth customer in batch i is in the system at epoch t, $I_j = 0$

otherwise. Let τ_i be the arrival epoch of batch i. Conditioning,

$E(I_j \mid \tau_i = t - u) = G^c(u)$ and, for $j \neq k$, $E(I_j I_k \mid \tau_i = t - u) = [G^c(u)]^2$,

from which we have

$$E(I_j) = \int_0^t G^c(u)\, du/t, \quad \text{and, for } j \neq k, \quad E(I_j I_k) = \int_0^t [G^c(u)]^2\, du/t. \tag{42}$$

From (41) and (42),

$$E(R_i) = v \int_0^t G^c(u)\, du/t, \quad \text{and} \tag{43}$$

$$E(R_i^2) = v \int_0^t G^c(u)\, du/t + E\{\mathscr{v}(\mathscr{v} - 1)\} \int_0^t [G^c(u)]^2\, du/t. \tag{44}$$

Substituting these expressions into (39) and (40), we have

$$E\{N(t)\} = \lambda v \int_0^t G^c(u)\, du, \quad \text{and} \tag{45}$$

$$V\{N(t)\} = \lambda v \int_0^t G^c(u)\, du + \lambda E\{\mathscr{v}(\mathscr{v} - 1)\} \int_0^t [G^c(u)]^2\, du. \tag{46}$$

As $t \to \infty$, the moments of $N(t)$ converge to the moments of N, and we have

$$E(N) = \lambda v/\mu = \mathscr{m}, \tag{47}$$

which is of course the offered load, and [substituting (38)] peakedness

$$\mathscr{z} = V(N)/E(N) = 1 + (\imath - 1)r, \tag{48}$$

where

$$r = \mu \int_0^\infty [G^c(u)]^2\, du. \tag{49}$$

For exponential service, these equations become

$$r = 1/2 \quad \text{and} \quad \mathscr{z} = (\imath + 1)/2. \tag{50}$$

For Poisson arrivals, $\imath = 1$ and, as (48) shows, $\mathscr{z} = 1$, and the service distribution plays no role. (Also recall Erlang's loss formula for the $M/G/\ell$ loss system. These are examples of service distribution *insensitivity*.)

For $\imath > 1$, peakedness depends on G through r, which is a rather unconventional measure of service time regularity, where increasing r means greater regularity. In fact, r is maximized ($r = 1$) uniquely for *constant* service, where $[G^c]^2 = G^c$ for all u. Thus for constant service,

$$r = 1 \quad \text{and} \quad \mathscr{z} = \imath, \tag{51}$$

which is nearly double (50) when $\imath$ is large.

Remark. This notion of regularity is consistent with other definitions for common distributions, e.g., hyper-exponential, exponential, Erlang, and constant are listed in order of increasing regularity.

When $\iota > 1$, a remarkable feature of (48) is that increasing r *increases* peakedness; constant service is the worst case! This *contrary* behavior (contrary to conventional wisdom) has an intuitive explanation: Irregular service spreads the departure epochs of customers in the same batch, permitting observance of a "partial" batch. With constant service, we find either "all" or "none." Thus we get larger deviations from fixed $E(N)$ and larger variance when service is regular.

7-7 THE $BM/G/\ell$ LOSS SYSTEM

This is the same model as in Section 7-6 except that now we have a loss system with $\ell < \infty$ channels.

Consider the following question: When an arriving batch at a loss system finds fewer idle channels than the batch size, is any portion of the batch served? Clearly, this will depend on potential applications of the model. We introduced *BM* arrivals in Section 7-6 because this allowed us to create an elementary model of a loss system with arbitrarily large peakedness, where this model can be used to approximate more complex situations, such as the superposition of overflow processes. For this purpose, batches represent closely spaced arrivals, not arrivals that actually occur simultaneously, where an arrival is served if it finds at least one idle channel.

Assumption. When a batch of size j finds i idle servers, $i < j$, i customers are served, and $j - i$ customers are lost.

In this section, we will work out two elementary examples for the purpose of (1) illustrating contrary behavior, and (2) comparing the EQRM and other approximations with our exact calculations.

In the first example, we compare two service distributions: constant (D) and exponential (M). The exponential case can be worked out easily with balance equations.

Constant service is more difficult except for the following special case: Batches are of constant size v, and ℓ/v is an integer. With constant service, customers in the same batch depart together. Groups of v servers become busy and idle together, and the $BM/D/\ell$ loss system behaves exactly like a $M/D/(\ell/v)$ loss system, where a channel is now a group of v servers. From Erlang's loss formula, the second system is equivalent to an $M/M/(\ell/v)$ loss system.

In Table 7-4, we list the blocking probability for each service distribution as a function of the offered load $m = \lambda v/\mu$ for the case: constant batch size $v =$

TABLE 7-4

$\mathscr{m}$	*BM/D/4* (exact)	*BM/D/4* (EQRM)	*BM/M/4* (exact)	*BM/M/4* (EQRM)	*BM/M/4* (HA)
.5	.024	.084	.020	.030	.009
1	.077	.142	.066	.071	.042
2	.200	.249	.184	.176	.149
3	.310	.346	.295	.278	.261
4	.400	.423	.387	.374	.357
6	.529	.540	.521	.510	.500
8	.615	.622	.610	.600	.595

2, and $\ell = 4$ channels. These exact results are compared with HA and EQRM. Notice that for constant service, $\mathscr{z} = \mathscr{i} = v$, and HA is exact! For exponential service, we used (35) to interpolate HA. For both cases, we used the Rapp approximation for EQRM, checking to see that (c, a) determined by Rapp gave (close to) the correct $(\mathscr{m}, \mathscr{z})$. In a few cases, minor adjustments were made.

The exact calculations exhibit the contrary behavior mentioned at the end of Section 7-6. The effect is small here, but it can be much larger for large batch size. For constant service and low blocking probabilities, the EQRM does poorly. For exponential service, it performs better than HA. However, this example is small, and the assumption of constant batch size with integer ℓ/v may cause quirky behavior. Furthermore, the EQRM was not developed to handle either batch effects or constant service.

We chose constant batch size above in order to illustrate contrary behavior, but this is not a reasonable way to model overflows from an $M/M/c$ loss system. Overflows occur only during intervals where the system is full. It is easy to see that the number of overflows during such an interval has a *geometric* distribution, and, for large c, these intervals are apt to be short. If there are many similar primary stations, we can model the combined overflow process as a batch Poisson process with a geometric batch size distribution

$$P(v_j = i) = x(1 - x)^i, \qquad i = 0, 1, \ldots .$$

For batch arrival rate λ, we have

$$\mathscr{m} = \lambda v = \lambda(1 - x)/x \quad \text{and} \quad \mathscr{i} = (2 - x)/x. \tag{52}$$

For exponential service at the second station,

$$\mathscr{z} = 1/x. \tag{53}$$

To compare with the EQRM, we start with a single primary station (c, a), generate $(\mathscr{m}, \mathscr{z})$ and then use (52) and (53) to determine λ and x ($\mu = 1$). For $a = 100$ and $c = 119$, we found $\mathscr{m} = .6867$, $\mathscr{z} = 5.147$, $x = .1943$, and $\lambda = .1656$. We then analyzed an $\ell = 4$ channel secondary station. The blocking probabilities turned out to be very close: $B(BM) = .4655$ and $B(\text{EQRM}) = .4511$. However,

TABLE 7-5

k	B_k	z_k	Katz
1	.0185	1	.0185
2	.0330	1.5	.0724
3	.0545	2	.1263
4	.0918	2.5	.1802
5	.2734	3	.2341
6	.3945	3.5	.2880
7	.4810	4	.3418

the corresponding fraction of time each system was blocked was substantially different: $B_T(BM) = .0185$ and $B_T(\text{EQRM}) = .0256$.

We now investigate the Katz approximation. Partition the BM arrival process into separate arrival processes for each fixed batch size $\nu = k$, $k = 0, 1, \ldots$. (Batches of size zero experience no losses.) The true blocking probability for each fixed batch size, B_k, is easily computed for the BM model. To investigate nonlinearity, we used $B(BM)$ and $B_T(BM)$ to fit (37). The results are in Table 7-5.

This is not to suggest that constant batch size is an appropriate model for overflows from a single primary station. We are simply combining independent flows that have different peakedness. Now suppose that in this example, there is some oddball substream with batch size distribution

$$P(\nu = 1) = .94 \quad \text{and} \quad P(\nu = 7) = .06.$$

This substream has mean batch size 1.36, peakedness 1.93, and, from Table 7-5, blocking probability

$$B_0 = [(.0185)(.94) + 7(.481)(.06)]/1.36 = .161.$$

Compared with the $k = 3$ case, this substream has lower peakedness, but nearly triple the blocking probability.

While showing only one example, and a special one at that, we have seen that $B_i(z_i)$ can be highly nonlinear and may not even be monotone. Furthermore, models that have very close overall blocking probabilities can have quite different time-average behavior.

7-8 LARGE TWO-STAGE MODELS; HEAVY TRAFFIC

In this section, we briefly discuss some recent work on large systems of the type treated in Sections 7-1 through 7-7. There are three reasons to study large systems: (1) Real ones are getting larger; (2) some of the recursion formulas in this chapter, (28) in particular, are now known to be somewhat unstable computationally (Fredericks [1986], personal communication); and (3) determining asymptotic properties (limiting behavior as one or more parameters approach ∞) often leads to useful approximations for large systems and improves our understanding of smaller ones as well.

Whitt [1984a] considers a $G/GI/\infty$ queue, where the first G means that the arrival process $\{\Lambda(t)\}$ is a stationary point process (see Section 2-9). The GI means that the service times are i.i.d. and independent of the arrival process. This is a modification of the notation introduced in Chapter 5. Let the arrival rate be $\lambda > 0$, and the service distribution be G, with service rate $\mu > 0$ and offered load $m = \lambda/\mu$.

Renewal arrivals with inter-arrival distribution A and arrival rate $\lambda > 0$ will be a stationary point process if the time until the first arrival has distribution A_e, the equilibrium distribution. Thus the arrival process considered here is a generalization of renewal input.

Whitt assumes that the arrival process is asymptotically normal (as in a central limit theorem), i.e., for some $i > 0$,

$$[\Lambda(t) - \lambda t]/(\lambda i t)^{1/2} \xrightarrow{D} N(0, 1), \tag{54}$$

as $t \to \infty$, where D denotes convergence in distribution.

For renewal input, $i = E(T^2)/E^2(T)$, where $T \sim A$ [see equation (215) in Chapter 2]. Related to but not quite implied by (54) is the linear growth of the variance of $\Lambda(t)$ as $t \to \infty$, and the asymptotic independence of the increments of $\{\Lambda(t)\}$. For example, the second property means that when properly centered and normalized [as in (54)], $\Lambda(t/2)$ and $[\Lambda(t) - \Lambda(t/2)]$ become independent as $t \to \infty$. Now Poisson and batch Poisson are the only arrival processes that have both stationary and independent increments. For these processes, it is easily shown that (54) holds; in particular, it holds for the BM process $\{C(t)\}$ in Section 7-6, where i is given by (38).

Let $N(m)$ be the stationary number of customers in the $G/GI/\infty$ system as a function of the offered load. Whitt fixes G and lets $m \to \infty$ by increasing λ. In Whitt's paper, *heavy traffic* means $m \to \infty$. What is going on is actually quite sophisticated. As λ increases, there is actually a sequence of arrival processes being considered, and instead of (54), the time scale for them is changed. He then quotes a heavy traffic limit theorem of Borovkov [1967], where, under a properly stated (54),

$$(N(m) - m)/\sqrt{m z} \xrightarrow{D} N(0, 1), \tag{55}$$

as $m \to \infty$, where z, now an *asymptotic* peakedness, is given by

$$z = 1 + (i - 1)r, \tag{56}$$

where r is given by (49)!

Thus the contrary behavior illustrated in Sections 7-6 and 7-7 is by no means confined to BM arrivals, and would be expected to hold in sufficiently heavy traffic whenever $i > 1$. The difference is that the results in Section 7-6 are exact and do not require that m be large. This phenomenon has not been discussed in many places. Independent of Borovkov, Haji and Newell in 1971 (summarized in

Newell [1982], p. 193) derived (56) as an approximation for the $GI/G/\infty$ queue, i.e., this model specialized to renewal input. Contrary behavior for other models, including the loss system example in Section 7-7, are illustrated in Wolff [1977b]; see also Problem 5-39.

Let N have stationary distribution $\{p_j\}$ in Section 7-2. From (55), the distribution of N is approximately normal with mean m and variance mz when m is large. Let N_ℓ, with distribution $\{p_j(\ell)\}$ in Section 7-3, be the stationary number of customers in system for the $GI/M/\ell$ loss system. To approximate the distribution of N_ℓ, Whitt proposes the heuristic

$$N_\ell \approx \sim (N \mid N \leq \ell), \tag{57}$$

that is, we truncate and re-normalize $\{p_j\}$. By combining these approximations, Whitt shows that

$$P(N_\ell = i) \approx (1/\sqrt{mz})\varphi\left(\frac{i - m}{\sqrt{mz}}\right) \Big/ \Phi\left(\frac{i - m}{\sqrt{mz}}\right), \tag{58}$$

where φ and Φ are the standard normal density and distribution functions, respectively. To approximate B_T, set $i = \ell$ in (58),

$$B_T = P(N_\ell = \ell). \tag{59}$$

Remark. It is easily shown that heuristic (57) is exact for the $BM/M/\ell$ loss system with an arbitrary batch size distribution. For geometric batch sizes, this model is a special case of $GI/M/\ell$.

Now the departure rate of served customers is $\mu E(N_\ell) = \lambda E(N_\ell)/m$, and hence the blocking probability is

$$B = 1 - E(N_\ell)/m. \tag{60}$$

Whitt uses (55), (57), and standard properties of the normal distribution to approximate (60),

$$B \approx \sqrt{z/m}\, \varphi\left(\frac{\ell - m}{\sqrt{mz}}\right) \Big/ \Phi\left(\frac{\ell - m}{\sqrt{mz}}\right), \tag{61}$$

where, from (58),

$$B \approx zB_T, \tag{62}$$

where (62) has been found (as an approximation) by others.

Whitt suggests approximations for a number of other quantities as well, but he presents numerical comparisons only for different approximations for B [(61), primarily with the EQRM and HA]. In general, the approximations were found to be consistent with each other.

Remark. A word of caution here. We should expect (61) to perform better than (58) because (61) depends on integrating the normal density function, rather than evaluating the density at a particular point. [The density in (61) is a consequence of the integration.] The justification for (62) given here is no better than (58). As our limited computational experience in Section 7-7 suggests, estimating B_T is more difficult than estimating B. Furthermore, quite apart from questions of computational stability, recursion expressions such as (28) are only approximations when there is more than one primary station. Fortunately, B rather than B_T is a measure of primary importance. On the other hand, B_T appears in the Katz and other approximations that can at times be very poor [caution × caution = (caution)^2].

Newell [1984a] analyzes the classic two-stage and sequential models in heavy traffic under the assumptions of Poisson arrivals and exponential service, i.e., the model treated in Section 7-4. Here, *heavy traffic* means large offered load a, and a large number of channels c, at the *primary* station. As $a \to \infty$, he also permits $c \to \infty$, and obtains asymptotic properties of the overflow process that depend crucially on the relative magnitudes of a and c. If this is done so that $\mathcal{m} \to \infty$ at the secondary station, other properties of the overflow process, $\mathcal{z}$ in particular, are also changing.

The analysis is much too involved for detailed discussion here. The important point is that while N will be approximately normal when $(a - c)/\sqrt{c} \gg 1$, it may be very skewed otherwise. This has important implications for the accuracy of both the EQRM and approximations based on the normal distribution.

When there are many primary stations, the central limit theorem suggests that the normal distribution might produce good estimates. Whether or not this is true, if we are in a range where N is skewed, we would expect EQRM blocking probability estimates to be higher in the extreme tail and lower in the middle.

The remarkable closeness of the estimates in Whitt shed no light on this question because even when the secondary station is lightly loaded in Whitt's Table 4, (a, c) is in a range where we expect N to be normal. We tried $a = 500$ and $c = 510$, and then computed $\mathcal{m} = 11.71$ and $\mathcal{z} = 11.30$ as input to a secondary station with ℓ channels. We then computed blocking probability estimates at the secondary station using the normal approximation (61) and the EQRM. Our results appear in Table 7-6. While $\mathcal{m}$ is small for a central limit approximation, we do not believe that this accounts for the difference.

As an alternative to the EQRM, Newell suggests a batch Poisson model with a geometric batch size for the case $c - a \gg \sqrt{a}$, where both c and a are large. In this case, visits to the full state c are very rare, but once this occurs, many additional visits may also occur in a short period of time. To account for this possibility, Newell approximates the number in system process as the difference between two Poisson processes, where for $t \geq 0$,

$$N(t) = \Lambda(t) - \Omega(t), \tag{63}$$

where the first process is Poisson at rate λ (the original arrival process), and the

TABLE 7-6

ℓ	Normal	EQRM
12	.768	.549
18	.477	.382
24	.259	.254
30	.117	.160
36	.043	.095
42	.012	.054

second is Poisson at rate $c\mu$ (representing service completions). Initially the system is full, i.e., $N(0) = c$.

We are permitting $N(t) > c$ here. Arrivals when the (real) system is full increase $N(t)$ and are overflows. Because $c\mu > \lambda$, (63) will approach $-\infty$ as $t \to \infty$. The number of overflows until this occurs,

$$\mathscr{v} = \max \{N(t) - c: t \geq 0\}, \tag{64}$$

is a proper random variable, and $\mathscr{v}$ is the batch size Newell proposes to use.

To find the distribution of $\mathscr{v}$, let $y = P(\mathscr{v} \geq 1)$, and u be the probability that the first transition in $N(t)$ is "up," i.e., an overflow, where

$$u = \lambda/(\lambda + c\mu) = a/(a + c) \quad \text{and} \quad d \equiv 1 - u = c/(a + c).$$

Because overflows occur only when the system is full,

$$P(\mathscr{v} \geq i + 1 \mid \mathscr{v} \geq i) = y, \qquad i = 1, 2, \ldots,$$

i.e., $\mathscr{v}$ has a geometric distribution. To find y, condition on whether the first transition is up or down. If down, $\{N(t)\}$ must climb up two "steps" for $\{\mathscr{v} \geq 1\}$ to occur, and $P(\mathscr{v} \geq 1 \mid \text{down}) = y^2$. We have

$$y = 1u + y^2 d. \tag{65}$$

Factoring out the root $y = 1$, we have

$$y = u/d = a/c, \tag{66}$$

the only root of (65) in the interval (0, 1). It follows that

$$P(\mathscr{v} = i) = (a/c)^i(1 - a/c), \qquad i = 0, 1, \ldots, \tag{67}$$

and a BM/M/∞ queue with batch size distribution (67) has peakedness

$$\mathscr{z} = c/(c - a). \tag{68}$$

Comparing (68) with (22), we see that in the proposed range $(c - a) \gg \sqrt{a}$, $\mathscr{m}_c \ll a$, and (68) will be larger. For $c = 600$ and $a = 500$, $\mathscr{m}_c = .0007$, and we get $\mathscr{z} = 6$ and $\mathscr{z} = 5.95$, respectively. However, we saw in Table 7-6 that it is not necessary to go out to the extreme tail for the EQRM to be highly skewed. In that example, $m_c = 11.7$, and (68) would be very poor. Note that in this case, we are well outside the range of validity for this approximation.

Clearly, more research needs to be done on the questions raised here.

7-9 RETRIALS AND THE ORBIT MODEL; THE RTA APPROXIMATION

In Sections 7-1 through 7-8, we assumed that blocked customers at the (any, if there are more than one) primary station would be served somewhere else, if at all. In this section, we assume instead that they may return to the primary station some random time later to try again (*retrials*), and will be served there if, on return, the primary station is not full. For brevity, we will call the primary station *the system.*

A customer may return to the system many times. Between trials, a customer is said to be *in orbit*. Either initially or on return from orbit, a customer who finds the system full may decide not to reenter orbit. These customers leave the system forever and are lost. Of course, they may get service somewhere else (by a competitor?), but this is outside our model. Customers receive no service while in orbit. See Figure 7-3.

We assume that arrivals who find the system full will enter orbit with probability $\alpha_1 > 0$, and leave the system forever (are lost) with probability $1 - \alpha_1$. Retrials who find the system full will reenter orbit with probability α_2, and leave the system forever with probability $1 - \alpha_2$, independent of the number of prior trials each customer has made. (More generally, we could let the probability that a customer enters orbit be some arbitrary function of the number of prior trials that customer has made.) We place no limit on the number of customers who may be in orbit at the same time.

Some of our results will hold only for two special cases of the above: When $\alpha_1 = \alpha_2 < 1$, we have *geometric* orbits, i.e., the number of retrials that a customer is willing to make has a geometric distribution. When $\alpha_2 = 1$, we have *infinite* orbits, i.e., once in orbit, a customer will try again without limit.

We assume that arrivals to the system (primary station) are Poisson at rate

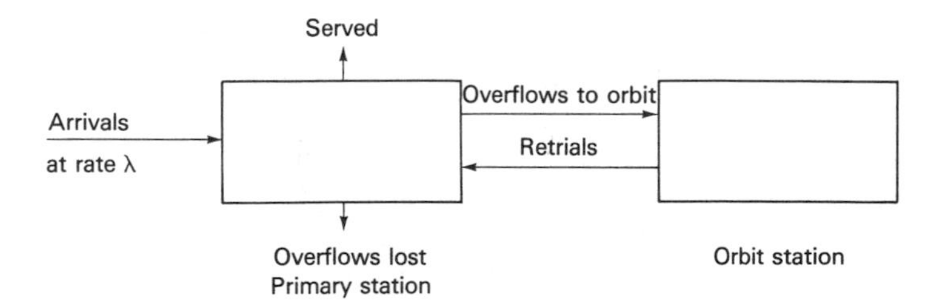

Figure 7-3

$\lambda > 0$, and that the system has a capacity of $k < \infty$ customers. Arrivals and retrials who find k customers in system (the system is full or at capacity) overflow and either are lost or go into orbit. We assume exponential service with (combined) service rate $\mu_i > 0$ when there are i customers in system. For example, if there are $c < k$ channels, we can let $\mu_i = i\mu$ for $i = 0, \ldots, c$, and $\mu_i = c\mu$ for $i = c + 1, \ldots, k$. Thus, our model is a (slight) generalization of a multichannel queue with queue limit; for $c = k$, we have a loss system, the basic model in Sections 7-1 through 7-8. Finally, we assume that customers in orbit behave independently of each other, and that the time a customer spends in orbit between tries is exponential at rate $\gamma > 0$. Thus when there are j customers in orbit, the retrial rate is $j\gamma$.

Our model is a continuous-time Markov chain with state space (i, j), $i = 0, \ldots, k$, and $j = 0, 1, \ldots$, where i is the number of customers in system, and j is the number of customers in orbit. It is easily shown (Greenberg [1986]) that when $\alpha_2 < 1$, the system is stable for all values of the other parameters, and when $\alpha_2 = 1$ and μ_k is the maximum service rate, the system is stable if and only if $\lambda\alpha_1 < \mu_k$. (We mean stability of the "enlarged" system that includes the orbit station, so that the chain is positive recurrent.)

Because our performance measures are averages, we will consider only stable systems. These measures may be found in terms of the steady-state probabilities $\{p_{ij}\}$ for the chain. Unfortunately, except for a few special cases, solving for these probabilities is intractable, and closed-form expressions for elementary measures such as fraction of customers lost do not exist.

We will approximate the distribution of the number of customers in system, where these probabilities are fractions of time,

$$q_i \equiv \sum_{j=0}^{\infty} p_{ij}, \qquad i = 0, \ldots, k.$$

First, let λ_R be the (at present, unknown) retrial rate of customers returning from orbit, and r_i be the fraction of retrials who find i customers in system, $i = 0, \ldots, k$. (Each return of a customer from orbit counts as a retrial.) Now equate transition rates between adjacent states,

$$\lambda q_i + \lambda_R r_i = \mu_{i+1} q_{i+1}, \qquad i = 0, \ldots, k - 1, \tag{69}$$

and the rates that customers enter and leave orbit,

$$\lambda q_k \alpha_1 + \lambda_R r_k \alpha_2 = \lambda_R. \tag{70}$$

To motivate the approximation, observe that customers enter orbit only when the system is full. If the return rate γ is large (relative to λ and μ), the system will have little time to change before customers return, and we expect that $r_k > q_k$. On the other hand, if γ is small, the system will have time to approach steady state when customers return, and we expect that for each i, $r_i \approx q_i$. Equality is the PASTA property, and is the basis for our

Approximation. Retrials see time averages (*RTA*), i.e.,

$$r_i = q_i \qquad \text{for } i = 0, \ldots, k. \tag{71}$$

From (69), (71), and the normalizing condition $\sum_i q_i = 1$, we can easily solve for (approximate) the q_i in terms of λ_R,

$$q_i \approx b_i \Big/ \sum_{i=0}^{k} b_i \qquad \text{for } i = 0, \ldots, k, \tag{72}$$

where $b_0 = 1$, and

$$b_i = (\lambda + \lambda_R)^i \Big/ \prod_{j=1}^{i} \mu_j \qquad \text{for } i = 1, \ldots, k. \tag{73}$$

From (70) and (71), we also have

$$q_k \approx \lambda_R/(\lambda\alpha_1 + \lambda_R\alpha_2). \tag{74}$$

Equating the right-hand side of (72) for $i = k$ with the right-hand side of (74), we have a polynomial in λ_R of order $k + 1$. When μ_j is nondecreasing in j, it can be shown (Greenberg) that the polynomial has a unique positive root. In this case, the root is easy to find numerically because as λ_R increases, (74) crosses (72) from below.

Once λ_R has been determined, various first-moment performance measures may be approximated. For example, we approximate the rate that customers are served in either of two equivalent ways,

$$(\lambda + \hat{\lambda}_R)(1 - \hat{q}_k) = \sum_{i=1}^{k} \mu_i \hat{q}_i, \tag{75}$$

and the average number of trials per customer (served or not) as

$$1 + \hat{\lambda}_R/\lambda, \tag{76}$$

where ^ denotes approximate quantities.

In the absence of retrials ($\alpha_1 = 0$), $\{q_i\}$ is insensitive to the service distribution G (Erlang's loss formula). Consequently, if the primary station is a c-channel loss system ($k = c$) with *general* service distribution G, we propose to use the *RTA approximation,* specialized to $\mu_i = i\mu$, $i = 0, \ldots, c$.

Now suppose that in the absence of retrials, the primary station is an $M/G/c$ queue with queue limit where $k > c$. In Section 11-8, we show how to incorporate retrials into approximations for this model, such as that of Nozaki and Ross [1978].

Notice that RTA does not depend on the retrial rate per customer in orbit, γ. The quality of the approximation obviously does; this is investigated in Section 7-12. We show in the next section that, at least for the special cases of geometric and infinite orbits, (75) is an upper bound on the true rate that customers are served.

Remark. Contrast this approximation with the behavior of overflow processes for the two-stage and sequential models in earlier sections. To use the approximation "overflows see time averages" there would lead to gross errors. See for example the last paragraph of Section 7-4.

7-10 MONOTONICITY AND AN UPPER BOUND

In this section, we show that for two special cases of the orbit model with exponential service, (75) is an upper bound on the rate that customers are served.

We will need a preliminary result for

$$L_i \equiv \sum_{j=0}^{\infty} jp_{ij}/q_i, \qquad i = 0, \ldots, k,$$

i.e., L_i is the expected number of customers in orbit when there are i customers in system. It is immediate that L, the average number of customers in orbit, can be written

$$L = \sum_{i=0}^{k} L_i q_i, \quad \text{and} \quad \lambda_R = \gamma L.$$

Similarly, by interpreting r_i as a ratio of rates, we have $r_i = L_i q_i / L$, and we can rewrite (69) and (70) as

$$(\lambda + \gamma L_i) q_i = \mu_{i+1} q_{i+1}, \qquad i = 0, \ldots, k-1, \text{ and} \tag{77}$$

$$(\lambda \alpha_1 + \gamma L_k \alpha_2) q_k = \gamma L. \tag{78}$$

Clearly, we now see that approximation (71) is equivalent to setting

$$L_i = L \qquad \text{for all } i.$$

We state our preliminary result as

Theorem 2. L_i is a strictly increasing function of i.

Sketch of Proof. Let $L_i^{(2)} = \sum_{j=0}^{\infty} j^2 p_{ij}/q_i$, the second moment of a conditional distribution, where clearly, $L_i^{(2)} > L_i^2$, $i = 0, \ldots, k$. Now write the balance equations for states $(0, j)$,

$$(\lambda + j\gamma) p_{0j} = \mu_1 p_{1j}, \qquad j = 0, 1, \ldots,$$

multiply by j and sum, obtaining

$$(\lambda L_0 + \gamma L_0^{(2)}) q_0 = \mu_1 L_1 q_1.$$

From (78) and the preceding second moment inequality, we have $L_0 < L_1$. The proof is completed by writing out the balance equations about states (i, j) for $j = 0, 1, \ldots,$ and $i = 1, \ldots, k-1$, and showing by induction that

$$(\lambda L_i + \gamma L_i^{(2)})q_i < \mu_{i+1} L_{i+1} q_{i+1}, \qquad i = 1, \ldots, k-1,$$

from which the result follows.

Remark. Notice that the balance equations about the "system full states" were not used to prove this result. The particular way we model the behavior of customers who find the system full (on arrival or as retrials) is irrelevant.

We are now ready for

Theorem 3: Geometric Orbits. When $\alpha_1 = \alpha_2 = \alpha < 1$, and the μ_i are nondecreasing in i,

$$\sum_{i=1}^{k} \mu_i q_i < \sum_{i=1}^{k} \mu_i \hat{q}_i,$$

i.e., the RTA rate that customers are served is strictly more than the true rate.

Proof. From (77) and (78), the left- and right-hand sums in the theorem are

$$(\lambda + \gamma L) - \gamma L/\alpha \quad \text{and} \quad (\lambda + \gamma \hat{L}) - \gamma \hat{L}/\alpha, \quad \text{respectively,}$$

and the theorem is equivalent to $L > \hat{L}$. Assume the contrary, i.e., that

$$L \leq \hat{L}, \quad \text{or, equivalently, that} \quad \sum_{i=1}^{k} \mu_i q_i \geq \sum_{i-1}^{k} \mu_i \hat{q}_i. \tag{79}$$

From Theorem 2, $L_k > L$, and, from (79),

$$q_k = \gamma L/(\lambda + \gamma L_k) < \gamma L/(\lambda + \gamma L) \leq \gamma \hat{L}/(\lambda + \gamma \hat{L}) = \hat{q}_k. \tag{80}$$

If μ_i is nondecreasing in i, the following is easily shown from (80) and the right-hand inequality in (79): There is some j in the range $1, \ldots, k-1$, such that

$$q_j > \hat{q}_j \quad \text{and} \quad q_{j+1} \leq \hat{q}_{j+1}, \tag{81}$$

and there is some i in the range $0, \ldots, j-1$, such that

$$q_i \leq \hat{q}_i \quad \text{and} \quad q_{i+1} > \hat{q}_{i+1}. \tag{82}$$

From (77) and (81)

$$q_{j+1}/q_j = (\lambda + \gamma L_j)/\mu_{j+1} < \hat{q}_{j+1}/\hat{q}_j = (\lambda + \gamma \hat{L})/\mu_{j+1},$$

and hence $L_j < \hat{L}$; similarly, $L_i > \hat{L}$, in contradiction with Theorem 2. This completes the proof.

Theorem 4: Infinite Orbits. When $\alpha_2 = 1$, and the μ_i are nondecreasing in i,

$$\sum_{i=1}^{k} \mu_i q_i \leq \sum_{i=1}^{k} \mu_i \hat{q}_i,$$

i.e., the RTA rate that customers are served is more than the true rate.

We omit a proof, which is very similar to the preceding one.

Remark. There is a curious difference in these results in that in Theorem 4, the inequality is not strict. In fact, equality holds when $k = 1$.

7-11 THE FREDERICKS AND REISNER APPROXIMATION FOR THE ORBIT MODEL

Unlike the RTA approximation, Fredericks and Reisner [1979] develop an approximation for this model that explicitly depends on γ. They consider only the geometric orbits case, but it is easy to modify their method for arbitrary α_1 and α_2.

In the notation of Sections 7-9 and 7-10, they define

$$\lambda_i' = \lambda_R r_i / q_i = \gamma L_i \quad \text{and} \quad \lambda_i = \lambda + \lambda_i', \qquad i = 0, \ldots, k, \tag{83}$$

and write (77) as

$$\lambda_i q_i = \mu_{i+1} q_{i+1}, \qquad i = 0, \ldots, k-1, \tag{84}$$

where (84) resembles the recursion formula for the state probabilities of a standard birth and death model. The λ_i here are not Markovian rates, however, but have meaning only as long-run averages.

Now let $N(t)$ be the number of customers in system at epoch $t \geq 0$, and for all states m and n, let

$$Q_{mn}(t) = P\{N(t) = n \mid N(0) = m\}, \tag{85}$$

where these transient probabilities satisfy the boundary conditions

$$Q_{mn}(0) = \delta_{mn},$$

where $\delta_{mn} = 1$ if $m = n$, $\delta_{mn} = 0$ otherwise.

Suppose that at epoch 0, either an arrival or a retrial finds the system full (state k), and enters orbit, an event that occurs at rate λ_R. This customer returns after a time that is distributed exp (γ) and is independent of the transient behavior of the system. It follows that the fraction of retrials who find i is

$$r_i = \int_0^\infty Q_{ki}(t) \cdot \gamma e^{-\gamma t}\, dt \equiv \gamma \tilde{Q}_{ki}(\gamma), \qquad i = 0, \ldots, k, \tag{86}$$

where $\tilde{Q}$ denotes the Laplace transform of Q. (Elsewhere in this book, $\tilde{Q}$ would denote the Laplace-Stieltjes transform of Q.) Now multiply (86) by $\lambda_R = (\lambda\alpha_1 + \lambda_k'\alpha_2)q_k$, and we have

$$\lambda_i' q_i = (\lambda\alpha_1 + \lambda_k'\alpha_2)q_k\gamma\tilde{Q}_{ki}(\gamma), \qquad i = 0, \ldots, k. \tag{87}$$

For $i = k$, we can "solve" for $\tilde{Q}_{kk}(\gamma)$,

$$\tilde{Q}_{kk}(\gamma) = (\lambda_k - \lambda)/[\lambda\alpha_1 + (\lambda_k - \lambda)\alpha_2]\gamma. \tag{88}$$

Technically, (85) through (88) ignore dependence on the unknown initial distribution of the number of customers in orbit, which, because of our initial conditions, is not a time average. Otherwise, we haven't made any assumptions. Fredericks and Reisner now make the following

Assumption. The system (primary station) is a "one-dimensional" birth and death process with transition rates given in (84).

Now the λ_i are unknown, but a transient analysis of the birth and death process determines them. We denote by *FRA* both this assumption and approximation that follows.

Using formulas for the $\tilde{Q}_{ki}(\gamma)$ developed in Section 5.3 of Riordan [1962] (or the Appendix in Fredericks and Reisner) for a one-dimensional birth and death process, we obtain a recursion for $\delta_i \equiv (\lambda_i - \lambda)/(\lambda_{i-1} - \lambda)$:

$$\begin{aligned} \delta_1 &= 1 + \gamma/\lambda_0, \\ \delta_{i+1} &= [(\gamma + \lambda_i + \mu_i)\delta_i - \mu_i]/\lambda_i\delta_i \qquad \text{for } i = 1, \ldots, k-1, \text{ and} \end{aligned} \tag{89}$$

$$\tilde{Q}_{kk}(\gamma) = \delta_k/[(\gamma + \mu_k)\delta_k - \mu_k]. \tag{90}$$

For any $\lambda_0 > \lambda$, we can iteratively evaluate (89), (90), and (88). As a function of λ_0, (88) crosses (90) from below, and equating these equations determines the λ_i.

7-12 QUALITY OF THE ORBIT MODEL APPROXIMATIONS

For exponential service, $\{q_i\}$ can be calculated fairly easily only for $k = 1$ in the geometric orbits case (Cohen [1957a]), and only up to $k = 2$ in the infinite orbits case (Keilson et al. [1968]). For general service, the only traceable case is infinite orbits with $k = 1$ (Keilson et al. [1968], Aleksandrov [1974], and Lubacz and Roberts [1984]). The first two papers assume $\alpha_1 = 1$ as well. The approach of Lubacz and Roberts makes clear why this case is "easy." (See Section 7-13.)

We compare exact server utilization (calculated or estimated by simulation) with the upper bound RTA and the FRA approximation. (This measure is equivalent to the fraction of customers ultimately served.) More extensive comparisons appear in the cited references. We also present a lower bound (LB) that is achieved when $\alpha_1 = 0$, or in the limit as $\gamma \to \infty$. Tables 7-7 through 7-10 are for conventional single or multichannel loss system or queue limit models determined by specifying c and k. In each case, the service rate per channel is $\mu = 1$, and we let $\alpha_1 = \alpha$.

In these examples, FRA does very well, and the results suggest that for exponential service it may also give an upper bound. (Should this not be the case, calculating simple examples more accurately than we have done may turn up a counterexample.) It tends to make larger errors for large α and in the intermediate range of γ. RTA does well when γ is small, as expected. Both methods, particularly RTA, do better for infinite orbits than for geometric. For infinite orbits, customers are lost only when (external) arrivals find the system full, and they do see time averages. Orbiting customers tend to find the system in states $k - 1$ and k, compared to time averages, and hence tend to keep the system full, increasing losses from new arrivals. For geometric orbits, orbiting customers are lost as well.

It is of interest to improve the lower bound. This can be done by truncating the state space by putting an upper limit on the number of customers in orbit. For some computational results using this approach, see Greenberg.

Comparison of Tables 7-7 and 7-8 show that constant service performs better than exponential, although the effect here is small. Simple examples have shown that hyper-exponential service performs worse. This is what we normally expect, and is certainly not the contrary behavior illustrated in Sections 7-6 and 7-7.

Now that we have become sensitized to this possibility, we might wonder why contrary behavior did not occur. The departure rate from the orbit station varies randomly with the number of customers in orbit. This similarity to a Poisson process with randomly varying rate (as in the superposition of overflow processes in the two-stage model) is misleading, however. The primary and orbit stations are closely linked; it is not correct to view the arrival process of customers from orbit as independent of the primary station.

TABLE 7-7

Case	λ	LB	α	RTA (UB)	$\gamma = 2$ True	$\gamma = 2$ FRA	$\gamma = 1$ True	$\gamma = 1$ FRA	$\gamma = .5$ True	$\gamma = .5$ FRA	$\gamma = .1$ True	$\gamma = .1$ FRA
Geometric orbits Exponential service $c = k = 1$ Exact results	.5	.333	.3	.359	.346	.346	.350	.350	.354	.354	.358	.358
			.7	.413	.381	.383	.392	.395	.401	.403	.410	.411
			.9	.461	.429	.436	.442	.447	.450	.453	.458	.459
	1	.500	.3	.545	.524	.525	.531	.532	.537	.537	.543	.543
			.7	.646	.595	.603	.615	.621	.629	.633	.642	.643
			.9	.758	.699	.723	.725	.741	.741	.750	.756	.758
	2	.667	.3	.718	.701	.701	.708	.708	.712	.713	.717	.717
			.7	.826	.796	.800	.808	.813	.817	.819	.824	.824
			.9	.921	.902	.912	.913	.917	.917	.919	.921	.921

TABLE 7-8

Case	λ	LB	α	RTA	γ = 2 True	γ = 1 True	γ = .5 True	γ = .1 True
Geometric orbits, Constant service, $c = k = 1$, Simulation results	.5	.333	.3	.3591	.349	.353	.356	.359
			.7	.4129	.390	.400	.406	.411
	1	.500	.3	.5445	.529	.536	.539	.543
			.7	.6461	.611	.627	.636	.644
	2	.667	.3	.7183	.706	.712	.716	.718
			.7	.8258	.806	.816	.821	.825

Remark. The simulation runs, which used the regenerative method discussed in Section 2-20, have estimated standard deviation less than .0005.

TABLE 7-9

Case	λ	LB	α	RTA (UB)	γ = 2 True	γ = 2 FRA	γ = 1 True	γ = 1 FRA	γ = .5 True	γ = .5 FRA	γ = .1 True	γ = .1 FRA
Infinite orbits, Exponential service, $c = k = 2$, Exact results	1	.400	.3	.423	.422	.422	.422	.422	.422	.423	.423	.423
			.7	.462	.460	.460	.461	.461	.461	.461	.462	.462
	2	.600	.3	.665	.663	.663	.664	.664	.664	.665	.665	.665
			.7	.801	.798	.800	.799	.801	.800	.801	.801	.801
	4	.792	.3	.887	.885	.886	.886	.887	.886	.887	.887	.887

Although not proposed for this purpose, FRA does well for constant service, if we use in Table 7-8 the FRA values from Table 7-7. In several cases, the "true" (simulated) value falls between RTA and FRA.

Because of the small size and limited number of models in this section, we must temper our observations. Numerical results for FRA for larger systems and a different measure of performance may be found in Fredericks and Reisner.

TABLE 7-10

Case	λ	LB	α	RTA (UB)	γ = 2 True	γ = 2 FRA	γ = 1 True	γ = 1 FRA	γ = .5 True	γ = .5 FRA	γ = .1 True	γ = .1 FRA
Infinite orbits, Exponential service, $c = 1$; $k = 2$, Exact results	.5	.429	.3	.446	.444	.444	.445	.445	.445	.445	.446	.446
			.7	.474	.471	.472	.472	.473	.473	.473	.474	.474
	1	.667	.3	.724	.716	.717	.718	.720	.721	.721	.723	.723
			.7	.838	.823	.829	.827	.833	.831	.835	.836	.837
	2	.857	.3	.948	.938	.943	.942	.945	.945	.947	.947	.948

7-13 THE $M/G/1$ LOSS SYSTEM WITH INFINITE ORBITS

We consider this model in more detail because (apparently) it is the only example of the models discussed in Section 7-9 that has closed-form solutions for all the usual first-moment performance measures.

Since $\alpha_2 = 1$ here, let $\alpha_1 = \alpha$, where $0 < \alpha \leq 1$. Let S_j be the service time of the jth customer served, $j = 1, 2, \ldots$, where the S_j are i.i.d., with mean $E(S) = 1/\mu < \infty$, and define $\rho = \lambda/\mu$. It can be shown that this model is stable if and only if

$$\alpha\rho < 1. \tag{91}$$

Assume this condition holds.

We use the notation in Sections 7-9 and 7-10 for time-average quantities $\{q_i\}$ and $\{L_i\}$, where $i = 0, 1$ here. Because $\alpha_2 = 1$ and the system is stable, any customer who enters orbit will be served. Let λ_s be the arrival rate of those (eventually) served, where (PASTA)

$$\lambda_s = \lambda q_0 + \lambda\alpha q_1 \quad \text{and} \quad (L = \lambda w), \tag{92}$$

$$q_1 = \lambda_s/\mu. \tag{93}$$

From (92), (93), and normalization, we have

$$q_1 = \rho/(1 + \rho - \alpha\rho) \quad \text{and} \quad q_0 = (1 - \alpha\rho)/(1 + \rho - \alpha\rho). \tag{94}$$

Thus server utilization and equivalent performance measures are known exactly for this model.

By equating rates into and out of orbit, we can now find L_0:

$$q_0\gamma L_0 = \lambda\alpha q_1, \text{ or}$$

$$L_0 = \lambda\alpha\rho/\gamma(1 - \alpha\rho). \tag{95}$$

We now find L_1 by a simplification of a clever method by Lubacz and Roberts: Assume *all* customers who get served enter orbit first, and customers in orbit are served FIFO. This in no way affects transition rates or the number of customers in orbit. All it does is identify *which* customer is served whenever a service begins.

Under this FIFO rule, let w_1 be the expected waiting time in orbit *while the server is busy*; excluded from w_1 is the waiting time in orbit that occurs while the server is idle. Now from $L = \lambda w$ we have

$$q_1 L_1 = \lambda_s w_1, \tag{96}$$

where λ_s is given by (92).

To get another equation relating L_1 and w_1, we consider the orbit station as found by an arrival who gets served. In terms of N_a, the number of customers in orbit found by such an arrival, the actual waiting time may be written

$$W_1 = \sum_{i=1}^{N_a} S_i + S_r, \text{ with mean} \tag{97}$$

$$w_1 = E(N_a)/\mu + E(S_r), \tag{98}$$

where the S_i are the service times of the customers found in orbit, and S_r is the remaining service time, if any, of the customer found in service. Now $S_r = 0$ if the system is found empty, and, for Poisson arrivals, $(S_r \mid S_r > 0) \sim S_e$, a random variable having the *equilibrium* service distribution.

The substream of served customers is not Poisson, and the expected values in (98) are time-average quantities appropriately weighted for this substream:

$$E(N_a) = q_0L_0/(q_0 + \alpha q_1) + \alpha q_1L_1/(q_0 + \alpha q_1), \text{ and} \tag{99}$$

$$E(S_r) = \alpha q_1 E(S_e)/(q_0 + \alpha q_1), \text{ where} \tag{100}$$

$$E(S_e) = \mu E(S^2)/2.$$

Combining these expressions, we can now find L_1 and $L = q_0L_0 + q_1L_1$,

$$L_1 = L_0 + \lambda\alpha E(S_e)/(1 - \alpha\rho), \text{ and} \tag{101}$$

$$L = L_0 + \lambda\alpha q_1 E(S_e)/(1 - \alpha\rho), \tag{102}$$

where L_0 and q_1 are found above. Thus L depends on the service time variance, but the q_i do not. Notice that $L_1 > L_0$, consistent with Theorem 2. Other first-moment measures such as the average waiting time in orbit w, of those who really enter orbit, are easily found,

$$w = L/\lambda\alpha q_1.$$

Remark. Because $\alpha_2 = 1$ here, customers in orbit are never lost, and this led to an elementary determination of $\{q_i\}$. This works for $c = k = 1$, but not otherwise; there are too many unknowns. This also helps to explain why the corresponding geometric orbits model is much more difficult to solve.

7-14 OTHER MODELS AND THE LITERATURE

The models in this chapter are queueing networks. They differ from the models in Chapter 6 in that here, the paths customers take through the network depend on the congestion they find. This difference makes the models in this chapter much harder to solve, even under the idealized assumption of Poisson arrivals and exponential service. We no longer have what were called product-form solutions to balance equations. An attempt to find approximate solutions with this property (or, what is in some cases equivalent, the PASTA property) would lead to gross errors for the models in Sections 7-4 and 7-5. On the other hand, we have seen that for retrials, this works well in some cases.

Nevertheless, the literature related to the models here goes back much further than that for Chapter 6. The reason is simple: With the invention of the telephone and the development of rudimentary communication networks early in this century, trunk lines were necessary, and equal access to them impractical. Servers (lines) were arranged in ordered groups (*graded* systems), and customers were served by the lowest group that had a free server. This idealized setup is the model treated in Sections 7-1 through 7-4. Of course, the order may be different for different subsets of customers. The model in Section 7-5 is one way to treat this case.

The early literature is summarized by Syski [1960]. Also see Riordan. We list some notable early contributions: The two-stage model with Poisson arrivals and an ∞-channel second stage was solved by Kosten [1937], with balance equations and generating functions; a formidable achievement! He obtained variance formula (23) and complex formulas for higher moments in terms of what are called *Charlier* polynomials. For renewal input, the factorial moment results in Sections 7-2 and 7-3 were obtained by Cohen [1957b], and at about the same time by Takács (see his book [1962]). When specialized to Poisson arrivals, their results can be used to obtain Kosten's formulas. Wilkinson is given credit for the equivalent random method, and published [1956] extensive charts for finding the inverse (c, a).

In spite of the enormous amount of work done in the "early days," this is an area of active research today; e.g., see the proceedings of any recent teletraffic conference. Many of these papers are concerned with estimating blocking probabilities for each of several primary groups, e.g., see the Katz approximation in Sections 7-5 and 7-7.

This is the main objective of Fredericks [1983], which we mention for another reason. He uses recursion formula (11) in that paper, which is equivalent to our (28), to derive an elementary upper bound on B_T, the fraction of time the secondary station is blocked. Derivations of (28) based on balance equations go back at least to Brockmeyer [1954].

Hayward's approximation has been around for a long time, but the first extensive study of it appears to be Fredericks [1980]. (Fredericks credits W. S. Hayward (c. 1959).) Fredericks also modifies the approximation and considers offered loads that have peakedness $z < 1$.

Retrials are even more difficult to handle. Cohen [1957(a)] considers the geometric orbits model treated in Sections 7-9 through 7-12 for loss systems with exponential service. He uses balance equations, but is unable to do much with them, except in limiting cases as parameters approach 0 or ∞. The most interesting case is for $\gamma \to 0$, where he obtains what amounts to the RTA approximation. Riordan asserts but does not prove the same limiting behavior for the corresponding infinite orbits model with $\alpha_1 = \alpha_2 = 1$.

While the retrial literature is much smaller, a survey of this literature by

Jonin [1984] lists 49 references, including three mentioned in Section 7-12. However, the important 1979 paper by Fredericks and Reisner is not among them.

The literature on multistage overflow processes is huge and scattered. There does not appear to be any self-contained, reasonably complete and up-to-date treatment of this topic. Sections 7-1 through 7-8 are intended to fill this need. In addition, we have constructed examples that are intended to raise questions about the accuracy of even the most esteemed approximation methods. The purpose is not to stop their use; design engineers have to use something. Instead, we hope to influence the course of future research in this area.

The retrial literature has much less to offer in terms of results. Our focus here is on new ways of looking at these models and both the rationale for and properties of approximation methods. Most of Sections 7-10 and 7-12 are from Greenberg; also see Greenberg and Wolff [1987] for more detailed proofs. An alternative to the state space truncation method mentioned in Section 7-12 is to put an upper limit on the combined return rate of all customers in orbit. When this is done, the matrix-geometric method of Neuts [1981] can be used to obtain numerical solutions.

One important feature of the orbit models is that queueing is permitted. For example, a reservation system may have more lines than operators. This can occur in sequential arrangements as well; unfortunately, the corresponding analysis would be difficult, particularly if we can simultaneously have queueing at a primary station and an idle server at the secondary station.

PROBLEMS

7-1. For a two-stage model with Poisson arrivals and $c = \ell = 1$, write the balance equations for the steady-state distribution $\{p_{ij}\}$, where p_{ij} is the fraction of time there are i customers at station 1 and j customers at station 2. Let f_k be the fraction of time station k is full, $k = 1, 2$. Show that

$$f_1 f_2 < p_{11}.$$

Interpret this result.

7-2. When an overflow occurs at the primary station in Figure 7-1, define a stochastic process at the primary station that regenerates. Does the corresponding process at the secondary station regenerate at these epochs as well? Why or why not?

7-3. For a two-stage model with Poisson arrivals, suppose the primary station operates as a c-channel queue with queue limit, where $k > c$ is the maximum number of customers permitted at the primary station. The secondary station operates as an ℓ-channel loss system.

(a) Suppose that whenever a departure occurs from a full secondary station, a customer in queue at the primary station (if any) will move immediately to the idle secondary server. Similarly, an arrival finding all c channels busy at the primary station will be served by an idle secondary server (if any), rather than queue.

Explain how you would calculate such quantities as the overall fraction of customers lost.

(b) Why is this problem more complicated if we do not make the simplifying assumptions in (a)?

7-4. Suppose station 1 is a c-channel loss system and station 2 is an ℓ-channel loss system, where each station has its own Poisson stream of primary customers, at rate λ_i for station i, $i = 1, 2$. Arrivals at their own full primary station overflow to the other station and are lost if it also is full. What fraction of customers is lost from each stream?

7-5. By a simplified version of the argument used to obtain (4), find (5).

7-6. Starting with the generating function of a binomial distribution, find the expression referred to in the footnote on page 348.

7-7. Find equation (16).

7-8. Show that the two equations in (17) are equivalent.

7-9. For the two-stage model with Poisson arrivals, show that m_c is a convex function of c.

7-10. For the $BM/M/\ell$ loss system, write the balance equations for the steady-state distribution of N_ℓ. Observe that ratios p_{n+1}/p_n, $n = 0, \ldots, \ell - 1$, are determined by the balance equations about states $0, \ldots, \ell - 1$. This implies that truncation approximation (57) is exact for this model.

7-11. Argue that when $\alpha_2 < 1$, the model in Section 7-9 is stable for all values of the other parameters.

8

THE $M/G/1$ AND $GI/M/c$ QUEUES

Chapters 5 and 6 were primarily concerned with queues that can be modeled as continuous-time Markov chains. General distributions could only be represented approximately in terms of those that are phase-type. In this chapter, we treat two kinds of queues: single channel with Poisson arrivals and general independent service times ($M/G/1$), and multichannel with general independent inter-arrival times (renewal arrivals) and independent exponential service times ($GI/M/c$).

As usual, we let $\{N(t): t \geq 0\}$ be the number of customers in system process. For each model, we will find that at certain epochs *embedded* in $[0, \infty)$, this process is a discrete-time Markov chain, called an *embedded chain*. The analysis of each chain and its relation to the original process is an important aspect of this chapter. We also used embedded chains to advantage in Sections 7-2 and 7-3, but this chapter may be read independently of that one.

We will also define and analyze the *busy period* and variations such as *exceptional first service* for the $M/G/1$ queue that are important in Chapter 10 for treating priority queues.

8-1 ANALYSIS OF AN EMBEDDED MARKOV CHAIN FOR THE $M/G/1$ QUEUE

Consider a single channel queue with Poisson arrivals at rate λ and general independent service with common distribution G, the $M/G/1$ queue. Let S denote a service time random variable with $E(S) = 1/\mu$, $\rho = \lambda/\mu$, and $E(e^{-sS}) = \tilde{G}(s)$. We obtained first moment results for this model in Section 5-13.

Phase-type G was treated earlier, where we defined an appropriate state space with a *countable* number of states such that the corresponding continuous-time process was a Markov chain. In particular, the time spent in each state between transitions was exponential. Balance equations were used to find steady-state probabilities.

For arbitrary G, we can define a continuous-time Markov process with state space $\{(n, x)\}$ say, where at any epoch t, n is the number of customers in system and x is either the attained or remaining service time of the customer in service. However, this state space is not countable because x varies over a continuum. Analyses of Markov processes of this kind require methods beyond our present scope and usually are more difficult.

In Chapter 4, a discrete-time Markov chain was defined on the epochs where transitions of a continuous-time chain occur. The general idea now is to select epochs (points) *embedded* in continuous time such that a sequence of random variables defined at these epochs is a Markov chain.

For the $M/G/1$ queue, we have to get rid of attained service time x. For $n = 1, 2, \ldots$, let

$$X_n = \text{number of customers in system left behind by departure } C_n, \text{ and}$$

$$Y_n = \text{number of arrivals that occur during the service time } S_n \text{ of } C_n.$$

Now X_{n+1} clearly depends on X_n, S_{n+1}, and the Poisson arrival process during S_{n+1}. It is easy to see that $\{X_n\}$ is a Markov chain.

We now represent X_{n+1} in terms of X_n. If $X_n > 0$, the next departure will decrease the number in system by 1. During S_{n+1}, the number in system will increase by Y_{n+1}. If $X_n = 0$, we must first have an arrival before the next service begins. Hence

$$\begin{aligned} X_{n+1} &= X_n - 1 + Y_{n+1} && \text{for } X_n \geq 1, \text{ and} \\ &= Y_{n+1} && \text{for } X_n = 0. \end{aligned}$$

It will be convenient to combine both cases into one equation,

$$X_{n+1} = X_n - 1 + \delta_n + Y_{n+1}, \tag{1}$$

where $\delta_n = 1$ if $X_n = 0$, $\delta_n = 0$ otherwise, and Y_{n+1} is *independent* of the other terms on the right-hand side. Note that

$$\delta_n^2 = \delta_n, \quad E(\delta_n) = P(\delta_n = 1), \quad \text{and} \quad X_n\delta_n = 0.$$

Now $\{Y_n\}$ is a sequence of i.i.d. random variables, where Y_n is the number of Poisson arrivals during S_n. Dropping subscripts, let Y be the number of arrivals during $S \sim G$. For later use, we obtain the mean, variance, and generating function $H(z) \equiv E\{z^Y\}$ of Y. We condition on S, where $(Y \mid S) \sim P(\lambda S)$, and therefore has the mean, variance, and generating function of a Poisson distribution with parameter λS. From $E(Y \mid S) = \lambda S$,

$$E(Y) = \lambda/\mu = \rho; \text{ from} \tag{2}$$

$$V(Y) = E\{V(Y \mid S)\} + V\{E(Y \mid S)\} = E(\lambda S) + V(\lambda S),$$

$$V(Y) = \rho + \lambda^2 V(S); \text{ and from} \tag{3}$$

$$E\{z^Y \mid S\} = e^{\lambda S(z-1)},$$

$$H(z) = E\{e^{\lambda S(z-1)}\} = \tilde{G}[\lambda(1 - z)]. \tag{4}$$

Let $P(Y = j) = a_j > 0, j = 0, 1, \ldots$, and denote the transition probability matrix of the chain by $P = (p_{ij})$, where $p_{ij} = P(X_{n+1} = j \mid X_n = i)$, for $i \geq 0$ and $j \geq 0$. From (1), it is easy to see that

$$P = \begin{pmatrix} a_0 & a_1 & a_2 & \cdots \\ a_0 & a_1 & a_2 & \cdots \\ 0 & a_0 & a_1 & \cdots \\ 0 & 0 & a_0 & \cdots \\ \vdots & \vdots & \vdots & \ddots \end{pmatrix}, \tag{5}$$

and that the chain is irreducible and aperiodic.

Without reference to periodicity, we know from Theorem 5 in Chapter 3 that all states in this chain are *positive* if and only if there exists a unique probability vector π that satisfies

$$\pi = \pi P, \tag{6}$$

where for each j, $\pi_j > 0$, and $1/\pi_j$ is the mean recurrence time of state j. We also have the time-average interpretation (for the discrete-time process): π_j is the *fraction of departing customers who leave j customers behind.*

A probability vector (distribution) π that satisfies (6) is called *stationary* because if X_0 has distribution π, $\{X_n\}$ is a stationary sequence of random variables, each $\sim \pi$. Equivalently, we will work with (1), where stationarity simply means that X_n and X_{n+1} have the same distribution, which implies that they have the same moments and generating function.

First, take the expected value of (1) and, under stationarity, cancel $E(X_{n+1}) = E(X_n)$:

$$E(X_{n+1}) = E(X_n) - 1 + E(\delta_n) + E(Y_{n+1}), \text{ and}$$

$$E(\delta_n) = 1 - E(Y_{n+1}) = 1 - \rho = P(X_n = 0) = \pi_0. \tag{7}$$

Similarly, take the variance of both sides of (1) and cancel $V(X_{n+1}) = V(X_n)$.

$$\begin{aligned} V(X_{n+1}) &= V(X_n + \delta_n) + V(Y_{n+1}) \\ &= V(X_n) + V(\delta_n) + 2 \operatorname{Cov}(X_n, \delta_n) + \rho + \lambda^2 V(S), \text{ where} \end{aligned}$$

$\operatorname{Cov}(X_n, \delta_n) = -E(X_n)E(\delta_n)$, $V(\delta_n) = \rho(1 - \rho)$, and we can solve for $E(X_n)$,

$$E(X_n) = \rho + [\rho^2 + \lambda^2 V(S)]/2(1 - \rho) = \rho + \lambda^2 E(S^2)/2(1 - \rho). \tag{8}$$

The generating function, $\Pi(z) \equiv \sum_{j=0}^{\infty} \pi_j z^j = E(z^{X_n}) = E(z^{X_{n+1}})$ is obtained in analogous fashion from (1),

$$E(z^{X_{n+1}}) = E(z^{(X_n - 1 + \delta_n + Y_{n+1})}) = H(z)E(z^{(X_n - 1 + \delta_n)}),$$

$$\Pi(z) = H(z)[E(z^{(X_n - 1)} \mid X_n > 0)(1 - \pi_0) + E(z^0 \mid X_n = 0)\pi_0].$$

From this expression and $\Pi(z) = E(z^{X_n} \mid X_n > 0)(1 - \pi_0) + \pi_0$, we find $\Pi(z)$,

$$\Pi(z) = \pi_0(1 - z)H(z)/[H(z) - z]. \tag{9}$$

We already know that $\pi_0 = 1 - \rho$, but this can be obtained independently from (9) by taking the limit of that expression as $z \to 1$, applying L'Hospital's rule, and normalizing: $\Pi(z) = 1$. In either case, and from (4), we obtain

$$\Pi(z) = \frac{(1 - \rho)(1 - z)\tilde{G}[\lambda(1 - z)]}{\tilde{G}[\lambda(1 - z)] - z} \qquad \text{for } \rho < 1. \tag{10}$$

Before we proceed, a few comments about our methods are in order. The validity of taking moments of both sides of (1) and canceling [e.g., of $E(X_n)$ to get (7) and $V(X_n)$ to get (8)] depends on the canceled terms being finite. Fortunately, the derivation of (10) does not suffer from this theoretical defect, provided that $\pi_0 = 1 - \rho$ is obtained by taking the limit rather than using (7). In fact, for $\rho < 1$, we have found the generating function of a probability distribution. For $\rho \geq 1$, we get $\pi_0 \leq 0$, which, whether true or nonsense ($\pi_0 < 0$), is not the positive case. Hence, all states are positive if and only if $\rho < 1$.

For the record, it turns out that $\rho = 1$ is the null case and $\rho > 1$ is the transient case. We discuss these cases in connection with the busy period in Section 8-4, and prove these results for the more general $GI/G/1$ queue in Chapter 9. In the context of Markov chains, Theorem 11 in Chapter 3 may also be used to establish these results.

We may obtain (8) and higher moments of (stationary) X_n by differentiating (10). For $\rho < 1$ and any positive integer k, this will also establish that

$$E(X_n^k) < \infty \quad \text{if and only if} \quad E(S^{k+1}) < \infty. \tag{11}$$

Hence, our original derivation of (8) implicitly required that $E(S^3) < \infty$, though the result is true even if $E(S^2) = \infty$.

Also observe that we made no mention of order of service. Each time a service begins, we need only assume that the service time is an independent draw from the service distribution. This is true for FIFO, LIFO, and random selection of the next customer served. Clearly, the order of service doesn't matter *from the server's point of view*. Similarly, the stochastic properties of the departure process (e.g., the probability of j departures in $[0, t]$) does not depend on order of service. (Priority rules such as "shortest job first" are clear exceptions to the statements made here because they induce dependence between successive service times.) Under the assumption that successive service times (in the order served) are i.i.d., the number of customers in system process $\{N(t): t \geq 0\}$, L, Q, w, and d are independent of the order of service.

From the *customer's* point of view, order of service does matter, i.e., the waiting time *distribution* is affected. We return to this point in Section 8-5.

In the stable (positive) case $\rho < 1$, the embedded Markov chain is a regenerative process with finite mean cycle length, where visits to any specific state may be chosen as regeneration points. It will be convenient to choose visits to state 0, i.e., departure epochs that leave the system empty. For the chain, departure epochs are integers, where epoch n simply means the nth departure. It is important to realize that many other processes will also be regenerative. For example, the sequence of waiting times $\{W_n\}$ is regenerative and has the same regeneration points. The continuous-time number-in-system process $\{N(t)\}$ is also regenerative, with regeneration points that are the *real time* epochs where the chain leaves the system empty. It is easy to show (e.g., from Wald's equation) that $\{N(t)\}$ has finite mean cycle length. Similarly, the work-in-system process $\{V(t)\}$ is regenerative with the same regeneration points as $\{N(t)\}$.

It follows that the time average (continuous or discrete) of each of these processes is also the mean of a corresponding limiting distribution, and that each of these processes has a "stationary version," meaning, among other things, that properties of each distribution may be found from a stationary analysis, as we did with $\pi = \pi P$ above. The practical significance of all this is that we routinely will equate a time average such as w with the mean of a corresponding stationary random variable, W.

8-2 FIFO WAITING TIMES FOR THE $M/G/1$ QUEUE

Assume $\rho < 1$. Let customer C_n wait W_n in system and leave X_n customers behind. Under FIFO, the customers left behind by a departure are precisely those who arrived while the departing customer was in the system. Thus X_n is the number of Poisson arrivals during W_n, an interval of random duration. The mathematical relationship of X_n to W_n is the same as that of Y to S in Section 8-1; see equations (2) through (4).

For example, from $E(X_n \mid W_n) = \lambda W_n$,

$$E(X_n) = \lambda E(W_n). \tag{12}$$

Now the distribution of W_n determines that of X_n, and conversely. If W_n is stationary, so is X_n, and in this case, we have from (8) and (12) that

$$E(X_n) = \lambda w, \tag{13}$$

$$w = 1/\mu + \lambda E(S^2)/[2(1 - \rho)], \text{ and} \tag{14}$$

$$d = w - 1/\mu = \lambda E(S^2)/[2(1 - \rho)]. \tag{15}$$

From inspection of (14) and (15), we see that w and d depend on the service distribution only through the first two moments, and, for fixed λ and μ, they are increasing functions of $V(S)$. Note the sensitivity of these results to ρ as $\rho \to 1$. It is easy to obtain corresponding results found in Chapter 5; e.g., for $M/E_k/1$, set $E(S^2) = (k + 1)/k\mu^2$.

From $L = \lambda w$, we have $E(X_n) = L$. At first this seems like a curious coincidence, because $L = \sum jp_j$ is a *time* average, where p_j is the fraction of time there are j customers in system, but $E(X_n) = \sum j\pi_j$ is an average over points in the embedded chain. In Section 8-3, we show that for this queue and this chain a deeper result holds: $\{\pi_j\} = \{p_j\}$.

To obtain the corresponding transform, let W_n have stationary distribution $W(t)$ and transform $E(e^{(-sW_n)}) = \tilde{W}(s)$. Conditioning in the same manner, we have

$$\Pi(z) = \tilde{W}[\lambda(1 - z)].$$

Now set $\lambda(1 - z) = s$, and (10) becomes

$$\tilde{W}(s) = \frac{(1 - \rho)s\tilde{G}(s)}{s - \lambda + \lambda\tilde{G}(s)}. \tag{16}$$

This result is called the *Pollaczek-Khintchine Formula*. [Some authors give this name to (14) instead.] To obtain the corresponding transform for the stationary delay in queue $\tilde{D}(s)$, observe that because the delay and service time of the same customer are independent,

$$\tilde{W}(s) = \tilde{D}(s) \cdot \tilde{G}(s), \text{ and}$$

$$\tilde{D}(s) = \frac{(1 - \rho)s}{s - \lambda + \lambda\tilde{G}(s)} = \frac{1 - \rho}{1 - \rho[\mu(1 - \tilde{G}(s))/s]} = \frac{1 - \rho}{1 - \rho\tilde{G}_e(s)}, \tag{17}$$

where $\tilde{G}_e(s)$ is the transform of the *equilibrium distribution*! The right-hand expression for $\tilde{D}(s)$ can be expanded as a geometric series,

$$\tilde{D}(s) = \sum_{j=0}^{\infty} (1 - \rho)[\rho\tilde{G}_e(s)]^j. \tag{18}$$

The representation of $\tilde{D}(s)$ in (18) was discovered before any physical interpretation of this form was known. As discussed in Chapter 6, we now know that under certain rules of operation, the stationary distribution of N, the number of customers in system in an $M/G/1$ queue, is insensitive to the service distribution. Two such rules are *processor sharing,* where the server is shared equally by all customers in system, and *preemptive LIFO,* where each customer begins service on arrival, preempting anyone in service, and, at all times, the most recent arrival who is still in system is being served.

Under either rule, N has distribution

$$P(N_{\mathrm{PL}} = j) = (1 - \rho)\rho^j, \qquad j = 0, 1, \ldots, \tag{19}$$

where for definiteness, PL denotes preemptive LIFO, so that we will not confuse this result with FIFO and standard LIFO, where the distribution of N clearly depends on G. Furthermore, stationary work in system has the representation

$$V = \sum_{i=1}^{N_{\mathrm{PL}}} S_{ri}, \tag{20}$$

where S_{ri} is the remaining service time of the ith customer in system, and given $N_{\mathrm{PL}} = j, j = 0, 1, \ldots,$ the S_{ri} are i.i.d. with $S_{ri} \sim G_e, i = 1, \ldots, j$.

From equations (18) through (20), it is easy to see that D and V have the same distribution. On the other hand, since D is the work found by an arrival and V is the work in system at a "random" time, we already know from PASTA that $D \sim V$. Thus under a rule such as PL, the right-hand side of (18) does have a physical interpretation in terms of the work found by an arrival.

As a check on what we have done, for the $M/M/1$ queue, $\tilde{G}(s) = \mu/(\mu + s)$, and (16) becomes

$$\tilde{W}(s) = (\mu - \lambda)/(\mu - \lambda + s),$$

a result obtained in Section 5-6.

8-3 TIME AND TRANSITION AVERAGES

We have made the observation several times in previous chapters that the transition rates into and out of each state in a discrete-state stochastic process must be equal. This is what we are doing in (6). More generally, we can partition the state space of a stochastic process (discrete state or not) arbitrarily into two disjoint subsets, A and A^c, say. The transition rate of $A \to A^c$ must equal the rate of $A^c \to A$.

For our $M/G/1$ queue, let $A = \{0, \ldots, j\}$ for arbitrary $j \geq 0$. Let λ^a be the transition rate for $A \to A^c$, where these transitions occur every time an arrival finds j customers in system. Let λ^d be the transition rate for $A^c \to A$, where these transitions occur every time a departure leaves j customers in system. We have $\lambda^a = \lambda^d$. Let $\pi_j^a = \lambda^a/\lambda$ be the *fraction of arrivals* who find j customers in system. If the system is stable ($\rho < 1$ here), the departure rate of served customers is also λ, and $\pi_j = \lambda^d/\lambda$, the *fraction of departures* who leave j customers in system, making $\pi_j^a = \pi_j$. Since j is arbitrary we have

$$\{\pi_j^a\} = \{\pi_j\}, \tag{21}$$

i.e., the *distributions* are the same.

For $j = 0, 1, \ldots,$ now let p_j be the fraction of *time* there are j customers in system. From PASTA, we have $\{\pi_j^a\} = \{p_j\}$, and from (21),

$$\{\pi_j\} = \{p_j\}. \tag{22}$$

While (22) depends on having Poisson arrivals, (21) is a very general result. It holds for batch arrivals provided we number arrivals sequentially, including customers in the same batch. (Think of customers within their own batch forming a line, where the kth customer in this line finds $k - 1$ customers from this batch in system, in addition to customers from prior batches.) It holds for batch departures, and also for the total number in system in an arbitrary queueing network, or for that matter a "black box" that customers enter and leave. Finally, no particular stochastic assumptions are needed provided the rates are well-defined limits and the overall arrival and departure rates are the same.

8.4 THE $M/G/1$ BUSY PERIOD

A typical service facility will have periods where one or more customers are present (it is busy) and periods with no customers present (it is empty or idle). A *busy period* is an interval that begins when an arrival finds the facility empty and ends when, for the first time after that, a departure leaves the system empty. *Idle periods* are the intervals between successive busy periods. Clearly, busy and idle periods alternate.

Not only are busy periods of direct interest, e.g., in relation to machine breakdowns or operator rest periods, but we will also find that they occur in the analysis of other quantities.

We need some notation. Let

B = duration of a busy period,

K = number of customers served during B, and

I = duration of an idle period.

Suppose a single-channel queue starts empty, and let $S_1, \ldots, S_K$ be the service times during the first busy period, B. Clearly,

$$B = \sum_{n=1}^{K} S_n. \tag{23}$$

For the $M/G/1$ (and even the $GI/G/1$) queue, the S_n are i.i.d., and K is a stopping time for this sequence, provided that K is a proper random variable. Applying Wald's equation, we have

$$E(B) = E(K)E(S). \tag{24}$$

However, busy periods need not end, i.e., K and B may be defective random variables. (They will be either defective or proper together.) For $M/G/1$ queues, we showed that all states are positive if and only if $\rho < 1$. Now K is the time (number of transitions) between visits to state 0, and when $\rho < 1$,

$$E(K) = 1/\pi_0 = 1/(1 - \rho). \tag{25}$$

The case $\rho = 1$ is mathematically interesting; B and K are proper, but $E(B) = E(K) = \infty$. See Chapter 9. Technically, Wald's equation requires $E(K) < \infty$. However, when summing nonnegative random variables, it may be extended to cover the infinite expectation case as well.

For the remainder of this section, assume $\rho < 1$. The sequence of busy and idle periods may be viewed as an alternating renewal process, and since the fraction of time busy is $\rho = \lambda E(S)$, it follows that

$$\rho = E(B)/[E(B) + E(I)], \tag{26}$$

where, because arrivals are Poisson, $E(I) = 1/\lambda$. Thus, from either (24) and (25), or (26), we have

$$E(B) = (1/\mu)/(1 - \rho) = 1/(\mu - \lambda). \tag{27}$$

Going beyond first moments requires a more detailed analysis. Let

$$S = \text{first service time during } B, \text{ and}$$

$$Y = \text{number of arrivals during } S.$$

Given $Y = 0$, $B = S$. Given $Y = 1$, $B - S = B_1$, say, where B_1 is an interval that begins with one customer in system who is just entering service. Because arrivals are Poisson, the distribution of B_1 does not depend on when this customer arrived, and it is easy to see that B_1 has the *same distribution* as an (unconditional) B.

More generally, suppose a busy period begins at epoch 0, and $Y = y \geq 1$. At epoch S^+ (just after the first departure), the number in system is y. Let $U_i \geq 0$, $i \in \{0, \ldots, y - 1\}$, be the departure epoch of the first departure who leaves i customers behind. Also let

$$B_y = U_{y-1} - S, B_i = U_{i-1} - U_i, i = 1, \ldots, y - 1, \text{ and write} \tag{28}$$

$$B = U_0 = S + B_y + \cdots + B_1. \tag{29}$$

Now B_1 begins with $(i - 1)$ customers in queue. For each fixed i, suppose we decide not to service any of them while any other customers are present. (Put them in our "back pocket" and forget about them.) For this order of service, which may be quite different from FIFO, B_i clearly has the same distribution as B. Furthermore, because arrivals are Poisson and service times are i.i.d., the B_i are i.i.d. However, as discussed in Section 8-1, *order of service doesn't matter*, i.e., altering the order of service as done here has no effect on the distribution of the B_i. This is merely a device to assist us in determining their properties.

We have shown that B can be written

$$B = S + \sum_{i=1}^{Y} B_i, \tag{30}$$

where, given $S = s$ and $Y = y \geq 1$, $B_1, \ldots, B_y$ are i.i.d. $\sim B$.

In principle, (30) is sufficient to determine the distribution of B. We will obtain information about B by conditioning on S and Y, using the conditional distribution $(Y \mid S) \sim P(\lambda S)$. To illustrate, we find (27),

$$E(B \mid S, Y) = S + YE(B),$$
$$E(B \mid S) = S + \lambda SE(B), \text{ and}$$
$$E(B) = 1/\mu + \rho E(B) = 1/(\mu - \lambda).$$

An expression for the transform $\tilde{B}(s) \equiv E(e^{-sB})$ is also easy to obtain,

$$E\{e^{-sB} \mid S, Y) = e^{-sS}[\tilde{B}(s)]^Y,$$
$$E\{e^{-sB} \mid S\} = e^{-sS}e^{\lambda S[\tilde{B}(s)-1]} = e^{-S[s+\lambda-\lambda\tilde{B}(s)]}, \text{ and}$$
$$\tilde{B}(s) = \tilde{G}[s + \lambda - \lambda\tilde{B}(s)] \qquad \text{for any } s \geq 0. \tag{31}$$

From (30) directly or by differentiating (31), we can also find the variance,

$$V(B) = \frac{\rho/\mu^2 + V(S)}{(1 - \rho)^3}\,. \tag{32}$$

While (31) is an implicit expression for $\tilde{B}(s)$, we have demonstrated that it determines moments of B. In fact, as we discuss later in this section, (31) uniquely determines the distribution of B.

To obtain information about K, let K_i be the number of departures during B_i, and write

$$K = 1 + \sum_{i=1}^{Y} K_i, \tag{33}$$

where, given $S = s$ and $Y = y \geq 1$, $K_1, \ldots, K_y$ are i.i.d. $\sim K$.

The generating function $\Gamma(z) \equiv E(z^K)$ is found by conditioning (33),

$$E\{z^K \mid S, Y\} = z[\Gamma(z)]^Y,$$
$$E\{z^K \mid S\} = ze^{\lambda S[\Gamma(z)-1]},$$
$$\Gamma(z) = z\tilde{G}[\lambda - \lambda\Gamma(z)] \qquad \text{for any } z \in [0, 1]. \tag{34}$$

From either (33) or (34), we can find the mean (25) and the variance of K,

$$V(K) = \frac{\rho + \lambda^2 V(S)}{(1 - \rho)^3}\,. \tag{35}$$

Observe that (32) and (35) have $(1 - \rho)^3$ in the denominator, but (25) and (27) have only $(1 - \rho)$ there. Hence, if we let $\lambda \to \mu$ $(\rho \to 1)$, not only do the means and variances grow large, but so do the corresponding coefficients of variation. By this measure, the distributions of B and K become very irregular as $\rho \to 1$.

We now discuss different conceptual representations for busy periods. The $\{U_i\}$ used to define (28) is an example of *ladder variables*. Think of walking down a ladder one step at a time. We may lose some ground from time to time (go back up a few steps). However, the only way to reach the bottom is to have been on each intermediate step previously. The recognition that $B_i \sim B$ and is independent of the past means that going down each step is of equivalent difficulty. The U_i may be regarded as renewal epochs. For a systematic exploitation of these ideas in the more general context of random walks and the *GI/G/*1 queue, see Chapter 9.

Perhaps a more natural (but less efficient) way to represent a busy period is as follows: Let $\beta_0 = S_1$, the service time of the first customer in a busy period, $\gamma_0 = 1$, and γ_1 be the number of arrivals during β_0. Let γ_2 be the number of arrivals during the interval

$$\beta_1 = \sum_{n=2}^{\gamma_1+1} S_n,$$

which is the total time to serve the γ_1 arrivals during β_0. Define β_j = total time to serve the γ_j arrivals during β_{j-1}, and γ_{j+1} = number of arrivals during β_j, $j = 1, 2, \ldots$. We have

$$B = \sum_{j=0}^{\infty} \beta_j, \text{ and} \tag{36}$$

$$K = \sum_{j=0}^{\infty} \gamma_j. \tag{37}$$

The utility of this representation becomes apparent when we recall that Y_1, $Y_2, \ldots$ are i.i.d., where Y_n is the number of arrivals during S_n. We can think of $K - 1$ as the number of "descendants" of the customer who begins the busy period and Y_n as the number of "progeny" of customer C_n. The number of descendants in generation j is γ_j. This has the mathematical structure of a *branching process*. (See Section 3-9.)

From (36) and (37), it is easy to find (27) and (25). Finding the corresponding transform expressions by using these ideas is more work. This was done for K in Chapter 3; (34) here is a special case of equation (105) in Chapter 3. This extra effort produces some useful results (Theorem 16 in Chapter 3): K is a proper random variable if and only if $E(Y) = \rho \le 1$, in which case, the generating function (and hence the distribution) of K is uniquely determined by (34).

For branching processes in general, busy periods have no physical meaning. Nevertheless, we can apply the methods of Section 3-9 to busy periods. Let $\tilde{B}_n(s)$ be the transform of $\sum_{j=0}^{n} \beta_j$. By conditioning on β_0 and γ_1, it is easily shown that for any $s \geq 0$,

$$\tilde{B}_{n+1}(s) = \tilde{G}[s + \lambda - \lambda \tilde{B}_n(s)], \; n = 0, 1, \ldots, \text{ and then that} \tag{38}$$

$$\lim_{n \to \infty} \tilde{B}_n(s) = \tilde{B}(s), \tag{39}$$

where $\tilde{B}(s)$ satisfies (31). Analogous to the proof of Theorem 16 in Chapter 3, it can be shown that (31) uniquely determines the distribution of B. Also see Feller [1971], p. 441.

8-5 THE $M/G/1$ QUEUE WITH EXCEPTIONAL FIRST SERVICE; LIFO WAITING TIMES

An important model in the analysis of priority queues is the $M/G/1$ queue with *exceptional first service*: That is, we have a single-channel queue with Poisson arrivals, where the first customer in a busy period has "exceptional" service time distribution G_a, and we denote an exceptional service time by $S_a \sim G_a$. Subsequent service times during the same busy period are independent of S_a, and are i.i.d. with service distribution G. As before, let λ be the arrival rate and $S \sim G$ denote an "ordinary" service time.

One interpretation of this model is that some kind of setup time is needed before an arrival who finds the system empty can begin service. In this context, S_a can be defined as the sum of the setup time and an ordinary service time. However, we will not assume any special structure of this sort; there are no constraints on the relative magnitudes of exceptional and ordinary service times.

Using an embedded Markov chain, where $\{X_n\}$ is defined as in Section 8-1, we obtain the FIFO waiting time transform in Problem 8-9. Our concern here is with busy periods. Let

B_a = duration of an exceptional first service busy period, and

Y_a = number of arrivals during S_a.

Assume $\rho = \lambda/\mu < 1$, where $1/\mu$ is the mean of G, and define $B_1, \ldots, B_y$ as in Section 8-4. Since these random variables are generated by service times that are distributed as G, they are still i.i.d., each distributed as an *ordinary* $M/G/1$ busy period. We now write

$$B_a = S_a + \sum_{i=1}^{Y_a} B_i, \tag{40}$$

where B_a is a proper random variable.

The first moment is again easily obtained,

$$E(B_a \mid S_a, Y_a) = S_a + Y_a E(B),$$

$$E(B_a \mid S_a) = S_a + \lambda S_a E(B), \text{ and}$$

$$E(B_a) = E(S_a)/(1 - \rho), \tag{41}$$

where as before, $E(B) = 1/(\mu - \lambda)$. Note that $E(B_a)$ is proportional to $E(S_a)$. Result (41) will turn out to be important in applications.

Similarly, from (40),

$$E\{e^{-sB_a} \mid S_a, Y_a\} = e^{-sS_a}[\tilde{B}(s)]^{Y_a},$$

$$E\{e^{-sB_a} \mid S_a\} = e^{-sS_a}e^{\lambda S_a[\tilde{B}(s)-1]}, \text{ and}$$

$$E\{e^{-sB_a}\} = \tilde{G}_a[s + \lambda - \lambda\tilde{B}(s)] \qquad \text{for any } s \geq 0, \tag{42}$$

where $\tilde{B}(s)$ is the transform of an ordinary $M/G/1$ busy period. Comparing these derivations with corresponding ones in Section 8-4, we see that up until the last step, they are identical. This is because the conditional distributions are identical. Only in the last step does it matter that S_a may have a different distribution.

Now let K_a be the number of customers served during B_a, define $K_1, \ldots,$ K_y as before, and we have

$$K_a = 1 + \sum_{i=1}^{Y_a} K_i. \tag{43}$$

The following are easily obtained:

$$E(K_a) = 1 + \lambda E(S_a)/(1 - \rho), \text{ and} \tag{44}$$

$$E\{z^{K_a}\} = z\tilde{G}_a[\lambda - \lambda\Gamma(z)] \qquad \text{for any } z \in [0, 1], \tag{45}$$

where $\Gamma(z)$ is the generating function of the number of customers served in an ordinary $M/G/1$ busy period.

As an application of busy periods, consider an ordinary $M/M/1$ queue with LIFO order of service, i.e., whenever a departure occurs, the next customer served is the most recent arrival, and let

$$D_L = \text{the stationary delay in queue.}$$

Because the stochastic behavior of the number in system process does not depend on this change (from FIFO), the results of Sections 8-1 and 8-4 apply. In particular, the event $\{D_L = 0\}$ means that an arrival finds the system empty, and

$$P(D_L = 0) = 1 - \rho. \tag{46}$$

Now suppose $D_L > 0$. Customers found in queue can be ignored under LIFO because they do not contribute to D_L. In fact, LIFO delay will end when the system first clears of the customer found in service *and* future arrivals. Thus for exponential service, $\{D_L \mid D_L > 0\}$ is stochastically equivalent to an ordinary $M/M/1$ busy period.

More generally, the LIFO $M/G/1$ queue can be approached in the same way. The only difference is that now $\{D_L \mid D_L > 0\}$ is stochastically equivalent to an exceptional first service busy period, where S_a is the *remaining* service of the customer found in service by a Poisson arrival who finds the server busy. Averaging over all such arrivals, it is an elementary exercise to show that S_a is distributed as the *equilibrium* distribution G_e, i.e., for any $t \geq 0$,

$$P(S_a \leq t) = G_e(t) \equiv \mu \int_0^t G^c(u)\, du. \tag{47}$$

Consequently, for ordinary $M/G/1$ queues with LIFO service,

$$E\{e^{-sD_L} \mid D_L > 0\} = \tilde{G}_e[s + \lambda - \lambda\tilde{B}(s)], \text{ and} \tag{48}$$

$$E\{e^{-sD_L}\} = 1 - \rho + \rho\tilde{G}_e[s + \lambda - \lambda\tilde{B}(s)] \qquad \text{for any } s \geq 0, \tag{49}$$

where $\tilde{G}_e(s) = \mu[1 - \tilde{G}(s)]/s$.

From the above it is easy to verify that

$$E(D_L) = E(D_F) = \lambda E(S^2)/2(1 - \rho), \tag{50}$$

where in this section, we denote FIFO delay by D_F.

However, higher moments are different. From (17) and (49), it may be shown that

$$E(D_F^2) = \frac{\lambda E(S^3)}{3(1 - \rho)} + \frac{\lambda^2 E^2(S^2)}{2(1 - \rho)^2}, \tag{51}$$

$$E(D_L^2) = \frac{\lambda E(S^3)}{3(1 - \rho)^2} + \frac{\lambda^2 E^2(S^2)}{2(1 - \rho)^3}. \tag{52}$$

Because $1 - \rho < 1$, the second moment and hence the variance is larger under LIFO than under FIFO, i.e., LIFO is less regular. (See Section 5-14.) There seems to be no ready explanation for the similar appearance of these expressions.

As $\rho \to 1$, LIFO and FIFO delay distributions behave very differently; $V(D_L)/d^2 \to \infty$ (like a busy period), but $V(D_F)/d^2 \to 1$. In fact, from (17), it is easily shown that as $\rho \to 1$ ($\lambda \to \mu$ with G fixed),

$$\lim_{\rho \to 1} E\{e^{-s(1-\rho)D_F}\} = [s\mu E(S^2)/2 + 1]^{-1} \qquad \text{for any } s \geq 0. \tag{53}$$

i.e., the distribution of $(1 - \rho)D_F$ converges to an exponential with mean $\mu E(S^2)/2$.

8-6 THE $GI/M/1$ QUEUE

We now consider a single-channel queue with i.i.d. inter-arrival times (renewal input) and i.i.d. exponential service times. Let $T_n \sim A$ be the time between arrivals n and $n + 1$, $\lambda = 1/E(T_n)$, $\tilde{A}(s) = E\{e^{-sT_n}\}$, μ be the service rate, and $\rho = \lambda/\mu$.

Because the non-Markovian feature of this model is the inter-arrival distribution, we embed points at arrival epochs. For $n = 1, 2, \ldots$, let

X_n = number of customers in system found on arrival by C_n, and

Y_n = number of Poisson "events" at rate μ that occur during T_n,

where, provided the system is busy, each event is a departure. It is convenient to permit events to occur at all times, so that the event process is Poisson. Events when the system is empty have no effect on it.

With this definition, the Y_n are i.i.d. with properties analogous to arrival process quantities in Section 8-1. Let $Y_n \sim \{r_j\}$, and note that

$$E(Y_n) = \mu/\lambda = 1/\rho, \text{ and} \tag{54}$$

$$E\{z^{Y_n}\} = \tilde{A}[\mu(1 - z)]. \tag{55}$$

It is easy to see that

$$X_{n+1} = \max\{X_n + 1 - Y_n, 0\}, \qquad n = 1, 2, \ldots, \tag{56}$$

where $\{X_n\}$ is a Markov chain with transition probability matrix

$$P = \begin{pmatrix} \sum_{i>0} r_i & r_0 & & \\ \sum_{i>1} r_i & r_1 & r_0 & \\ \vdots & \vdots & \ddots & \ddots \end{pmatrix}. \tag{57}$$

Because the r_j are strictly positive, the chain is irreducible.

The form of the stationary vector π turns out to be very simple: $\pi_j = (1 - \alpha)\alpha^j$, $j = 0, 1, \ldots$, for some $\alpha \in (0, 1)$, as can be verified by plugging into (57) and applying the argument that follows (61).

Simply doing this is not very enlightening. To understand this result, recall that for a positive irreducible chain, π_j is the frequency of visits to state j, and for every pair of states, π_j/π_i is a relative frequency, i.e.,

$$\pi_j/\pi_i = E(V_{ij}) \qquad \text{for all } i \text{ and } j, \tag{58}$$

where V_{ij} is the number of visits to j between returns to i. (See Section 3-5.)

Consider sequences of transitions that generate $V_{0,1}$, e.g., for $0 \to 1 \to 2 \to 3 \to 1 \to 0$, $V_{0,1} = 2$. Every sequence of states of form $s_1 \to s_2 \to \cdots \to s_n$, where $s_1 = s_n = 0$, and $s_j > 0$ for $j = 2, \ldots, n - 1$, will generate a value of $V_{0,1}$.

Now consider sequences that generate $V_{j,j+1}$ for any $j \geq 1$. From (57), note that any transition to a state $s < j$ stops the process because the chain cannot visit state $j + 1$ again without first returning to state j.

A sequence $s_1 \to \cdots \to s_{n-1} \to s_n$ generates $V_{0,1} = k$ if and only if $s_1 + j \to \cdots \to s_{n-1} + j \to \{0, \ldots, j\}$ generates $V_{j,j+1} = k$. Furthermore, from (57), these two sequences *have the same probability*. Consequently, the distribution of $V_{j,j+1}$ is independent of j, and, from (58).

$$\pi_{j+1}/\pi_j = \alpha, \tag{59}$$

a constant independent of j. Clearly, $\alpha > 0$. Normalization, which is possible only if $\alpha < 1$, gives us

$$\pi_j = (1 - \alpha)\alpha^j, \qquad j = 0, 1, \ldots, \quad \text{for some } \alpha \in (0, 1). \tag{60}$$

Thus, if $\pi = \pi P$ has a solution, it must be of form (60). We now try to find it. For $j = 1$, $\pi_1 = (1 - \alpha)\alpha = \sum_{i=0}^{\infty} r_i(1 - \alpha)\alpha^i = (1 - \alpha)R(\alpha)$. The choice of j doesn't matter; it is easy to show that a solution exists if and only if α satisfies

$$\alpha = R(\alpha) \qquad \text{for some } \alpha \in (0, 1), \text{ where} \tag{61}$$

$$R(z) \equiv \sum_{j=0}^{\infty} r_j z^j = E\{z^{Y_n}\}.$$

Note that $R(z)$ is a strictly convex function. Finding solutions to (61) is a classic problem that arises in the analysis of branching processes. In Section 3-9, we show that (61) has a unique solution if $E(Y_n) = 1/\rho > 1$, and no solution otherwise. (Of course, uniqueness also follows from the theory of arbitrary irreducible chains.)

Summarizing our results and recalling (55), $\pi = \pi P$ has a solution if and only if $\rho < 1$, in which case, it is of form (60), where $\alpha \in (0, 1)$ is the unique solution to

$$\alpha = \tilde{A}[\mu(1 - \alpha)]. \tag{62}$$

It is also true that the chain is null if $\rho = 1$, and transient if $\rho > 1$. For the remainder of this section, assume $\rho < 1$.

Because α is the proportion of arrivals who find the server busy, we should get $\alpha = \rho$ for the $M/M/1$ queue. This is easily verified; from (62),

$$\alpha = \lambda/[\lambda + \mu(1 - \alpha)],$$

an equation with two roots, 1 and $\alpha = \rho$.

For Erlang arrivals, it may be shown that $\alpha < \rho$, i.e., the proportion of arrivals who find the server busy is less than the proportion of time it is busy. On the other hand, for hyper-exponential arrivals, $\alpha > \rho$.

Denote a stationary X_n by N_a, the number of customers in system found by an arrival, a random variable with distribution (60). We can represent a stationary waiting time as the sum

$$W = \sum_{i=1}^{N_a+1} S_i, \tag{63}$$

where, because service is exponential, the S_i are i.i.d. $\sim$ exp (μ).

From (60) and (63), it is an elementary exercise to show that

$$E\{e^{-sW}\} = (1 - \alpha)\mu/[(1 - \alpha)\mu + s], \tag{64}$$

i.e., the waiting time distribution is exponential,

$$W \sim \exp\,[(1 - \alpha)\mu]. \tag{65}$$

The corresponding first moments are

$$w = E(W) = 1/(1 - \alpha)\mu, \text{ and} \tag{66}$$

$$L = \lambda w = \rho/(1 - \alpha). \tag{67}$$

Now let N be the stationary number of customers in system for the continuous-time process $\{N(t)\}$, where for $j = 0, 1, \ldots, P(N = j) = p_j$ is the proportion of time there are j customers in system. For arbitrary $j \geq 1$, partition the state space into $\{0, \ldots, j - 1\}$ and $\{j, j + 1, \ldots\}$, and equate transition rates between these subsets. From the Markov chain, the "up" rate is $\lambda\pi_{j-1}$, and from PASTA, the "down" rate is μp_j. We have

$$\lambda\pi_{j-1} = \mu p_j, \text{ or} \tag{68}$$

$$p_j = \rho\pi_{j-1} = \rho(1 - \alpha)\alpha^{j-1}, \qquad j = 1, 2, \ldots. \tag{69}$$

Of course, consistent with $L = \lambda w$, we still have

$$p_0 = 1 - \sum_{j \geq 1} p_j = 1 - \rho. \tag{70}$$

For fixed ρ, N is *stochastically* increasing and L is increasing in α. Similarly, for fixed λ and μ, W is stochastically increasing and w is increasing in α. By any of these measures, Erlang arrivals perform better than Poisson, and hyper-exponential arrivals perform worse. In particular, using E, M, and H to denote these cases, we have

$$L_E < L_M < L_H.$$

In contrast with the $M/G/1$, these inequalities depend on α through (62), not simply on the mean and variance of A. Furthermore, $L \neq E(N_a) = \alpha/(1 - \alpha)$ when $\alpha \neq \rho$. In particular, $E(N_a) < L$ for Erlang arrivals and $E(N_a) > L$ for hyper-exponential arrivals. Regular arrivals tend to see the system *less congested* than the corresponding time average.

8-7 THE *GI/M/c* QUEUE

We now extend some $GI/M/1$ results to c channels. The same notation will be used except that service times are now distributed exp (μ/c). Thus the departure rate is μ only when all channels are busy.

Now $\{X_n\}$ is still a Markov chain, but some of the transition probabilities have changed. On the range $\{X_n \geq c - 1, X_{n+1} \geq c\}$,

$$X_{n+1} = X_n + 1 - Y_n, \tag{71}$$

i.e., the departure process behaves like a Poisson process at rate μ provided that all channels remain busy. From (71), the transition probabilities are given by

$$p_{ij} = r_{j-i+1} \qquad \text{for } i \geq c - 1 \quad \text{and} \quad j \geq c. \tag{72}$$

These are identical to those in (57) on the same range.

By the same argument as in Section 8-6, the distribution of $V_{j,j+1}$ for $j \geq c - 1$ here is the same as for $V_{0,1}$ in the single-channel case. Consequently, in the c-channel case,

$$\pi_{j+1}/\pi_j = \alpha, \qquad j = c - 1, c, \ldots . \tag{73}$$

As in the single-channel case, $\alpha \in (0, 1)$ is the unique solution to

$$\alpha = \tilde{A}[\mu(1 - \alpha)]$$

if and only if $\rho = \lambda/\mu < 1$. The ratio of terms in (73) will not be α for $j < c - 1$; because there are only a finite number of such terms, normalization can be done. Hence, this is the positive case. (The transient and null cases are also the same as before.) Assume $\rho < 1$ in the remainder of this section.

From (73), $(N_a - c \mid N_a \geq c)$ has a geometric distribution, where $\{N_a \geq c\}$ is equivalent to the event $\{D > 0\}$. Now $(D \mid D > 0)$ is the sum of $N_a - c + 1$ inter-departure times that are i.i.d. $\sim$ exp (μ). It is easily shown that

$$(D \mid D > 0) \sim \exp\,[(1 - \alpha)\mu], \tag{74}$$

where (74) holds for any $c = 1, 2, \ldots .$

The unconditional delay distribution does depend on the number of channels. To facilitate comparison, denote by D_1 and D_c the respective single- and c-channel delays. By unconditioning (74), the delay distribution is a mixture of a unit step at 0 and an exponential. For $c = 1$,

$$D_1 \sim (1 - \alpha)U_0 + \alpha \exp\,[(1 - \alpha)\mu]. \tag{75}$$

For $c \geq 2$, we have not explicitly represented $\pi_0, \ldots, \pi_{c-2}$. If it were true that $\pi_0 = \cdots \pi_{c-2} = 0$, D_c would have the same distribution as D_1. However, $\pi_j > 0$ for all j. From (73), the ratio of successive terms is fixed on the range $j \geq c - 1$, and having $\pi_i > 0$ for $i < c - 1$ can only make all the π_j, $j \geq c - 1$, smaller by some fixed factor. We have shown that $\beta \equiv P(D_c > 0) < \alpha$. Hence for $c \geq 2$,

$$D_c = (1 - \beta)U_0 + \beta \exp[(1 - \alpha)\mu], \tag{76}$$

where $\beta < \alpha$.

We could explicitly represent β and, in particular cases, compute β. This is not necessary for the comparisons we have in mind, e.g., from (75) and (76), D_c is *stochastically smaller* than D_1. In particular, for $c \geq 2$,

$$d_c < d_1. \tag{77}$$

Our results are consistent with comparisons of $M/M/1$ and $M/M/c$ queues that we made in Chapter 5. For Poisson arrivals and exponential service, we also found that when the respective service times are included, $w_c > w_1$. This turns out to be true for the models treated here as well. A proof may be found in Section 11-7.

PROBLEMS

8-1. Specialize (15) for an $M/E_k/1$ queue. Compare with Chapter 5 results.

8-2. For a stable $M/G/1$ queue, let Y_n be the number of arrivals during service time S_n, $n = 1, 2, \ldots$. Let $P(S_r \leq t)$ be the fraction of these arrivals who find that the *remaining* service time of the customer in service does not exceed t. By a direct argument, show that

$$P(S_r \leq t) = G_e(t) = \mu \int_0^t G^c(u)\,du, \qquad t \geq 0,$$

the equilibrium distribution. If you get stuck, this is a minor modification of Example 2-7. What customers are *not* included in this average?

8-3. *The $M/G/1$ queue with dissatisfied customers.* On completion of service, suppose that with probability $(1 - \alpha)$, customers are dissatisfied with the service received and join the end of the queue to be served again, independent of whether they were dissatisfied before. Assume all service times are i.i.d.

(a) An alternative way of operating would be to serve dissatisfied customers again immediately, until they are satisfied. How would this change affect the number in system process? Under what conditions is the system stable? Assume these conditions hold below.

(b) How would the change proposed in (a) affect L, w, and d? Find an expression for d in terms of λ, α, and properties of the original service time distribution. Find Q.

(c) How would the change proposed in (a) affect the waiting time distribution? What performance measure would improve?

8-4. *Continuation.* Assuming dissatisfied customers join the end of the queue, let D_1 be the delay in queue of an arrival until service begins *for the first time*. Represent D_1 as a sum and find its mean d_1. (You will need the result in Problem 8-2.) Find Q_1, the average number of customers in queue who are waiting to be served for the first time.

8-5. *The BM/G/1 queue.* We have batch Poisson arrivals (BM), with batch arrival rate λ and batch size distribution $P(\nu = j) = \alpha_j, j = 0, 1, \ldots$. Customers are still served one at a time; $S \sim G$ denotes a customer service time. Assume $\rho \equiv \lambda E(\nu)/\mu < 1$, and that ν and S have finite second moments.

(a) What fraction of time is the server busy serving batches of size j?

(b) Given a batch of size j is being served, what is the expected number of customers *from the same batch* in queue, for $j = 0, 1, \ldots$?

(c) Assume FIFO. We say a batch is in service if any of its customers are. Let Q_b be the average number of batches in queue. Use (b) to show that

$$Q = Q_b E(\nu) + \lambda E\{\nu(\nu - 1)\}/2\mu.$$

8-6. *Continuation.* We look at the same model in a different way. Using the same convention as above, let D_b be the delay in queue of a batch, S_b be the total time to serve all customers in a batch, and $d_b = E(D_b)$. Show that

(a) $d_b = \lambda E(S_b^2)/[2(1 - \rho)]$,

where $E(S_b^2) = E(\nu)V(S) + E(\nu^2)E^2(S)$.

(b) Clearly, all customers in the same batch have the same D_b. Added to this is delay caused by others in their own batch. Independent of 8-5(c), show that

$$d = d_b + E\{\nu(\nu - 1)\}/2\mu E(\nu).$$

Show that this expression is equivalent to that found in Problem 8-5(c).

8-7. *Continuation.* It is easy to express D in terms of D_b,

$$D = D_b + D_w, \qquad \text{where} \quad D_w = \sum_{i=1}^{J-1} S_i,$$

J is the position in line of a randomly selected customer within that customer's own batch, the S_i are the i.i.d. service times of the customers in front of this one, and D_b is independent of D_w. Think of D_w as the "within batch" delay.

(a) What is the distribution of J?

(b) Show that $J - 1$ has generating function

$$E(z^{J-1}) \equiv \beta(z) = [1 - \alpha(z)]/(1 - z)E(\nu),$$

where $\alpha(z) = E(z^{\nu})$.

(c) Show that D_w has transform

$$E(e^{-sD_w}) = \beta[\tilde{G}(s)].$$

(d) Let $\tilde{G}_b(s)$ be the transform of the total time to serve all customers in a batch. Show that

$$\tilde{G}_b(s) = \alpha[\tilde{G}(s)].$$

(e) Find the transforms of D_b and D.

8-8. *Continuation.* For yet another way to look at the $BM/G/1$ queue, let X_n be the number of customers in system left behind by departure C_n, where $\{X_n\}$ is a Markov chain.

One can find stationary $E(X_n)$ by modifying the stationary chain analysis in the chapter, but it is messy. Instead, show that

(a) $E(X_n) = L + E\{\nu(\nu - 1)\}/2E(\nu)$, and also that

(b) $E(X_n) = \lambda w E(\nu) + E\{\nu(\nu - 1)\}/2E(\nu)$.

These expressions are obviously equal, but argue each one independently.

8-9. *The M/G/1 queue with exceptional first service.* (Welch [1964]) We now do an embedded chain analysis of the model in Section 8-5. Let $S_a \sim G_a$ be an exceptional first service, and $S \sim G$ be an ordinary service time, where $E(S_a) < \infty$, and $\lambda E(S) = \rho < 1$. Define $\{X_n\}$, $\{Y_n\}$, and π as in Section 8-1.

(a) Write X_{n+1} as a function of X_n and Y_{n+1}. Take the expected value of this expression, cancel first moments, and show that

$$\pi_0 = (1 - \rho)/[1 - \rho + \lambda E(S_a)].$$

(b) Use (44) to find π_0.

(c) Modify the argument leading to (9) and (10); show that π has generating function

$$\Pi_a(z) = \frac{\pi_0[\tilde{G}[\lambda(1 - z)] - z\tilde{G}_a[\lambda(1 - z)]]}{\tilde{G}[\lambda(1 - z)] - z}.$$

(d) Give two explanations for why $E\{X_n\} = \lambda w$.

(e) Argue that the waiting time transform satisfies $\tilde{W}_a[\lambda(1 - z)] = \Pi_a(z)$.

8-10. Suppose a stable $M/M/1$ queue is observed "at random" and found to be busy, and we want an expression for the distribution of the remaining busy period. From one point of view, it should be the equilibrium busy period distribution $B_e(t)$, say. Given that n customers are found in system, it would be the n-fold convolution of the busy period distribution $B^{(n)}(t)$. Show that these points of view are consistent, i.e., show that

$$B_e(t) = \sum_{n=1}^{\infty} B^{(n)}(t)P(N = n \mid N \geq 1).$$

8-11. For a stable $M/G/1$ queue under FIFO, why is it true that

(a) $d = Q/\mu + E(S_r)$,
where S_r is the remaining service of the customer found in service, if any?

(b) Find $E(S_r)$ and equation (15).

8-12. *The M/G/1 queue with server breakdown.* Suppose that the time until breakdown of the server is distributed exp (ν), where the server breaks down only when it is busy. Breakdown times and repair times of the server are each i.i.d., and independent of each other. Let R denote a repair time. Assume FIFO, $E(R^2) < \infty$, and that no work is performed on customers while the server is down. Define delay D_n as the time between the arrival of customer C_n and the initiation of service on C_n, *completion time* $\mathscr{C}_n$ as the time between the initiation and the completion of service on C_n, and $W_n = D_n + \mathscr{C}_n$. From the customers' point of view, the $\mathscr{C}_n$ are service times.

(a) Argue that the $\mathscr{C}_n$ are i.i.d., $n = 1, 2, \ldots$.

(b) For any fixed n, represent $\mathscr{C}_n$ as a sum of random variables; find $E(\mathscr{C}_n)$.

(c) When is this system stable? Assume these conditions hold below.

(d) Find $E(\mathscr{C}_n^2)$ and argue that expected delay is

$$d = \lambda E(\mathscr{C}^2)/2[1 - \lambda E(\mathscr{C})].$$

Remark. Allowing the server also to break down while idle is a more difficult problem. It will be treated by more refined methods in Chapter 10.

8-13. Use representations (36) and (37) to derive $E(B)$ and $E(K)$. [*Hint:* Argue that $E(\gamma_{n+1} \mid \gamma_n) = \rho\gamma_n$.]

8-14. For a FIFO $M/G/1$ queue, use the right-hand representation in (17) to show that as $\rho \to 1$ ($\lambda \to \mu$ with G fixed)

$$\lim_{\rho\to 1} E\{e^{-s(1-\rho)D_F}\} = [2/\mu E(S^2)]/[2/\mu E(S^2) + s],$$

which implies that $(1 - \rho)D_F$ converges in distribution to an exponential with mean $\mu E(S^2)/2$.

8-15. Assume the arrival of cars at a stoplight is Poisson at rate λ. When the light turns red, the cars form a single line during the red light period, a constant R. The first car in line starts τ_1 seconds after the light turns green, and for $i > 1$, the ith car starts τ_i seconds after the car in front. The τ_i are i.i.d. with finite mean $1/\mu$. Measured from when the light turns green, let T be the time until the line of cars hits 0. New arrivals join the line only while it is strictly positive. Assume cars have zero length, the green period is of infinite length, and ignore acceleration effects.
(a) When is T a proper random variable? Assume this is true.
(b) Find expressions for $E(T)$ and $E(e^{-sT})$.

8-16. *The $M/M/1$ queue under work policy v.* The service facility is turned off whenever it is empty. It is turned on again when the amount of *work* in system exceeds a predetermined quantity v. For an initially empty system, the server is turned on by arrival number $X = x$ if

$$\sum_{n=1}^{x-1} S_n \le v < \sum_{n=1}^{x} S_n.$$

Once on, the server remains on until the system is empty. Assume the system is stable.

(a) How is $\sum_{n=1}^{X} S_n - v$ distributed?

(b) Find $E\left\{\sum_{n=1}^{X} S_n\right\}$ and $E(X)$.

(c) Find the expected time from when the server is turned on until the system is empty.

8-17. Derive (38).

8-18. *The $M/G/1$ queue under preemptive LIFO.* Under *preemptive* LIFO, customers begin service immediately on arrival, interrupting anyone found in service. When a service is completed, the most recent arrival in queue resumes service. Assume that the work-in-system process is not affected by this rule. Let W_{PL} be the waiting time in system and N_a be the (stationary) number of customers in system found by an arrival.
(a) Identify W_{PL} with a standard $M/G/1$ quantity and find $E(W_{PL})$.

(b) Find $P(N_a \geq 1)$.

(c) Argue that for any $j < k < n$, the events {arrival n interrupts arrival k} and {arrival k interrupts arrival j} are independent. What is $P(N_a \geq 2)$?

(d) Argue that N_a has a geometric distribution.

8-19. Find $E(D_L)$ by differentiating (49).

8-20. Find (51) from the right-hand side of (17).

8-21. Let N_a be the number of customers in system found by an arrival at an $E_2/M/1$ queue. Find

$$L_{E_2/M/1} = 2\rho/(1 - 4\rho + \sqrt{1 + 8\rho}),$$

and show that

$$P(N_a > 0) < \rho.$$

8-22. *Continuation.* For an $H/M/1$ queue, show that

$$P(N_a > 0) > \rho.$$

8-23. Derive (74).

8-24. We now use $L = \lambda w$ to derive (69). Assume FIFO and for $j = 1, 2, \ldots$, let

$W(j)$ = time a customer spends in system while "jth in line," counting the customer in service, and

$L(j)$ = expected number of customers in system who are jth in line.

(a) Argue that $L(j) = \sum_{i \geq j} p_i$.

(b) Argue that the fraction of customers who, on arrival or later, are jth in line is $\sum_{i \geq j-1} \pi_i$, and that

$$L(j) = \sum_{i \geq j} p_i = \lambda \sum_{i \geq j-1} \pi_i E\{W(j)\}, \qquad j = 1, 2, \ldots.$$

(c) Now $E\{W(j)\} = 1/\mu$. Complete the argument. [What is $L(j) - L(j-1)$?]

8-25. For a stable $GI/M/1$ queue under LIFO, relate $(D_L \mid D_L > 0)$ to B, and find

(a) the mean duration of a busy period,

(b) the mean duration of an idle period, and

(c) the expected number of customers served during a busy period.

Remark. LIFO here and FIFO in Problem 8-24 were chosen for convenience. The quantities found do not depend on these assumptions.

8-26. For the same ρ, compare the $E_2/M/1$ and $M/E_2/1$ queues and show that

$$L_{M/E_2/1} > L_{E_2/M/1} \qquad \text{for all } \rho \in (0, 1),$$

where $L_{E_2/M/1}$ was found in Problem 8-21.

8-27. For the same ρ, compare the $H/M/1$ and $M/H/1$ queues for the special case $H =$

$(1 - \beta)U_0 + \beta$ exp, and show that

$$L_{H/M/1} > L_{M/H/1} \qquad \text{for all } \rho \in (0, 1).$$

Remark. Both here and in Problem 8-26, we are better off with the less regular distribution as the service distribution.

9

RANDOM WALKS AND THE $GI/G/1$ QUEUE

In this chapter, we give up the exponential distribution entirely. Both inter-arrival and service times are permitted to have arbitrary distributions. The methods in Chapters 5 through 8 that were based on discrete- or continuous-time Markov chains no longer apply. However, renewal theory turns out to be a valuable tool, and the queue processes in this chapter will be regenerative (when their cycles are proper random variables).

The focus of our analysis also shifts from the number of customers in system (or queue) to delay in queue. The behavior of the delay-in-queue process depends on the behavior of a sequence of partial sums known as a random walk. After showing this connection, we develop the theory of random walks in Sections 9-2 through 9-7, and then apply these results to the $GI/G/1$ queue in Sections 9-8 through 9-10. We relate delay to work in system in Section 9-11. The Wiener-Hopf factorization and Spitzer's identity are presented in Section 9-12.

9-1 THE $GI/G/1$ QUEUE

As in Chapter 5, T_n is the time between the arrival of customers C_n and C_{n+1}, and S_n is the service time of C_n, $n = 1, 2, \ldots$. The $GI/G/1$ queue is a single-channel queue where $\{T_n\}$ and $\{S_n\}$ are independent sequences of i.i.d. random variables. The inter-arrival and service time distributions are denoted by $A(t) = P(T_n \leq t)$ and $G(t) = P(S_n \leq t)$, $t \geq 0$, respectively. Let $E(T_n) = 1/\lambda$ and $E(S_n) = 1/\mu$, where, as before, λ is the arrival rate and μ the service rate.

Assume FIFO order of service, and let D_n be the delay in queue of C_n.

Thinking of D_n as *work-in-system* found by C_n, we have

$$D_{n+1} = D_n + S_n - T_n \qquad \text{if } D_n + S_n \ge T_n,$$
$$= 0 \qquad \text{if } D_n + S_n < T_n.$$

Combining these two cases, we have

$$D_{n+1} = \max(D_n + S_n - T_n, 0), \qquad n = 1, 2, \ldots. \tag{1}$$

Observe that given D_n, D_{n+1} depends only on the *difference* of S_n and T_n. To take advantage of this simplification, it is convenient to define

$X_n = S_n - T_n$, and (1) becomes

$$D_{n+1} = \max(D_n + X_n, 0) \equiv (D_n + X_n)^+, \qquad n = 1, 2, \ldots, \tag{2}$$

where, as defined in Chapter 1, $(\cdot)^+$ is called the *positive part* of whatever is inside the parentheses. Similarly, $(\cdot)^-$ will denote the *negative part*.

Using (2) recursively, we can write

$$D_{n+1} = \max(D_1 + X_1 + \cdots + X_n, X_2 + \cdots + X_n, \ldots, X_n, 0), \qquad n \ge 1. \tag{3}$$

For each fixed n, equation (3) shows that aside from initial conditions (the value of D_1), D_{n+1} depends on the partial sums of $X_1, \ldots, X_n$, *summed in reverse order*. The study of the behavior of partial sums of i.i.d. random variables is called *fluctuation theory* or the theory of *random walks*.

We now begin our treatment of random walks, which are important in their own right and have applications outside of queueing theory.

9-2 RANDOM WALKS AND LADDER VARIABLES; BASIC TERMINOLOGY AND NOTATION

Let $X_1, X_2, \ldots$ be a sequence of i.i.d. random variables, unrestricted in sign, with some common distribution F. To avoid trivialities, we assume that F has positive mass on both sides of the origin, i.e., $F(0) < 1$ and $F(0^-) > 0$.

The sequence of partial sums

$$Z_0 = 0, \quad Z_n = X_1 + \cdots + X_n, \qquad n = 1, 2, \ldots \tag{4}$$

is called the *random walk* generated by $\{X_n\}$. We will think of Z_n as the *position* or *height* of the random walk at *epoch n*. If $\{Z_n \in I\}$ for some interval I and some $n \ge 1$, we say the random walk *visits* or *hits* I at epoch n.

A useful way to think about random walks is to suppose that the X_n are the outcomes (gains or losses) from a sequence of bets, where for each n, Z_n is the net cumulative gain after n bets. Of particular interest in this context is whether $Z_n > 0$ for any n, and if so, those epochs n where $Z_n > Z_j$ for all $j < n$.

Formally, we call $n \geq 1$ a *(strict ascending) ladder epoch* if

$$Z_n > Z_j \qquad \text{for all } j < n. \tag{5}$$

Let $\mathscr{K}_1$ be the first epoch at which (5) occurs, where $\mathscr{K}_1$ is a (possibly defective) random variable. Given $\mathscr{K}_1 = k$ for each $k \geq 1$, let $k + \mathscr{K}_2$ be the second epoch at which (5) occurs. Because the X_n are i.i.d., the partial sum process $\{Z_{n+\mathscr{K}_1} - Z_{\mathscr{K}_1}; n \geq 0\}$ "starts over" at epoch k. Hence, $\mathscr{K}_2$ has the same distribution as $\mathscr{K}_1$, and is independent of $\mathscr{K}_1$. In this way, we define a sequence of (possibly defective) *ladder epochs*

$$\mathscr{K}_1, \mathscr{K}_1 + \mathscr{K}_2, \ldots, \tag{6}$$

which are the random epochs where (5) occurs, where the $\mathscr{K}_i$ are i.i.d.

Similarly, we define a sequence of (strict ascending) *ladder heights*

$$\mathscr{H}_1 = Z_{\mathscr{K}_1}, \mathscr{H}_1 + \mathscr{H}_2 = Z_{\mathscr{K}_1+\mathscr{K}_2}, \ldots, \tag{7}$$

where the $\mathscr{H}_i$ are i.i.d. and possibly defective. In fact, $\mathscr{H}_1$ occurs (is finite) if and only if $\mathscr{K}_1$ occurs. It will be convenient to place the defective mass of both distributions at $+\infty$,

$$P(\mathscr{H}_1 < \infty) = P(\mathscr{K}_1 < \infty) = \sum_{k=1}^{\infty} P(\mathscr{K}_1 = k) \leq 1. \tag{8}$$

The pairs $(\mathscr{K}_1, \mathscr{H}_1)$, $(\mathscr{K}_1 + \mathscr{K}_2, \mathscr{H}_2 + \mathscr{H}_2)$, . . . are called *ladder points,* where the $(\mathscr{K}_i, \mathscr{H}_i)$ are i.i.d. Clearly, ladder points can be defined in several ways, depending on the inequality in (5):

$>$ strict ascending (*sa*),

$\geq$ weak ascending (*wa*),

$\leq$ weak descending (*wd*), and

$<$ strict descending (*sd*).

If F is continuous, the corresponding strict and weak cases are equivalent. However, there are important special cases where the distinction is essential, e.g., for $P(X_n = 1) = p$ and $P(X_n = -1) = q$, where $p + q = 1$. We plot a realization of $\{Z_n\}$ for this case in Figure 9-1, where the ladder points of each type are denoted by the preceding abbreviations. Note that a strict ladder point is also weak, where both are either ascending or descending.

For general F, let $\mathscr{H}_1^w, \mathscr{H}_1^w + \mathscr{H}_2^w, \ldots$ denote the sequence of weak ascending ladder heights, and let the distributions of the first strict and weak ladder heights be

$$H(t) = P(\mathscr{H}_1 \leq t) \quad \text{and} \quad H_w(t) = P(\mathscr{H}_1^w \leq t), \qquad t \geq 0,$$

respectively. In order to relate these distributions, consider the event

$$E = \{\mathscr{H}_1^w = 0\},$$

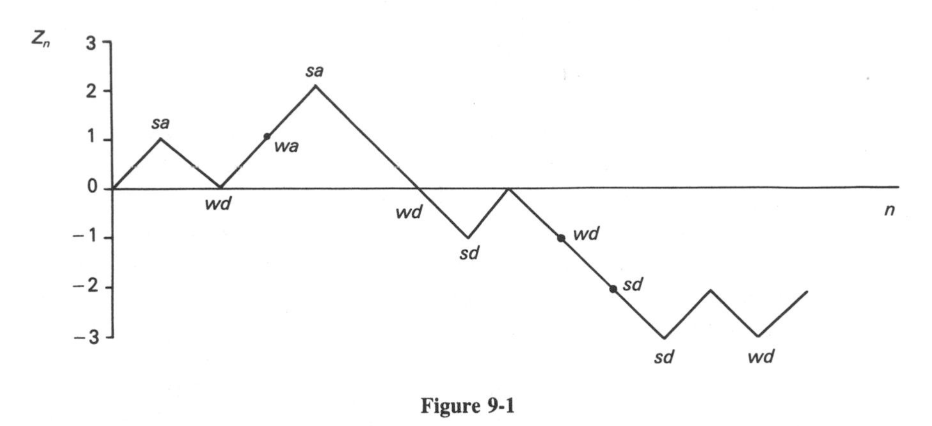

Figure 9-1

i.e., the event that the first weak ladder height occurs *and is not strict*. Define

$$\beta = P(E) = \sum_{n=1}^{\infty} P(Z_1 < 0, \ldots, Z_{n-1} < 0, Z_n = 0), \tag{9}$$

where, because $P(X_1 > 0) > 0$,

$$\beta < 1. \tag{10}$$

Given that E occurs, the random walk starts over, and the distribution of $\mathscr{H}_1$ is unaffected, i.e., $\mathscr{H}_1$ and E are independent. It follows that $\mathscr{H}_1$ and E^c are also independent! If E^c occurs, $\mathscr{H}_1^w = \mathscr{H}_1$ ($= \infty$ if $\mathscr{H}_1$ does not occur). Hence, by conditioning on whether E occurs, we have

$$H_w(t) = \beta U_0(t) + (1 - \beta)H(t), \qquad t \geq 0, \tag{11}$$

where U_0 is the unit step at the origin.

From (10) and (11), we make the important observation that the strict and weak ladder variables of the same type (ascending or descending) are either *both proper* or *both defective*. When defective, of course, the defects (mass at infinity) may be different.

In terms of convolutions of H and H_w, we define (ordinary) *renewal functions*

$$m(t) = \sum_{n=1}^{\infty} H^{(n)}(t) \quad \text{and} \quad m_w(t) = \sum_{n-1}^{\infty} H_w^{(n)}(t), \qquad t \geq 0, \tag{12}$$

where, for example, $m_w(t)$ is the expected number of weak ladder heights that fall in $[0, t]$.

From (11), there is an easy relationship between these renewal functions,

$$m_w(t) = [m(t) + \beta]/(1 - \beta). \tag{13}$$

[See Problem 9-1 for a probabilistic derivation of (13).]

From (13), or the observation above that H and H_w are either proper or defective together,

$$\lim_{t\to\infty} m_w(t) = \lim_{t\to\infty} m(t) = \infty \tag{14}$$

in the proper case. In the defective case, it is easily shown (Section 2-10) that the corresponding limits are finite (but different when $\beta \neq 0$); e.g.,

$$\lim_{t\to\infty} m_w(t) = H_w(\infty)/[1 - H_w(\infty)] < \infty, \tag{15}$$

when these distributions are defective. We call the renewal process *recurrent* when H is proper, and *transient* otherwise.

Notation for other ladder variables will be needed occasionally. For example, let $\mathscr{K}_1^w$ be the first weak ascending ladder epoch, and $(\overline{\mathscr{K}}_1, \overline{\mathscr{H}}_1)$ be the first strict descending ladder point. Corresponding notation for weak descending variables is obvious, but we shall avoid its use.

9-3 DUALITY

Joint distributions of partial sums of i.i.d. random variables are independent of the order in which these variables are summed. The first application of this observation is called *duality*, where random variables are summed in *reverse* order.

For each *fixed* $n \geq 1$, number $X_1, \ldots, X_n$ in reverse order,

$$X_1^* = X_n, \ldots, X_n^* = X_1, \text{ with partial sums}$$

$$Z_j^* = \sum_{i=1}^{j} X_i^* = Z_n - Z_{n-j}, \qquad j = 0, \ldots, n, \tag{16}$$

where (16) defines a one-to-one mapping in n-dimensional space from points $(Z_1, \ldots, Z_n) = (z_1, \ldots, z_n)$ to $(Z_1^* = z_n - z_{n-1}, Z_2^* = z_n - z_{n-2}, \ldots, Z_n^* = z_n)$.

Now consider the event

$$\{(Z_1, \ldots, Z_n) \in A\}, \tag{17}$$

where A is an (essentially arbitrary) region in n-dimensional space. The mapping (16) and A determines a region A^* such that

$$\{(Z_1^*, \ldots, Z_n^*) \in A^*\} \tag{18}$$

is the *same event* as (17). Corresponding to event (17), we also define *dual event*

$$\{(Z_1, \ldots, Z_n) \in A^*\}. \tag{19}$$

Since (17) and (18) are the same event, they have the same probability. On the other hand, because the vectors in (18) and (19) have the same joint distribution, the probability that they each fall in the same region A^* must be equal.

Combining, we have

$$P[(Z_1, \ldots, Z_n) \in A] = P[(Z_1, \ldots, Z_n) \in A^*], \tag{20}$$

i.e., *dual events have the same probability.*

For example, consider the event that (n, Z_n) is a (strict ascending) ladder point, i.e., that for some fixed n, (5) holds. From (16),

$$\{Z_j^* > 0, j = 1, \ldots, n\}$$

is the same event. From (16) and duality, we have that for every fixed $n \geq 1$,

$$P(Z_n > Z_j, \text{ for all } j < n) = P(Z_j > 0, j = 1, \ldots, n). \tag{21}$$

More generally, for every $n \geq 1$ and $t > 0$, the following are dual events:

$$\{Z_n > Z_j \text{ for all } j < n;\ Z_n \in (0, t]\}, \text{ and} \tag{22}$$

$$\{Z_j > 0, j = 1, \ldots, n;\ Z_n \in (0, t]\}. \tag{23}$$

For each $n \geq 1$, let

$$\begin{aligned} J_n &= 1 \qquad \text{if (23) occurs,} \\ &= 0 \qquad \text{otherwise, where} \end{aligned}$$

$$\sum_{n=1}^{\infty} J_n = \text{number of visits of } \{Z_n\} \text{ to } (0, t] \text{ prior to the first visit to } (-\infty, 0].$$

On the other hand, from (22) and duality

$$E(J_n) = P(\text{a ladder height falls in } (0, t] \text{ at epoch } n). \tag{24}$$

Summing (24) on n, we have

Theorem 1: Duality Theorem. For every $t > 0$, $m(t)$ has two interpretations:

(i) $m(t)$ is the expected number of visits of Z_n to $(0, t]$ prior to the first visit to $(-\infty, 0]$, and

(ii) $m(t)$ is the expected number of strict ascending ladder heights in $(0, t]$.

Example 9-1: Coin Tossing

Let $P(X_n = 1) = p$, $P(X_n = -1) = q$; $p + q = 1$. If $\mathscr{H}_1$ occurs, we clearly have $\mathscr{H}_1 = 1$. Similarly, $\mathscr{H}_1 + \mathscr{H}_2 = 2$ (if both occur). From the theory of Markov chains (or Theorem 3), we know that $\mathscr{H}_1$ is proper if $p \geq q$, and it is easily shown (see Problem 9-2) that $P(\mathscr{H}_1 < \infty) = p/q$ if $p < q$. Hence $m(1) = P(\mathscr{H}_1 \text{ occurs})$ and, for every integer $k \geq 1$,

$$\begin{aligned} m(k) - m(k-1) = P(\mathscr{H}_1 + \cdots + \mathscr{H}_k = k) &= 1 \qquad \text{if } p \geq q, \\ &= (p/q)^k \qquad \text{if } p < q. \end{aligned}$$

For a fair coin ($p = q$), the Duality Theorem gives us this surprising result: For every $k \geq 1$, the expected number of visits of $\{Z_n\}$ to k prior to the first visit to $(-\infty, 0]$ is equal to 1! Clearly, the probability of reaching k prior to visiting $(-\infty, 0]$ decreases with k. However, if this event occurs, the random walk now is at height k, and (for large k) is likely to visit k many times before visiting $(-\infty, 0]$.

The Duality Theorem is the basis for

Theorem 2. There exists only two types of random walks:

(i) *Oscillating*. Both the ascending and descending renewal processes are recurrent ($\mathscr{K}_1$ and $\overline{\mathscr{K}}_1$ are both proper), $\{Z_n\}$ oscillates w.p.1 without bound between $-\infty$ and $+\infty$, and

$$E(\mathscr{K}_1) = E(\overline{\mathscr{K}}_1) = \infty. \tag{25}$$

(ii) *Drift*. $\{Z_n\}$ drifts to either $+\infty$ (*positive drift*) or $-\infty$ (*negative drift*), e.g., for negative drift, $Z_n \to -\infty$ w.p.1 as $n \to \infty$, and reaches a finite maximum. Thus for negative drift,

$$\mathscr{M} \equiv \max\{Z_0, Z_1, \ldots\} < \infty, \text{ w.p.1}, \tag{26}$$

the ascending renewal process is transient ($\mathscr{K}_1$ is defective), the descending renewal process is recurrent ($\overline{\mathscr{K}}_1$ is proper), and

$$E(\overline{\mathscr{K}}_1) < \infty. \tag{27}$$

Proof. For every $n \geq 1$, observe that $\{Z_j \geq 0, 0 \leq j \leq n\} = \{\overline{\mathscr{K}}_1 > n\}$. From duality and that $P(Z_0 \geq 0) = 1$,

$$\begin{aligned}
E\{\overline{\mathscr{K}}_1\} &= \sum_{n=0}^{\infty} P(\overline{\mathscr{K}}_1 > n) \\
&= \sum_{n=0}^{\infty} P(Z_j \geq 0 \quad \text{for } 0 \leq j \leq n) \\
&= 1 + \sum_{n=1}^{\infty} P(Z_n \geq Z_j \quad \text{for } 0 \leq j \leq n) \\
&= 1 + m_w(\infty).
\end{aligned} \tag{28}$$

Thus from (28), (27) holds if and only if the ascending process is transient. Clearly, (26) is the height of the last ascending ladder point that occurs, which is finite w.p.1. Similarly, after each descending ladder height $\overline{\mathscr{H}}_1, \overline{\mathscr{H}}_1 + \overline{\mathscr{H}}_2, \ldots$, the random walk visits $[\overline{\mathscr{H}}_1, \infty), [\overline{\mathscr{H}}_1 + \overline{\mathscr{H}}_2, \infty), \ldots$ a finite number of times. Now

$\sum_{i=1}^{k} \overline{\mathscr{H}}_1 \to -\infty$, as $k \to \infty$, and we have negative drift. Conversely, we have positive drift. Because it is impossible for both the ascending and descending processes to terminate, the only alternative is for both to be recurrent. In this case, (28) implies (25). Because there is no last ladder height in either direction, $\{Z_n\}$ oscillates without bound between $-\infty$ and $+\infty$.

Remark. Duality in general and the proof of Theorem 2 in particular illustrate the connection between strict ladder variables in one direction and weak ladder variables in the other direction. This is why we introduced four types of ladder points in Section 9-3. Nevertheless, the basic properties of the weak and strict processes in the same direction are the same.

We now tie in Theorem 2 with properties of $\{X_n\}$ and F. Suppose that $E(X) \equiv E(X_n)$ is well defined, i.e., either finite, $+\infty$, or $-\infty$. [This rules out $E(X^+) = E(X^-) = \infty$.] By the strong law of large numbers,

$$\lim_{n\to\infty} Z_n/n = E(X) \quad \text{w.p.1.} \tag{29}$$

If $E(X) > 0$, (29) implies that w.p.1, $Z_n < 0$ for at most finitely many n; this is positive drift. Conversely, for $E(X) < 0$, we have negative drift.

Now suppose $E(X) = 0$. At least one renewal process, ascending say, must be recurrent. For every $k \geq 1$, let $n(k)$ be the kth ladder epoch. That is,

$$\frac{Z_{n(k)}}{n(k)} = \frac{\mathscr{H}_1 + \cdots + \mathscr{H}_k}{\mathscr{K}_1 + \cdots + \mathscr{K}_k} = \frac{(\mathscr{H}_1 + \cdots + \mathscr{H}_k)/k}{(\mathscr{K}_1 + \cdots + \mathscr{K}_k)/k}, \tag{30}$$

where the $\mathscr{H}_i$ and the $\mathscr{K}_i$ are i.i.d. As $k \to \infty$, $n(k) = \mathscr{K}_1 + \cdots + \mathscr{K}_k \to \infty$ w.p.1. The strong law implies that subsequences converge, i.e.,

$$\lim_{k\to\infty} Z_{n(k)}/n(k) = E(X) = 0 \quad \text{w.p.1.} \tag{31}$$

On the other hand, as $k \to \infty$, the right-hand side of (30) converges to

$$E(\mathscr{H}_1)/E(\mathscr{K}_1), \tag{32}$$

where $E(\mathscr{H}_1) > 0$ (possibly $+\infty$). This is possible only if $E(\mathscr{K}_1) = \infty$. This implies that the descending renewal process must also be recurrent; we have oscillation. [If $E(\mathscr{H}_1) = \infty$, (32) is undefined, but this does not change the conclusion.] When $E(X) > 0$, (30) through (32) imply

$$E(\mathscr{H}_1) = E(X)E(\mathscr{K}_1), \tag{33}$$

where $E(\mathscr{K}_1) < \infty$. We have proven

Theorem 3. When $E(X) = 0$, we have oscillation; $\mathscr{H}_1$ and $\mathscr{K}_1$ are proper, and $E(\mathscr{K}_1) = \infty$. We have positive drift when $E(X) > 0$, and negative drift when

$E(X) < 0$. When $E(X) > 0$, $E(\mathscr{K}_1) < \infty$, (33) holds, and

$$E(\mathscr{H}_1) < \infty \tag{34}$$

if and only if we also have $E(X) < \infty$.

Remarks. Of theoretical but limited practical interest, the case $E(X^+) = E(X^-) = \infty$ is not covered by Theorem 3. For examples of this type, e.g., when F is a Cauchy distribution, see Feller [1971]. For $0 < E(X) < \infty$, (33) is a special case of *Wald's equation.*

From Theorems 2 and 3 applied to

$$\mathscr{M} = \max(Z_0, Z_1, \ldots), \text{ we have}$$

$$\begin{aligned} \mathscr{M} &= \infty \quad \text{w.p.1} \qquad \text{if } E(X) \geq 0, \\ &< \infty \quad \text{w.p.1} \qquad \text{if } E(X) < 0. \end{aligned} \tag{35}$$

In the latter case, $\mathscr{M}$ is the sum of the $\mathscr{H}_i$ that occur (the sum is the same for the strict and weak cases), i.e.,

$$\mathscr{M} = \sum_{i=1}^{\mathscr{N}} (\mathscr{H}_i \mid \mathscr{H}_i < \infty), \tag{36}$$

where $\mathscr{N}$, the number of ladder heights that occur, has geometric distribution

$$\begin{aligned} P(\mathscr{N} = n) &= P(\mathscr{H}_i < \infty, \quad i = 1, \ldots, n; \mathscr{H}_{n+1} = \infty) \\ &= [1 - H(\infty)][H(\infty)]^n, \qquad n = 0, 1, \ldots . \end{aligned} \tag{37}$$

Given $\mathscr{N} = n$, the $\mathscr{H}_i$ in (36) are i.i.d. with distribution

$$P(\mathscr{H}_i \leq t \mid \mathscr{H}_i < \infty) = H(t)/H(\infty), \qquad t \geq 0. \tag{38}$$

Combining (37) and (38), we have that when $E(X) < 0$,

$$\begin{aligned} P(\mathscr{M} \leq t) &= 1 - H(\infty) + \sum_{n=1}^{\infty} [1 - H(\infty)]H^{(n)}(t) \\ &= [1 - H(\infty)][m(t) + 1], \qquad t \geq 0. \end{aligned} \tag{39}$$

9-4 A REPRESENTATION OF THE LADDER HEIGHT DISTRIBUTION

There is an important relation between the ascending ladder height distribution and the descending ladder height renewal function. As in Section 9-3, we pair weak ascending with strict descending.

For $t \geq 0$, define

$$R_n(t) = P(Z_1 < 0, \ldots, Z_{n-1} < 0, Z_n > t),\ n = 1, 2, \ldots,$$

$$R(t) = \sum_{n=1}^{\infty} R_n(t) = P(t < \mathcal{H}_1^w < \infty).$$

$$\overline{m}_n(t) = P[Z_1 < 0, \ldots, Z_{n-1} < 0, Z_n \in (-t, 0)],\ n = 1, 2, \ldots,$$

$$\overline{m}(t) = \sum_{n=1}^{\infty} \overline{m}_n(t),$$ and for compactness below let

$$\overline{m}_0(t) = 1 \quad \text{and} \quad g(t) = \overline{m}(t) + \overline{m}_0(t).$$

$\overline{m}(t)$ = expected number of visits of $\{Z_n; n \geq 1\}$ to $(-t, 0)$ prior to the first visit to $[0, \infty)$, and, by duality,

= expected number of strict descending ladder heights in $(-t, 0)$.

Except that renewals *at t* are excluded, $g(t)$ is a general (or delayed) *renewal function.*

Conditioning on X_{n+1} yields

$$P(Z_1 < 0, \ldots, Z_n < 0, Z_{n+1} > t \mid X_{n+1} = u, u > t) = \overline{m}_n(u - t),$$ and

$$R_{n+1}(t) = \int_t^{\infty} \overline{m}_n(u - t)\, dF(u), \qquad n = 0, 1, \ldots, \tag{40}$$

and summing over n, we have

$$R(t) = \int_t^{\infty} g(u - t)\, dF(u). \tag{41}$$

In the case of negative drift, the first ascending ladder height may never occur. To account for this, let $\gamma = P(\mathcal{H}_1^w < \infty)$, where $0 < \gamma < 1$ if $E(X) < 0$, and $\gamma = 1$ if $E(X) = 0$. In either case, $\gamma > 0$, and we write

$$R(t)/\gamma = \int_t^{\infty} g(u - t)\, dF(u)/\gamma \tag{42}$$

$$= P(t < \mathcal{H}_1^w < \infty \mid \mathcal{H}_1^w < \infty), \tag{43}$$

which is the *tail* of a proper distribution function.

Remark. Equation (41) may be regarded as a "disguised version" of Equation 3.7(a) on page 399 of Feller [1971]. As we see in the next section, however, (41) leads to a remarkably easy proof of conditions for finite ladder height moments. In Section 9-10, these conditions are used for an easy proof of conditions for finite delay moments for the $GI/G/1$ queue.

9-5 CONDITIONS FOR FINITE LADDER HEIGHT MOMENTS

Our strategy is to obtain linear bounds on renewal function $g(t)$ which, when substituted into (42), produce the desired results.

From the elementary renewal theorem,

$$\lim_{t \to \infty} g(t)/t = -1/E(\overline{\mathscr{H}}_1) \equiv \ell . \tag{44}$$

For $E(X) < 0$, it follows from Theorem 3 that

$$\ell > 0 \quad \text{if and only if} \quad E(X) > -\infty. \tag{45}$$

Conditions for $\ell > 0$ when $E(X) = 0$ will be determined later.

From (44), we can find linear upper and lower bounds on $g(t)$,

$$a + bt < g(t) < c + et \qquad \text{for all } t \geq 0, \tag{46}$$

where $e > 0$, and, when $\ell > 0$, $b > 0$.

Let $J = (\mathscr{H}_1^w \mid \mathscr{H}_1^w < \infty)$. For every $k > 0$, the kth moment of J can be found from (42).

$$\begin{aligned} \gamma E(J^k)/k &= \gamma \int_0^\infty t^{(k-1)} P(J > t)\, dt \\ &= \int_0^\infty t^{(k-1)} \int_t^\infty g(u - t)\, dF(u)\, dt \\ &= \int_0^\infty \int_0^u t^{(k-1)} g(u - t)\, dt\, dF(u). \end{aligned} \tag{47}$$

Replacing $g(t)$ in (47) by the upper bound in (46), we get

$$\begin{aligned} \gamma E(J^k)/k &\leq \int_0^\infty \int_0^u t^{(k-1)}[c + e(u - t)]\, dt\, dF(u) \\ &\leq \int_0^\infty [cu^k/k + eu^{(k+1)}/k(k + 1)]\, dF(u). \end{aligned} \tag{48}$$

From (48) and $e > 0$,

$$E\{(X^+)^{(k+1)}\} < \infty \Rightarrow E(J^k) < \infty. \tag{49}$$

Similarly, whenever $\ell > 0$, $b > 0$, and, from the lower bound in (46),

$$E(J^k) < \infty \Rightarrow E\{(X^+)^{(k+1)}\} < \infty. \tag{50}$$

Note that (49) and (50) hold when $\mathscr{H}^w$ and J are replaced by *strict* ladder heights. Furthermore, when $E(X) = 0$, the counterpart of (49) holds for descending ladder heights. In particular,

$$E\{(X^-)^2\} < \infty \Rightarrow -E(\mathscr{H}_1) < \infty, \tag{51}$$

i.e., $\ell > 0$, and (50) holds!

By combining the above, we have

Theorem 4. For a random walk with $-\infty \leq E(X) \leq 0$, and every $k > 0$,

$$E\{(X^+)^{(k+1)}\} < \infty \Rightarrow E(J^k) < \infty,$$

where $J = (\mathscr{H}_1^w \mid \mathscr{H}_1^w < \infty)$ has distribution (42). If, in addition,

either (i) $-\infty < E(X) < 0$,

or (ii) $E(X) = 0$ and $E\{(X^-)^2\} < \infty$,

$$E(J^k) < \infty \Rightarrow E\{(X^+)^{(k+1)}\} < \infty.$$

Remark. Equation (45) is easily extended to other moments. That is, under negative drift,

$$E\{(X^-)^k\} < \infty \Leftrightarrow E\{|\mathscr{H}_1|^k\} < \infty \qquad \text{for } k > 0. \tag{52}$$

We will show (52) in Section 9-10.

9-6 A COMBINATORIAL RESULT

We will determine the distribution of $\mathscr{H}_1$, the first (strict ascending) ladder epoch, by a method that depends on a combinatorial result about the cyclic arrangement of real numbers.

The terminology is the same as in Section 9-2, where $x_1, x_2, \ldots$ are now arbitrary real numbers, with partial sums

$$z_0 = 0, \quad z_n = \sum_{j=1}^{n} x_j, \qquad j = 1, 2, \ldots .$$

As in (5), we say $n \geq 1$ is a ladder epoch if $z_n > z_j$ for all $j < n$.

For fixed $n \geq 1$, define and number the n cyclic *arrangements* of $(x_1, \ldots, x_n)$:

number	*arrangement*
0	$(x_1, \ldots, x_n)$
1	$(x_2, \ldots, x_n, x_1)$
$\vdots$	$\vdots$
$n-1$	$(x_n, x_1, \ldots, x_{n-1})$

Let $z_j^{(k)}$ be the sum of the first j elements in arrangement k, where it is easy to see that

$$z_j^{(k)} = \begin{cases} z_{k+j} - z_k & \text{for } j = 0, \ldots, n-k, \\ z_n - z_k + z_{j-n+k} & \text{for } j = n-k+1, \ldots, n. \end{cases} \tag{53}$$

We now state an elementary but important combinatorial result.

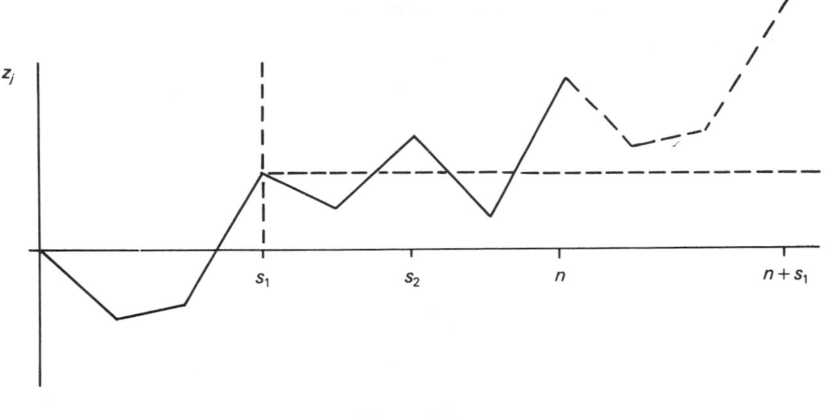

Figure 9-2

Theorem 5. For arbitrary fixed $n \geq 1$, let r be the number of arrangements for which n is a ladder epoch. Then for each of these arrangements, n is the rth ladder epoch. Furthermore, $r \geq 1$ if and only if $z_n > 0$.

Proof. Suppose $z_n > 0$, for otherwise n cannot be a ladder epoch for any arrangement. Let s be the largest ladder epoch for arrangement 0, in the range $1, \ldots, n$. From (53), we see that n is a ladder epoch for arrangement s. Thus $r \geq 1$. (Also see Figure 9-2 and how it is used in the remainder of the proof.) Now relabel arrangement s as arrangement 0, i.e., suppose this was the initial arrangement. For some $r \geq 1$, suppose n is the rth ladder epoch, where the earlier ladder epochs occur at $s_1 < \cdots < s_{r-1}$. Now consider each arrangement in turn, where arrangements are represented by shifting the origin (up or down) and over. In Figure 9-2, where $n = 7$ and $r = 3$, we have shown the shift corresponding to s_1 by dashed lines. It is easy to see that only for arrangements $0, s_1, \ldots, s_{r-1}$ will n be a ladder epoch, where in each case, n is the rth ladder epoch.

Remark. We have stated Theorem 5 for the strict ascending case. It holds for the weak case as well; the only change is that $r \geq 1$ if and only if $z_n \geq 0$.

9-7 THE DISTRIBUTION OF LADDER EPOCHS

We now revert to earlier notation, where uppercase letters denote random variables. Denote the distribution of $\mathscr{K}_1$ by

$$\alpha_n = P(\mathscr{K}_1 = n) = P(Z_1 \leq 0, \ldots, Z_{n-1} \leq 0, Z_n > 0), \qquad n = 1, 2, \ldots,$$

with generating function

$$E(z^{\mathscr{K}_1}) = \sum_{n=1}^{\infty} \alpha_n z^n \equiv \mathscr{A}(z), \qquad |z| \leq 1.$$

Recall properties of generating functions: When $\mathscr{K}_1$ is proper,

$$\mathscr{A}(z) \to \mathscr{A}(1) = 1, \text{ as } z \to 1; \text{ when } \mathscr{K}_1 \text{ is defective, } \mathscr{A}(1) < 1.$$

We now state this section's main result, where ln denotes natural log.

Theorem 6

$$\ln \frac{1}{1 - \mathscr{A}(z)} = \sum_{n=1}^{\infty} z^n P(Z_n > 0)/n. \tag{54}$$

Proof. For the cyclic arrangements of $(X_1, \ldots, X_n)$, and each $r \geq 1$, let

$$\begin{aligned} J_s &= 1 \quad \text{if } n \text{ is the } r\text{th ladder epoch for arrangement } s, \\ &= 0 \quad \text{otherwise,} \quad \text{for } s = 0, \ldots, n-1. \end{aligned}$$

Now

$$E(J_0) = P(J_0 = 1) = P(\mathscr{K}_1 + \cdots + \mathscr{K}_r = n) \equiv \alpha_n^{(r)},$$

where $\{\alpha_n^{(r)}, n \geq 1\}$ denotes the r-fold convolution of $\{\alpha_n\}$, and has generating function

$$\sum_{n=1}^{\infty} \alpha_n^{(r)} z^n = [\mathscr{A}(z)]^r, \qquad r \geq 1. \tag{55}$$

Furthermore, since the X_j are i.i.d., the J_s are identically distributed, and, from Theorem 5, the sum

$$J_0 + \cdots + J_{n-1}$$

can take on only two possible values, 0 and r. Hence,

$$\begin{aligned} \alpha_n^{(r)} &= E(J_0) = E(J_0 + \cdots + J_{n-1})/n \\ &= rP(J_0 + \cdots + J_{n-1} = r)/n. \end{aligned} \tag{56}$$

Dividing (56) by r and summing, where $r \geq 1$ if and only if $Z_n > 0$, we have

$$\sum_{r=1}^{\infty} \alpha_n^{(r)}/r = \sum_{r=1}^{\infty} P(J_0 + \cdots + J_{n-1} = r)/n = P(Z_n > 0)/n. \tag{57}$$

From (55), we get (54) by multiplying (57) by z^n and summing on n:

$$\sum_{n=1}^{\infty} \sum_{r=1}^{\infty} z^n \alpha_n^{(r)}/r = \sum_r \sum_n = \sum_{r=1}^{\infty} [\mathscr{A}(z)]^r/r = \ln \frac{1}{1 - \mathscr{A}(z)} = \sum_{n=1}^{\infty} z^n P(Z_n > 0)/n. \tag{58}$$

Theorem 6 is remarkable in that the distribution of $\mathscr{K}_1$ depends only on the distribution of the Z_n at the origin, as we now illustrate.

Example 9-2

Let F be symmetric and continuous, so that $P(Z_n > 0) = 1/2$ for all n. Equation (54) becomes

$$\ln \frac{1}{1 - \mathscr{A}(z)} = \sum_{n=1}^{\infty} z^n/2n = (1/2) \ln [1/(1 - z)], \text{ or}$$

$$\mathscr{A}(z) = 1 - \sqrt{1 - z}, \tag{59}$$

independent of the particular F with this property!

Another application of Theorem 6 is

Theorem 7. A random walk drifts to $-\infty$ if and only if

$$\sum_{n=1}^{\infty} P(Z_n > 0)/n < \infty. \tag{60}$$

Proof. A random walk has negative drift if and only if $\mathscr{K}_1$ is defective, in which case $\mathscr{A}(1) < 1$, and the left-hand side of (54) converges as $z \to 1$.

9-8 STABILITY CONDITIONS FOR THE *GI/G/1* QUEUE; ASSOCIATED QUEUES

For the remainder of this chapter, assume $-\infty \le E(X) < \infty$. For every random walk, we now introduce an *associated* single-channel FIFO queue.

Let $\{(S_n, T_n); n \ge 1\}$ be an i.i.d. sequence of pairs of nonnegative random variables, and

$$X_n = S_n - T_n, \qquad n = 1, 2, \ldots, \tag{61}$$

where the X_n are i.i.d., and generate the random walk in Section 9-2. The S_n and T_n are the respective service times and inter-arrival times at a single-channel FIFO queue; they generate delay sequence $\{D_n\}$ in Section 9-1.

Note that we do *not* require that S_n and T_n be independent. [For some distributions of X_n, this requirement is impossible to meet, e.g., when $P(X_n = j) > 0$ for every integer j, except that $P(X_n = 0) = 0$.] Nevertheless, we continue to use our standard notation; $E(T) \equiv E(T_n) = 1/\lambda$, and $E(S) \equiv E(S_n) = 1/\mu$, and $\rho = \lambda/\mu$, where λ and μ are arrival and service rates. In particular, it is always possible to define an associated queue for any distribution F by

$$S_n = X_n^+ \quad \text{and} \quad T_n = X_n^-, \qquad n = 1, 2, \ldots. \tag{62}$$

When $\{S_n\}$ and $\{T_n\}$ are independent sequences, we have the standard *GI/G/1* queue defined in Section 9-1. Thus an associated queue is a minor generalization of the *GI/G/1* queue, which we can define in this manner because $\{D_n\}$ depends only on $\{X_n\}$.

With this definition, we also ensure that an associated queue exists for every random walk. In fact, there are many, e.g., for any constant $c > 0$, let $S_n = X_n^+ + c$, and $T_n = X_n^- + c$, $n \geq 1$. More generally, we could add the same nonnegative random variable to both X_n^+ and X_n^-. To avoid trite technicalities, we assume $E(S) < \infty$.

In the $GI/G/1$ case, the distributions F, A, and G, are related by

$$F(t) = \int_0^\infty G(t + u)\, dA(u) \qquad \text{for all } t. \tag{63}$$

If $D_1 = 0$ in (3), we have as an elementary application of duality that

$$D_{n+1} \sim M_n \equiv \max(Z_0, \ldots, Z_n), \qquad n = 1, 2, \ldots. \tag{64}$$

As $n \to \infty$, $M_n \to \mathscr{M}$ w.p.1, where $\mathscr{M}$ has properties (35), and, for $E(X) < 0$, distribution (39). From (64), $\{D_n\}$ *converges in distribution* to random variable D, say, where $D \sim \mathscr{M}$.

Now suppose $D_1 > 0$, where its value is independent of $\{X_n\}$. This can only make every D_n larger, and will not affect convergence in distribution (to $+\infty$) when $E(X) \geq 0$. When $E(X) < 0$, $Z_n \to -\infty$, as $n \to \infty$, and from (3), $\{D_n\}$ converges in distribution to D for every finite value of D_1. We have proven

Theorem 8. For every initial value of D_1, $\{D_n\}$ for an associated queue converges in distribution to D. For $E(X) < 0$ ($\rho < 1$), D is proper and has distribution (39). For $E(X) \geq 0$ ($\rho \geq 1$), $D = \infty$.

Remarks. The senses in which $\{D_n\}$ and $\{M_n\}$ converge are quite different. From (64), realizations $\{M_n(\omega)\}$ are monotone nondecreasing as $n \to \infty$. From (2), $\{D_n\}$ does not have this property. For $E(X) \leq 0$, $\{D_n\}$ does not converge to a random variable; in fact, in Section 9-9 we find that it is a regenerative process. For $E(X) < 0$, the mean cycle length is finite, and we call $\{D_n\}$ *stable.* For $E(X) \geq 0$, we call $\{D_n\}$ *unstable.*

9-9 BUSY PERIODS

For an associated queue generated by $\{(S_n, T_n)\}$, let customer C_1 arrive at epoch 0 and find the system empty. Thus $D_1 = 0$, and C_1 initiates a *busy period,* a time interval during which the server is continuously busy. Define

$$Y_n = (D_n + S_n - T_n)^- = (D_n + X_n)^-, \qquad n = 1, 2, \ldots. \tag{65}$$

When $Y_n > 0$, it is called an *idle period*, and is the time between the departure of C_n and the arrival of C_{n+1}. When $Y_n = 0$, C_n departs at or after the arrival epoch of C_{n+1}. In either case, Y_n is the *idle time of the server* between the departure of C_n and the commencement of service on C_{n+1}.

By combining (65) with (2), we have

$$D_{n+1} - Y_n = D_n + S_n - T_n = D_n + X_n, \qquad n = 1, 2, \ldots. \tag{66}$$

Summing (66), we have

$$D_{n+1} = \sum_{i=1}^{n} S_i - \left(\sum_{i=1}^{n} T_i - \sum_{i=1}^{n} Y_i\right) = Z_n + \sum_{i=1}^{n} Y_i, \qquad n = 1, 2, \ldots, \tag{67}$$

where $\sum S_i$ is the cumulative work brought by arrivals prior to C_{n+1}, $\sum T_i$ is the corresponding elapsed time, and $\sum Y_i$ is the cumulative prior idle time of the server. Thus $\sum T_i - \sum Y_i$ is the prior *work performed,* and the middle expression in (67) has the interpretation:

$$\text{FIFO delay} = \text{work found} = \text{cumulative work} - \text{work performed}. \tag{68}$$

Remark. For $D_1 > 0$, D_1 would be added to the middle and right-hand expressions in (67). If we call it *initial work* and include it in cumulative work, (68) is still true.

For some $k \geq 1$, consider the event

$$\{Y_1 = 0, \ldots, Y_{k-1} = 0, Y_k > 0\}. \tag{69}$$

From (67), (69) is equivalent to $\{D_{n+1} = Z_n \geq 0,\ n < k, \text{ and } D_{k+1} = 0 > Z_k\}$, which is the event

$$\{Z_n \geq 0,\ n = 1, \ldots, k - 1;\ Z_k < 0\}. \tag{70}$$

That is, for i.i.d. $\{X_n\}$, (69) is equivalent to $\overline{\mathcal{K}}_1 = k$ and $\overline{\mathcal{H}}_1 = Z_k = -Y_k < 0$.

On the other hand, (69) also means that k customers are served during the first busy period, Y_k is the duration of the first idle period, and C_{k+1} initiates the second busy period. Busy and idle periods were defined in Chapter 8. Henceforth, we adopt the notation used there, i.e., $K_1 = \overline{\mathcal{K}}_1$, and $I_1 = -\overline{\mathcal{H}}_1$. For $i \geq 1$, let

$$K_i = \text{number of customers served in the } i\text{th busy period},$$

$$I_i = \text{duration of the } i\text{th (subsequent) idle period, where}$$

$$B_1 = \sum_{n=1}^{K_1} S_n = \text{duration of the first busy period, and}$$

$$B_1 + I_1 = \sum_{n=1}^{K_1} T_n = \text{duration of the first busy cycle.}$$

The K_i are i.i.d. When K_1 is proper, $\{D_n\}$ is a discrete-time regenerative process, with regeneration points K_1, $K_1 + K_2, \ldots$, and the number (of customers) in system process $\{N(t)\}$ is regenerative, with regeneration points $(B_1 + I_1)$, $(B_1 + I_1) + (B_2 + I_2), \ldots$. Furthermore, K_1 is a stopping time for both $\{S_n\}$ and $\{T_n\}$. From Theorem 3 and Wald's equation, we have

Theorem 9. For an associated queue, K_1, B_1, and I_1 are proper if and only if $E(X) \leq 0$ $(\rho \leq 1)$.

$$\text{For } E(X) = 0 \ (\rho = 1), \qquad E(K_1) = E(B_1) = \infty.$$

$$\text{For } E(X) < 0 \ (\rho < 1), \qquad E(K_1) < \infty,$$

$$E(B_1) = E(K_1)/\mu < \infty \quad \text{and} \quad E(B_1) + E(I_1) = E(K_1)/\lambda,$$

where $E(I_1) < \infty$ if and only if $\lambda > 0$ $(0 < \rho < 1)$.

Thus $\{D_n\}$ has finite mean cycle length where $\rho < 1$, and $\{N(t)\}$ has finite mean cycle length when $0 < \rho < 1$.

Analogous to Theorem 6, we have an expression for the generating function $\Gamma(z) = E\{z^{K_1}\}$:

Theorem 10.

$$\ln \frac{1}{1 - \Gamma(z)} = \sum_{n=1}^{\infty} z^n P(Z_n < 0)/n. \tag{71}$$

Important Remarks. Consider the work-in-system process $\{V(t); t \geq 0\}$ first introduced in Section 5-13. $\{V(t)\}$ is regenerative whenever $\{N(t)\}$ is, and has the same regeneration points. While equations such as (66) depend on FIFO, the results in this section do not. However $\{S_n\}$ and $\{T_n\}$ are defined, we assume that the sample paths of $\{V(t)\}$ are determined by $\{(S_n, T_n)\}$, as in Figure 5-15, and we say that *work is conserved*. Thus the distribution of K_1 depends only on $\{X_n\}$. The distributions of B_1 and I_1 depend on $\{(S_n, T_n)\}$, but not order of service. Also notice that we have chosen to define K_1, B_1, and I_1 in terms of *strict* descending ladder variables. In particular, idle periods are strictly positive. This was done for convenience and clarity. We could have chosen weak ladder variables, which would change the distributions but not the fundamental properties of these quantities. The sequence $\{D_n\}$ is not affected by this choice.

9-10 CONDITIONS FOR FINITE DELAY MOMENTS

Consider a single-channel FIFO queue associated with a random walk. Assume $E(X) < 0$, so that $\{D_n\}$ is stable, and D is proper and has distribution (39). In this section, we obtain conditions for D to have finite moments.

Now D (technically, $\mathscr{H}$) has representation

$$D = \sum_{i=1}^{\mathscr{N}} J_i, \tag{72}$$

where $\mathscr{N}$ is the number of strict ascending ladder heights that occur, and has geometric distribution (37), with finite moments of all orders. In this section, $J_i \equiv (\mathscr{H}_i \mid \mathscr{H}_i < \infty)$, but they have the same moment properties as the weak ladder height called J in Theorem 4.

From (72) and Minkowski's inequality (Kingman and Taylor [1966], p. 184),

$$E(D^k \mid \mathscr{N}) \leq \mathscr{N}^k E(J^k) \qquad \text{for } k \geq 1. \tag{73}$$

Since the kth power is a concave function for $0 < k < 1$,

$$E(D^k \mid \mathscr{N}) \leq \mathscr{N} E(J^k), \qquad 0 < k < 1. \tag{74}$$

From (37) and (72) through (74),

$$H(\infty)E(J^k) \leq E(D^k) \leq a(k)E(J^k), \qquad k > 0, \tag{75}$$

where $a(k) = E(\mathscr{N}^k)$ for $k \geq 1$, and $a(k) = E(\mathscr{N})$ for $0 < k < 1$. Because $\mathscr{N}$ has finite moments of all orders, (75) implies that

$$E(D^k) < \infty \quad \text{if and only if} \quad E(J^k) < \infty \qquad \text{for every } k > 0. \tag{76}$$

From (76) and Theorem 4, we have

Theorem 11. For an associated queue generated by $\{(S_n, T_n)\}$, with

$$X_n = S_n - T_n, \qquad n = 1, 2, \ldots,$$

where the X_n are i.i.d., with mean $-\infty \leq E(X) < 0$, and every $k > 0$,

$$E\{(X^+)^{(k+1)}\} < \infty \Rightarrow E(D^k) < \infty.$$

If in addition, $E(X) > -\infty$ ($\lambda > 0$),

$$E(D^k) < \infty \Rightarrow E\{(X^+)^{(k+1)}\} < \infty.$$

For the $GI/G/1$ case, we may replace $E\{(X^+)^{(k+1)}\} < \infty$ by $E\{S^{(k+1)}\} < \infty$.

Important Remark. Theorem 11 is consistent with properties of delay for the $M/G/1$ and $GI/M/1$ queues found in Chapter 8. Note that aside from $\lambda > 0$, the inter-arrival distribution plays no role in this result. This suggests that when $\{S_n\}$ and $\{T_n\}$ are independent sequences, the conclusions of this theorem may hold for very general arrival processes. If true, there are important implications for tandem and network queues. While appropriate conditions are an open question at this time, some additional restrictions on the arrival process will be necessary. For example, when batch arrivals are permitted, there is a "batch size biasing" effect, and higher moments of the batch size distribution play a role.

We now determine conditions for finite idle period moments when $E(X) < 0$. In this case, $\{Y_n\}$ is regenerative with finite mean cycle length. Let Y_∞ have its limiting distribution, which from (65) exists in a pointwise sense as $n \to \infty$. In fact, we can use (65) to represent Y_∞.

$$Y_\infty = (D + X_n)^- \le X_n^-, \tag{77}$$

where D, the stationary delay, has distribution (39). From (77), Y_∞ has finite kth moment when X_n^- does, and (condition on D), Y_∞ has infinite kth moment when X_n does. The subsequence of strictly positive Y_n are the idle periods, and it follows that an idle period has the representation

$$I = (Y_\infty \mid Y_\infty > 0), \tag{78}$$

and hence the same moment properties as Y_∞. Thus when $E(X) < 0$ $(\lambda > 0)$

$$E\{(X^-)^k\} < \infty \Leftrightarrow E(I^k) < \infty \qquad \text{for } k > 0, \tag{79}$$

which is equivalent to (52).

9-11 PREEMPTIVE LIFO AND WORK

Now consider an associated single-channel queue under *preemptive LIFO* (*PL*). Every customer begins service on arrival, interrupting any customer being served. On service completion, the most recent arrival in queue resumes service. [Now $P(T_n = 0) > 0$ is possible, and we define the "most recent" arrival as the highest numbered arrival in queue.] The remaining service when a customer resumes service is what it was at the point of interruption, so that *work is conserved,* and the sample paths of $\{V(t)\}$ are the same under PL as under FIFO.

Work found by arrival $n \ge 1$ is also the same under PL and FIFO, and we still denote it by D_n, the FIFO delay. When $P(T_n > 0) = 1$, D_n is simply

$$D_n = V(t_n^-),$$

where t_n is the arrival epoch of C_n. Thus PL will turn out to be a tool for analyzing FIFO, which is our main interest in it.

Assume $-\infty < E(X) < 0$, so that $\{V(t)\}$ has finite mean cycle length. We now relate the distributions of stationary FIFO delay D to stationary work in system V. Under PL, we can split up $V(t)$,

$$V(t) = D_{\varkappa(t)} + S_{\varkappa(t),r}, \qquad t \ge 0, \tag{80}$$

where $\varkappa(t)$ is the arrival number of the customer in service at epoch t, $D_{\varkappa(t)}$ is the work found by that customer on arrival, and $S_{\varkappa(t),r}$ is the remaining service of that customer. If the system is empty at epoch t, we set $D_{\varkappa(t)} = S_{\varkappa(t),r} = 0$. Note that (80) *holds on sample paths.*

For every fixed $n \ge 1$, consider the set $\{t\colon \varkappa(t) = n\}$ in (80). If $S_n \le T_n$, this set is an interval of length S_n. Otherwise, this set is the union of intervals of *total length* S_n, all within the same busy period. For all points in this set, $D_{n(t)} = D_n$, where D_n and S_n *are independent*.

With this observation, we now find the joint stationary distribution of the right-hand quantities in (80) by finding the corresponding fraction of time: Let D_n and S_{nr} denote these stationary quantities, where for $x, y \geq 0$,

$$I_{xy}(t) = 1 \qquad \text{if } D_{n(t)} > x \quad \text{and} \quad S_{n(t),r} > y,$$

$$= 0 \qquad \text{otherwise, and}$$

$$P(D_n > x, S_{nr} > y) = \lim_{t\to\infty} \int_0^t I_{xy}(u)\, du/t. \tag{81}$$

We find limit (81) as an application of Theorem 5 in Section 5-15 ($\overline{H} = \lambda\overline{G}$). [It is easily shown that (198) in Section 5-15 holds.] The contribution of C_n to the integral in (81) is represented in terms of the indicator of the event $\{D_n > x\}$,

$$I_{\{D_n > x\}}(S_n - y)^+, \tag{82}$$

where, averaged over customers, $D_n \sim D$. From this and independence, we have

$$P(D_n > x, S_{nr} > y) = \lambda E\{I_{\{D_n>x\}}(S_n - y)^+\} = \lambda P(D > x) \int_y^\infty P(S_n > u)\, du. \tag{83}$$

Now $P(S \leq t) = G(t)$, $t \geq 0$, and let $S_e \sim G_e$,

$$P(S_e \leq y) = \mu \int_0^y G^c(t)\, dt = G_e(y), \qquad y \geq 0, \tag{84}$$

the familiar *equilibrium* service distribution, and (83) becomes

$$P(D_n > x, S_{nr} > y) = \rho P(D > x) P(S_e > y), \qquad x, y \geq 0. \tag{85}$$

From (80), (81), and (85), we have

Theorem 12. For an associated single-channel queue with $-\infty < E(X) < 0$, the stationary distribution of work in system V has representation

$$P(V > v) = \rho P(D + S_e > v), \qquad v \geq 0, \tag{86}$$

where D and S_e are independent, D is the stationary delay under FIFO, and S_e has equilibrium service distribution (84).

Remark. The formality of the proof obscures the intuitive content of this theorem. At a random point in time, the remaining service of anyone found in service is $\sim G_e$. Under PL, the additional work present is that found on arrival by the customer in service at this random time, and is the delay this customer would have under FIFO. Because D_n and S_n are independent, there are no hidden length-biasing effects, and these quantities are independent.

The equilibrium distribution has the property

$$E\{S^{(k+1)}\} < \infty \Leftrightarrow E(S_e^k) < \infty \qquad \text{for } k > 0.$$

Now $E(S^k) < \infty \Rightarrow E\{(X^+)^k\} < \infty$. From Theorems 11 and 12, we have

Theorem 13. For an associated single-channel queue with $-\infty < E(X) < 0$,

$$E\{S^{(k+1)}\} < \infty \Leftrightarrow E(V^k) < \infty \qquad \text{for } k > 0. \tag{87}$$

Important Generalization. We have shown Theorem 12 for an associated queue, a generalization of $GI/G/1$ that permits some dependence between S_n and T_n. There is an important generalization in a different direction: Let $\{S_n\}$ and $\{T_n\}$ be independent sequences, where the S_n are i.i.d., but the T_n need not be. Suppose $\{T_n\}$ is either regenerative with finite mean cycle length, or ergodic and stationary, with arrival rate λ, where $0 < \rho < 1$. In either case (because of the existence of a stationary version, regenerative is a special case) it has long been known (Loynes [1962]) that $\{D_n\}$ has a stationary ergodic version and a unique limiting distribution. Theorem 12 holds for this case, and our proof carries over. Note that D_n and S_n are still independent.

For a standard $GI/G/1$ queue with $0 < \rho < 1$, stationary work in system also has the representation

$$V \sim (D + S - T_e)^+, \tag{88}$$

where D, S, and T_e are independent. D is the stationary FIFO delay, $S \sim G$, and $T_e \sim A_e$, the *equilibrium* inter-arrival distribution. See Problem 9-9.

Remarks. The intuitive content of (88) is this: At a random moment, the time since the last arrival is $\sim T_e$. That arrival finds work D and brings work S. The remaining work at the random moment is $D + S$ minus elapsed time T_e, if positive. Again there are no hidden length-biasing effects. While we may write (88) for more general single-channel queues, it will not have the same meaning; the random variables on the right are dependent: For an associated queue where S_n and T_n are dependent, S and T_e are dependent. For the important generalization following Theorem 13, D and T_e are dependent.

9-12 WIENER-HOPF FACTORIZATION; THE DELAY DISTRIBUTION TRANSFORM

Relation (41) between the ascending and descending ladder height processes, the form needed in Section 9-5, is only one of several ways of expressing essentially the same information.

We now derive another version of this result, again pairing weak ascending with strict descending. This time we relate the ascending renewal function to the descending ladder height distribution.

To steamline the equations that follow, we also introduce notation for set functions (measures) that correspond to distributions and renewal functions, e.g., $F\{J\} = P(X_n \in J)$ for any interval J.

For $A = [0, \infty)$ and any interval J, let

$$m_n^w\{J\} = P(Z_1 \geq 0, \ldots, Z_{n-1} \geq 0, Z_n \in JA), \qquad n \geq 0, \text{ and}$$

$$\overline{H}_n\{J\} = P(Z_1 \geq 0, \ldots, Z_{n-1} \geq 0, Z_n \in JA^c), \qquad n \geq 1,$$

where $m_0^w\{J\}$ is the measure corresponding to the unit step at the origin. Conditioning on $\{Z_1 \geq 0, \ldots, Z_n \geq 0\}$ and Z_n, we have for $n \geq 0$ that

$$m_{n+1}^w\{J\} = \int_0^\infty F\{J - x\}\, dm_n^w(x) \qquad \text{for } J \subset A, \text{ and} \tag{89}$$

$$\overline{H}_{n+1}\{J\} = \int_0^\infty F\{J - x\}\, dm_n^w(x) \qquad \text{for } J \subset A^c, \tag{90}$$

where $m_n(x) = m_n\{(0, x]\}$. Thus for arbitrary J, $J = JA \cup JA^c$, and the above equations can be combined and written as

$$m_{n+1}^w\{J\} + \overline{H}_{n+1}\{J\} = \int_0^\infty F\{J - x\}\, dm_n^w(x), \qquad n \geq 0, \tag{91}$$

where the integral is a convolution that we write as $F * m_n^w\{J\}$. Summing on n, where by duality $\sum_{n=1}^{\infty} m_n^w = m_w$ as defined in (12),

$$m_w\{J\} + \overline{H}\{J\} = F\{J\} + F * m_w\{J\}. \tag{92}$$

Convolving (92) with H_w and using $m_w * H_w = m_w - H_w$,

$$m_w - H_w + \overline{H} * H_w = F * m_w,$$

which, when combined with (92) yields

$$F = H_w + \overline{H} - \overline{H} * H_w, \tag{93}$$

a remarkable relation between the measure F and the corresponding ascending and descending ladder height measures. By setting $J = (-\infty, t]$, (93) becomes a relation between point functions that is of the same form.

Taking the transform of (93), we have

$$\tilde{F}(s) = \tilde{H}_w(s) + \tilde{\overline{H}}(s) - \tilde{\overline{H}}(s)\tilde{H}_w(s). \tag{94}$$

To ensure the existence of these transforms, s is restricted to be imaginary, i.e., (94) is a relation between *characteristic functions*.

Now (92), (93), and (94) are equivalent; largely because of (94) and similar expressions, they are called *Wiener-Hopf factorizations*. It can be shown (Feller [1971], p. 402) that they are unique. In particular, this means that if G^+ and G^- are (possibly defective) probability measures concentrated on A and A^c, respectively, such that

$$F = G^+ + G^- - G^+ * G^-,$$

then $G^+ = H_w$ and $G^- = \overline{H}$. (Why can't they both be defective?)

Important Remark. Equation (93) holds with strict ascending and weak descending ladder height distributions, as does (92) for the corresponding processes, with strict inequalities replacing weak ones in the derivation, and $A = (0, \infty)$.

Now suppose we have negative drift, so that the ascending process terminates and the corresponding associated queue is stable. An idle period $I = -\overline{\mathscr{H}}$, and the transform of (92) may be written

$$\tilde{m}(s) + 1 = [1 - \tilde{I}(-s)]/[1 - \tilde{F}(s)].$$

Stationary delay $D \sim \mathscr{M}$, and the transform of the weak-ascending version of (39) is

$$\tilde{D}(s) = [1 - H_w(\infty)][1 - \tilde{I}(-s)]/[1 - \tilde{F}(s)]. \tag{95}$$

From (15) and (28), (95) can be written

$$\tilde{D}(s) = a_0[1 - \tilde{I}(-s)]/[1 - \tilde{F}(s)], \tag{96}$$

where $a_0 = 1/E(K_1)$. Equation (96) is derived independently as equation (14) in Chapter 11.

For completeness, we state without proof another classical result:

Theorem 14: Spitzer's Identity. For $|r| < 1$ and imaginary s,

$$\sum_{n=0}^{\infty} r^n E(e^{-sM_n}) = \exp\left\{\sum_{n=1}^{\infty} r^n E(e^{-sZ_n^+})/n\right\}, \tag{97}$$

and, under negative drift,

$$\tilde{D}(s) = E(e^{-s\mathscr{M}}) = \exp\left\{\sum_{n=1}^{\infty} [E(e^{-sZ_n^+}) - 1]/n\right\}. \tag{98}$$

Spitzer's identity has been used to derive our next result, but this involves a rather delicate differentiation. (There are similar problems with respect to differentiating (96); see technical remarks following equation (14) in Chapter 11.) Instead we present a direct derivation based on Asmussen [1987], page 177. In expressions below $\{X; E\}$ denotes the random variable

$$\{X; E\}(\omega) = \begin{cases} X(\omega) & \text{if } \omega \in E, \\ 0 & \text{otherwise.} \end{cases}$$

For $n \geq 1$, write

$$M_n = \{M_n; Z_n > 0\} + \{M_n; Z_n \leq 0\}$$
$$= \{X_1; Z_n > 0\} + \{\max_{1 \leq i \leq n} (Z_i - X_1); Z_n > 0\} + \{M_{n-1}; Z_n \leq 0\}.$$

Now the middle term above $\sim \{M_{n-1}; Z_n > 0\}$, $\{X_1; Z_n > 0\} \sim \{X_i; Z_n > 0\}$ for $1 \leq i \leq n$, and $Z_n^+ = \sum_{i=1}^{n} \{X_i; Z_n > 0)$. From these observations,

$$E(M_n) = E(Z_n^+)/n + E(M_{n-1}), \qquad n \geq 1,$$

and we have

$$E(M_n) = \sum_{i=1}^{n} E(Z_i^+)/i, \; n \geq 1. \tag{99}$$

From (99) and monotone convergence, it is immediate that

$$d = E(\mathscr{M}) = \sum_{n=1}^{\infty} E(Z_n^+)/n. \tag{100}$$

In spite of the simple appearance of (100) and the elegance of Spitzer's identity, these expressions are awkward computationally. Notice that conditions for $d < \infty$ are not apparent from (100). Even for finite d, convergence of (100) will be slow for ρ near 1. We can find lower bounds on d that have some utility in light traffic. This is considered in Chapter 11.

On the other hand, when either the ascending or the descending ladder height distribution is known, Wiener-Hopf factorization determines the other, and various distributions of interest can be computed. This occurs in important special cases, e.g., when F has an exponential tail (see Problems 9-4 and 9-5). This is illustrated by Example 9-3 and also by Problem 9-17, which generalizes the $GI/M/1$ queue.

When X is integer valued, another special case occurs when X is *skipfree*. We say X is skipfree *to the right* if $P(X \geq 2) = 0$, and skipfree *to the left* if $P(X \leq -2) = 0$. When X is skipfree to the right, $(\mathscr{H}_1 \mid \mathscr{H}_1 < \infty) = 1$. Problems 9-3 and 9-4 are particularly easy because X is skipfree in both directions.

Example 9-3: The $M/G/1$ Queue

For a stable $M/G/1$ queue, we know that $-\overline{\mathscr{H}} = I \sim \exp(\lambda)$, and $\tilde{F}(s) = \tilde{G}(s)\tilde{A}(-s)$, where $A(s) = \lambda/(\lambda + s)$. From (94)

$$s\tilde{H}_w(s) = \lambda[1 - \tilde{G}(s)],$$
$$\tilde{H}_w(s) = \rho\mu[1 - \tilde{G}(s)]/s = \rho\tilde{G}_e(s),$$

where $\tilde{G}_e$ is the transform of the *equilibrium* service distribution. (Note that a characteristic function determines the corresponding distribution even when restricted to the imaginary s.) Thus $(\mathscr{H}_i^w \mid \mathscr{H}_i^w < \infty) \sim G_e$, and $P(\mathscr{H}_i^w < \infty) = \rho$. It is a simple matter to work out

$$\tilde{D}(s) = (1 - \rho)/[1 - \rho\tilde{G}_e(s)], \tag{101}$$

the form encountered in equation (17) in Chapter 8.

It is easily shown that I is exponential for an associated queue (not necessarily $M/G/1$) when F has exponential left tail, and $\tilde{D}(s)$ is easy to work out in this case as well.

9-13 LITERATURE NOTES

The literature on random walks is vast, with connections to several branches of mathematics. We have emphasized combinatorial methods, which provide a number of elementary and elegant proofs of important results; see Chapter 12 of Feller [1971] for a more complete treatment and references to both the development of combinatorial methods and other approaches. These methods are combined with Fourier analysis in Chapter 18 of Feller, where a proof of Spitzer's identity may be found. Also see Asmussen [1987] and the references therein, and Prabhu [1980].

Random variables $X_1, \ldots, X_n$ are said to be *exchangeable* if each of the $n!$ permutations of these variables has the same joint distribution. Sequence $\{X_n\}$ is exchangeable if $X_1, \ldots, X_n$ are exchangeable for every n. Clearly, i.i.d. is a special case, and i.i.d. random variables become exchangeable when subjected to a symmetric constraint, e.g., $(X_1, \ldots, X_n \mid Z_n > 0)$. Some of our combinatorial methods and results are valid for exchangeable variables; e.g., see Problem 9-13 and page 423 of Feller.

It is now generally agreed that Lindley [1952] was the first to recognize the connection between random walks and the $GI/G/1$ queue. He derived a Wiener-Hopf integral equation for the stationary delay distribution, which, if we chose to write it out, would be the positive part of equation (4) in Chapter 11.

Conditions for finite delay moments for the $GI/G/c$ queue go back to a classic paper by Kiefer and Wolfowitz [1956]. Several simpler proofs for the $GI/G/1$ case have appeared since, e.g., Lemoine [1976], which unnecessarily restricts the interarrival distribution to be nonlattice. There is a separate literature on ladder height moments for oscillating random walks; see for example Chow and Lai [1979] and Doney [1980]. Theorem 4 for the case $E(X) = 0$ was first shown by Doney.

Our approach to moment conditions in Sections 9-5 and 9-10, which unifies the oscillating and drift cases, is based on Wolff [1984]; also see Asmussen. In Chapter 11, we use conditions for finite $GI/G/1$ delay moments to obtain the Kiefer and Wolfowitz conditions for finite $GI/G/c$ delay moments.

For the $GI/G/1$ queue, relation (86) between the stationary work and delay distributions was first derived by Takács [1963]; see Lemoine [1974] for a simpler proof. Relation (88) was first derived by Hook [1969]; see Harrison and Lemoine [1976] for a simpler proof. See Takács, Cohen [1969], and Lemoine for derivations of the number-of-customers-in-system distribution in Problem 9-10. These proofs all require a nonlattice inter-arrival distribution. Our proofs, which remove this restriction, are simpler still.

The important generalization of (86) to stationary and ergodic arrivals is by Miyazawa [1979]. Theorem 13 follows easily from his results.

There are several papers on the $GI/G/1$ queue under preemptive LIFO, e.g., Fakinos [1981] and Yamazaki [1984], which are primarily concerned with processes generated by the PL rule. See Problems 9-14 and 9-15. This rule can also be used as we have done to derive other results that hold under FIFO.

PROBLEMS

9-1. **(a)** What is the distribution of the number of weak ascending ladder heights that occur before either the first strict one occurs or the ascending process terminates?

(b) Argue that $m_w(0) = \beta/(1 - \beta)$.

(c) Given $\mathcal{K}_1 = n$ and $\mathcal{H}_1 = u$, what is the expected number of weak ladder heights that fall on u?

(d) Derive (13).

9-2. Suppose $P(X = 1) = p$ and $P(X = -1) = q = 1 - p$, and let $y = P(\mathcal{H}_1 < \infty)$.

(a) What is $P(\mathcal{H}_1 + \mathcal{H}_2 < \infty)$?

(b) Given $\mathcal{H}_1$ occurs, what is its value?

(c) What is $P(\mathcal{H}_1 < \infty \mid X_1 = -1)$?

(d) Show that y satisfies the equation

$$(y - 1)(qy - p) = 0.$$

(e) For $p < q$, why must $y = p/q$?

(f) Without using Theorem 3, why must $y = 1$ when $p \geq q$?

9-3. *Continuation.* For $|z| < 1$, let $k(z) = E(z^{\mathcal{K}_1})$. Show that

$$k(z) = [1 - \sqrt{1 - 4pqz^2}]/2qz.$$

9-4. **(a)** For the $M/G/1$ queue, show that F has *exponential left tail,* i.e., $F(x)$ is of the form

$$\text{(i)} \qquad F(x) = qe^{\lambda x} \qquad \text{for } x < 0,$$

where $q = \int_0^\infty e^{\lambda t}\, dG(t)$.

(b) For the $GI/M/1$ queue, show that F has *exponential right tail,* i.e.,

$$\text{(ii)} \qquad F(x) = 1 - pe^{-\mu x} \qquad \text{for } x > 0,$$

where $p = \int_0^\infty e^{-\mu t}\, dA(t)$.

Remark. We say F has exponential left (or right) tail whenever (i) [or (ii)] holds; the other tail is arbitrary. We require only that $q = F(0^-)$ and $p = F^c(0)$.

9-5. *Continuation.*

(a) Let F have exponential right tail. For $t > 0$, use (41) to show that $P(t < \mathcal{H}_1^w < \infty)$ is proportional to $e^{-\mu t}$, and hence that $(\mathcal{H}_1 \mid \mathcal{H}_1 < \infty) \sim \exp(\mu)$. [For *GI/M/*1, F is continuous, but more generally, F could have a discontinuity at the origin. In this case, $P(\mathcal{H}_1^w = 0) > 0$.]

(b) If F has exponential left tail, argue by symmetry that $(-\overline{\mathcal{H}}_1 \mid \overline{\mathcal{K}}_1 < \infty) \sim \exp(\lambda)$.

(c) If F has exponential left tail and $E(X) \le 0$, why is $\overline{m}(t) = \lambda t$?

9-6. Let $b_n = P(Z_1 < 0, \ldots, Z_{n-1} < 0, Z_n = 0)$, and $\mathcal{B}(z) = \sum_{n=1}^{\infty} b_n z^n$.

(a) For every n, show that $b_n = P(Z_1 > 0, \ldots, Z_{n-1} > 0, Z_n = 0)$.

(b) Show that

$$\ln \frac{1}{1 - \mathcal{B}(z)} = \sum_{n=1}^{\infty} z^n P(Z_n = 0)/n.$$

Why does the sum converge as $z \to 1$?

9-7. *Continuation of Problem 9-6 specialized to Problems 9-2 and 9-3.* Argue that

$$b_n = qP(\mathcal{K}_1 = n - 1) \qquad \text{for } n \ge 2,$$

$$\mathcal{B}(z) = qzk(z), \text{ and}$$

$$\sum_{n=1}^{\infty} z^n P(Z_n = 0)/n = \ln [2/(1 + \sqrt{1 - 4pqz^2})].$$

9-8. From (64), argue that when $D_1 = 0$,

$$P(D_{n+1} > x) \ge P(D_n > x) \qquad \text{for all } n \text{ and } x,$$

i.e., $\{D_n\}$ is a stochastically increasing sequence of random variables.

9-9. *Proof of (88).* For arbitrary fixed $v \ge 0$ and every $t \ge 0$, let

$$I(t) = 1 \qquad \text{if } V(t) > v,$$

$$= 0 \qquad \text{otherwise, and}$$

$$P(V > v) = \lim_{t\to\infty} \int_0^t I(u)\, du/t.$$

Let the contribution to this integral by C_n be made during T_n, and argue that

$$\int_{t_n}^{t_{n+1}} I(t)\, dt = \min [(W_n - v)^+, T_n],$$

where $W_n = D_n + S_n$ and T_n are independent. For stationary W and T, write

$$\lambda E\{\min [(W - v)^+, T] \mid W - v = x\} = \lambda \int_0^x A^c(u)\, du = P(T_e \le x) = P(T_e < x).$$

By unconditioning and applying $\overline{H} = \lambda \overline{G}$, we have (88),

$$P(V > v) = P(W > T_e + v), \qquad v \ge 0,$$

where $W = D + S$; D, S, and T_e are independent.

9-10. *The stationary distribution of the number of customers in system.* For a $GI/G/1$ FIFO queue with $0 < \rho < 1$, let $N(t)$ be the number of customers in system at epoch t, N_{an} be the number of customers (in system) found by customer C_n on arrival, and N_{dn} be the number of customers left behind by C_n on departure. Processes $\{N(t)\}$, $\{N_{an}\}$, and $\{N_{dn}\}$ are regenerative with proper limiting [in a (continuous or discrete) time-average sense] distributions. Let N, N_a, and N_d be the corresponding stationary random variables, $N \sim \{p_j\}$, and $N_a \sim \{\pi_j\}$, where these distributions have frequency interpretations, e.g., p_j is the fraction of time there are n customers in system.

Arrivals and departures can occur simultaneously, and we adopt these conventions: Arrivals after C_n are those with higher arrival numbers, and N_{dn} is the number of them who have arrived by $t_n + W_n$. Similarly, N_{an} is the number of arrivals before C_n who depart on or after t_n. The convention we adopt may affect $\{\pi_j\}$ but will not affect $\{p_j\}$. Our convention is consistent with the earlier one that an arrival finds the system empty only if it has been empty a positive length of time. (If this is too confusing, assume F is continuous.)

(a) Why is $N_d \sim N_a$? (If necessary, refer to Section 8-3.)

(b) Let I_n be the indicator of the event $\{N_{dn} \geq j\}$, where $\{I_n\}$ is also regenerative. The stationary version of $\{W_n\}$ induces stationary versions of $\{N_{dn}\}$ and $\{I_n\}$. From this, argue that

$$P(N_d \geq j) = \int_0^\infty A^{(j)}(t)\, dW(t), \qquad j \geq 1,$$

where $A^{(j)}$ is the j-fold convolution A, and $W(t)$ is the stationary distribution of waiting time in system.

(c) From (a) and (b), what is π_j?

(d) Why is $p_0 = 1 - \rho$?

(e) There are j customers in system at some random moment if and only if the server is busy and there have been $j - 1$ arrivals while the customer in service has been in the system. With this idea in mind, let $I(t)$ be the indicator of the event $\{N(t) \geq j\}$, for arbitrary fixed $j \geq 1$. Let the contribution to the area under $\{I(t)\}$ by C_n be made between departures n and $n + 1$, and argue that

$$\int_{t_n + D_n}^{t_n + W_n} I(t)\, dt = \{S_n - [T_n^{(j-1)} - D_n]^+\}^+,$$

where $T_n^{(0)} = 0$, $T_n^{(j)} = T_n + \cdots + T_{n+j-1}$ for $j \geq 1$, $T_n^{(j)} \sim A^{(j)}$, and the quantities S_n, D_n, and $T_n^{(j-1)}$ are independent.

(f) Now show that

$$\lambda E\{\{S_n - [T_n^{(j-1)} - D_n]^+\}^+ \mid [T_n^{(j-1)} - D_n]^+ = x\} = \rho G_e^c(x) = \rho P(S_e > x).$$

(g) From (f), $\overline{H} = \lambda \overline{G}$, and that $\overline{G}$ is the expected value of the corresponding stationary version, show that

$$P(N \geq j) = \rho P[D + S_e > T^{(j-1)}], \qquad j \geq 1,$$

where D, S_e, and $T^{(j-1)}$ are independent. What is $\{p_j\}$?

(h) For the $M/M/1$ queue, $D + S_e \sim W$, where we know that $W \sim \exp(\mu - \lambda)$ (from earlier chapters or by specializing Example 9-3). Use (g) to show that

$$p_j = (1 - \rho)\rho^j, \qquad j \geq 0.$$

9-11. Derive (99) by "naively" differentiating (97).

9-12. Derive (101).

9-13. *Exchangeable random variables.* Suppose $X_1, \ldots, X_n$ are exchangeable, as defined in Section 9-13. Argue that (20) and the right-hand side of (56) are still valid. [However, the $\mathscr{K}_i$ are no longer i.i.d., and the left-hand side of (56) is not a convolution.]

9-14. *Preemptive LIFO.* (Fakinos [1981]) Suppose a $GI/G/1$ queue actually operates under PL. Now adopt the convention that an arrival finds a customer in service only if the remaining service is strictly positive. Think of the queueing process starting at $-\infty$, rather than at epoch 0. For all integers $i < n$, let A_{ni} be the event that customer C_n interrupts C_i, which depends only on inter-arrival times $T_i, \ldots, T_{n-1}$ and service times $S_i, \ldots, S_{n-1}$. Because A_{ni} and A_{ij} depend on disjoint sets of random variables, they are independent events. Let $B_n = \bigcup_{i<n} A_{ni}$ be the event that C_n finds the system busy.

(a) Show that $P(A_{ni}) = P(\mathscr{K}_1 = n - i)$, and that $P(B_n) = P(\mathscr{K}_1 < \infty)$.

(b) Let S_{nr} be the *remaining service time* of the customer found in service by C_n, where $S_{nr} = 0$ if the system is found empty. Show that

$$(S_{nr}, A_{ni}) \sim (\mathscr{H}_1, \mathscr{K}_1 = n - i).$$

(c) Argue that given (S_{nr}, A_{ni}),

$$(S_{ir}, A_{ij}) \sim (\mathscr{H}_1, \mathscr{K}_1 = i - j).$$

(d) For $\rho < 1$, let N_{an} be the number of customers in system found by C_n. Argue that

$$N_{an} \sim \mathscr{N},$$

the number of strict ascending ladder heights that occur, with geometric distribution (37), and that the *work found* by C_n has distribution (39).

(e) What is the distribution of waiting time in system under PL?

Remarks. PL generates a ladder variable process with the time index reversed. The same analysis works for what we called an associated queue.

9-15. *Continuation.* (Yamazaki [1984]) Let N_{PL} be the stationary number of customers in system at a random point in time.

(a) What is $P(N_{\text{PL}} = 0)$?

(b) Let $I(t)$ be the indicator of the event $\{N_{\text{PL}}(t) = j\}$, for arbitrary fixed $j \geq 1$. While C_n is in service, the number of *additional* customers in system equals the number found by C_n on arrival, N_{na}, where N_{na} is independent of S_n. From this and $\overline{H} = \lambda\overline{G}$, show that

$$P(N_{\text{PL}} = j) = \rho P(\mathscr{N} = j - 1), \qquad j \geq 1.$$

(c) Check that $E(N_{\text{PL}}) = \lambda E(W_{\text{PL}})$.

Remark. Unlike FIFO, N_{PL} and W_{PL} will depend on the way N_{an} is defined when the probability of a simultaneous arrival and departure is positive.

9-16. *Wiener-Hopf factorization.* If $E(X) = 0$ and $E\{|X|^3\} < \infty$, show that

$$E(X^2) = -2E(\mathscr{H}_1^w)E(\overline{\mathscr{H}}_1).$$

Verify for these cases: $M/M/1$, and $P(X = 1) = P(X = -1) = 1/2$.

9-17. Let $E(X) < 0$, and suppose that $(X \mid X > 0) \sim \exp(\mu)$, i.e., F has exponential right tail with density $F^c(0)\mu e^{-\mu t}$. We know from Problem 9-5 that $\mathscr{H}_1$ has density of form $\alpha\mu e^{-\mu t}$, where $0 < \alpha < 1$.

(a) Either from (36) through (39) or by inverting the transform $\tilde{m}(s)$, show that $m(t)$ has *derivative*

$$\alpha\mu e^{-(1-\alpha)\mu t}, \qquad t > 0.$$

(b) Find α by evaluating the strict-ascending, weak-descending version of (92) for $J = (-\infty, 0]$, obtaining (integrate by parts)

$$1 = F(0) + \alpha\mu \int_{-\infty}^{0} F(t) e^{(1-\alpha)\mu t}\, dt$$

$$= F(0)/(1 - \alpha) - \alpha \int_{-\infty}^{0} e^{(1-\alpha)\mu t}\, dF(t)/(1 - \alpha).$$

Now

$$\int_0^\infty e^{(1-\alpha)\mu t}\, dF(t) = \mu F^c(0) \int_0^\infty e^{-\alpha\mu t}\, dt = F^c(0)/\alpha,$$

and combining these expressions, we have that α satisfies

$$\int_{-\infty}^{\infty} e^{(1-\alpha)\mu t}\, dF(t) = \tilde{F}[(\alpha - 1)\mu] = 1, \qquad \alpha \in (0, 1),$$

where it can be shown that the solution is unique.

(c) For the corresponding associated queue, show that

$$W \sim \exp[(1 - \alpha)\mu].$$

(d) For the $GI/M/1$ queue, specialize $\tilde{F}(s)$ and show that the above equation for α becomes $\alpha = \tilde{A}[(1 - \alpha)\mu]$, which is equation (62) in Chapter 8.

9-18. Generalize $GI/G/1$ to batch arrivals; apply Problems 8-5 through 8-7.

10

WORK CONSERVATION AND PRIORITY QUEUES

Queues form because the resources available to satisfy customer service requirements are limited. A *priority rule* determines the allocation of resources to customers. Thus FIFO and LIFO are priority rules where, on a service completion, the next customer to begin service was the first (or last) to join the queue. More generally, a rule may take into account other factors such as differences in delay costs or service times from one customer to another. These factors may be explicitly modeled, or may be implicit in the assignment of customers to *priority classes*, where some classes receive better service than others.

As rules become complex, so will their analysis and implementation, and the list of possible rules is endless. Whole books could be (and have been) written on this topic.

In this chapter, we deal primarily with single-channel queues and will treat some of the better known rules. Our objective is twofold: (1) to develop general concepts and methods for priority queues and related models, and (2) to explain why and in what sense some rules are better than others.

The most important concept in this chapter is *work conservation*, which means that customer service times are not affected by the particular priority rules under consideration. We can think of each service time as being drawn from a (possibly class-dependent) distribution on arrival.

To get closed form expressions for such quantities as expected delay, we will have to make simplifying assumptions, which usually will be to assume that under standard FIFO, we have an $M/G/1$ queue. When this is done, PASTA and properties of an exceptional first service busy period (treated in Section 8-5, and abbreviated here by *EFSBP*) will be needed.

10-1 WORK CONSERVATION AND OTHER CONSERVATION PRINCIPLES

We restrict our attention to single- or multiple-channel queues (in parallel), but the concept of work applies to queueing networks as well.

The service time S of a customer is an amount of work, in worker-hours say, that must be done. Some priority rules permit interrupting customers during service, so that a customer in queue may have received some service previously. Some rules permit customers to share servers. We say a customer with service time S has *attained service* a at epoch t if the total server hours allocated to that customer by t is a.

Definition. A priority rule is called *work-conserving* if for all S, a, and t, the *remaining service time* (*RST*) of a customer at epoch t is $S - a$. If $a = S$, the customer has completed service and has departed.

We define the *work-in-system* process $\{V(t)\}$ by: For every epoch t,

$$V(t) = \text{the sum of the RSTs of all customers in system at epoch } t. \tag{1}$$

Now suppose we have a *single-channel* queue with arrival epochs t_j and service times S_j, $j = 1, 2, \ldots$. If work is conserved (and the server is busy whenever there is work to do), the *sample paths* of $\{V(t)\}$ are determined by $\{t_j, S_j\}$, as in Figure 5-15, for *all* priority rules. In particular, we have

Theorem 1. For a single-channel work-conserving queue,

$$E(V) \equiv \lim_{t\to\infty} \int_0^t V(u)\, du/t, \tag{2}$$

whenever (2) exists, and $\{V(t)\}$ are the same for all rules.

We will find this invariance principle to be very useful. Busy and idle periods [determined by epochs where $V(t) > 0$ and $V(t) = 0$] are also invariant.

It is easy to see that other measures of performance such as L and Q are not invariant; e.g., we show in Section 10-3 that if the server is always allocated to the shortest *remaining* service, the number-in-system process $\{N(t)\}$ is minimized *on all sample paths*. Thus priorities can affect L and Q, and this is one of the main reasons to consider them.

On the other hand, for a stable $c \geq 1$ channel queue with arrival rate λ, service rate μ, and $\rho = \lambda/c\mu < 1$, the expected number of customers in service (the expected number of busy servers, and for $c = 1$, the fraction of time the server is busy),

$$L_s = \lambda/\mu = c\rho, \tag{3}$$

is invariant.

We will partition work in system by priority class, and also by work associated with customers in queue (*queue* work) and work associated with customers in service (*service* work), where, with the obvious time-average notation for the latter partition,

$$E(V) = E(V_q) + E(V_s). \tag{4}$$

Preemptive priority rules permit interrupting customers during service (returning them to queue); *nonpreemptive* or *postponable* rules do not, so that once service begins on a customer, that customer is served to completion.

For a stable $c \geq 1$ channel queue, we wish to represent $E(V_s)$ for some unspecified priority rule. First assume the rule is postponable. A realization of service work $\{V_s(t)\}$ is represented for $c = 1$ in Figure 10-1. The priority rule rearranges the order that customers enter service, which rearranges the order of the triangular contributions to the area under $\{V_s(t)\}$. If the average waiting time w is finite (over all customers for each priority rule under consideration), and $\lim_{n\to\infty} \sum_{j=1}^{n} S_j^2/n \equiv E(S^2) < \infty$, Theorem 5 in Section 5-15 applies to $\{V_s(t)\}$, and we have

$$E(V_s) = \lambda E(S^2)/2, \tag{5}$$

which is also invariant, and therefore (when $c = 1$), so is $E(V_q)$!

If $c > 1$, several triangular contributions may be in progress simultaneously, but Theorem 5 still applies and (5) holds. If preemptions are permitted, a triangle may be split into pieces, as in the sloped portion of Figure 10-2, but the *area* is the same, and (5) still holds.

We will not be able to do much with work conservation at this level of generality, and from now on, assume that we have a $GI/G/c$ queue with $\rho < 1$ and $E(S^2) < \infty$. Under these conditions, the standard FIFO model is stable with finite expected delay; see Chapter 11.

For $c = 1$, (5) has an interesting interpretation. Suppose we observe a $GI/G/1$ queue "at random." With probability ρ, the system is busy, and, given it is busy, the remaining service of the customer found in service has equilibrium distribution G_e, corresponding to service distribution G. Let S_r be the remaining service time of the customer found in service, where $S_r = 0$ if the system is

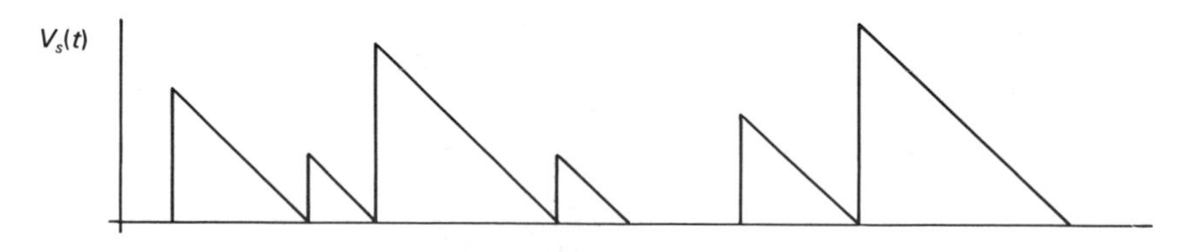

Figure 10-1

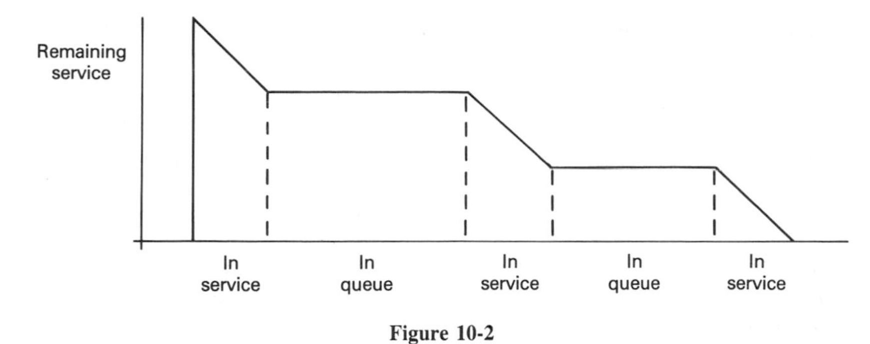

Figure 10-2

empty. Denote $(S_r \mid S_r > 0) = S_e \sim G_e$, and we have

$$E(S_r) = \rho E(S_e) = (\lambda/\mu)\mu E(S^2)/2 = \lambda E(S^2)/2. \tag{6}$$

We also remark that for $\rho < 1$, a $GI/G/1$ queue operating under a work-conserving priority rule is a regenerative process with finite mean cycle length, and when we also have $E(S^2) < \infty$, time averages of portions of work in system exist and are finite.

FIFO and LIFO are examples of postponable rules where order of service is independent of the service times of customers in queue. Under rules with this property, the service time of a customer about to enter service is independent of past history, and the number-in-system process $\{N(t)\}$ and the departure process are *stochastically* invariant, i.e., these processes under different rules are stochastically equivalent. If the randomness of service is a property of *servers* rather than customers, rearranging customers has no effect on the sequence of service times, and both $\{N(t)\}$ and the departure process are invariant on *sample paths*. (Essentially the same idea is used in Chapter 11 to obtain important upper bounds on various quantities for the $GI/G/c$ queue.)

10-2 FIXED PRIORITY CLASSES

Now assume that we have a $GI/G/1$ queue where arrivals are partitioned into a finite or countable number of priority classes, numbered 1, 2, ... For brevity, we call customers in class j *j-jobs*.

For $j = 1, 2, \ldots$, let p_j be the probability that an arrival is a j-job, independent of all else. Jobs within class are served FIFO and have i.i.d. service times with generic j-job service time $S_j \sim G_j$, and $\rho_j = \lambda p_j E(S_j)$. Let S denote a generic "overall" service time, where $S \sim G = \sum p_j G_j$, and $\rho = \lambda E(S) = \sum \rho_j$. Let Q_j be the average number of j-jobs in queue, and d_j be the average j-job delay in queue, where the corresponding overall quantities are $Q = \sum Q_j$ and $d = \sum p_j d_j$.

For any *postponable* rule, a j-job's contribution to time-average work in queue is a rectangle of height S_j, length D_j, and, from independence, expected area

$$d_j E(S_j). \tag{7}$$

Hence, from (7) and $L = \lambda w$, time-average j-job work in queue is

$$\lambda p_j d_j E(S_j) = \rho_j d_j = Q_j E(S_j). \tag{8}$$

The right-hand expression, which is also the expected value of the *sum* of the service times of all j-jobs in queue, is intuitive and easy to remember. It follows that we can write $E(V)$ in either of two ways,

$$E(V) = \sum \rho_j d_j + \lambda E(S^2)/2 = \sum Q_j E(S_j) + \lambda E(S^2)/2. \tag{9}$$

From work conservation and (7), we have

Theorem 2.

$$E(V_q) = \sum \rho_j d_j = \sum Q_j E(S_j) = \rho d_F, \text{ a constant,} \tag{10}$$

independent of the priority rule, where d_F is the average delay under FIFO.

Thus, except in trivial cases such as $E(S_j) = 0$ for some j, a priority rule can decrease expected delay for one class (compared with some alternative) only by *increasing* expected delay for some other class. There is no free lunch! (When preemptions are allowed, this conclusion need *not* hold.)

Remark. Theorem 2 and the conclusions that follow do not require renewal arrivals. See equations (2) and (5). We call (10) a *conservation law*.

Thus far we have been intentionally vague about defining specific rules in order to emphasize the generality of the above concepts and results. Now suppose we order the priority classes, where class 1 is the highest, class 2 is next, etc., where jobs in a higher class have either preemptive or postponable priority over jobs in lower classes. Under preemptive priorities, a job preempts (interrupts) on arrival any lower priority job found in service. Postponable priorities operate only at service completion epochs, where the next job served is drawn from the highest class present in queue.

In the examples that follow, we also assume Poisson arrivals.

Example 10-1: The $M/G/1$ Queue with Postponable Priorities

We now obtain an expression for the expected delay of j-jobs by considering work found by an arrival. From PASTA, $E(V)$ and the portions of time-average work represented in (9) are also arrival averages. The delay of an arriving 1-job is the sum of the service times of the 1-jobs found in queue and the remaining service time of any job found in service. Hence, the expected 1-job delay is the sum of the relevant terms in (9),

$$d_1 = \rho_1 d_1 + \lambda E(S^2)/2, \text{ and solving for } d_1,$$

$$d_1 = \lambda E(S^2)/2(1 - \rho_1), \tag{11}$$

which is similar to the standard $M/G/1$ expected delay, where now the numerator includes all customers, but the denominator has ρ_1 rather than ρ.

Let R be the relevant work found by a 2-job, where R is the sum of the service times of all 1- and 2-jobs found in queue and the remaining service of any customer found in service, with

$$E(R) = \rho_1 d_1 + \rho_2 d_2 + \lambda E(S^2)/2. \tag{12}$$

The arriving 2-job's delay D_2 will be at least R, and will exceed R if a 1-job arrives while R is being performed. In fact, D_2 will be the time until R has been performed *and* the system is clear of 1-jobs. Notice that the order of service within R is irrelevant, and that R may be split into subintervals. This occurs whenever a 1-job arrives while a 2-job composing R is in queue.

Thus D_2 has the structure of an *exceptional first service busy period* (*EFSBP*) of an $M/G/1$ queue composed of 1-jobs only, where R has the role of the exceptional first service. It follows that

$$d_2 = E(R)/(1 - \rho_1), \tag{13}$$

and we can now solve for d_2. For arbitrary class j, there are two ways to proceed: One is by induction, where we redefine R for each j, and observe that D_j is the time until R has been performed and the system is clear of all jobs of higher priority. A quick but tricky alternative is to combine all classes higher that j into a new class 1; class j is now new class 2. (Why is it true that *on sample paths*, combining classes in this manner has no effect on class j-job delays?) Either way, it is easily shown that for $j = 1, 2, \ldots,$

$$d_j = \frac{\lambda E(S^2)}{2(1 - \sum_{i<j} \rho_i)(1 - \sum_{i \le j} \rho_i)}. \tag{14}$$

Remark. Compare this analysis with that in Section 5-14 for the same model, where we had not yet introduced the concept of an exceptional first service busy period. For the earlier analysis, there must be some "lowest" class.

Example 10-2: The $M/G/1$ Queue with Preemptive-Resume Priorities

Preemptive-resume means that on reentering service, an interrupted customer resumes service at the point of interruption, with no gain or loss, i.e., the rule is work conserving.

As far as 1-jobs are concerned, this model is a standard $M/G/1$ queue composed only of 1-jobs, and

$$d_1 = \lambda_1 E(S_1^2)/2(1 - \rho_1). \tag{15}$$

For j-jobs, higher class jobs can be combined into a new class 1, and lower class jobs are irrelevant. Thus the problem reduces to the 2-class case. Suppose this has been done, and for the rest of this example, suppose there are only 2 classes.

Let D_{2f} be delay of a 2-job prior to entering service *for the first time*, and let V be the work in system found by a 2-job arrival, where $E(V)$ is the time-average work in system, which is also the average delay of the corresponding $M/G/1$ queue without priorities, i.e.,

$$E(V) = \lambda E(S^2)/2(1 - \rho).$$

Now D_{2f} is the time until V has been performed and the system is clear of 1-jobs. Thus D_{2f} is an EFSBP, with expected value

$$d_{2f} = E(V)/(1 - \rho_1) = \lambda E(S^2)/2(1 - \rho_1)(1 - \rho). \tag{16}$$

Define the *completion time* $\mathscr{C}$ of a 2-job to be the time from when service begins on the 2-job until service is completed. Thus the 2-job waiting time in system is $W_2 = D_{2f} + \mathscr{C}$. A completion time can begin only when the system is clear of 1-jobs, and ends when service time S_2 has been completed and the system is clear of 1-jobs. Thus $\mathscr{C}$ is also an EFSBP, where

$$E(\mathscr{C}) = E(S_2)/(1 - \rho_1). \tag{17}$$

Although our primary emphasis will be on finding first-moment performance measures such as expected delay, in this case the corresponding distributions, or at least their transforms, are also easy to find. First observe that $\mathscr{C}$ depends on S_2 and service times from a Poisson arrival stream, and therefore is independent of D_{2f}. Thus

$$E(e^{-sW_2}) = E(e^{-sD_{2f}})E(e^{-s\mathscr{C}}). \tag{18}$$

The transform of an EFSBP is given in equation (42) in Chapter 8 in terms of the corresponding transform of exceptional first service and an ordinary busy period. The transforms of V, S_2, and (appropriate) ordinary $M/G/1$ busy periods are known here.

Example 10-3: The $M/G/1$ Queue with Exceptional First Service

We now find the expected delay for this model (without priorities). The methods we are developing are by no means restricted to priority queues. (Also see Problem 8-9 for corresponding transform results.)

In the notation of Section 8-5, S_a is an exceptional first service and K_a is the number of customers served in the corresponding busy period, where

$$E(K_a) = 1 + \lambda E(S_a)/(1 - \rho).$$

Thus $1/E(K_a)$ is the fraction of customers who are exceptional, $\lambda_a = \lambda/E(K_a)$ is the arrival rate of exceptional customers, and $f_a = \lambda_a E(S_a)$ is the fraction of time an exceptional service is in progress. Similarly, $f_o = (\lambda - \lambda_a)E(S)$ is the fraction of time an ordinary service is in progress.

Delay is the sum of the (ordinary) service times of customers found in queue and the remaining service time of any customer found in service. From PASTA we can write

$$d = QE(S) + f_a E(S_{ae}) + f_o E(S_e), \tag{19}$$

where e denotes the appropriate equilibrium distribution, with $E(S_{ae}) =$

$E(S_a^2)/2E(S_a)$ and $E(S_e) = E(S^2)/2E(S)$. From (19) and $L = \lambda w$,

$$d = [f_a E(S_{ae}) + f_o E(S_e)]/(1 - \rho). \tag{20}$$

Once preemptions are allowed, work conservation may seem to be an unreasonable assumption. There may be some setup that would have to be repeated, or perhaps the entire attained service of an interrupted job would have to be done over. This point of view regards service times as attributes of particular jobs. Alternatively, the randomness of service within each job class may be an attribute of the server. With these possibilities in mind, we now name two specific alternatives to the resume assumption in Example 2.

A preemptive rule is called *repeat-identical* if on reentry into service, an interrupted job's remaining service is what it was at the point of interruption *plus* that job's attained service. That is, remaining service is equal to the original service time. Thus work performed on that job is wasted and must be done over. A rule is called *repeat-different* if the interrupted job's remaining service is a random draw from the service distribution for that class of jobs. Note that for constant service, these rules are the same.

Compared with resume, performance under repeat-identical will be worse, but this is not necessarily true under repeat-different. For example, how do resume and repeat-different compare when service is exponential?

Define completion times as in Example 10-2. For the $M/G/1$ model, it is easy to see that the 2-job completion times are i.i.d. for resume, the preceding repeat rules, and virtually any assumption we might care to make. For the next example, we assume i.i.d. completion times for some unspecified rule, where $E(\mathscr{C})$ and $E(\mathscr{C}^2)$ are known or can be computed, and to ensure stability, that $\lambda_2 E(\mathscr{C}) < 1$.

Example 10-4: The $M/G/1$ Queue with General Preemptive Rules

Consider the 2-class case with i.i.d. 2-job completion times. Now 1-jobs are not affected by preemption assumptions, and (15) still holds. A completion time is part of each 2-job's waiting time, and we can *define* it as that job's service time, where D_{2f} in Example 10-2 is now a conventional delay. With this definition, we have accounted for all 1-jobs that arrive during completion times, and (almost) reduced the 2-job delay problem to a conventional $M/G/1$ queue.

We have *not* accounted for 1-jobs that arrive when no completion time is in progress. An arriving 2-job who finds no 2-jobs in system will not begin service immediately if a 1-job busy period is in progress. If this delay is added to this 2-job's service time, this 2-job begins an EFSBP, and the model in Example 10-3 applies. This approach is not convenient, however, because 2-jobs who begin busy periods are "special" and lack the PASTA property; the added delay is difficult to deal with directly.

Instead we consider the congestion found by a "random" 2-job arrival. Delay D_{2f} is the sum of the completion times of 2-jobs found in queue, the remaining completion time of any completion time found in progress, and the remaining 1-job busy period of any busy period found in progress that began when the system was empty. The last two terms can't be positive simultaneously, but we can still add expected values.

Clearly, from PASTA, $1 - \lambda_2 E(\mathscr{C})$ is the probability of *not* finding a 2-job completion time in progress, and ρ_1 is the probability of finding a 1-job busy period in progress. It is easily shown (see Problem 10-12) that the joint probability of both events is their product

$$\rho_1[1 - \lambda_2 E(\mathscr{C})], \tag{21}$$

which is the probability that the last term included in D_{2f} is positive.

It follows that

$$d_{2f} = Q_{2f}E(\mathscr{C}) + \lambda_2 E(\mathscr{C})E(\mathscr{C}_e) + \rho_1[1 - \lambda_2 E(\mathscr{C})]E(B_{1e}), \text{ and} \tag{22}$$

$$d_{2f} = \frac{\lambda_2 E(\mathscr{C}^2)}{2[1 - \lambda_2 E(\mathscr{C})]} + \rho_1 E(B_1^2)/2E(B_1), \tag{23}$$

where B_1 is an ordinary 1-job busy period. As a check on what we have done, it can be shown that when specialized to resume, (23) becomes (16).

While work conservation does not hold for this model, our methods still apply.

10-3 OVERALL DELAY AND COST CONSIDERATIONS

Suppose we have a single-channel queue and a finite set of n jobs to do, with known service times $S_1, \ldots, S_n$, where all jobs are present initially (no future arrivals). The average delay (or equivalently the total delay) of the n jobs is minimized if they are served in the order of shortest to longest, i.e., by this criterion, the rule *shortest-job-first* is optimal. If the cost of delay is proportional to delay, where the cost rate of job j is c_j, total cost is minimized by the *$c\mu$-rule*, where $\mu_j = 1/S_j$ here, and the rule means that we serve in the order of largest to smallest $c_j\mu_j$.

If the S_j are independent random variables with known means, where only the completed service times are known when the next selection is made, and we rule out preemptions, the $c\mu$-rule minimizes expected cost, where now $\mu_j = 1/E(S_j)$. These conclusions are easily shown by interchanging adjacent pairs of jobs that are out of order.

Now return to Example 10-1 and suppose that jobs are partitioned into identifiable classes. We want to decide which class should have the highest priority (be class 1), and so forth. For the same linear cost model, the average cost per customer is of form

$$\sum c_j p_j d_j, \tag{24}$$

which is equivalent (multiply by λ) to the cost per unit time

$$\sum c_j Q_j. \tag{25}$$

Notice that in (14), the d_j all have the same numerator. Using an interchange argument, is easily shown (see Problem 5-42) that the $c\mu$-rule minimizes (24) and (25), where again $\mu_j = 1/E(S_j)$. The special case where the c_j are all equal is also of interest, where we do the shortest *expected* service time first.

These results can be generalized in several respects for an $M/G/1$ queue with postponable rules. The $c\mu$-rule is optimal even when we permit the rule to depend on the queue lengths of each class. The number of classes need not be countable, e.g., we can order by actual service times, where the service distribution may be continuous. In this case, it can be shown that shortest-job-first minimizes overall expected delay.

Now suppose we have a linear cost model for a $GI/G/1$ queue, postponable priorities, and two priority classes denoted by a and b. Which class should have higher priority? We want to minimize

$$c_a Q_a + c_b Q_b, \tag{26}$$

subject to [from (10)]

$$Q_a/\mu_a + Q_b/\mu_b = \text{a constant.} \tag{27}$$

Giving class a higher priority decreases Q_a and increases Q_b. Cost is reduced if and only if

$$c_a\mu_a > c_b\mu_b, \tag{28}$$

which again is the $c\mu$-rule. (We don't even need renewal arrivals for this conclusion, but only that certain limits exist and are finite.)

The conclusion (28) depends on the severe restrictions imposed when we have only two classes; the $c\mu$-rule is not necessarily optimal when there are as few as three classes. Poisson arrivals are needed for the more general statements made previously. When arrivals are *not* Poisson, we are able to somewhat "anticipate" when future arrivals will occur, and this will sometimes affect what actions we should take.

Nevertheless, the $c\mu$-rule should be regarded as a good rule of thumb for postponable priorities with linear costs. Preemptive priorities are quite a different story, even in the resume case. Notice that $E(S_1^2)$ appears in (15), and some dependence on second moment properties is inevitable.

Now consider the *shortest-remaining-service-time* rule (*SRST*), i.e., at all times, the server is allocated to the job with the shortest remaining service time. This is a preemptive rule, and we assume work conservation (resume). For this rule, let $l_1(t)$ be the length of the *longest* remaining service time at epoch t, where $l_1(t) = 0$ if the system is empty. The process $\{l_1(t)\}$ can only decrease at epochs where the number in system $N(t) = 1$. The process will jump to S_j at arrival epoch t_j if $S_j > l_1(t_j^-)$, and is left-continuous at t_j otherwise.

Now let $\{l_1'(t)\}$ and $\{N'(t)\}$ be the corresponding processes *under any alternative rule* (called *rule* A). Start off with any initial set of remaining service times at epoch 0. Between 0 and the first arrival epoch t_1, $\{l_1(t)\}$ is constant or, if $\{N(t)\}$ hits 1, $l_1(t) = V(t)$ from then on. Clearly, for any rule A, $l_1'(t) \leq l_1(t)$ for $0 \leq$

$t \leq t_1$, and repeating the argument for successive arrival epochs, we have that for all realizations,

$$l_1'(t) \leq l_1(t) \qquad \text{for all } t \geq 0. \tag{29}$$

Now let $l_j(t)$ and $l_j'(t)$ be the respective *sum* of the j longest remaining service times at epoch t under SRST and rule A, $j \geq 2$. On $[0, t_1)$, $\{l_j(t)\}$ is constant or, if either $N(0) \leq j$ or $\{N(t)\}$ hits j, $l_j(t) = V(t)$ from then on, and $l_j'(t) \leq l_j(t)$. By induction, suppose we know

$$l_i'(t_1) \leq l_i(t_1) \qquad \text{for } i < j. \tag{30}$$

If S_1 is not included in $l_j'(t_1)$, (30) clearly holds for $i = j$ from left-continuity. If S_1 is included.

$$l_j'(t_1) = l_{j-1}'(t_1) + S_1 \leq l_j(t_1),$$

whether or not S_1 is included in $l_j(t_1)$, so that (30) holds for $i = j$. Repeating the argument, we have that for all realizations,

$$l_j'(t) \leq l_j(t) \qquad \text{for all } t \geq 0 \quad \text{and} \quad j \geq 1. \tag{31}$$

From work conservation, the corresponding work-in-system processes are identical, where each process is a sum of remaining service times,

$$V(t) = l_{N(t)}(t) = l'_{N'(t)}(t) = V'(t) \qquad \text{for all } t \geq 0. \tag{32}$$

From (31) and (32), Theorem 3 is immediate.

Theorem 3. For every single-channel queue, SRST minimizes the number-in-system process, i.e., for all realizations

$$N(t) \leq N'(t) \qquad \text{for all } t \geq 0, \tag{33}$$

where $\{N'(t)\}$ is corresponding process under *any* work-conserving alternative.

It is easy to see that (33) does not hold for $c \geq 2$ channels, e.g., if for $c = 2$, we begin with service times 2, 1, and 1, and no arrivals occur, all three jobs are done by epoch $t = 2$ if we begin serving the longest job immediately, but they are not done until $t = 3$ under SRST. This occurs because one server is idle while there is work to do. This is not, in the terminology of Section 5-4, an efficient allocation of workers to work.

If idle servers could help out busy ones, with no loss in efficiency (*job sharing*), SRST would have property (33) for $c \geq 2$ channels. In fact, under this assumption, (2) is invariant for $c \geq 2$ channels as well, because for work in system, a multichannel queue is equivalent to a "fast" single channel queue. Even for conventional multichannel queues without job sharing, we would expect SRST to perform well for measures such as L, particularly for ρ near one.

Example 10-5: The $M/G/1$ Queue under SRST

We now find $w_r = E\{W_r\}$ for an $M/G/1$ queue under SRST where W_r is the stationary waiting time of a customer with service time $S = r$. As in Examples 10-2 and 10-4, we write

$$W_r = D_r + \mathscr{C}_r, \quad \text{with expectations} \quad w_r = d_r + c_r,$$

where D_r is the delay until service begins for the first time, and $\mathscr{C}_r$ is a completion time made up of service time r and busy periods generated by $Y \sim P(\lambda r)$ arrivals during r. An arrival with service time S at epoch $r - u$ will generate a busy period if $S < u$. Service times during this busy period must have the same property, and we have a standard $M/G/1$ busy period with

$$\rho(u^-) = \lambda E(S(u)) = \lambda \int_0^{u^-} t\, dG(t),$$

$$\text{where} \quad S(u) = S \text{ if } S < u,\ S(u) = 0 \text{ otherwise},$$

and let $\rho(u)$ be the corresponding expression when $\leq$ replaces $<$. Conditioning on the Y unordered arrival epochs during r, it is easily shown that

$$c_r = \int_0^r du/(1 - \rho(u^-)) = \int_0^r du/(1 - \rho(u)) = \int_0^r \frac{du}{1 - \lambda \int_0^u t\, dG(t)}, \quad r \geq 0. \tag{34}$$

To find d_r, let V_r be the (stationary) sum of the remaining service times of all customers in system with remaining service times that are $\leq r$, as found by an arrival. D_r is an EFSBP generated by V_r, in terms of arrivals during the busy period with service times that are $< r$. We have

$$d_r = E(V_r)/[1 - \rho(r^-)]. \tag{35}$$

To find $E(V_r)$, we change the rule: For any fixed r, customers with remaining service times that are $\leq r$ are served FIFO, and have preemptive priority over customers with remaining service times that are $> r$. This means that customers with original service $S \leq r$ are served FIFO, and customers with original service $S > r$ are allowed to complete service *once their remaining service reaches* r. This rule has no effect on the stochastic process $\{V_r(t)\}$. Let D_{Fr} be the delay of a customer with service time $S \leq r$ under this mixed preemptive-postponable rule, where $D_{Fr} = V_r$, and relevant customers in queue have service times $\leq r$. A customer with service time S who is found in service will cause delay if $S \leq r$, or, if $S > r$ and the final portion of length r is being performed. By combining these cases, the customer found in service will be allowed to complete the remainder of $S_r' \equiv \min(S, r)$, which is the corresponding equilibrium random variable S_{re}', with mean

$$E(S_{re}') = E\{(S_r')^2\}/2E(S_r') = \int_0^r tG^c(t)\, dt \Big/ \int_0^r G^c(t)\, dt.$$

We now use PASTA and $L = \lambda w$ in standard ways to write

$$d_{Fr} = Q_{Fr}\rho(r)/\lambda + \lambda E(S_r')E(S_{re}') \text{ and}$$

$$d_{Fr} = E(V_r) = \lambda \int_0^r tG^c(t)\, dt/[1 - \rho(r)], \tag{36}$$

from which we get

$$d_r = \frac{\lambda \int_0^r tG^c(t)\,dt}{[1 - \rho(r)][1 - \rho(r^-)]}, \qquad r \geq 0. \tag{37}$$

Now the average waiting time under this rule is $w_{\text{SRST}} = \int_0^\infty w_r\, dG(r)$, which can easily be computed for particular distributions and compared with FIFO. The improvement over FIFO can be very large for ρ near 1, e.g.,

$$w_{\text{SRST}}/w_F = .176 \text{ for } M/M/1 \text{ with } \rho = .99,$$

but is easily checked [evaluate (34) and (37) at $r = E(S)$] that $w_{\text{SRST}} = w_F$ for an $M/D/1$ queue.

We have primarily discussed first moment measures of performance in this section. Priority rules can substantially increase delay and waiting time variability (variance in particular), when compared with FIFO, particularly for ρ near one. For example, comparing FIFO and LIFO for an $M/G/1$ queue, we have [see (51) and (52) in Section 8-5],

$$E(D_L^2) = E(D_F^2)/(1 - \rho),$$

even though the corresponding means are the same. A priority rule that yields a small gain by some first-moment measure may do very poorly in terms of other measures of performance.

10-4 PRIORITY RULES BASED ON ATTAINED SERVICE; TAGGING JOBS

As we have seen, shortest-job-first and SRST reduce overall expected delay, and it is easy to construct examples where the reduction is substantial when compared with FIFO. Unfortunately, the information about service time durations needed to implement these rules often is not available.

We now treat rules that give better service to short jobs than to long jobs without having to know or estimate job durations. Rules based on *attained* service have this property. The rules we are about to discuss are widely known in the computer science literature, and variations of them have been used to determine access of jobs (processes) to computers and other equipment (processors).

Assume that we have a work-conserving $GI/G/1$ queue with arrival rate λ and generic service time $S \sim G$. (Later we specialize to $M/G/1$.) On entering service, a job is allocated a positive block of time (a *quantum*) in which to complete service, and is interrupted and removed from the server if service is not completed during that time. Thus the same job may make many passes through the server.

Let δ_i be the quantum size allocated to a job on its ith pass through the server, $i = 1, 2, \ldots$. (All jobs are treated alike.) It is sometimes assumed that the δ_i are all equal, or monotone, $\delta_1 \leq \delta_2 \leq \ldots$, but this is not required in the analysis. Assume $\sum \delta_i = \infty$. We usually assume the individual δ_i are finite, but by setting $\delta_i = \infty$ for some i, jobs are assured to complete service in at most i passes.

A job with service time S completes service on its first pass if $S \leq \delta_1$, and is interrupted otherwise. In either event, the time actually used is min $\{S, \delta_1\}$. Let

$$\Delta_{j+1} = \sum_{i=1}^{j} \delta_i, \qquad j = 1, 2, \ldots, \text{ with } \Delta_1 \equiv 0^-,$$

be the total time allocated to a job on its first j passes, where a job entering service for the jth time has *attained service* Δ_j. A job will complete service on the its jth pass if $\Delta_j < S \leq \Delta_{j+1}$.

We now define two specific rules:

1. *Feedback* (*FB*) to lower priority queues. Priority classes are numbered 1, 2, . . . , where 1 is the highest, 2 is next, etc. On arrival, all jobs are put in class 1. A job that is interrupted at the end of its jth pass is put in class $j + 1, j = 1, 2, \ldots$. On the completion of a pass (either a service completion or interruption), the next job to enter service is drawn from the highest class present in queue. Within class, the queue operates FIFO.
2. *Round Robin* (*RR*). There is a single queue that operates FIFO. Arrivals and interrupted jobs join the end of the queue.

These are by no means the only possible rules. To analyze these and other rules of this type, we will need to represent average work in system.

First write

$$E(V) = E(V_q) + \lambda E(S^2)/2. \tag{38}$$

A job that has completed $j - 1$ passes and is either in queue or in service we call a *j-pass*, $j = 1, 2, \ldots$. The expected remaining service of a j-pass in queue is

$$v_j \equiv E(S - \Delta_j \mid S > \Delta_j) = \int_{\Delta_j}^{\infty} G^c(t)\, dt/G^c(\Delta_j). \tag{39}$$

Now let Q_j be the average number of j-pass jobs in queue, and $E(V_{qj})$ be the average work in queue associated with these jobs, where

$$E(V_{qj}) = Q_j v_j, \text{ and}$$

$$E(V_q) = \sum Q_j v_j = Q_F E(S), \tag{40}$$

where Q_F is the average queue length under FIFO, and (40) is invariant with respect to these rules.

A job cannot be a j-pass, $j > 1$, unless it previously was an i-pass, $1 \leq i < j$. Thus for rules based only on attained service and, as in round robin, on factors independent of remaining service, it is clear that short jobs get better service than long jobs. In fact, with all else fixed, the delay of every job is an increasing function of its own service time. Does this mean that these rules necessarily perform better than FIFO?

The answer is no! The service distribution plays a crucial role here, and we first compare these rules with FIFO under the assumption that

$$E(S - t \mid S > t) \leq E(S) \qquad \text{for all } t \geq 0, \tag{41}$$

which means that for any attained service t, the mean *remaining* service is bounded above by the original mean service. Clearly, Erlang and constant service have property (41), and it is easily shown that (41) holds as an equality for all t if *and only if* S is exponential. This and other "tail" properties of distributions are discussed in some detail in Chapter 11.

From (39) and (40), (41) implies

$$Q_F E(S) = \sum Q_j v_j \leq \sum Q_j E(S) = QE(S), \text{ or}$$

$$Q_F \leq Q \text{ (and } d_F \leq d), \tag{42}$$

where Q is the expected number in queue (over all passes) for *any* attained service rule. Thus for these measures, FIFO is better than any attained service rule when service is *regular* in the sense of (41).

In fact, under either FB or RR with constant service, where $\delta_1 < S$, *every* job departs later than it would under FIFO (*strictly* later, except for those ending busy periods).

Reversing the inequality in (41) (as is the case for the hyper-exponential distribution) reverses (42), and for the same measures, FIFO is worse than any of these rules. For *exponential* service,

$$Q_F = Q \tag{43}$$

for all attained service rules, i.e., Q is *invariant* under these rules. In fact, the number-in-system process

$$\{N(t)\} \tag{44}$$

is invariant in a stochastic sense.

Notice that these conclusions are *independent* of the arrival process.

There are several possible rationales for implementing an attained service rule, e.g., for diverse groups of users, their composite service distribution may indeed be highly irregular. Another is that the rule may induce desirable behavior on the part of some users (from the point of view of system performance or other users), such as encouraging them to submit short jobs during busy times of the day. Of course, priority rules of the type treated in Section 10-2 may be used in combination with those discussed here, with higher rates charged for better service.

To analyze attained service rules in more detail, let r_j be the expected delay in queue of a job between making passes $j - 1$ and j, where

$$Q_j = \lambda G^c(\Delta_j) r_j, \qquad j = 1, 2, \ldots, \tag{45}$$

and let d_j be the total expected delay of a job that completes service on its jth pass, where

$$d_j = \sum_{i=1}^{j} r_i, \qquad j = 1, 2, \ldots. \tag{46}$$

A useful concept for analyzing specific rules is *tagging* a job. This means that we keep track of the progress of a specific job from arrival until departure, including relevant changes in the system such as arrivals and departures of other jobs. This idea is helpful for finding r_j, which will depend on changes in the system that occur after the arrival of the tagged job.

We can split a service time S into the work that will be performed on it on each pass, $\min(S, \Delta_2)$, $\min(S, \Delta_3) - \min(S, \Delta_2)$, etc. Similarly, we can split the workload $V(t)$ at epoch t into the portion of it that eventually will be performed on jobs making their jth pass, which we call *j-work*.

FB has a special property with regard to j-work that we use as follows: For every realization and every $j \geq 2$, the server can begin to do $\{i: i \geq j\}$-work only at epochs where $(i: i < j\}$-work hits zero. Thus if we combine classes $1, \ldots, j - 1$ into new highest class $J \equiv \{1, \ldots, j-1\}$ (set $\delta_J = \Delta_j$), this has no effect on $r_j, r_{j+1}, \ldots$.

Example 10-6: The $M/G/1$ Queue with Feedback

We now find r_j. For any $j \geq 2$, combine classes $1, \ldots, j - 1$ into new highest class J as described above and write

$$E(V) = Q_J v_J + \sum_{i \geq j} Q_i v_i + \lambda E(S^2)/2, \tag{47}$$

where $v_J = E(S)$, and Q_J is the expected number of 1-pass jobs in queue when classes are combined.

From (45), we can write (47) as

$$E(V) = \rho d_J + \lambda \sum_{i \geq j} r_i G^c(\Delta_i) v_i + \lambda E(S^2)/2, \tag{48}$$

where d_J is the expected first-pass delay that would occur if classes were combined.

Specializing to Poisson arrivals, d_J is the relevant portion of (47),

$$d_J = Q_J \int_0^{\Delta_j} G^c(t)\, dt + \lambda E(S^2)/2 - O_J, \tag{49}$$

where the integral is $E\{\min(S, \Delta_j)\}$, which is the expected work performed on a job on pass J, and $\lambda E(S^2)/2 - O_J$ is the expected time until any job found in service completes the pass in progress. Now $\lambda E(S^2)/2$ is the expected remaining service of any job found in service, and thus O_J, the *expected overage*, is the expected remaining service of that job *when it is interrupted*.

The actual overage is zero unless for $k = J, j, j + 1, \ldots$, a job requiring more than k passes is found making pass k. The probability this occurs is the corresponding fraction of time:

$$\begin{aligned} &\lambda G^c(\Delta_j)\Delta_j && \text{for } k = J, \\ &\lambda G^c(\Delta_{k+1})\delta_k && \text{for } k = j, j + 1, \ldots. \end{aligned} \tag{50}$$

Similarly, the expected remaining work of an interrupted job that was earlier found making pass k is v_j for $k = J$, and v_{k+1} for $k = j, j + 1, \ldots$. From (39) and (50), we have

$$O_J = \lambda \left[\Delta_j \int_{\Delta_j}^{\infty} G^c(t)\, dt + \sum_{k \geq j} \delta_k \int_{\Delta_{k+1}}^{\infty} G^c(t)\, dt \right], \qquad j = 1, 2, \ldots. \tag{51}$$

From (49) and $Q_J = \lambda d_J$, we can now solve for d_J,

$$d_J = [\lambda E(S^2)/2 - O_J] \left[1 - \lambda \int_0^{\Delta_j} G^c(t)\, dt \right]^{-1} \qquad \text{for } j = 2, 3, \ldots, \tag{52}$$

where O_J is given in (51), and for the original class 1, $d_1 = d_{\{1\}}$.

Now *tag* a job on arrival that requires at least j passes, and compare the tagged job's delay under two ways of combining classes, $J + 1 = \{1, \ldots, j\}$ and J. We relate d_{J+1} to $d_J + r_j$. On *sample paths*, total $\{i\colon i \leq j\}$-work is the same for J and $J + 1$, but the split between $\{i\colon i < j\}$-work and j-work is different. For every $i > j$, i-work is also the same. Under J, random variable $D_J + R_j + \Delta_j$ is the time until relevant work $D_{J+1} + \Delta_j$ is performed, *and* the system is clear of $\{i\colon i < j\}$-work, another EFSBP, with expectation

$$d_J + r_j + \Delta_j = (d_{J+1} + \Delta_j) \left[1 - \lambda \int_0^{\Delta_j} G^c(t)\, dt \right]^{-1}.$$

Solving for r_j, we have

$$r_j = \frac{d_{J+1} + \Delta_j}{1 - \lambda \int_0^{\Delta_j} G^c(t)\, dt} - d_J - \Delta_j \qquad \text{for } j = 2, 3, \ldots. \tag{53}$$

Example 10-7: The *M/G/1* Queue with Round Robin

Unlike FB, there is no useful way of combining classes. While we will not obtain an explicit formula for the r_j here, the approach is the same, and its implementation is more straightforward.

We tag a job on arrival and track its progress from pass to pass. Delay on the first pass is the relevant work found by the tagged job on arrival, which is the time for all customers in queue to complete their next pass plus the time to complete any remaining pass in service. From PASTA, r_1 is the appropriate portions of (38), (40), and (51)

$$r_1 = \sum_{i \geq 1} Q_i \omega_i + \lambda E(S^2)/2 - \lambda \sum_{i \geq 1} G^c(\Delta_{i+1})\delta_i v_{i+1}, \tag{54}$$

where the term on the far right is the expected overage (51) when $j = 1$, and

$$\omega_i = \int_{\Delta_i}^{\Delta_{i+1}} G^c(t)\, dt / G^c(\Delta_i),$$

which is the expected work performed on a job on its ith pass, given $S > \Delta_i$.

The tagged job's delay R_2, between its first and second pass, is the time for arrivals during $R_1 + \delta_1$ to make their first pass plus the time for jobs initially present (in queue and in service) to make an additional pass, if necessary. We have

$$r_2 = \lambda(r_1 + \delta_1)\omega_1 + \sum_{i \geq 1} Q_i G^c(\Delta_{i+1})\omega_{i+1}/G^c(\Delta_i) + \lambda \sum_{i \geq 1} G^c(\Delta_{i+1})\delta_i \omega_{i+1}. \tag{55}$$

For R_3, we need to account for the first pass by new arrivals during $R_2 + \delta_2$ and the next pass, as necessary, by jobs that arrived earlier. Continuing this way and from (45), we obtain after a little algebra the following system of linear equations for $\{r_j\}$:

$$\begin{aligned} r_1 &= \lambda \sum_{i \geq 1} r_i a_i + \lambda E(S^2)/2 - \lambda \sum_{i \geq 1} \delta_i \int_{\Delta_{i+1}}^{\infty} G^c(t)\, dt, \\ r_j &= \lambda \sum_{i=1}^{j-1} (r_i + \delta_i) a_{j-i} + \lambda \sum_{i \geq 1} r_i a_{i+j-1} + \lambda \sum_{i \geq 1} \delta_i a_{i+j-1}, \quad j = 2, 3, \ldots, \end{aligned} \tag{56}$$

where

$$a_i = G^c(\Delta_i)\omega_i = \int_{\Delta_i}^{\Delta_{i+1}} G^c(t)\, dt, \qquad i = 1, 2, \ldots,$$

which is the (unconditional) expected work performed on a job on its ith pass.

It can be shown (Wolff [1970b]) that (56) has a unique solution if either the δ_i are finite and *bounded*, or that a job is served in at most a finite number of passes k, which will be the case if we set $\delta_k = \infty$.

FB and RR can be mixed, e.g., pass 1 jobs have priority over all others, but lower passes are RR. More generally, there are limiting versions of attained service rules, as the quanta size $\to 0$.

The *least-attained-service* (*LAS*) rule allocates the server to the job with the *least* attained service. Thus a tagged job begins service immediately on arrival. Should the tagged job's attained service become equal to that of one or more other jobs, these jobs share the server equally, i.e., while there are n jobs in system that have least attained service, the remaining service of each of these jobs decreases at rate $1/n$. This is a limiting version of FB. When arrivals are Poisson, this model is easy to analyze; see Problem 10-17.

For RR, suppose $\delta_i = \delta$, a constant for all i, and $\delta \to 0$. In the limit, all jobs in system share the server equally, i.e., while the number of jobs in system is n, the remaining service of each job decreases at rate $1/n$. This rule is called *processor sharing* (*PS*). In Chapter 6, we found that for Poisson arrivals, PS is an example of a model where the stationary distribution of the number of customers in system is *insensitive* to the service distribution.

For some unspecified work-conserving rule, let job C_n have service time S_n, and let $W_n(a)$ be C_n's waiting time in system until its attained service reaches a, where $W_n(S_n)$ is C_n's entire waiting time. In terms of these quantities, attained service rules have a property that we now define for a queue with mutually independent service times that are also independent of the arrival process.

Definition. A rule is called *service-time-independent* if for every job C_n and every $t \geq 0$, $\{W_n(a): a \leq t\}$ and $\{S_n \mid S_n > t\}$ are independent.

Consider the experience of a tagged job with service time S and attained service $W(a)$ under a service-time-independent rule, where

$$w(a) = E\{W(a)\}, \qquad a \geq 0, \tag{57}$$

is called the *response time* function, which is regarded as an important performance measure for attained service rules. For Examples 10-6 and 10-7, (57) becomes

$$w(a) = \sum_{i=1}^{j} r_i + a = d_j + a \qquad \text{for } a \in (\Delta_j, \Delta_{j+1}], \qquad j = 1, 2, \ldots .$$

We now do a stationary analysis on the tagged job in order to prove the following conservation law for the response time function.

Theorem 4. For a stable $GI/G/1$ queue with $E(S^2) < \infty$, under any work-conserving rule that is also service-time-independent, the response time function $w(a)$ satisfies

$$\lambda \int_0^\infty w(a) G^c(a)\, da = E(V), \tag{58}$$

the time-average work in system.

Proof. Equation (58) follows from the generalization of $L = \lambda w$ stated as Theorem 5 in Chapter 5, which is (informally)

$$(\text{time average}) = (\text{arrival rate}) \cdot (\text{arrival average}),$$

where the time average here is $E(V)$. Let A_j be the contribution to the area under $\{V(t)\}$ by C_j. Because $\{V(t)\}$ is regenerative, so is $\{A_j\}$, and the arrival average of $\{A_j\}$ is $E(A)$, where A is the contribution of the tagged customer. Thus (58) amounts to showing that $E(A)$ is equal to the integral above.

The contribution of the tagged customer is an area such as in Figure 10-3, where we have plotted $S - a$ and $W(a)$ as a varies from 0 to S. For every sample point, $W(a)$ is a monotone function of a, and the area A is a Stieltjes integral that we integrate by parts,

$$A = \int_0^S (S - a)\, dW(a) = \int_0^S W(a)\, da = \int_0^\infty W(a)I(a)\, da, \tag{59}$$

where for every a, $I(a) = 1$ if $S > a$, and $I(a) = 0$ otherwise. Taking the expected value of the right-hand integral, interchanging E and $\int$, and using the independence of $W(a)$ and $I(a)$, we have

$$E(A) = \int_0^\infty w(a)G^c(a)\, da, \tag{60}$$

which proves (58).

Remarks. We don't really need renewal arrivals for (58), e.g., we might have stationary or regenerative input, provided the appropriate limits exist and we can justify a stationary analysis. For $c \geq 2$ channels and general service, (58) will hold for trivial comparisons such as FIFO and LIFO, where $E(V)$ is invariant. For the attained-service rules we have discussed, however, it is easy to see that even with work conservation, $\{V(t)\}$ (and hence $E(V)$) is rule dependent. Consequently, while (58) holds for each rule, it is not a conservation law in this case. On the other hand, for $c \geq 2$ channels and *exponential* service, the number-in-system process $\{N(t)\}$ is stochastically invariant, remaining service times are exponential, and $E(V)$ is invariant, even though sample paths are not.

Example 10-8: The *M/G/*1 Queue with Processor Sharing

The response time function has a remarkably simple form for this model that we now find using (see Section 6-8) these results: The stationary number of jobs in system N_{PS} is $\sim N_{M/M/1}$, and the attained service time of each of these jobs is $\sim G_e$. It follows that

$$L_{PS}(a) = L_{M/M/1}G_e(a) = \lambda \int_0^a G^c(t)\, dt/(1 - \rho), \tag{61}$$

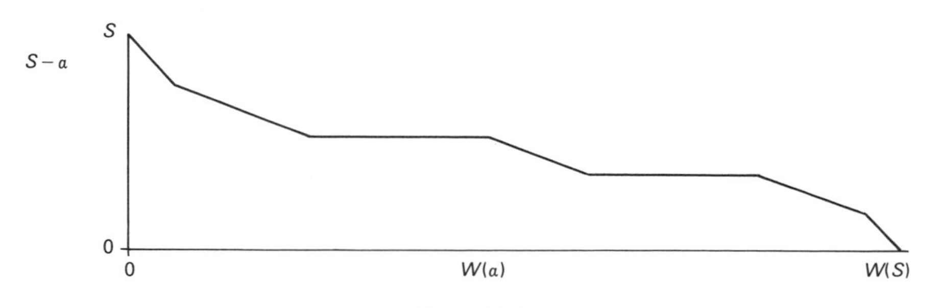

Figure 10-3

where $L_{PS}(a)$ is the expected number of jobs in system under PS that have attained service $\le a$. From $L = \lambda w$, we also have

$$L_{PS}(a) = \lambda \int_0^a G^c(t)\, dw_{PS}(t). \tag{62}$$

Equating these expressions, and from $w_{PS}(0) = 0$, we have

$$w_{PS}(a) = a/(1 - \rho), \qquad a \ge 0. \tag{63}$$

Remark. Applying the same argument to the $M/G/c$ PS queue, (61) becomes

$$L_{PS}(a) = L_{M/M/c} G_e(a),$$

(62) is unchanged, and (63) becomes

$$w_{PS}(a) = a L_{M/M/c}/\rho.$$

Example 10-9: The $M/G/1$ Queue with Preemptive LIFO (PL)

Recall that under PL, every arrival begins service immediately, interrupting any job that may be in service; on service completion, the most recently interrupted job resumes service. This is a work-conserving rule that is also service-time-independent.

To find $w_{PL}(a)$, the argument in Example 10-8 applies, but an even easier one is at hand: $W_{PL}(a)$ is an EFSBP with exceptional service a; hence

$$w_{PL}(a) = a/(1 - \rho). \tag{64}$$

In terms of the response time function, the overall average waiting time is

$$w = \int_0^\infty w(a)\, dG(a), \tag{65}$$

and is by no means conserved under these rules.

Even when (65) is not the major concern, there is more to choosing a good rule than simply comparing response time functions. For example, comparing FIFO and PS for $M/D/1$ (constant service), we have

$$w_F(a) = d + a = \frac{\rho E(S)}{2(1 - \rho)} + a,$$

and, comparing this with (63), PS has a lower response time function only for $a < E(S)/2$, so that PS is worse for all response times experienced by real jobs. This conclusion would hold for service times that are random over some narrow range. Thus, contrary to the remarks following equation (10), one rule outperforms another for *all* priority classes. (Class "a" here means service time $S = a$.) This illustrates again the poor performance of attained service rules when service times are regular. (The least-attained-service rule is even worse than PS.)

Also note that $w(a) = E(W \mid S = a)$ is an expectation, which does not account for the effect of a rule on variance, and low variance is usually achieved by FIFO.

The models in this section assume work conservation. This may not be strictly true, e.g., swapping one job for another on a central processor may result in a positive "swap time," which is a positive length of wasted CPU time. If this always occurs for an incomplete job at the end of a pass, whether or not other jobs are present, swap times can be incorporated into the definition of a service time, and the analysis in Examples 10-6 and 10-7 is still valid. Differential rules should be regarded as approximations, e.g., PS for RR. Unless swap times are small relative to quanta size (the meaning of "small" depends in part on server utilization), however, these approximations may be poor.

10-5 THE $M/G/1$ QUEUE WITH GENERALIZED VACATIONS

In Sections 10-1 through 10-4, we developed efficient methods for analyzing and comparing priority rules, primarily with respect to first-moment performance measures. Unfortunately, higher moments and distributions of (say) delay are often complex expressions that make comparisons difficult, even when they can be found. In this section, we obtain representations for certain stationary distributions for a broad group of models. In many cases, these representations are very useful for finding these other measures.

From the point of view of some class of customers, the server can be thought of as unavailable and "on vacation" when in reality it is serving some other class of customers. Similarly, with or without priorities, the server may actually take breaks or, from time to time, it may break down.

For Poisson arrivals, most of the specific models we have considered generate a generalized type of $M/G/1$ queue that we will define. We show in Theorem 6 that N, the stationary number of customers in system for this model, has a representation as the sum of two independent random variables, one of which is the corresponding number of customers in system for the standard $M/G/1$ queue. From this result, which is of considerable interest in itself, we will find a general expression for (the transform of) the stationary waiting time distribution.

The *(generalized) vacation model* is the result of the interaction of a *standard* $M/G/1$ queue with arrival rate λ, service rate μ, and server utilization $\rho < 1$, and periods called *vacations* during which the server provides no service.

The model has these properties: At all times, the server is either busy serving customers or on vacation. (For the standard $M/G/1$ queue, the idle periods are vacations.) A vacation can begin only at a service completion, and will begin whenever the departing customer leaves the system empty (of customers). A vacation can begin in other ways, e.g., after the completion of service of k customers since the last vacation, for some fixed k. Vacations begin and end according to well-defined rules that may depend on either the current state or the past evolution of the queue. (They may *not* depend on future increments of the Poisson arrival process; this is the "lack of anticipation" property needed in the proof of PASTA.) Service times are i.i.d., independent of all else. In many applications, the sequence of vacation times will be i.i.d., but this can be weakened. We will assume that the entire model is regenerative with finite mean cycle length, which justifies a stationary analysis.

Order of service is nonpreemptive and independent of customer service times, e.g., FIFO or LIFO, such that the customer number-in-system process $\{N(t)\}$ is stochastically invariant with respect to order of service. With no loss of generality, we assume *LIFO* in the following analysis.

Customers who arrive during a vacation are called *vacation* customers, and customers who arrive while the server is busy are called *regular*. For each vacation customer, let the *first-generation* customers be the regular customers who arrive during this vacation customer's service time, the *second-generation* customers be the arrivals during the service times of first-generation customers, and so on. The *sum* of the service times of the vacation customer and of customers on all subsequent generations is the "branching process way" of defining a standard $M/G/1$ busy period. (See equation (36) in Chapter 8.) Thus every customer falls into a well-defined busy period generated by a vacation customer, where the other customers in the period are regular. Note that subsequent vacations may cause the *performance* of a busy period to be split into intervals. We assume that when vacations are included, the $M/G/1$ queue is still stable; in particular cases, this will require verification.

Let

$\mathscr{u}_n$ = number of customers in system at the beginning of vacation n, and

$\mathscr{v}_n$ = number of vacation customers who arrive during vacation n,

where $\{(\mathscr{u}_n, \mathscr{v}_n)\}$ is a stationary (ergodic) sequence, and we make the additional assumption that $E(\mathscr{v}) \equiv E(\mathscr{v}_n) < \infty$. The joint distribution of $(\mathscr{u}_n, \mathscr{v}_n)$ is arbitrary. Let their (marginal) generating functions be

$$\alpha(z) \equiv E(z^{\mathscr{v}_n}) \quad \text{and} \quad \gamma(z) \equiv E(z^{\mathscr{u}_n}). \tag{66}$$

An arrival at the end of vacation n is included in $\mathscr{v}_n$.

Now tag a customer *on departure*. The tagged customer belongs to some busy period which is generated by a vacation customer, which in turn arrived during some (*tagged*) vacation. The set of arrivals during the tagged vacation is called the *batch*, where the vacation customer is a member of that batch. Customers present when the tagged vacation began are called *early* customers.

Under LIFO, the customers left behind by the tagged customer are early customers, members of the batch, or regular customers belonging to the tagged customer's busy period. Let the number of customers left behind by the departing tagged customer be

$$N_d = N_{de} + N_{dv} + N_{dr}, \text{ where} \tag{67}$$

N_{de} = number of early customers,

N_{dv} = number of remaining vacation customers from the batch, and

N_{dr} = number of remaining regular customers from the tagged customer's busy period.

We will determine the (stationary) joint distribution of these quantities.

For a standard $M/G/1$ queue, the customers left behind are never those who start busy periods. Because the busy periods here are equivalent to standard ones, and regular customers left behind are from the same busy period as the departing customer, it follows that

$$N_{dr} \sim N_{ds}, \tag{68}$$

where N_{ds} is the stationary number of customers left behind by a departure in a *standard* $M/G/1$ queue, for which we found the generating function

$$\Pi(z) \equiv E(z^{N_{ds}}) \tag{69}$$

as equation (10) in Chapter 8.

For the distributions of other quantities in (67), we first assume that the tagged customer is a vacation customer, and then we show that the distributions when the tagged customer is regular are the same.

To find the distribution of N_{dv}, let $\mathcal{v}_n$ have distribution

$$\alpha_j = P(\mathcal{v}_n = j), \qquad j = 0, 1, \ldots, \tag{70}$$

and recall the analysis of batch effects in Section 2-5. We have a standard example of length biasing or "batch size biasing." We want to find the "position" of a "randomly selected" vacation customer within that customer's batch, where we let $\mathcal{B}$ be the (stationary) number of vacation customers left behind (from the same batch) by the vacation customer, which, in the notation of that section, has the same distribution as $J - 1$, the number of vacation customers who are served in front of the vacation customer. Thus $\mathcal{B}$ has distribution

$$\beta_j \equiv P(\mathcal{B} = j) = P(\mathcal{v}_n > j)/E(\mathcal{v}) = \sum_{i>j} \alpha_i/E(\mathcal{v}), \qquad j = 0, 1, \ldots. \tag{71}$$

Remarks. LIFO within the batch is irrelevant; someone is served first, someone second, and so on. The results in Section 2-5 were derived under the assumption that the $\mathcal{v}_n$ are i.i.d. As will be shown, extension to so-called stationary ergodic sequences is straightforward. While not the definition, a stationary sequence with finite expectation is *ergodic* if the arithmetic average of the first n terms converges w.p.1 to this expectation, as $n \to \infty$. In the i.i.d. case, this is the strong law of large numbers. If $\{(\mathcal{u}_n, \mathcal{v}_n)\}$ is regenerative, there will be a stationary ergodic version of this sequence.

We now find the distribution of $\mathcal{U}$, the stationary number of early customers left behind by the tagged customer. Note that under LIFO, *all* vacation customers from the same vacation leave the *same early customers*, and hence the same *number* of early customers behind.

We are sampling from $\{\mathcal{U}_n\}$ in a particular way, and we again have a length biasing effect. This is easily taken into account by letting

$$\mathcal{V}_{ni} = \text{number of vacation customers during vacation } n \text{ who leave } i \text{ early customers behind,}$$

i.e., $\mathcal{V}_{ni} = \mathcal{V}_n$ if $\mathcal{U}_n = i$, $\mathcal{V}_{ni} = 0$ otherwise. Now $\{\mathcal{V}_{ni}\}$ is a stationary ergodic sequence with expectation

$$E(\mathcal{V}_{ni}) = \sum_j jP(\mathcal{U}_n = i, \mathcal{V}_n = j), \tag{72}$$

and it follows immediately that the *fraction* of vacation arrivals who leave i early customers behind is

$$\lim_{n\to\infty} \sum_{j=1}^{n} \mathcal{V}_{ji} \Big/ \sum_{j=1}^{n} \mathcal{V}_j = E(\mathcal{V}_{ni})/E(\mathcal{V}). \tag{73}$$

Now $\mathcal{U}$ is a representative of a stationary sequence of the number of early customers left behind by departing vacation customers, and it follows that

$$P(\mathcal{U} = i) = E(\mathcal{V}_{ni})/E(\mathcal{V}) = \sum_j jP(\mathcal{U}_n = i, \mathcal{V}_n = j)/E(\mathcal{V}), \qquad i = 0, 1, \ldots. \tag{74}$$

Of greater interest in applications is

$$\mathcal{Y} = \mathcal{U} + \mathcal{B},$$

the stationary sum of these quantities. For vacation n, there will be *one* vacation customer who leaves a total of y early and vacation customers behind if and only if $\mathcal{U}_n \leq y$ and $\mathcal{U}_n + \mathcal{V}_n > y$. By the argument used for (74), we have

$$P(\mathcal{Y} = y) = \sum_{i=0}^{y} \sum_{j>y-i} P(\mathcal{U}_n = i, \mathcal{V}_n = j)/E(\mathcal{V}), \qquad y = 0, 1, \ldots. \tag{75}$$

Similarly, the joint distribution of $(\mathcal{U}, \mathcal{B})$, and of course (71), may be found by the same method.

From their distributions above, moments of $\mathcal{B}$ and $\mathcal{U}$ may be expressed in terms of (factorial) moments of $\mathcal{V}_n$ and expectations of form $E(\mathcal{U}_n^i \mathcal{V}_n)$. First moments are

$$E(\mathcal{U}) = E(\mathcal{U}_n\mathcal{V}_n)/E(\mathcal{V}), E(\mathcal{B}) = E\{\mathcal{V}_n(\mathcal{V}_n - 1)\}/2E(\mathcal{V}), E(\mathcal{Y}) = E(\mathcal{U}) + E(\mathcal{B}). \tag{76}$$

Now suppose the tagged customer is regular. The number of early and vacation customers left behind is the same for all departures from the same busy period, independent of the number of customers served in that period and the position (service order) of the tagged customer within the period, where the number served in successive busy periods are i.i.d., independent of $\{(u_n, v_n)\}$. Thus for every busy period generated by a vacation customer for each vacation n, the number of departures from that period who leave k (say) regular customers behind is independent of (u_n, v_n) and the order of service of the vacation customer within that customer's batch. From these observations, it is easily shown that the distribution of (N_{de}, N_{dv}) is the same for regular customers as for vacation customers, so for all departures, $(N_{de}, N_{dv}) \sim (\mathcal{U}, \mathcal{B})$, and that N_{dr} is independent of $(\mathcal{U}, \mathcal{B})$.

If u_n and v_n are *independent*, (74) becomes the distribution of u_n, and (75) becomes the *convolution* of this distribution with (71).

Putting the pieces together, we have

Theorem 5. For the $M/G/1$ vacation model, the stationary number of customers left behind by a departure has representation

$$N_d = \mathcal{Y} + N_{ds}, \tag{77}$$

where $\mathcal{Y}$ and N_{ds} are independent, $\mathcal{Y}$ has distribution (75), and N_{ds} is the stationary number of customers in system left behind for the corresponding standard $M/G/1$ queue. If u_n and v_n are independent, (77) can be written

$$N_d = \mathcal{U} + \mathcal{B} + N_{ds}, \tag{78}$$

where $\mathcal{U}$, $\mathcal{B}$, and N_{ds} are independent, $\mathcal{U} \sim u_n$, $\mathcal{B} \sim$ (71), and $N_{ds} \sim$ as in (77).

By the elementary arguments used in Section 8-3, N_d has the same distribution as the number of customers in system *found by an arrival*, and from PASTA, the second distribution is the corresponding time average. Thus

$$N_d \sim N, \tag{79}$$

the stationary number of customers in system for the vacation model at a random time point, and of course,

$$N_{ds} \sim N_s, \tag{80}$$

the stationary number of customers in system for the standard $M/G/1$ queue. We have shown the following remarkable result:

Theorem 6. The stationary number of customers in system for the $M/G/1$ vacation model at a random time point has the representation

$$N = \mathcal{Y} + N_s, \tag{81}$$

where $\mathcal{Y}$ and N_s are independent, $\mathcal{Y}$ has distribution (75), and N_s is the stationary number of customers in system for the standard $M/G/1$ queue. If u_n and

v_n are independent, (81) can be written

$$N = \mathcal{U} + \mathcal{B} + N_s, \tag{82}$$

where $\mathcal{U}$, $\mathcal{B}$, and N_s are independent, $\mathcal{U} \sim u_n$, $\mathcal{B} \sim$ (71), and $N_s \sim$ as in (81).

Theorems of this kind are usually stated in terms of generating functions, and are called *factorization* theorems. Thus, combining (77) and (81), we have

$$\Gamma(z) \equiv E(z^N) = E(z^{N_d}) = \chi(z)\Pi(z), \tag{83}$$

where $\chi(z)$ is the generating function of distribution (75), and $\Pi(z)$ is the generating function of N_s, given by equation (10) in Chapter 8. If u_n and v_n are independent, (83) becomes

$$\Gamma(z) = \gamma(z)\beta(z)\Pi(z), \tag{84}$$

where $\gamma(z)$ is defined in (66), and

$$\beta(z) \equiv E(z^{\mathcal{B}}) = [1 - \alpha(z)]/(1 - z)E(v), \tag{85}$$

where $\alpha(z)$ is defined in (66), and the relation between $\beta(z)$ and $\alpha(z)$ in (85) follows in a straightforward manner from (71) when the order of summation is interchanged.

Remark. We can also write $N = N_e + N_v + N_r$, where these quantities are the respective stationary number of early, vacation, and regular customers (from the same busy period as the customer in service) in the system at a random time point. Theorem 6 does *not* determine their distribution. In general, N_r is *not* independent of N_e and N_v, and their marginal distributions differ from those for N_{dr} and N_{de} and N_{dv}. See Example 10-10(a).

Example 10-10

(a) For the ordinary $M/G/1$ queue, the number of arrivals during (at the end of) the nth idle period is $v_n = 1$, $u_n = 0$, $\mathcal{B} = 0$, and $N \sim N_s$. Note that $P(N_v = 1) > 0$, so that N_v is stochastically larger than $\mathcal{B}$, and N_r is stochastically smaller than N_s.

(b) For the exceptional-first-service model in Example 10-3, suppose an exceptional service $S_a = S + R$, say, where S is an ordinary service and R is a nonnegative random variable called *added delay*, independent of S, and, without loss of generality, the R-portion of service is performed first. Define the nth vacation period to be an idle period plus the next R-period, $v_n = 1 +$ (the number of arrivals during the nth R period), and $u_n = 0$. If R_n is the added delay at the beginning of busy period n, where the R_n are i.i.d. with distribution function A, v_n has generating function

$$\alpha(z) = E\{E(z^{v_n} \mid R_n)\} = z\tilde{A}[\lambda(1 - z)].$$

If S_a does not have this representation, e.g., S_a is *smaller* than S, the exceptional-first-service model is outside the scope of the vacation model.

(c) For the preemptive model in Example 10-4, let the ordinary $M/G/1$ queue be the 2-job queue, where 2-job service times are now their *completion times*. The exceptional first service is a completion time plus the remainder of any 1-job busy period found in progress by an arriving 2-job that begins a 2-job busy period. This is an example of added delay in (b), and fits the vacation model. Vacation periods begin when 2-job busy periods end. There is more than one way to define the v_n here. The simplest is to let $v_n = 1$ if the next arrival is a 2-job, and let v_n be the number of 2-job arrivals during a 1-job busy period otherwise. (If $v_n = 0$, we begin another vacation.) It is easy to obtain the generating function of v_n as a function of the 1-job busy period transform. There is another obvious way to define v_n. What is it, and why do both definitions result in the same $\mathcal{B}$? Note that in (a) through (c), the v_n are i.i.d., and vacations begin only when the system is empty.

(d) For the postponable model in Example 10-1, let the ordinary $M/G/1$ queue be the j-jobs. First suppose j is the *lowest* class, and combine all other jobs into a new class 1. A j-job will enter service only if there are no 1-jobs present. Vacations can begin with j-jobs in queue, so that $u_n > 0$ for some n. A vacation will begin with $u_n > 0$ only if there was at least one 1-job arrival during the preceding j-job service, so that u_n and v_n are dependent, but the form of the dependence is fairly simple. If j is not the lowest class, a lower class can enter service when $u_n = 0$, and the dependence is more complicated.

We now continue the analysis of the vacation model by supposing that customers are in fact served FIFO. Observe that this change from the LIFO rule used to derive Theorems 5 and 6 has no effect on the distribution of N and N_d. For the next result, it is necessary to make the following

Assumption. The waiting time of every customer is independent of the arrival process after that customer arrived.

Under FIFO, N_d is the number of arrivals during W, the waiting time of the tagged job, where N_d has generating function (83). From the assumption above, $(N_d \mid W)$ is distributed Poisson with parameter λW. By the standard argument used in Section 8-2, we have

Theorem 7. The stationary waiting time for the $M/G/1$ vacation model under FIFO has transform

$$E(e^{-sW}) = \chi(1 - s/\lambda)\tilde{W}_s(s), \tag{86}$$

where $\tilde{W}_s(s)$ is the standard waiting time transform in equation (16) in Chapter 8, and (86) can be expressed in terms of (84) and (85) when u_n and v_n are independent.

Remark. The preceding assumption clearly does not hold under LIFO, and may not hold under FIFO, e.g., for the so-called N-policy, where vacations begin when the system empties, and end when the number of customers in system reaches N. It does hold for the models in Example 10-10. For a priority queue,

Theorem 7 will follow if this assumption holds for the Poisson arrival process of the particular class we define as the standard $M/G/1$.

There are an enormous number of specific variations of the vacation model in the literature for which the results of this section provide a method of attack. In a typical application, however, $\{(u_n, v_n)\}$ is a process generated by the interaction of a standard $M/G/1$ queue, possibly other stochastic processes, and specified rules of operation. The joint distribution of (u_n, v_n) is not a "given," but must be found. In some cases, such as in Example 10-10(b,c), this is easy to do. In others, this can be very difficult to do, and other methods may work better.

10-6 RELATED LITERATURE

The literature on priority queues and related vacation models is enormous; Jaiswal [1968] and Kleinrock [1976] have extensive treatments and references, but Jaiswal in particular is rather dated.

The methods in the first two sections of this chapter and conclusion (28) are based on Wolff [1970a], but results derived in the examples are much older. Cobham [1954] was the first to derive (14). Gaver [1962] was the first to introduce the concept of a completion time, which he used to derive transform expressions corresponding to results in Examples 10-2 and 10-4. Similar ideas are found in Avi-Itzhak and Naor [1963]. Kleinrock used the concept of work conservation earlier, but in a more restricted way.

Sample-path arguments for time-average work in system in this chapter and in Section 5-13 owe their validity to Heyman and Stidham [1980].

Cox and Smith [1961] proved the optimality of the $c\mu$-rule for an $M/G/1$ queue with fixed postponable priority classes by the interchange argument suggested in Section 10-3. These results have been generalized in several ways e.g., the $c\mu$-rule is optimal when we permit the rule to depend on the queue length of each class, but these generalizations are all restricted to the $M/G/1$ postponable model.

The optimality of SRST in the sense of (33) was first shown by Schrage [1968]. The proof given here is a simplification of one by Smith [1978]. Schrage and Miller [1966] analyzed the SRST $M/G/1$ queue, assuming a continuous service time distribution; they also made numerical comparisons with FIFO and other alternatives. Our derivation of (37) is new.

Section 10-4 through Example 10-7 is based on Wolff [1970b]. Attained service rules and response time functions for $M/G/1$ queues are treated at length in Kleinrock.

The conservation law stated as Theorem 4 was shown to hold for $M/G/1$ queues by Kleinrock (see p. 197 of [1976]), and extended to $GI/G/1$ queues by O'Donovan [1974]. The proof here is similar to that in Heyman and Sobel [1982], but their claimed generalization to $GI/G/c$ is false; see Remarks following the proof of this theorem.

Processor sharing result (63) was first shown by Sakata et al. [1969], and many elementary proofs have since been devised such as the one here and that of Tsoucas and Walrand [1983]. Their proof is for a station in a quasi-reversible network (see Chapter 6); the proof here extends to these networks. Some of these simple "proofs" are fallacious (not the cited ones, of course) and have led to claimed generalizations that are false.

The representation in Theorems 10-5 and 10-6 has been discovered many times for particular special cases of the vacation model; see the references in the next two papers. For the special case where $u_n = 0$, i.e., vacations begin only when the system is empty, the first simple and direct proof is by Fuhrmann [1984]. Fuhrmann and Cooper [1985] present a simple and direct proof for the general case where $\{(u_n, v_n)\}$ is a stationary ergodic sequence, but they determine the distribution of $\mathcal{Y}$ only for independent u_n and v_n. Thus the explicit results (74), (75), and (76) are new. Section 10-5 borrows LIFO and the assignment of customers to busy periods from Fuhrmann and Cooper, but a different technique is used to obtain these more explicit results.

PROBLEMS

10-1. *The shortest-job-first rule*. Consider an $M/G/1$ queue, where, on service completion, the next customer served is the one in queue with the shortest service time (FIFO breaks ties). Let $d(t)$ be the average delay in queue of a job with service time $S = t$.

(a) For fixed t, find $d(t)$ by defining three priority classes defined by $\{S_j < t\}$, $\{S_j = t\}$, and $\{S_j > t\}$. Consider the cases where G is continuous at t, and where it is not.

(b) Find d, the average delay over all jobs.

10-2. Consider an $M/M/c$ queue with two priority classes. Class 1 has *preemptive* priority over class 2. Both classes have the *some* service distribution, exp (μ). Work is conserved. Let class i have arrival rate λ_i, and $\lambda = \lambda_1 + \lambda_2$. Assume all standard $M/M/c$ results are known, and express your answers below in terms of them.

(a) Find w_1, the average waiting time in system of 1-jobs.

(b) Find w_2, the average waiting time in system of 2-jobs.

10-3. For a $GI/G/1$ queue, suppose that with probability β, a customer is dissatisfied with service on service completion, and joins the end of the queue, independent of whether this customer had been dissatisfied earlier. Subsequent service times, as needed, are random draws from G. Call this *Model 1*. *Model 2* is the same, except that a dissatisfied customer reenters service immediately, as many times as necessary, until satisfied.

(a) Comparing these models, what quantities (and processes) are invariant? Stochastically invariant?

(b) Find the average delay d_1 for Model 1 in terms of the average delay for a well-defined *standard* $GI/G/1$ queue.
(c) Compare waiting time variances $V(W_i)$, $i = 1, 2$.
(d) For $M/G/1$ and Model 1, find the average delay of a customer from arrival until service begins *for the first time*.

10-4. Consider an $M/G/1$ queue with two classes of jobs, where class 2 service $S_2 \sim G_2 = E_2$, a 2-Erlang distribution with mean $1/\mu_2$. Class 1-jobs have preemptive priority over 2-jobs found in their first phase of service, but only postponable priority over 2-jobs found in their second phase of service. Assume work is conserved, and FIFO within class.
(a) Write an expression for time-average work in system.
(b) Find the average delay of a 1-job, d_1.
(c) Find the expected waiting time in system of a 2-job, w_2.
(d) Instead of work conservation, suppose that when an interrupted 2-job reenters service, the service time is a random draw from E_2. How would this assumption affect your answers above?

10-5. Consider an $M/G/1$ queue with two classes of customers, where 1-jobs have preemptive priority over 2-jobs found in service that have *attained* service $\leq \alpha$, and postponable priority over 2-jobs with attained service $> \alpha$. Assume work is conserved.
(a) Represent time-average work in system and the portion of it that is relevant for 1-job delay.
(b) Find the average 1-job delay, d_1.
(c) Find the average 2-job waiting time in system, w_2.

10-6. *Continuation.* For what service distributions G_2 might a rule of the type in Problem 10-5 make sense? When might it make sense to reverse the preemption and postponable portions of S_2?

10-7. For an $M/G/1$ queue, suppose there is a start-up cost of $\$C$ whenever a busy period begins (the server begins work). In order to reduce start-up cost per unit time, it has been proposed that the queue be allowed to build before work begins. Here are two ways to do this:

(1) Under an *N-policy*, which controls the queue directly, the server begins work when the number of customers in system reaches some constant $N \geq 1$, where for $N = 1$, we have the standard $M/G/1$ queue.
(2) Under a *T-policy*, the server begins work when the idle time since the last busy period reaches a constant $T > 0$. If no arrivals occur during T, no start-up cost is incurred, and another T-period begins.

Under both policies, the server continues to work until the system is empty.
(a) Under both policies, what fraction of time is the server busy?
(b) Under a T-policy, suppose an arrival occurs during a T-period. What is the expected time until the T-period ends?
(c) Under a T-policy, justify the following:

$$d_T = (1 - \rho)T/2 + Q_T E(S) + \lambda E(S^2)/2,$$

and solve for d_T. (Subscripts denote policy.)

(d) For an N-policy, justify the following and solve for d_N:

$$d_N = (1 - \rho)(N - 1)/2\lambda + Q_N E(S) + \lambda E(S^2)/2.$$

(e) If the cost of delay is \$$h$/customer/unit time, write an expression for the (total cost)/(unit time) under both policies.

Remark. For the cost structure in (e), it can be shown that the optimal N-policy is better than any T-policy.

10-8. *Continuation.* An alternative to (1) and (2), called a *D-policy,* controls the queue in terms of work, and is a bit messy to carry out in detail for an arbitrary service distribution. To illustrate, but also to keep the algebra simple, suppose

$$P(S = 1) = P(S = 3) = 1/2.$$

The server begins to work when work in system exceeds some fixed $v > 0$, i.e., the server begins service with the kth arrival since the last busy period if

$$\sum_{j=1}^{k-1} S_j \leq v < \sum_{j=1}^{k} S_j, \; k = 1, 2, \ldots .$$

Let $v = 1.5$ in the calculations that follow.

In the analysis, it is convenient to identify three cases: *empty* means no customers in system, *busy* means server working, and *idle-not-empty* means the server is idle, but customers are present.

(a) What fraction of time is the server busy?
(b) What is the mean time between arrivals who find the system empty?
(c) What fraction of arrivals find the system empty?
(d) What is the expected delay of an arrival who finds the system empty?
(e) Find the average delay in queue.

10-9. Cars arriving at a single-channel automobile clinic are Poisson at rate λ. Checkup times are i.i.d. With probability α, a car is found to need repair. Repair times are i.i.d. and independent of checkup times. (Conclusions in (f) and (g) are not changed if thc chcckup and repair times for the same car are dependent.) Checkups have postponable priority over repairs. Otherwise, order of service is FIFO. Denote a checkup time by S_1, a repair time by S_2, and the total service time (checkup, and, if necessary, repair) by S.

(a) In terms of moments of S_1 and S_2, find $E(S)$ and $E(S^2)$.
(b) Represent $E(V)$, time-average work in system.
(c) What fraction of time is (i) a checkup in progress? What fraction of time is (ii) a repair in progress?
(d) Find d_1, the expected delay before a checkup.
(e) Find d_2, the expected delay between checkup and repair, for a car needing repair.
(f) Let d_R be the total expected delay under this rule. Compare this rule with pure FIFO, where customers in need of repair are repaired immediately after checkup. By equating time-average work, show that

$$d_R < d_F \quad \text{if and only if} \quad E(S_1) + \alpha E(S_2) < E(S_2),$$

i.e., the mean service time is less than the mean *remaining* service time of a car found to need repair.

(g) Why does your conclusion in (f) *not* depend on Poisson arrivals?

(h) Compare (f) with the corresponding result in Problem 5-4.

10-10. We now compare performance of a $GI/G/1$ queue under the shortest-job-first (SJF) rule to FIFO, where we rule out constant service.

(a) Show that D_{SJF} and S *for the same job* are positively correlated.

(b) Argue that $E(D_{\text{SJF}}S) = E(D_F S)$, and, from (a),

$$d_{\text{SJF}} < d_F.$$

10-11. In Example 10-4, compare resume and repeat-different when service is exponential.

10-12. In Example 10-4, argue that the fraction of 1-job arrivals who find a completion time in progress is $\lambda_2 E(\mathscr{C})$. These arrivals are served during the same completion time. Now argue that the fraction of time that a completion time is in progress *and* a 1-job is in service is

$$[\lambda_1 \lambda_2 E(\mathscr{C})]E(S_1) = \rho_1 \lambda_2 E(\mathscr{C}),$$

which is equivalent to (21).

10-13. This model is from Greenberg, Leachman, and Wolff [1988]. Suppose we have a narrow one-lane bridge on a two-lane road; see Figure 10-4. Arriving cars in each direction are Poisson at rate λ/minute from the right (r-cars), and at rate μ/minute from the left (ℓ-cars). Once a car begins to cross the bridge in one direction, arrivals from the other direction must wait, forming a queue. Arrivals from the same direction as cars on the bridge cross the bridge without delay. Each car takes T *minutes* to cross the bridge, a constant. The number of cars (from the same direction) on the bridge at one time can be large; we regard it as unlimited. (Note that the bridge is not drawn to scale.)

Let B_r and B_ℓ be the durations of a period during which the bridge is continuously occupied by r-cars and ℓ-cars, respectively. Once a period of one type ends, cars of the other type in queue, if any, initiate a period of the other type, where we make the approximation that cars in queue *cross together* in T minutes. If the other queue is empty when a B period ends, the bridge is empty until next arrival, which initiates a new (r or ℓ)-period.

(a) Relate B_r and B_ℓ to quantities from a more familiar queueing model.

(b) Find

$$E(B_r) = (e^{\lambda T} - 1)/\lambda \quad \text{and} \quad E(B_\ell) = (e^{\mu T} - 1)/\mu.$$

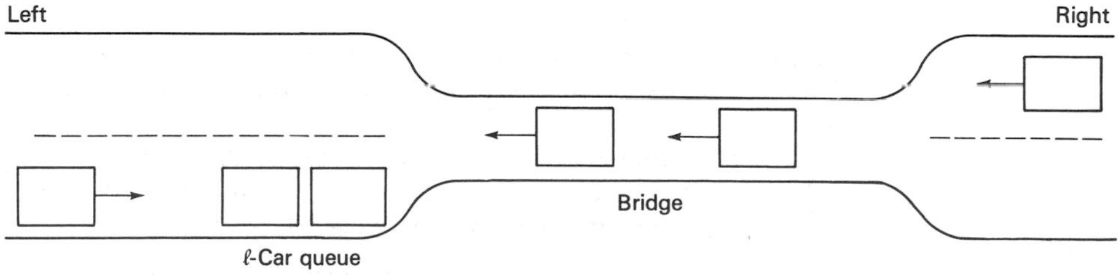

Left
Right
Bridge
ℓ-Car queue

Figure 10-4

(c) Let p_r be the proportion of time an r-period is in progress, f_r be the rate that r-periods begin, and q_r be the probability that when an ℓ-period ends, the r-queue is positive. Define ℓ-car quantities analogously, and let $p_0 = 1 - p_r - p_\ell$. Show that

$$\text{(i) } f_r = \lambda p_0 + q_r f_\ell \quad \text{and} \quad \text{(ii) } f_r E(B_r) = p_r.$$

(d) Show that

$$q_r = a/(1 - b), \text{ where}$$

$$a = \lambda[1 - e^{-(\lambda+\mu)T}]/(\lambda + \mu) \quad \text{and} \quad b = \mu[1 - e^{-(\lambda+\mu)T}]/(\lambda + \mu).$$

(*Hint:* Condition on "what happens" during the first T minutes of B_ℓ.)

(e) Obtain corresponding expressions for ℓ-car quantities, and solve for $p_0 = (\lambda + \mu)/D$, $p_r = (e^{\lambda T} - 1)[\mu e^{(\lambda+\mu)T} + \lambda]/D$, and $p_\ell = (e^{\mu T} - 1)[\lambda e^{(\lambda+\mu)T} + \mu]/D$, where

$$D = \lambda + \mu + (e^{\lambda T} - 1)[\mu e^{(\lambda+\mu)T} + \lambda] + (e^{\mu T} - 1)[\lambda e^{(\lambda+\mu)T} + \mu].$$

(f) Argue that the expected delay of an r-car has the form

$$d_r = p_\ell E(B_\ell^2)/2E(B_\ell).$$

(g) Because T is constant, the distribution of the number of ℓ-car arrivals during B_ℓ is easily shown to be geometric. Write down this distribution and show that

$$E(B_\ell^2) = 2e^{\mu T}[(e^{\mu T} - 1)/\mu - T]/\mu.$$

Remarks. The analysis is easily modified to permit the bridge crossing times in each direction to be different. Parts (a) through (e) may easily be generalized when T is random; results for these parts are valid when T is replaced by $E(T)$. Part (f) would still be valid; however, the second moment in (g) would change, and finding it would be very difficult.

10-14. Consider a $GI/G/1$ queue under preemptive LIFO (PL), and assume that work is conserved. Compare this rule with FIFO in (a) through (d) below. Is the quantity the same or different for the two rules, and, if different, for which rule is it greater? Indicate reasoning.

(a) Busy periods and work in system.

(b) For exponential service, expected waiting time.

(c) For exponential service, the stationary waiting time variance.

(d) For $E(S - t \mid S > t) \le E(S)$ for all $t \ge 0$, expected waiting time.

10-15. *Alternating priorities.* (Avi-Itzhak, et al. [1965]) An auto paint shop, operating as a single-channel work-conserving queue, will paint a car either red (r) or silver (s). Arrivals are Poisson at rate λ. With probability p_i, $i \in \{r, s\}$, an arrival selects color i, which has painting time $S_i \sim G_i$, and $\rho_i = \lambda p_i E(S_i)$.

When switching colors, dirty nozzles are replaced by clean ones, which is instantaneous, but the dirty nozzles must be cleaned. In order to reduce cleaning costs, the shop operates under this postponable rule: Paint cars the same color until no cars to be painted that color are present, then switch to the other color. Continue until empty. Arrivals finding the system empty are served immediately, whether or not a color change is required.

To analyze this problem, split work in system by color, where $E(V) = E(V_{\varkappa}) + E(V_{\varsigma})$. For all pairs of colors (i, j), let

$$E(V_{ij}) = \text{expected } j\text{-work in system given } i\text{-work in service.}$$

(a) What fraction of time is the server doing i-work, $i \in \{\varkappa, \varsigma\}$?
(b) Find an expression for $\rho_{\varkappa} d_{\varkappa} + \rho_{\varsigma} d_{\varsigma}$.
(c) For $i \neq j$, show that

$$d_i = E(V_i) + \rho_j E(V_{jj})/(1 - \rho_j).$$

(d) Let B_i denote the duration of an i-work busy period, where B_i begins either with the termination of a j-work busy period when i-work is present $(i \neq j)$, or with an i-arrival finding the system empty. Why is

$$E(B_i^2)/2E(B_i) = E(V_{ii})/(1 - \rho_i)?$$

(e) From (d), and by considering the j-work that arrived during the age of the current i-work busy period, show that for $i \neq j$,

$$E(V_{ij}) = \rho_j E(V_{ii})/(1 - \rho_i).$$

We now have enough information to solve for the d_i.
(f) Suppose $G_{\varkappa} = G_{\varsigma}$, i.e., service times are independent of color. Find

$$p_{\varkappa} d_{\varkappa} + p_{\varsigma} d_{\varsigma}$$

by a direct argument, independent of the analysis above.

10-16. Why is (44) invariant in a stochastic sense?

10-17. *The M/G/1 queue under the least-attained-service rule (LAS).* LAS is the differential version of the feedback rule analyzed in Example 10-6. Service begins on arrival and continues until either service is complete, a new arrival occurs, or the attained service of the arrival reaches that of one or more previous arrivals. In the latter case, jobs with the same (least) attained service share the server equally; the attained service of each job increases at a rate inversely proportional to the number of jobs sharing the server. Work is conserved.

For arbitrary $x \geq 0$, split service times S_j into "early" and "late" portions,

$$S_j = \min(S_j, x) + \max(S_j - x, 0) \qquad \text{for every } j,$$

and let $V_x(t)$ be the *early work* in system at epoch t, i.e., it is the sum of the remaining early service time portions of all customers in system. Clearly, early work has preemptive priority over late work.

(a) Show that time-average early work is

$$E(V_x) = \lambda E\{[\min(S_j, x)]^2\}/2[1 - \lambda E\{\min(S_j, x)\}] = \lambda \int_0^x t G^c(t)\, dt/(1 - \rho_x),$$

where $\rho_x = \lambda \int_0^x G^c(t)\, dt$.

(b) Let W_x be the waiting time in system of a job with service time $S = x$. Show that

$$w_x = E(W_x) = [x + E(V_x)]/(1 \quad \rho_x).$$

10-18. Recall the dissatisfied customer model, as in Problem 10-3. Instead, consider an $M/M/1$ queue where *every* customer joins the end of the queue after the first service completion, and leaves the system after the second. Thus every customer makes two passes through the service facility. (If the probability of dissatisfaction is $\beta = 1/2$ in the earlier model, the *expected* number of passes per customer is two.) Let r_i be the expected delay between passes $i - 1$ and i, and Q_i be the expected number of customers waiting to make pass i, $i = 1, 2$.

(a) Relate r_i to Q_i, $i = 1, 2$.

(b) What fraction of time is the system busy?

(c) Tag a customer on arrival and find r_1 in terms of Q_1 and Q_2.

(d) Find r_2 for the tagged customer in terms of the system as found on arrival, and changes that occur later.

(e) For $\rho = 2\lambda/\mu$, show that

$$Q_1 = Q_2 = \rho^2/2(1 - \rho).$$

10-19. *An insensitivity result*. This is a continuation of Problem 10-18. Let v_j be the number of passes made by customer j, where the v_j are i.i.d., with

$$P(v_j = i) = \alpha_i, \qquad i = 1, 2, \ldots,$$

with mean $E(v) = \sum_i i\alpha_i$, where $\rho \equiv \lambda E(v)/\mu < 1$. As noted, a customer who requires an additional pass joins the *end* of the queue. By the method of tagging a customer, obtain a set of linear equations for the Q_i, and show that

(a) $Q_i = \rho^2 P(v \geq i)/[(1 - \rho)E(v)]$, and

(b) $Q = \sum_i Q_i = \rho^2/(1 - \rho)$,

i.e., Q is *insensitive* to $\{\alpha_i\}$, depending only on the mean of this distribution.

10-20. Consider an attained service rule that is a mixture of FB and RR in Examples 10-6 and 10-7, where 1-pass jobs have priority over all others, and the others are served RR. To keep the algebra down, set $\delta_3 = \infty$. Find a set of linear equations for $\{r_1, r_2, r_3\}$ that uniquely determines these quantities.

10-21. Does (63) require Poisson arrivals; e.g., can $w_{PS}(a)$ be proportional to a for (say) the $GI/G/1$ queue? To investigate, consider the corresponding random variable $W_{PS}(a)$ for very large a. Ignoring work found on arrival,

$$W_{PS}(a) = a + \text{work to arrive during } W_{PS}(a).$$

(a) Argue (loosely but correctly by letting $a \to \infty$) that $w_{PS}(a)$ is proportional to a only if

$$w_{PS}(a) = a/(1 - \rho).$$

(b) Now plug result in (a) into (58), and conclude that for a stationary single-channel queue, $w_{PS}(a)$ is proportional to a only if

$$E(V) = \lambda E(S^2)/2(1 - \rho) = E(V_{M/G/1}).$$

10-22. Either by specializing (75) or directly, derive (71).

10-23. Derive (85) and find $E(\mathcal{B})$.

10-24. Find an expression for the generating function of $\mathscr{v}_n$ for the preemptive model in Example 10-10(c).

10-25. Analyze the N-policy and the T-policy in Problem 10-7 from the point of view of the vacation model in Section 10-5.

(a) Find the distribution of $\mathscr{v}_n$ under each policy.

(b) Does Theorem 7 apply to either policy?

11

BOUNDS AND APPROXIMATIONS

As we found in Chapter 6, queueing models of great complexity often turn out to have remarkably simple time-average properties, when they have Poisson arrivals and exponential service. (Even with these assumptions, the models in Chapter 7 are difficult to solve.) With phase-type distributions, we were able to extend these results to general service distributions, provided that stations with these distributions operate as *symmetric queues*. The results obtained were said to be *insensitive* to these distributions.

FIFO is not a symmetric rule, and measures such as L in Chapters 8 and 9 clearly are not insensitive to service (or inter-arrival) distributions. Standard FIFO versions of $GI/G/1$, $GI/G/c$, and even $M/G/c$ queues are difficult to solve for these measures. General distributions can of course be approximated by those of phase-type, but in the absence of insensitivity, this often leads to computational difficulties. Furthermore, simply grinding out numerical solutions, even when possible, does little to increase our understanding of these models.

In this chapter we present bounds and approximations on performance measures for FIFO queues and some tandem queues. Some results are in terms of means and variances of (unspecified) distributions. We also compare the performance of queueing models, e.g., single-channel queues with different inter-arrival and/or service distributions, and multichannel with single. Approximations under both heavy and light traffic are presented, and the effect of dependent service times on performance is briefly treated.

11-1 *GI/G/1* DELAY IN QUEUE

We use the notation in Chapter 9. In particular, we have a FIFO single-channel queue, where T_n is the inter-arrival time between customers C_n and C_{n+1}, S_n is the service time of C_n, $X_n = S_n - T_n$, and $\{T_n\}$ and $\{S_n\}$ are independent sequences of i.i.d. random variables. Let inter-arrival distribution A have mean $1/\lambda$ and *variance* σ_a^2, service distribution G have mean $1/\mu$ and *variance* σ_g^2, and $0 < \rho = \lambda/\mu < 1$. Thus the queue is stable. Let $c_a^2 = \lambda^2\sigma_a^2$ and $c_g^2 = \mu^2\sigma_g^2$, where c is called the *coefficient of variation.*

For D_n, the delay in queue of C_n, write equation (2) in Chapter 9 as

$$D_{n+1} = (D_n + X_n)_+, \qquad n = 1, 2, \ldots, \tag{1}$$

where *in this chapter only,* we denote positive and negative parts by subscripts, e.g., $X_+ = X^+ = \max(X, 0)$, and $X_+^2 = (X_+)^2$. We write equation (65) in Chapter 9 as

$$Y_n = (D_n + X_n)_-, \qquad n, 1, 2, \ldots, \tag{2}$$

where Y_n is the idle time of the server between the departure of C_n and the commencement of service on C_{n+1}. Combining these expressions, we have

$$D_{n+1} - Y_n = D_n + X_n, \qquad n = 1, 2, \ldots, \tag{3}$$

where D_n and X_n are independent, and $Y_n D_{n+1} = 0$.

We show in Section 9-8 that $\{D_n\}$ converges in distribution (in the conventional pointwise sense) to random variable D, and hence from (2), $\{Y_n\}$ converges in distribution to Y_∞. Letting $n \to \infty$ in (3), we can write

$$D_{n+1} - Y_\infty = D_n + X, \tag{4}$$

where D_n and D_{n+1} are both $\sim D$, but we have retained their subscripts because they are distinct random variables. Now (4) is simply a random-variable representation for an equation that the (limiting or stationary) distributions of D and Y_∞ must satisfy. Furthermore, because $\{D_n\}$ is regenerative,

$$d \equiv \lim_{n\to\infty} \sum_{i=1}^{n} D_i/n = E(D), \tag{5}$$

and (4) now will be used to obtain information about d.

Taking the expected value of (4) and canceling $E(D_n) = E(D_{n+1})$, we get

$$E(Y_\infty) = -E(X) = 1/\lambda - 1/\mu. \tag{6}$$

Squaring (4), taking expectations, and canceling $E(D_n^2) = E(D_{n+1}^2)$, we get $E(Y_\infty^2) = 2dE(X) + E(X^2)$, and, combining with (6),

$$d = -E(X^2)/2E(X) - E(Y_\infty^2)/2E(Y_\infty). \tag{7}$$

Thus (7) would determine d exactly if we knew the second term on the right.

[Note that $E(X)$ is negative, so that the first term in (7) is positive.] From (2), $E(X^2) < \infty \Rightarrow E(Y_\infty^2) < \infty$, and (7) is valid and finite whenever $E(X^2) < \infty$; see the technical remarks that follow.

Instead of (7), we use $X^2 = X_+^2 + X_-^2$ to also write

$$d = [E(X_+^2) + E(X_-^2) - E(Y_\infty^2)]/2E(-X). \tag{8}$$

As observed in Section 9-8, $(Y_\infty \mid Y_\infty > 0) \sim I$, a *GI/G/1 idle period*. Let

$$a_0 = P(Y_\infty > 0),$$

which is the proportion of arrivals who find the system empty, i.e., $a_0 = 1/E(K)$, where K is the number of customers served in a busy period. (Note that for convenience, we have adopted the convention that idle periods are strictly positive, and we say that an arrival finds the system empty only if it has been empty for a positive length of time.) Thus

$$E(Y_\infty) = a_0 E(I) = 1/\lambda - 1/\mu, \tag{9}$$

$E(Y_\infty^2) = a_0 E(I^2)$, and we rewrite (7) as

$$d = -E(X^2)/2E(X) - E(I^2)/2E(I). \tag{10}$$

The form of the term on the far right in (10) is intriguing—it is the mean of the *equilibrium* idle distribution, $E(I_e)$!

A stationary inter-departure time, denoted by τ, has representation

$$\tau = Y_\infty + S, \tag{11}$$

where Y_∞ and S are independent. From (6) and (11),

$$E(\tau) = 1/\lambda,$$

which of course means that for a stable *GI/G/1* queue, the departure rate is equal to the arrival rate.

It is a few lines of algebra to represent

$$\begin{aligned} c_\tau^2 \equiv V(\tau)/E^2(\tau) &= \lambda^2[V(Y_\infty) + V(S)] \\ &= 2\lambda(1 - \rho)[E(I^2)/2E(I)] - (1 - \rho)^2 + \rho^2 c_g^2. \end{aligned} \tag{12}$$

For an *M/G/1* queue, $E(I^2)/2E(I) = 1/\lambda$, and (12) becomes

$$c_\tau^2 = 1 + \rho^2(c_g^2 - 1). \tag{13}$$

We can also use (4) to obtain a representation for the transform of D. Let $D \sim D(t)$, $I \sim I(t)$, $X \sim F(t)$, with corresponding Laplace-Stieltjes transforms $\tilde{D}$, $\tilde{I}$, and $\tilde{F}$. Taking the transforms of both sides of (4), it is a few lines of algebra to show that

$$\tilde{D}(s) = a_0[1 - \tilde{I}(-s)]/[1 - \tilde{F}(s)]. \tag{14}$$

This is a more elementary derivation of (14) than the one in Section 9-12.

To check what we have done, consider the $M/G/1$ queue, where $a_0 = 1 - \rho$, $I \sim \exp(\lambda)$, $\tilde{I}(s) = \lambda/(\lambda + s)$, $\tilde{F}(s) = \tilde{G}(s)\tilde{A}(-s) = \lambda\tilde{G}(s)/(\lambda - s)$, and (10) and (14) become the familiar Pollaczek-Khintchine formulas found in Section 8-2.

Remark. With the exception of (12), the results of this section hold for the (slightly) more general associated queue defined in Section 9-8, where the X_n are i.i.d., but S_n and T_n may be dependent.

Technical Remarks. Expressions for higher moments of D can be obtained by taking the expected value of higher powers of (4). To be valid, however, our derivation of (6) and (7) requires that canceled delay moments be finite. While (6) can be derived in other elementary ways, we see that from Theorem 11 in Chapter 9, (7) would appear to require that $E(S^3) < \infty$. We sketch a *truncation* method by Kingman [1970] that takes care of the cancellation problem: Replace S_n by $S_n(y) = \min(S_n, y)$, $n \geq 1$, for some constant y, and let $D(y)$ be the stationary delay for the queue with these tuncated service times. It is easily shown (see Section 11-5) that $D(y)$ is stochastically smaller than the original stationary D, and that $D(y)$ is stochastically increasing in y, and converges in distribution to D as $y \to \infty$. Since $S_n(y)$ has finite moments of all orders, the corresponding delay moments are finite, and cancellation is valid in the truncated case. By monotone convergence, the moments of $D(y)$ converge to the moments of D, and we have that (7) is valid and finite whenever $E(X^2) < \infty$. An alternative that completely bypasses cancellation is to derive expressions for d and higher moments from (14). However, without making additional assumptions, (14) is valid only for *imaginary* s, i.e., (14) should be viewed as a characteristic function. The reason for this is that a transform of a nonnegative random variable, e.g., $E(e^{-sS})$, exists for complex s with real part $R(s) \geq 0$. For a nonpositive random variable, we are restricted to $R(s) \leq 0$, and for $X = S - T$, to $R(s) = 0$. L'Hospital's rule does not necessarily hold on the complex plane (Rudin [1964], p. 97), and obtaining (7) by differentiating (14) and naively taking a limit is not valid. Nevertheless, we state without proof that (7) may also be obtained from (14) under the condition that $E(X^2) < \infty$.

11-2 BOUNDS ON EXPECTED DELAY FOR GENERAL *GI/G/*1 QUEUES

From (6), (7), and $E(Y_\infty^2) \geq E^2(Y_\infty)$, we obtain after a few lines of algebra what is known as *Kingman's upper bound,*

$$d \leq \lambda(\sigma_a^2 + \sigma_g^2)/2(1 - \rho) \equiv J/\lambda, \text{ where} \tag{15}$$

$$J = (c_a^2 + \rho^2 c_g^2)/2(1 - \rho). \tag{16}$$

Notice that while this bound involves the variance of the inter-arrival distribution,

d is *finite* even when $\sigma_a = \infty$. When σ_a is finite, we will soon find that this bound is very good for ρ near 1 (*heavy traffic*). It is tight (equality holds) in the deterministic $D/D/1$ case.

From (16), we see that the variability of the inter-arrival and service distributions enter in the bound in an asymmetric way. Suppose these distributions are interchanged and rescaled so that ρ remains the same. We get a smaller upper bound when the service distribution is the one that is *less regular* (in terms of c). It is interesting to speculate whether the true d is smaller for this arrangement as well. For the pair of distributions (E_2, M) and for a special case of the pair (H, M), it is shown in Problems 8-26 and 8-27 that d is indeed smaller! It is not known how general this conclusion is, but because d depends on more than the first two moments, there certainly will be counterexamples. Also note that if one of the distributions has an infinite variance, d is finite only when the service distribution is the one that is *more regular*.

We now derive (19) and (21), sharper upper bounds on d by Daley [1977], who used the following moment inequality for an arbitrary nonnegative random variable R, say,

$$E\{(R - x)_+^\alpha\} \geq [E(R^\alpha)/E^\alpha(R)]E^\alpha\{(R - x)_+\} \quad \text{for } x \geq 0 \text{ and } \alpha \geq 1. \tag{17}$$

Daley presents a direct analytic derivation of (17). We need only the case $\alpha = 2$ below, and derive (17) for that case in Problem 11-5. We remark here that (17) is equivalent to having $E\{(R - x)_+^\alpha\}/E^\alpha\{R - x)_+\}$ nondecreasing in x. For $\alpha = 2$, this means that the coefficient of variation of $(R - x)_+$ is nondecreasing in x. The moments in (17) are assumed to be finite.

For $Y_\infty = (D + S - T)_- = (T - W)_+$, where W is the stationary waiting time in system, we first apply (17) by conditioning on W and then use $E\{E(Y_\infty^2 \mid W)\} \geq E^2\{E(Y_\infty \mid W)\} = E^2(Y_\infty)$:

$$E(Y_\infty^2 \mid W) \geq [E(T^2)/E^2(T)]E^2(Y_\infty \mid W), \text{ and hence}$$

$$E(Y_\infty^2) \geq (1 + c_a^2)E^2(Y_\infty) = (1 - \rho)^2E(T^2). \tag{18}$$

From (6), (7), and (18),

$$d \leq [E(X^2) - (1 - \rho)^2E(T^2)]/2E(-X) = \lambda[\sigma_a^2 + \sigma_g^2 - (1 - \rho)^2\sigma_a^2]/2(1 - \rho), \text{ or}$$

$$d \leq J/\lambda - (1 - \rho)c_a^2/2\lambda, \tag{19}$$

an improvement over (15) that is of particular interest for low to moderate ρ and large c_a^2. As $\rho \to 1$, $J \to \infty$, and the ratio of the two bounds $\to 1$.

Without making more assumptions, (19) may be the best attainable upper bound on d in terms of the first two moments of S and T. However, we can sharpen (19) by applying (17) to $Y_\infty = (-X - D)_+ = (X_- - D)_+$, and conditioning on D:

$$E(Y_\infty^2 \mid D) \geq [E(X_-^2)/E^2(X_-)] \cdot E^2(Y_\infty \mid D),$$

$$E(Y_\infty^2) \geq [1 + c^2(X_-)] \cdot E^2(Y_\infty), \tag{20}$$

and we get

$$d \le J/\lambda - (1 - \rho)c^2(X_-)/2\lambda. \tag{21}$$

Applying (17) to $X_- = (T - S)_+$, we have $c(X_-) \ge c_a$, and (21) is indeed sharper than (19).

To get a lower bound, observe that from (2), X_- is stochastically larger than Y_∞, and has larger moments. Hence from (8),

$$d \ge E(X_+^2)/2E(-X) = \lambda E(X_+^2)/2(1 - \rho). \tag{22}$$

For ρ near 1 and symmetric X, $E(X_+^2) \approx E(X^2)/2 \approx (\sigma_a^2 + \sigma_g^2)/2$, and the lower bound in (22) is approximately 1/2 the upper bound in (15). In Section 11-10, we will find that our upper bounds are very good in heavy traffic; hence this lower bound will be poor in this range. Marshall [1968a] obtained a more complicated lower bound on d that depends on the actual distribution of X; it is even worse than (22) in heavy traffic, but is usually better for ρ near 0 (*light traffic*).

Consider the $M/M/1$ queue, where all the above bounds are readily computed. In particular,

$$E(X_+^2) = 2\rho/\mu(\lambda + \mu), \text{ and} \tag{23}$$

$$c^2(X_-) = 1 + 2\rho. \tag{24}$$

We compare the lower bound (22) and the best of the upper bounds (21) for $Q = \lambda d$,

$$\rho^3/(1 - \rho)(1 + \rho) \le Q = \rho^2/(1 - \rho) \le \rho^2(2 - \rho)/(1 - \rho). \tag{25}$$

The upper bound is asymptotically sharp (in a ratio sense) in heavy traffic ($\rho \to 1$), but off by a factor of 2 in light traffic. As expected, the lower bound is off by a factor of 2 in heavy traffic, and by this measure it is very poor in light traffic.

Remarks. In light traffic, the ratio of a bound to the corresponding true value is the appropriate measure, rather than their difference, because light traffic queueing is of practical interest only when delay is expensive. A bound is *asymptotically sharp* (in either light or heavy traffic) if the ratio approaches 1 as ρ approaches either 0 or 1. While (15), (19), and (21) are all asymptotically sharp in heavy traffic, their absolute differences can be large, even for moderately large ρ. Equation (15) is of great historic interest, but it is clearly superseded by (19) because both bounds require the same information and are easy to calculate.

As poor as it is, (22) is better than bounds found in some queueing textbooks! It should not be discarded, however, because it is easier to compute than most alternatives, and it provides useful information for moderate size ρ.

To improve on (22), we use the well-known result that when $D_1 = 0$, $\{D_n\}$ is stochastically increasing. (See Problem 9-8 and also Section 11-5.) Thus $D_2 \sim$

$X_+ \overset{st}{\leq} D$. Now write (7) as

$$2E(-X)d = E(X^2) - E\{(X + D)_-^2\} = E(X^2) - E\{(X_- - D)_+^2\},$$

and, replacing D by D_2, we have

$$2(1 - \rho)d/\lambda \geq E(X^2) - E\{(X_- - X'_+)_+^2\}, \tag{26}$$

where $X'_+ \sim X_+$, but it is independent of X_-.

From $X^2 = X_+^2 + X_-^2$, it is easy to see that (26) is better than (22); it still does poorly in heavy traffic, but usually does very well in light traffic. For the $M/M/1$ queue (26) becomes

$$(1 - \rho)Q \geq 1 - \rho + \rho^2 - (1 + 2\rho)/(1 + \rho)^3.$$

In this case, (26) is asymptotically sharp in light traffic, and approaches 5/8 of the true value in heavy traffic.

In light traffic, we can approximate d from (4) by $E(X_+)$, i.e.,

$$d \geq E(X_+), \tag{27}$$

which is the first term in equation (100) in Chapter 9, but this usually will be worse than (26). We obviously do better by including the first two terms,

$$d \geq E(X_+) + E\{(X_1 + X_2)_+\}/2. \tag{28}$$

For the $M/M/1$ queue, (28) is better than (26) for $\rho \leq 0.2$. Irregular service [large $c^2(X_+)$] favors (26) over (28).

For sufficiently irregular arrivals, (26) and (28) can be poor even in light traffic, but at least in Example 11-1, Daley's upper bounds are good.

Example 11-1

Let $S = 1$, $P(T = 0) = 0.9$, and $P(T = 100) = 0.1$. We have introduced a strong batch-arrivals effect, and d is almost entirely the result of other customers within the same batch. The batch size distribution is

$$P(\nu = j) = .1(.9)^{j-1}, \qquad j \geq 1,$$

The "within batch" expected delay is $d_{wb} = E\{(\nu)(\nu - 1)\}/2E(\nu) = 9$. We also have $\lambda = \rho = 0.1$, $c_g^2 = 0$, $c_a^2 = c^2(X_-) = 9$, $E(X_+) = 0.9$, $E\{(X_1 + X_2)_+\} = 1.62$, $E(X^2) = 981$, and $E\{(X_- - X'_+)_+^2\} = 962.37$. Under each equation number below, we list the numerical value of that bound:

(27)		(26)		(28)				(21)		(19)		(15)
0.9	<	1.035	<	1.71	<	d ≈ 9	<	9.5	=	9.5	<	50

Because Kingman's upper bound ignores arrival process variability, it performs poorly here. Daley's bounds, on the other hand, approximate $c^2(Y_\infty)$ very well.

The results in this section hold for all *GI/G/*1 queues. In Section 11-4, we obtain bounds on d from (10) by making assumptions under which we can bound $E(I^2)/2E(I)$.

11-3 TAIL PROPERTIES OF DISTRIBUTIONS

The bounds in Section 11-4 hold when the inter-arrival distribution has certain tail properties. These properties are collected here for easy reference.

Let $H \sim K(t)$ be a nonnegative random variable, with $K(0) < 1$, and let the *support* of H be the smallest interval $\mathcal{T}$ such that $P(H \in \mathcal{T}) = 1$. Let b be the right-hand end point of $\mathcal{T}$, where $b = \infty$ when H is unbounded. In reliability theory terminology, suppose H is the life of a component, and let

$$(H - t \mid H > t)$$

be the *remaining life* of the component given it has survived *age t*. If for some constant γ,

$$E\{(H - t \mid H > t)\} = \int_t^\infty K^c(x)\,dx/K^c(t) \le \gamma \qquad \text{for all } t < b, \tag{29}$$

we say H (and K) has *bounded mean residual life* (*from above*) by γ, denoted by *BMRL*-$\overline{\gamma}$, where we must have $\gamma \ge E(H)$.

If (29) holds for $\gamma = E(H)$, we say H has the property *new better than used in expectation* (is *NBUE*). If the inequality in (29) is reversed, we have BMRL *from below,* denoted by *BMRL*-$\underline{\gamma}$, and if it also holds for $\gamma = E(H)$, we have *new worse than used in expectation* (*NWUE*).

These are examples of tail properties of distributions that are useful in reliability, queueing, and other areas. The term NBUE comes from this stronger property: We say H is *new better than used* (*NBU*) if

$$(H - t \mid H > t) \le_{\text{st}} H \qquad \text{for all } t < b, \tag{30}$$

and we say H is *new worse than used* (*NWU*) if the stochastic inequality in (30) is reversed. (In this chapter, the notation $\le_{\text{st}}$ replaces $\overset{\text{st}}{\le}$.)

We say H has *decreasing mean residual life* (is *DMRL*) if it is NBUE and the expectation in (29) is nonincreasing in t, $t < b$. Similarly, we have *increasing mean residual life* (*IMRL*).

We say H has *increasing failure rate* (is *IFR*) if (30) holds and

$$(H - t \mid H > t) \text{ is stochastically decreasing as } t \text{ increases, } t < b. \tag{31}$$

Now (30) implies that $K(0) = 0$, and it can be shown that (31) implies that $K(t)$ has density function $k(t)$ for all $t < b$. It is easily shown that (31) is equivalent to

$$r(t) \equiv k(t)/K^c(t) \text{ is nondecreasing in } t \qquad \text{for } t < b, \tag{32}$$

where $r(t)$ is called the *failure rate function,* which accounts for the term IFR. Similarly, we say H has *decreasing failure rate* (is *DFR*) if $(H - t \mid H > t)$ is stochastically increasing in t, and this is equivalent to nonincreasing $r(t)$. IFR distributions are continuous, except that in the bounded case, $P(H = b) > 0$ is possible. DFR distributions are necessarily unbounded and continuous, except that $P(H = 0) > 0$ is possible.

It is easily shown that these properties are related as follows:

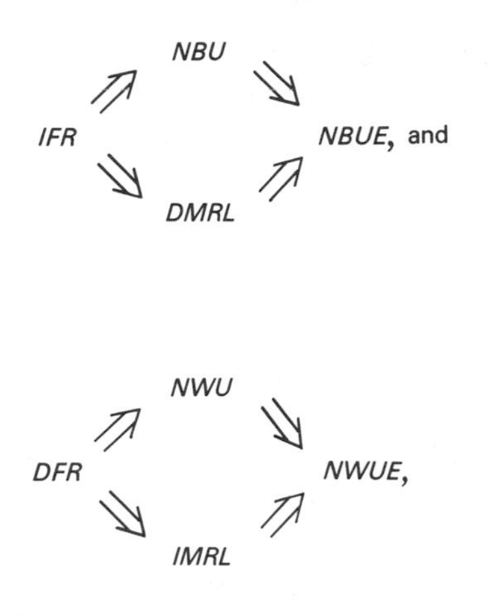

meaning that a distribution with one of these properties has all the properties to the right.

It is easily shown that

$$\textit{mixtures of DFR distributions are DFR,} \tag{33}$$

and it may also be shown (Barlow and Proschan [1981], p. 100) that

$$\textit{convolutions of IFR distributions are IFR.} \tag{34}$$

In fact, (34) is the continuous analog of the convolution property (see equation (38) in Chapter 6) of log-concave distributions.

For exponential H, (30) holds as an equality; H has constant failure rate and is both IFR and DFR. From (34), Erlang distributions are IFR, and from (33), hyper-exponential distributions are DFR.

11-4 *GI/G/1* BOUNDS BASED ON TAIL PROPERTIES OF THE INTER-ARRIVAL DISTRIBUTION

An idle period is the remainder of some inter-arrival time when a busy period ends, i.e., an *excess*. Its distribution is complicated, but may be represented in terms of the time since the last arrival, an *age*. Unfortunately, the age distribution is also complicated. Our approach will be to make assumptions about the inter-arrival distribution that allow us to obtain bounds on $E(I^2)/2E(I)$ that do not require knowledge of the age distribution.

Let $H \sim K$ and Z be independent nonnegative random variables, R =

$(H - Z \mid H > Z)$, and $P(Z \leq u \mid H > Z + v) = F_v(u)$. Our bounds will be a consequence of this elementary result:

$$H \text{ is BMRL-}\overline{\gamma}(\underline{\gamma}) \Rightarrow R \text{ is BMRL-}\overline{\gamma}(\underline{\gamma}). \tag{35}$$

Proof. We have that for $0 \leq v \leq t < \infty$.

$$P(R > t \mid R > v) = P(H > Z + t \mid H > Z + v) = \int_0^\infty \frac{K^c(u + t)}{K^c(u + v)}\, dF_v(u), \tag{36}$$

and

$$E(R - v \mid R > v) = \int_v^\infty P(R > t \mid R > v)\, dt.$$

Interchange the order of integration, bound the inside integral by γ, and we have the result.

Note that we require no assumptions about the distribution of Z. To apply (35), let W be the stationary waiting time (in system) of customer C, and let T be the inter-arrival time that begins when C arrives. Obviously W and T are independent. Furthermore, $(T - W \mid T > W)$ is a stationary idle period I, and $(W \mid T > W)$ is the time since the last arrival when a busy period ends. An immediate consequence of (35) is

$$T \text{ is BMRL-}\overline{\gamma}(\underline{\gamma}) \Rightarrow I \text{ is BMRL-}\overline{\gamma}(\underline{\gamma}). \tag{37}$$

When $\gamma = 1/\lambda$ in (37), T is NBUE (NWUE). An immediate consequence of (9) and (37) is

$$T \text{ is NBUE (NWUE)} \Rightarrow a_0 = 1/E(K) \geq (\leq)\ 1 - \rho. \tag{38}$$

The meaning of (38) is that when T is NBUE, the fraction of arrivals who find the system empty is greater than the corresponding fraction of time, and when T is NWUE, the converse is true. Recall that Erlang is NBUE and hyper-exponential is NWUE.

Now the fraction of time busy is related to busy and idle periods by $\rho = E(B)/[E(B) + E(I)]$, and from (38) it is immediate that

$$T \text{ is NBUE (NWUE)} \Rightarrow E(B) \leq (\geq)\ 1/(\mu - \lambda). \tag{39}$$

Recall that $E(B) = 1/(\mu - \lambda)$ for the $M/G/1$ queue.

To get a lower bound on d from (10), we need an upper bound on $E(I^2)/2E(I)$, and the converse. To do this, recall that in renewal theory, the limiting distribution (in the time-average sense) of both the excess and the age distribution is the *equilibrium distribution*, $K_e(t)$, corresponding to inter-event time $H \sim K$, where provided that $E(H) < \infty$,

$$H_e \sim K_e(t) = \int_0^t K^c(u)\, du/E(H),$$

and $E(H_e) = E(H^2)/2E(H)$.

At some random moment, both the excess and age are $\sim K_e$, but given the age equals u, say, the excess is $\sim(H - u \mid H > u)$. It follows that $K_e^c(t)$ may also be written

$$K_e^c(t) = \int_t^\infty \frac{K^c(u + t)}{K^c(u)}\, dK_e(u), \tag{40}$$

which can be verified analytically if desired.

Note that the form of (40) is a special case of (36) with $v = 0$, and that (35) does not depend on F_v. By the same argument used to prove equation (35), we have

$$H \text{ is BMRL-}\overline{\gamma}(\underline{\gamma}) \Rightarrow E(H^2)/2E(H) \le (\ge)\ \gamma. \tag{41}$$

For $\gamma = E(H)$, it follows immediately from (41) that

$$H \text{ is NBUE (NWUE)} \Rightarrow c(H) \le (\ge)\ 1. \tag{42}$$

From (37) and (41), it follows immediately that

$$T \text{ is BMRL-}\overline{\gamma}(\underline{\gamma}) \Rightarrow E(I^2)/2E(I) \le (\ge)\ \gamma. \tag{43}$$

From (43), (10), and a little algebra, we have that when T is NBUE, $E(I) \ge 1/\lambda$, and

$$d \ge J/\lambda - (1 + \rho)/2\lambda. \tag{44}$$

To appreciate the quality of this bound, we convert it to Q and combine it with (19),

$$J - (1 + \rho)/2 \le Q \le J - (1 - \rho)c_a^2/2, \tag{45}$$

where under NBUE, $c_a^2 \le 1$. The *width* of the bounds (upper–lower) in (45) is less than 1, i.e., we have determined Q to *within one customer*.

This is very impressive in heavy traffic, where J is large and $\to \infty$ as $\rho \to 1$. We can narrow the width further if we are able to compute $c^2(X_-)$ and use (21). Note that the lower bound in (44) is an equality for the $M/G/1$ queue. In light traffic, this bound may be either better or worse than (26) and (28), and it is easier to compute.

If T is NWUE, the inequality in (44) is reversed,

$$Q \le J - (1 + \rho)/2, \tag{46}$$

and may be either better or worse than (19) and (21).

We can improve on (44) and (46) by making the stronger assumption that T

has DMRL (IMRL). We do this by representing $E(I_e)$ and $E(T_e)$ in a particular way. From (40),

$$E(T_e) = \int_0^\infty A_e^c(t)\,dt = \int_0^\infty g(u)\,dA_e(u), \tag{47}$$

where

$$g(u) = E(T - u \mid T > u) = \int_0^\infty [A^c(u + t)/A^c(u)]\,dt.$$

Now I_e is the remainder of an idle period measured from some random epoch, given that the system is empty at that epoch, but it is also the remainder of the *inter-arrival time* in which the idle period falls. Let T_a be the age of that inter-arrival time and $T_a \sim A_a$, a distribution different from both A and A_e. Conditioning on T_a, we can write

$$E(I_e) = \int_0^\infty g(u)\,dA_a(u). \tag{48}$$

To compare A_a and A_e, we now select a point at random without any conditioning, and let $T_s \sim A_s$ be the length of the inter-arrival interval in which this point falls, where A_s has the corresponding *spread* distribution. The work present at the beginning of this interval is independent of interval length, and hence W is independent of T_s. Let $U \sim$ uniform on (0, 1), independent of all else, and let the actual location of our random point within T_s be $UT_s \sim A_e$, the age of T_s, and $(1 - U)T_s \sim A_e$ be the excess. If $UT_s > W$, which implies that $T_s > W$, we have fallen into a length-biased idle period with excess $(1 - U)T_s$ and age $UT_s - W$, both now $\sim I_e$, and we have generated $UT_s = T_a \sim A_a$. The occurrence of the event $\{UT_s > W\}$ can only make UT_s stochastically larger than it would be otherwise, i.e., we have shown that

$$A_e \leq_{\text{st}} A_a. \tag{49}$$

Remark. Equation (49) can be shown in more conventional ways; it does not depend on the distributions of T and W.

Now $g(u)$ is monotone nonincreasing (nondecreasing) if A is DMRL (IMRL). From (49) and the definition of stochastic ordering (57) in Section 11-5,

$$T \text{ has DMRL (IMRL)} \Rightarrow E(I_e) \leq (\geq)\, E(T_e). \tag{50}$$

Remark. By the same methods, it can be shown that

$$T \text{ has IFR (DFR)} \Rightarrow I_e \leq_{\text{st}} (\geq_{\text{st}})\, T_e.$$

From (10) and (50), we have

$$T \text{ is DMRL (IMRL)} \Rightarrow Q \geq (\leq)\, J - (c_a^2 + \rho)/2. \tag{51}$$

Under DMRL, the improved lower bound (51), when paired with (19), reduces the width of the bounds to $\rho(1 + c_a^2)/2$. Under IMRL, upper bound (51) is better than (19), and may be either better or worse than (21).

11-5 COMPARISON METHODS WITH APPLICATIONS TO THE *GI/G/*1 QUEUE

In this section, we compare performance measures for the *GI/G/*1 queue when the inter-arrival and/or service distribution is changed. When service times become larger, inter-arrival times become smaller, or either one becomes less regular, we expect performance to be worse. Indeed, this will often be the case, and we first develop concepts and methods for obtaining results of this kind.

In this section, let $X \sim F$ and $Y \sim G$ be arbitrary random variables. We now introduce several definitions of what we mean by smaller or more regular.

The strongest sense in which X is smaller than Y would be for this to hold for every point ω in the sample space, i.e.,

$$X(\omega) \leq Y(\omega) \qquad \text{for every } \omega \in \Omega. \tag{52}$$

Virtually equivalent is to assume that (52) holds on a set $\{\omega\}$ with probability one. (This distinction can be important when X and Y are defined as w.p.1 limits of sequences of random variables. Otherwise you can (*should*) forget about it!)

Our definition of stochastic ordering in Section 1-9 was motivated by (52), for clearly, if (52) holds,

$$F(t) \geq G(t) \qquad \text{for all real } t. \tag{53}$$

When (53) holds, we say X (F) is *stochastically smaller* than Y (G), denoted by either $X \leq_{st} Y$ or $F \leq_{st} G$. (Again, $\leq_{st}$ replaces $\overset{st}{\leq}$.)

While (53) is clearly weaker than (52), we can use (52) to give elementary proofs of properties of this ordering. The key is to observe that (53) is a relation between marginal distributions, and we can choose the corresponding joint distribution any way we wish. We generate X and Y as functions of U, a random variable uniformly distributed on (0, 1): Define inverse functions of F and G, $F^{-1}(u) \equiv \min\{x: F(x) \geq u\}$ and $G^{-1}(u)$, and let

$$X = F^{-1}(U) \quad \text{and} \quad Y = G^{-1}(U), \tag{54}$$

where it is easily shown that if X and Y are generated by (54),

$$X \sim F \quad \text{and} \quad Y \sim G. \tag{55}$$

Remark. In simulation, (54) is one of the standard ways of generating values of random variables with arbitrary (but specified) distributions from values of uniformly distributed random variables. The generation of random variables that have (or appear to have) a uniform distribution has been the subject of a great deal of research, and methods for doing this are widely available.

Thus if X and Y satisfy (53), they can always be put on a sample space such that they also satisfy (52), and in this sense, (52) and (53) are equivalent. This allows us to use *sample-path* arguments. To illustrate, let $g(u)$ be an arbitrary nondecreasing function. If X and Y satisfy (52), then

$$g(X(\omega)) \leq g(Y(\omega)) \qquad \text{for all } \omega \in \Omega, \text{ and}$$
$$E\{g(X)\} \leq E\{g(Y)\}, \tag{56}$$

provided these expectations are well defined (i.e., not of form $\infty - \infty$). Now suppose (56) holds for functions of form $g(u) = 1$ for $u > t$, $g(u) = 0$ otherwise, for every fixed t. We have

$$F^c(t) \leq G^c(t) \qquad \text{for every } t,$$

which is (53). It follows that

$$X \leq_{\text{st}} Y \Leftrightarrow E\{g(X)\} \leq E\{g(Y)\}, \tag{57}$$

for all nondecreasing functions g, whenever these expectations are well defined. Thus we have an equivalent definition of a stochastic ordering. The right-hand inequality in (57) is reversed when g is nonincreasing.

Using the same ideas, it is easily shown that if $X_2 \sim F_2$ is independent of $X_1 \sim F_1$, and $Y_2 \sim G_2$ is independent of $Y_1 \sim G_1$, where $X_i \leq_{\text{st}} Y_i$, $i = 1, 2$,

$$X_1 + X_2 \sim F_1 * F_2 \leq_{\text{st}} Y_1 + Y_2 \sim G_1 * G_2, \tag{58}$$

and we say that stochastic ordering is *closed under convolution*.

It is immediate that if $X \leq_{\text{st}} Y$, we have $-Y \leq_{\text{st}} -X$,

$$X_+ \leq_{\text{st}} Y_+ \quad \text{and} \quad Y_- \leq_{\text{st}} X_-, \tag{59}$$

i.e., stochastic ordering is *closed under positive part*.

It is also easily shown that if $\{X_n\}$ converges in distribution to X, $\{Y_n\}$ converges in distribution to Y, and $X_n \leq_{\text{st}} Y_n$ for every n, then

$$X \leq_{\text{st}} Y, \tag{60}$$

i.e., stochastic ordering is *closed under convergence in distribution*.

Finally, it is also immediate that

$$X \leq_{\text{st}} Y \Rightarrow X + c \leq_{\text{st}} Y + c \text{ for every constant } c, \text{ and}$$
$$X \leq_{\text{st}} Y \Rightarrow cX \leq_{\text{st}} cY \text{ for every constant } c > 0,$$

i.e., $\leq_{\text{st}}$ is closed under *shift* and *scaling*, respectively, and that $\leq_{\text{st}}$ has the following three named properties, where F, G, and H are distributions:

$$\text{for all distributions } F,\ F \leq_{\text{st}} F, \qquad (\textit{reflexivity})$$
$$F \leq_{\text{st}} G \text{ and } G \leq_{\text{st}} H \Rightarrow F \leq_{\text{st}} H, \text{ and} \qquad (\textit{transitivity})$$
$$F \leq_{\text{st}} G \text{ and } G \leq_{\text{st}} F \Rightarrow F = G. \qquad (\textit{antisymmetry})$$

Orderings with these three properties are called *partial orderings*. (An ordering would be called *complete* if it also has the property that all pairs of distributions are ordered, but this is not true for either $\leq_{st}$ or the other orderings we will consider.)

Remark. In random-variable notation, the third property should be written $X =_{st} Y$, which means that X and Y have the same distribution, *not* that they are the same random variable.

Example 11-2

We now compare two $GI/G/1$ queues, $A_1/G_1/1$ and $A_2/G_2/1$, where $G_1 \leq_{st} G_2$ and $A_2 \leq_{st} A_1$. Recall that $D_{n+1} = (D_n + S_n - T_n)_+$, and let $D_n^{(i)}$ be the delay of customer n for the queue with distributions A_i and G_i, $i = 1, 2$. Suppose both queues begin empty, so that $D_1^{(i)} = 0$. From (58) and (59), it is immediate that

$$D_2^{(1)} \leq_{st} D_2^{(2)}, \text{ and, by induction on } n,$$

$$D_n^{(1)} \leq_{st} D_n^{(2)} \qquad \text{for every } n. \tag{61}$$

As $n \to \infty$, these random variables converge in distribution to $D^{(i)}$ when $\rho_i < 1$, and from (60) we have

$$D^{(1)} \leq_{st} D^{(2)}. \tag{62}$$

Recall that these limiting distributions are independent of initial conditions.

Remark. By the same methods, we also have that for a $GI/G/1$ queue with $D_1 = 0$,

$$D_n \leq_{st} D_{n+1}, \qquad n \geq 1.$$

Compare this argument with that in Problem 9-8.

Example 11-3

Many well-known families of distributions are stochastically ordered. By either scaling or direct observation,

$$\exp(\mu) \leq_{st} \exp(\lambda) \qquad \text{when } \mu > \lambda.$$

For independent $X \sim P(\lambda)$ and $Y \sim P(\mu - \lambda)$, $X(\omega) \leq X(\omega) + Y(\omega) \sim P(\mu)$; we have

$$P(\lambda) \leq_{st} P(\mu) \qquad \text{when } \mu > \lambda.$$

By shifting, normal distributions with the same variance are stochastically ordered by their means. By convolution and the stochastic ordering of the exponential, Erlang distributions with the same number of phases are stochastically ordered by their means.

Stochastic ordering is too strong a property for the comparisons we would like to make, e.g., Erlang with exponential service, with the same service rate. Toward this end, we now introduce the following

Definition. X (F) has *smaller mean residual life* than Y (G), denoted by

$$X \leq_c Y \quad \text{or} \quad F \leq_c G, \text{ if}$$

$$E\{(X - t)_+\} \leq E\{(Y - t)_+\} < \infty \qquad \text{for all real } t. \tag{63}$$

We call $\leq_c$ a *convex* ordering because it has this characterization that we state here without proof (see Stoyan [1983], p. 9, and also Problem 11-13):

$$X \leq_c Y \Leftrightarrow E\{g(X)\} \leq E\{g(Y)\}, \tag{64}$$

for all *nondecreasing convex* functions g for which these expectations are finite. (*Convex* is defined in Problem 5-40.)

Now $g(x) = (x - t)_+$ is nonincreasing and convex, and (64) simply enlarges the class for which the right-hand inequality holds.

If $g(x)$ is an arbitrary convex function and $E(X)$ is finite, we can find a linear function $\ell(x)$ such that $g(x) \geq \ell(x)$ for all x, and $g(E(X)) = \ell(E(X))$. From this and $E\{\ell(X)\} = \ell(E(X))$, it follows immediately that

$$E\{g(X)\} \geq g(E(X)), \tag{65}$$

a classical and very useful result known as *Jensen's inequality*.

Stochastic ordering is clearly a stronger property than $\leq_c$, i.e.,

$$X \leq_{\text{st}} Y \quad \text{and} \quad E(Y_+) \leq \infty \Rightarrow X \leq_c Y, \tag{66}$$

and there is no sample-path interpretation of (63).

To give intuitive meaning to (63), shift X, i.e., let $Y = X + c$ for $c > 0$, and (63) will hold. Thus (63) provides some information about the relative magnitude or location of X and Y. On the other hand, if $E(X) = E(Y)$, (63) provides information about the relative *variability* of X and Y, with Y being more variable. For example, if $X = a$, a constant, it is easy to see that

$$X = a \leq_c Y \tag{67}$$

for all random variables Y with $E(Y) = a$. Thus (63) combines these notions in what will turn out to be a very useful, if at first unconventional, way.

An immediate consequence of (64) is that

$$X \leq_c Y \Rightarrow E(X_+^r) \leq E(Y_+^r) \qquad \text{for } r \geq 1, \tag{68}$$

which of course are moments of X and Y when they are nonnegative random variables. In this case,

$$V(X) \leq V(Y) \tag{69}$$

when $E(X) = E(Y)$, but (69) may not hold when $E(X) < E(Y)$.

Important Remark. When X and Y are non-negative and have the same mean, it is easily shown that

$$X \leq_c Y \Leftrightarrow X_e \leq_{\text{st}} Y_e, \tag{70}$$

but when the means are not equal, it is possible for both

$$X \leq_c Y \quad \text{and} \quad X_e \geq_{st} Y_e$$

to hold in a strict sense. For example, let $P(X = 0) = 1 - \alpha$, and $P(X = Y + c) = \alpha$, where $Y \sim \exp(\lambda)$ and $\alpha < e^{-\lambda c}$.

We now determine other properties of $\leq_c$. From (63), it is easy to see that

$$\leq_c \text{ is closed under } \textit{shift, scaling, and positive part,} \tag{71}$$

and

$$\leq_c \text{ is a } \textit{partial ordering.} \tag{72}$$

Now suppose $X_2 \sim F_2$ is independent of $X_1 \sim F_1$, and $Y_2 \sim G_2$ is independent of $Y_1 \sim G_1$, where $X_i \leq_c Y_i$, $i = 1, 2$. From shift,

$$(X_1 + X_2 \mid X_2) \leq_c (Y_1 + X_2 \mid X_2),$$

and hence $\leq_c$ is closed under *random shift,* i.e.,

$$X_1 + X_2 \leq_c Y_1 + X_2 \quad \text{and} \quad Y_1 + X_2 \leq_c Y_1 + Y_2,$$

and from transitivity,

$$X_1 + X_2 \leq_c Y_1 + Y_2, \text{ i.e.,}$$

$$\leq_c \text{ is } \textit{closed under convolution.} \tag{73}$$

Now suppose that $\{X_n\}$ converges in distribution to X and $\{Y_n\}$ converges in distribution to Y, and $X_n \leq_c Y_n$ for every n. If $E(X)$ and $E(Y)$ are finite, it is easily shown that $X \leq_c Y$, i.e.,

$$\leq_c \text{ is } \textit{closed under convergence in distribution.} \tag{74}$$

Finally, write

$$(X - t)_+ = X - t + (X - t)_- = X - t + (-X - (-t))_+,$$

and do the same for Y. If we have *finite* $E(X) = E(Y)$, it follows immediately from (63) that

$$X \leq_c Y \Leftrightarrow -X \leq_c -Y. \tag{75}$$

Remark. In this case, (64) holds for convex functions that need not be monotone.

We are now ready for an important application of $\leq_c$.

Example 11-4

As in Example 11-2, we compare two *GI/G/*1 queues, $A_1/G_1/1$ and $A_2/G_2/1$, where now $G_1 \leq_c G_2$, $A_1 \leq_c A_2$, and the A_i have the same finite mean, i.e., the queues have the same arrival rate. We use the same notation and initial conditions, $D_1^{(i)} = 0$. From (75) applied to $T_n^{(i)}$, convolution, positive part, and induction, we immediately have

$$D_n^{(1)} \le_c D_n^{(2)} \qquad \text{for every } n. \tag{76}$$

If we also have $\rho_i < 1$, and that the service distributions have finite variance (so that the $d^{(i)}$ are finite), then, by convergence in distribution,

$$D^{(1)} \le_c D^{(2)}. \tag{77}$$

From (68) and (77), we have

$$d_1 \le d_2, \tag{78}$$

and more generally that $D^{(1)}$ also has smaller rth moment, $r \ge 1$.

Example 11-4 is important because pairs of distributions that are prominent in queueing theory satisfy $\le_c$. We state without proof some well-known results for nonnegative X and Y with $E(X) = E(Y) < \infty$:

$$X \sim \text{IFR} \quad \text{and} \quad Y \sim \exp \Rightarrow X \le_c Y, \text{ and} \tag{79}$$

$$X \sim \exp \quad \text{and} \quad Y \sim \text{DFR} \Rightarrow X \le_c Y. \tag{80}$$

Recall that Erlang is IFR, and hyper-exponential is DFR. (A crossing property of tails of distributions that we used in Section 6-5 accounts for these results.) Within the Erlang family, we have

$$X \sim E_r \quad \text{and} \quad Y \sim E_k,\ 1 \le k < r \Rightarrow \qquad X \le_c Y. \tag{81}$$

Stochastic ordering within families extends these results to $E(X) < E(Y)$.

For $M/G/1$ queues, we get an even stronger result, as shown in Example 11-5.

Example 11-5

We now compare two $M/G/1$ queues with the same arrival and service rates, and $G_1 \le_c G_2$. From (70), $G_{1e} \le_{st} G_{2e}$, and recall that for $M/G/1$ queues, stationary D has the representation

$$D^{(i)} = \sum_{j=1}^{\mathcal{N}} S_{ej}^{(i)},$$

where the $S_{ej}^{(i)} \sim G_{ie}$ are i.i.d. Because both queues have the same ρ, $P(\mathcal{N} = n) = (1 - \rho)\rho^n$, $n \ge 1$, is also the same. It is immediate that

$$(D^{(1)} \mid \mathcal{N}) \le_{st} (D^{(2)} \mid \mathcal{N}),$$

and hence

$$D^{(1)} \le_{st} D^{(2)}, \tag{82}$$

a stochastic ordering! From Example 11-2, (82) also holds when $\lambda_1 < \lambda_2$.

Remark. With a bit more effort, it can be shown that (82) holds even when the service rates are different.

Example 11-4 also answers the following questions: For all *GI/G/*1 queues with either

1. fixed inter-arrival distribution A and service rate μ, or
2. fixed service distribution G and arrival rate λ,

what inter-arrival (or service) distribution minimizes d? From (67) and (78), the answer is

$$\textit{constant}\text{ inter-arrival (service) times.} \tag{83}$$

We now ask whether the comparison methods we have developed can be applied to the *GI/G/c* queue. In Section 11-6 we will show that the stochastic ordering results in Example 11-2 are still valid; however, this is not the case for $\le_c$, and in fact, the conclusion in (83) can be incorrect. In Problem 5-39 we show that for a particular 2-channel queue, d is smaller for exponential service than for constant service, at least in light traffic! This is an example of what we call *contrary behavior*. The example has batch arrivals and is not a *GI/G/c* queue; however, with a bit more effort, it can be modified to reach this conclusion.

Even when we have (77), as nice as it is, this does not necessarily mean that we should prefer $A_1/G_1/1$ to $A_2/G_2/2$, even in terms of delay. Now $\le_c$ is not a stochastic ordering, and (77) does *not* imply that $P(D^{(1)} = 0) \ge P(D^{(2)} = 0)$. (The probability of zero delay, or that it is very short, is an important performance measure for emergency service systems.) As discussed earlier, $\le_c$ combines aspects of both magnitude and variability.

Can variability be great enough that the left tails cross? The answer turns out to be yes, as the following result by Whitt [1984c] shows, and we state without proof: For two *GI/G/*1 queues with the *same* hyper-exponential inter-arrival distribution, and service distributions $G_1 \le_c G_2$ with the *same* service rate,

$$P(D^{(1)} = 0) \le P(D^{(2)} = 0), \tag{84}$$

and from (67), (84) is *minimized* for constant service.

Notice that this result relies on having irregular arrivals, as do other examples of contrary behavior in this book. (Contrary behavior can occur in other situations as well.) To illustrate (84), we now construct an example where this probability can be computed.

Example 11-6

As a limiting case of the hyper-exponential, let inter-arrival distribution A be of form

$$A = (1 - \alpha)U_0 + \alpha \exp(\alpha\lambda),$$

where U_0 is the unit step at the origin, and let the service distribution be of form

$$G = (1 - \beta)U_0 + \beta G_{\beta\mu},$$

where $G_{\beta\mu}$ is any distribution with mean $1/\beta\mu$, and no mass at the origin. Now we have batch Poisson arrivals with batch size denoted by υ. The probability that a *batch* finds the system empty is $1 - \rho$, independent of α and β, where $\rho = \lambda/\mu$. For a customer to have zero delay, the batch must find the system empty, and customers in front of this customer from the same batch must have zero service time, where we denote the number of these customers by J. Hence,

$$P(D = 0) = (1 - \rho) \sum_{j=0}^{\infty} P(J = j)(1 - \beta)^j, \tag{85}$$

which is easily computed from the known distribution of J. Notice that this probability is independent of $G_{\beta\mu}$, and increases with $1 - \beta$. If we pick a scaled parametric family, e.g., $G_{\beta\mu} = \exp(\beta\mu)$, $P(D = 0)$ *increases* with the coefficient of variation of G. For fixed A and μ, however, (85) depends only on β, which shows that c_g is not the appropriate measure of variability for this phenomenon. This example is too simple to pick up that constant service minimizes (84).

For nonnegative random variables X and Y, we now introduce a *transform* ordering: We write $X \leq_{\text{L}} Y$ if

$$E\{e^{-sX}\} \geq E\{e^{-sY}\} \qquad \text{for all } s > 0, \tag{86}$$

which are finite Laplace-Stieltjes transforms.

As functions of x, $-e^{-sx}$ is increasing, and e^{sx} is increasing and convex. It follows from (57) that

$$X \leq_{\text{st}} Y \Rightarrow X \leq_{\text{L}} Y, \tag{87}$$

and from (64) that

$$-X \leq_{\text{c}} -Y \Rightarrow Y \leq_{\text{L}} X. \tag{88}$$

Recall that when $E(X) = E(Y)$, $-X \leq_{\text{c}} -Y \Leftrightarrow X \leq_{\text{c}} Y$.

Important Remark. While ordering $\leq_{\text{c}}$ combines aspects of (relatively) small and regular, $\leq_{\text{L}}$ combines small and *irregular*. For example, when comparing Erlang and exponential distributions *with the same mean,* we have

$$E_k \leq_{\text{c}} M, \qquad \text{but} \quad M \leq_{\text{L}} E_k.$$

While small and regular are aspects of service times that (usually) improve performance, small and irregular are aspects of inter-arrival times that make performance worse.

We apply $\leq_{\text{L}}$ in the next example.

Example 11-7

Consider two $GI/M/1$ queues with service rate μ and $\rho_i < 1$, where $A_2 \leq_L A_1$. We know from Section 8-7 that for queue i, waiting time in system

$$W^{(i)} \sim \exp\,[(1 - \alpha_i)\mu],$$

where $\alpha_i \in (0, 1)$ is the unique solution to

$$\alpha_i = \tilde{A}_i[\mu(1 - \alpha_i)], \qquad i = 1, 2. \tag{89}$$

From (86), $\tilde{A}_1(s)$ lies below $\tilde{A}_2(s)$ for all s, and from (89), this implies

$$\alpha_1 \leq \alpha_2, \text{ and} \tag{90}$$

$$W^{(1)} \leq_{\text{st}} W^{(2)}. \tag{91}$$

From Example 11-2, or directly from (89), (91) holds when $\mu_1 < \mu_2$.

11-6 THE $GI/G/c$ QUEUE—BASIC PROPERTIES

We use the same notation as before: $\{T_n\}$ and $\{S_n\}$ are independent sequences of i.i.d. random variables, where $S_n \sim G$ is the service time of customer C_n, and $T_n \sim A$ is the time between the arrival of C_n and C_{n+1}. Let t_n be the arrival epoch of C_n. We have $c \geq 1$ channels (servers) *fed by a single FIFO queue,* arrival rate λ, service rate μ, and now define $\rho = \lambda/c\mu$.

Think of a bank or a post office, where arrivals who find all servers busy join a single FIFO queue. An equivalent arrangement is a retail shop where customers take a number on arrival. Let V_n be the work found by C_n on arrival. As defined in Chapter 5 and elsewhere.

$$V_n = \text{sum of the remaining service times of all customers } C_j \text{ in system, } j < n, \text{ at epoch } t_n. \tag{92}$$

The remaining service time of a customer in queue is the service time.

Suppose we know the remaining service times of all customers in service and in queue at epoch t_n. (Think of the service times as being drawn from G on arrival.) From this, it is easy to figure out where everyone in front of C_n will eventually get served. In fact, we can partition V_n into

$$V'_{ni} = \text{the portion of } V_n \text{ eventually performed at channel } i,$$

for $i = 1, \ldots, c$, and represent D_n, the delay of C_n, as

$$D_n = \min_i V'_{ni}.$$

Actually, there is some ambiguity in this description, e.g., where are customers served who find more than one idle server? It turns out to be more convenient to *order* the V'_{ni} from smallest to largest, where $V_{n1} \leq \cdots \leq V_{nc}$ are the corresponding ordered quantities, and to define $\mathbf{V}_n$ as the *vector* of these quantities,

$$\mathbf{V}_n = (V_{n1}, \ldots, V_{nc}), \tag{93}$$

where now

$$D_n = V_{n1} \quad \text{and} \quad V_n = \sum_{i=1}^{c} V_{ni}, \qquad n = 1, 2, \ldots . \tag{94}$$

We now show how to construct $\mathbf{V}_{n+1}$ from $\mathbf{V}_n$. First add S_n to V_{n1}. Although we have lost the identity of the servers, it is still true that between arrivals, positive work at each server decreases with slope -1. Let $\mathbf{V}''_{n+1}$ be the vector

$$\mathbf{V}''_{n+1} = ((V_{n1} + S_n - T_n)_+, (V_{n2} - T_n)_+, \ldots, (V_{nc} - T_n)_+), \tag{95}$$

which would be $\mathbf{V}_{n+1}$ except that the components of (95) may not be ordered. Now let

$$\mathbf{V}_{n+1} = \mathcal{R}(\mathbf{V}''_{n+1}), \tag{96}$$

where $\mathcal{R}$ is simply the rearrangement of the components of (95) that puts them in order.

Given sequence $\{T_n, S_n\}$ and any initial work vector $\mathbf{V}_0$ independent of $\{T_n, S_n\}$, where $\mathbf{V}_0 = (0, \ldots, 0) \equiv \mathbf{0}$ means that the system starts empty, $\{\mathbf{V}_n\}$ is a well-defined stochastic process. Unfortunately, however, it is not easy to deal with, and we state without proof this important result from a classic paper by Kiefer and Wolfowitz [1955].

Theorem 1. When $0 < \rho < 1$, $\{\mathbf{V}_n\}$ converges in distribution to $\mathbf{V}$, which has a unique stationary distribution that is independent of the initial conditions, and the queue is stable.

When $\rho > 1$, or when $\rho = 1$ and $P(S_n = cT_n) < 1$, Kiefer and Wolfowitz show that the queue is unstable, and that a stationary distribution does not exist.

One of the advantages of the functional definition of stochastic ordering (57) is that it extends to vectors: We say vectors $\mathbf{X} \leq_{st} \mathbf{Y}$ if for all nondecreasing (in every component of the vector) real-valued functions g,

$$E\{g(\mathbf{X})\} \leq E\{g(\mathbf{Y})\}, \tag{97}$$

whenever these expectations are well defined. Sample-path arguments are often used to show (97), by showing that $\mathbf{X}_n(\omega) \leq \mathbf{Y}_n(\omega)$ component-wise. In fact, it can be shown that $\mathbf{X} \leq_{st} \mathbf{Y}$ if and only if there exists a sample space on which the corresponding inequality holds on sample paths. (See Kamae et al. [1977], and for other characterizations of stochastic ordering, see Shanthikumar [1987].)

We now illustrate this approach for the $GI/G/c$ queue: Suppose $\mathbf{V}_0$ is fixed but arbitrary. From (95) and (96), it is easy to see that for every n, $\mathbf{V}_n(\omega)$ is minimized (component-wise) when $\mathbf{V}_0 = \mathbf{0}$, and however the system starts, $\mathbf{V}_{n+1}(\omega)$ is a nondecreasing function of $\mathbf{V}_n(\omega)$. Clearly, $\mathbf{V}_0 = \mathbf{0} \leq_{st} \mathbf{V}_1$. Now suppose $\mathbf{V}_0 \sim \mathbf{V}_1$, but independent of $\{T_n, S_n\}$. We have

$$(\mathbf{V}_1 \mid \mathbf{V}_0 \sim \mathbf{V}_1) \sim (\mathbf{V}_2 \mid \mathbf{V}_0 = \mathbf{0}) \geq_{st} (\mathbf{V}_1 \mid \mathbf{V}_0 = \mathbf{0}),$$

and continuing in this manner.

$$(\mathbf{V}_n \mid \mathbf{V}_0 = \mathbf{0}) \text{ is } \textit{stochastically increasing in } n. \tag{98}$$

Analogous to Example 11-2, we compare two $GI/G/c$ queues, $A_1/G_1/c$ and $A_2/G_2/c$, where $G_1 \leq_{st} G_2$ and $A_2 \leq_{st} A_1$. Generate the two $\{T_n, S_n\}$ sequences by $T_n^{(i)} = A_i^{-1}(U_{2n-1})$ and $S_n^{(i)} = G_i^{-1}(U_{2n})$, where the U_n are i.i.d. $\sim$ uniform on (0, 1). We get a sample-path comparison of the two $\{\mathbf{V}_n\}$ sequences. It is immediate that for any initial vector $\mathbf{V}_0^{(1)} = \mathbf{V}_0^{(2)}$,

$$\mathbf{V}_n^{(1)} \leq_{st} \mathbf{V}_n^{(2)} \qquad \text{for every } n, \tag{99}$$

and from the derivation, (99) also holds as a sample-path result.

As remarked earlier, when the preceding distributions are ordered by $\leq_c$, this does not imply that the means of the components of (99) are ordered.

Kiefer and Wolfowitz also have a rather intricate proof of the following result, which is needed for their proof of Theorem 1: For a $GI/G/c$ queue with $\rho < 1$, and any fixed initial vector $\mathbf{V}_0 = \mathbf{v}$, there is a proper random variable M such that

$$(\mathbf{V}_n \mid \mathbf{V}_0 = \mathbf{v}) = (\mathbf{V}_n \mid \mathbf{V}_0 = \mathbf{0}) \qquad \text{for all } n > M. \tag{100}$$

This means that the effect of initial conditions completely dies out in finite time. This result also follows from the theory related to the following remarks.

Important Remarks. The stable $GI/G/c$ queue has been analyzed by many famous mathematicians. It is a classic *hard problem*. One reason for this is that if $P(T \leq S) = 1$ and $\rho < 1$, which can easily occur when $c \geq 2$, the system will never empty! Whitt [1972] shows that if $P(T > S) > 0$ and $\rho < 1$, the system will empty over and over again (*infinitely often*). Whenever this occurs, the next arrival will find the system empty, $\{\mathbf{V}_n\}$ starts over, and in fact, $\{\mathbf{V}_n\}$ is a regenerative process with finite mean cycle length. Continuous-time processes such as the number-of-customers-in-system process $\{N(t)\}$ will also regenerate, and they have finite mean cycle length when $\lambda > 0$. Is $\{\mathbf{V}_n\}$ regenerative when $P(T \leq S) = 1$ and $\rho < 1$? Kiefer and Wolfowitz, who bound $\{\mathbf{V}_n\}$ by a positive recurrent Markov chain, don't answer this question, because $\{\mathbf{V}_n\}$ does not necessarily regenerate when the chain does.

It turns out that $\{\mathbf{V}_n\}$ is indeed regenerative when $\rho < 1$, but only in what we called the Asmussen sense in Section 2-21, i.e., while cycle lengths are i.i.d. with finite mean, the cycles themselves are identically distributed, but *one-dependent*. The first *correct* proof of this is by Charlot et al. [1978], but they rely on complicated constructions in Kiefer and Wolfowitz. There have been many attempts to find simpler proofs of this property, which with aperiodicity implies Theorem 1, and also (independently) of Theorem 1. The simplest of these use stochastic upper bound (123) on total work $V(t)$ derived in Section 11-7. See Asmussen [1987], Section 11-2, and also Sigman [1988a], who extends this result to multichannel queues in tandem. The mathematics in this work is beyond the scope of this book, and it turns out that even the description of what constitutes regeneration points in these models is quite complicated.

The time-average properties of the "classical" regenerative processes in Chapter 2 hold for these more general one-dependent processes, and hence for the $GI/G/c$ queue. See Sigman [1989].

11-7 BOUNDS FOR THE *GI/G/c* QUEUE; DELAY MOMENTS

In spite of the mathematical difficulties discussed in Section 11-6, we now obtain elementary and useful bounds on various quantities for the $GI/G/c$ queue, in terms of corresponding quantities for one or more $GI/G/1$ queues.

First, we compare the $GI/G/c$ queue generated by $\{T_n, S_n\}$ with a corresponding *fast single server,* i.e., a $GI/G/1$ queue generated by $\{T_n, S_n/c\}$. Thus both queues have the same arrival process, but customers at the single server queue are served c times as fast. Let $V_c(t)$ and $V_1(t)$ be the work in system at epoch $t \geq 0$ for these two queues, where the subscript denotes the number of channels. If $A(0) = 0$ (no batches), we have

$$V_c(t_n^-) = V_n,$$

as defined in (92). Assume for convenience that both queues are empty at epoch 0. We subscript other familiar quantities in the same way, e.g., d_1 and d_c.

There is some ambiguity about how to define work here, because the two queues work (serve customers) at different rates. We have to decide on a consistent unit of work. Suppose work is measured in worker hours, say, in the c-server system. Thus, $V_c(t)$ is defined as usual, but now $V_1(t)$ is the total worker hours of work that remain to be done at epoch t, *if done at the c-server queue.*

With this convention, stochastic processes $\{V_c(t)\}$ and $\{V_1(t)\}$ have jumps of height S_n at epoch t_n, for every n (with multiple jumps at some points when $A(0) > 0$). Between arrivals, $V_c(t)$ decreases with slope $-j$ when there are j busy servers, $1 \leq j \leq c$, but $V_1(t)$ decreases with slope $-c$ when positive. Thus when positive, $V_1(t)$ is decreasing *at least as fast as* $V_c(t)$. It follows immediately that for every point ω in the sample space,

$$V_1(t, \omega) \leq V_c(t, \omega) \qquad \text{for all } t \geq 0, \tag{101}$$

i.e., (101) *holds on sample paths.*

Now suppose $0 < \rho < 1$ (ρ is the same for both queues). From (101), corresponding time-average work satisfies the same inequality,

$$E(V_1) \leq E(V_c), \tag{102}$$

and we know from equation (87) in Chapter 9 that $E(V_1) < \infty$ when $E(S^2) < \infty$. As a consequence of upper bounds, all the usual first-moment performance measures for the c-server queue are finite under the same condition.

We now convert (102) into a lower bound on d_c, using equation (152) in Chapter 5 for d_c. Recall that even for multiserver queues, the contribution of C_n to the area under $\{V_c(t)\}$ is a rectangular area of form $S_n D_n$ plus a triangular area of form $S_n^2/2$. For our fast single server, the form of the rectangular area is the same (with a different D_n), but because of our work convention, the corresponding triangle has height S_n, length S_n/c, and *area* $S_n^2/2c$. Writing out the two expressions for time-average work, (102) becomes

$$E(V_1) = \lambda d_1/\mu + \lambda E(S^2)/2c \leq E(V_c) = \lambda d_c/\mu + \lambda E(S^2)/2, \text{ or}$$

$$d_c \geq d_1 - (c - 1)(c_g^2 + 1)/2\mu c. \tag{103}$$

Converting (103) into a bound on w_c, we have

$$w_c \geq w_1 + (c - 1)(1 - c_g^2)/2\mu c. \tag{104}$$

From (104), we see that a sufficient condition for

$$w_c \geq w_1 \tag{105}$$

is $c_g^2 \leq 1$, which is consistent with results in earlier chapters for the *M/M* and *GI/M* cases, where $c_g = 1$. While this condition is certainly not necessary (these inequalities are in fact strict), we show as Example 5-8 that reversals of (105) can occur when the service distribution is sufficiently irregular. (When $c \geq 2$, customers pass a "plugged up" server.)

Remark. We expect the fast single server to be "better" because it uses the service capacity more efficiently. A precise formulation of this idea is (101) and (102). To go beyond this to conclude that all performance measures are better is simply loose reasoning, and can lead to incorrect conclusions.

Important Remarks. Because (101) is a sample-path result free of stochastic assumptions, so is (102), provided only that the time averages exist and are finite. For (103) through (105), we need $E(SD) = dE(S)$, for which having $\{S_n\}$ i.i.d. and independent of $\{T_n\}$ is sufficient, and that certain averages exist and are finite. Thus *we do not require renewal arrivals for these results.*

For *GI/M* (see equation (6) in Chapter 8, Brumelle [1973], and Cox and Smith [1961], Sec. 3.2),

$$D_c \leq_{\text{st}} D_1, \text{ and hence} \tag{106}$$

$$d_c \leq d_1. \tag{107}$$

Another way to make these comparisons is to let $\{T_n, S_n\}$ generate the single server queue, and $\{T_n, cS_n\}$ generate the c-server queue. In this way, we get a sequence of queues for $c = 1, 2, \ldots$. With this approach, Mori [1975] shows that for GI/M,

$$D_c \text{ is } \textit{stochastically decreasing in } c, \tag{108}$$

a generalization of (106). He claims to show (108) for constant service (GI/D) as well, but his proof of this is incorrect, and in fact we now know that (108) does not hold for GI/D (see Daley and Rolski [1984]). However, Mori does show (106) for GI/D.

Important Remark. For GI/G, Suzuki and Yoshida [1970] show that for $c \geq 2$,

$$d_c \leq \lambda(\sigma_a^2 + \sigma_g^2/c^2)/2(1 - \rho),$$

but only for $\rho \leq 1/c$, where this bound is (15) applied to d_1. While of little use in this range, this bound has been widely conjectured to hold for $\rho \in (0, 1)$. When true, of course, (107) is stronger, and (19) applies.

We now obtain upper bounds on performance measures for general $GI/G/c$ queues. A well-known proposal for generating bounds is to *cyclically assign* arrivals to the servers, i.e., serve arrivals $i, i + c, i + 2c, \ldots$ at server i, $i = 1, \ldots, c$. An alternative that is more convenient for some mathematical purposes is to *randomly assign* customers to servers, i.e., with probability $1/c$, assign arrival n to server i, $i = 1, \ldots, c$, independent of all else.

Both of these rules are independent of the congestion at the servers and are clearly "dumb"—and therefore worse than the original $GI/G/c$ queue—right?

As with the validity of the lower bounds, we need to be precise about the senses in which these dumb rules are in fact bad (and hence lead to upper bounds). An *essential observation* is this: Under these rules, *it is possible to have a positive queue and an idle server simultaneously,* whereas this is not the case for a standard $GI/G/c$ system fed by a single queue.

The Model. We actually bound a more general model: Let the arrival process be arbitrary with arrival epochs $0 \leq t_1 \leq t_2, \ldots$, and let $\{S_n\}$ be a sequence of i.i.d. service times, independent of the arrival process. Call a standard c-server model fed by a single FIFO queue the *original system*. We will bound performance measures for this system by corresponding ones for a *modified system* that assigns arrivals to servers in some arbitrary manner that is *independent* of $\{S_n\}$. Assume both systems start empty.

Definition. We define S_n to be the service time of the nth customer to *enter* service. For the original system, this is the service time of the nth arrival, but this is not necessarily the case for the modified system.

Remarks. Conventionally, we would associate a service time random variable with each arriving customer, independent of where that customer is assigned. For the purpose of obtaining bounds that hold on sample paths, the futility of doing this is well known. For example, the modified system can be empty at an epoch where the original system is busy. (See also Stoyan [1976] or p. 115 of Stoyan (1983].) In earlier chapters, we have on occasion found it convenient to associate service times with servers rather than customers, e.g., when comparing LIFO with FIFO. Here we think in terms of simulating these systems, where we generate service times in the order in which their values need to be known. Because our service times are i.i.d. and independent of all else, both the original and modified systems are each stochastically equivalent under these variations, but their *interaction* is not.

For the original system, let J_n be the nth ordered departure epoch from the queue. $\{(J_n + S_n)\}$ be the set of unordered departure epochs from the system, and K_n be the nth ordered departure epoch from the system, $n \geq 1$. Let J'_n and K'_n be the corresponding quantities for the modified system.

Our main result is that departures occur sooner in the original system. For every fixed ω in the sample space,

$$K_1 = \min_{i \geq 1} (J_i + S_i) = \min_{1 \leq i \leq c} (t_i + S_i), \text{ and in general}$$

$$K_n = \text{the } n\text{th-order statistic from } \{J_i + S_i\text{: } i < n + c\} \text{ and} \tag{109}$$

$$J_n = \max (t_n, K_{n-c}), \qquad n = 1, 2, \ldots, \tag{110}$$

where $K_j \equiv 0$ for $j \leq 0$. For the modified system,

$$K'_n = \text{the } n\text{th-order statistic from } \{J'_i + S_i\text{: } i < n + c\}, \text{ but} \tag{111}$$

$$J'_n \geq \max (t_n, K'_{n-c}), \qquad n = 1, 2, \ldots, \tag{112}$$

rather than (110), because we can have a positive queue and an idle server simultaneously. From our definition of $\{S_n\}$, the service times in (109) and (111) are the *same random variables*. It is now an easy inductive proof (Wolff [1977a]) that *for all realizations,*

$$J'_n(\omega) \geq J_n(\omega), \qquad n = 1, 2, \ldots. \tag{113}$$

It follows immediately that for all realizations,

$$K_n \leq K'_n, \qquad n = 1, 2, \ldots, \tag{114}$$

$$Q(t) \leq Q'(t), \qquad t \geq 0, \text{ and} \tag{115}$$

$$N(t) \leq N'(t), \qquad t \geq 0, \tag{116}$$

where Q and N are the number of customers in queue and in system, respectively. If S_n is defined as the service time of the nth arrival, the distribution of each of these quantities remains the same, and (113) through (116) hold as stochastic orderings.

To deal with work in system, let $S_{pn}(t)$ be the amount of work performed on S_n by epoch t in the original system, i.e., for $t \geq 0$,

$$S_{pn}(t) = [\min (t - J_n, S_n)]_+ \quad \text{and} \quad S'_{pn}(t) = [\min (t - J'_n, S_n)]_+ . \tag{117}$$

For each system, the corresponding *cumulative work performed* by epoch t is

$$C_p(t) = \sum_{n=1}^{\infty} S_{pn}(t) \quad \text{and} \quad C'_p(t) = \sum_{n=1}^{\infty} S'_{pn}(t), \qquad t \geq 0. \tag{118}$$

From (113), it is immediate that for all realizations,

$$C_p(t) \geq C'_p(t), \qquad t \geq 0. \tag{119}$$

For each system, the *cumulative work to arrive* by epoch t is

$$C(t) = \sum_{\{n:t_n \leq t\}} S_n \quad \text{and} \quad C'(t) = \sum_{\{n:t'_n \leq t\}} S_n, \tag{120}$$

where t'_n is the arrival epoch of the customer who turns out to have service time S_n in the modified system. The *work in system* for each system is

$$V(t) = C(t) - C_p(t) \quad \text{and} \quad V'(t) = C'(t) - C'_p(t). \tag{121}$$

The set of service times included in the sums in (120) may be different (*will* be different if for some n, $t_n \leq t$ but $t'_n > t$). Thus the fact that *more* work has been performed does not imply that *less* remains to be done, and we *do not* have a sample path inequality for work in system. The problem is the same for work in queue, $V_q(t)$.

Nevertheless, we will obtain stochastic bounds on these quantities. First, notice that while our definition of $\{S_n\}$ induces a relabeling of $\{t_n\}$ to produce $\{t'_n\}$, it is not necessary to relabel the entire sequence to determine the behavior of the modified system over any finite interval.

Instead, we proceed as follows: Fix $t \geq 0$. For every n where $J'_n \leq t$, define S_n as before. The remaining arrivals in the modified system are in queue. To them, we assign the service times of the arrivals in the *original system* that have not yet been assigned. We call this procedure *partial relabeling* (*PR*). Now *PR* does not change $\{C'_p(u): 0 \leq u \leq t\}$, but the cumulative work to arrive by epoch t in the modified system is now $C(t)$ rather than $C'(t)$.

Let $V''(t) = C(t) - C'_p(t)$, which is the work in system at epoch t for the modified system under PR. From (119) and (121),

$$V(t) \leq V''(t) \tag{122}$$

holds on for all ω.

At epoch t, PR's only effect on the modified system is the identity of the service times associated with customers in queue. Because these service times are random draws from the service distribution,

$$V''(t) =_{\text{st}} V'(t).$$

From (122), and because this argument works for every fixed t, we have

$$V(t) \leq_{\text{st}} V'(t), \qquad t \geq 0, \tag{123}$$

and by the same argument,

$$V_q(t) \leq_{\text{st}} V_q'(t), \qquad t \geq 0. \tag{124}$$

Equations (123) and (124) remain valid when $\{S_n\}$ is defined in the usual way.

Notice that as t varies, the assignment of service times to customers under PR will change, and (122) will not hold simultaneously for all t. Thus, we do not have a conventional sample-path result, but we do have very useful stochastic orderings.

What about delay? For $n = 1, 2, \ldots$, define

$$D_n = J_n - t_n \quad \text{and} \quad D_n' = J_n' - t_n, \tag{125}$$

where D_n is the delay in queue of the nth arrival in the original system. However, D_n' does not have the same interpretation in the modified system because the nth departure from queue may not be the nth arrival. Nevertheless, we have from (113) that for all realizations,

$$D_n(\omega) \leq D_n'(\omega), \qquad n = 1, 2, \ldots. \tag{126}$$

Let D_n'' be the actual delay of the nth arrival in the modified system. For every realization, departure epochs are a sequence of real numbers; the actual delays are determined by appropriately relabeling the J_n' in (125). Thinking of the modified system as a "black box", $\{D_n'\}$ is the delay sequence we would get *if* it operated FIFO, and $\{D_n''\}$ is the delay sequence we actually get for some unspecified alternative.

If all the queues in the modified system happen to be empty at epoch $J_k(\omega)$, the first k relabeled J_n' are some reordering of $J_1', \ldots, J_k'$, and

$$\sum_{n=1}^{k} D_n'(\omega) = \sum_{n=1}^{k} D_n''(\omega). \tag{127}$$

Otherwise, some of the relabeled departure epochs may be larger than J_k', so that in any event,

$$\sum_{n=1}^{k} D_n'(\omega) \leq \sum_{n=1}^{k} D_n''(\omega) \qquad \text{for all } k. \tag{128}$$

Just as we did in Section 5-14, Kingman [1962c] uses (127) to show that FIFO has smaller variance than LIFO and other similar alternatives. This is true because the sum of squares corresponding to (127) is minimized when the labeling is FIFO, which is how we defined the D_n'. We extended this result to the expected value of nondecreasing convex functions of delay in Problem 5-41. We do the same here.

For any nondecreasing convex function g, we compare $\sum g(D_n)$. Under either (127) or (128), the sum is minimized under FIFO, i.e.,

$$\sum_{n=1}^{k} g[D'_n(\omega)] \le \sum_{n=1}^{k} g[D''_n(\omega)] \qquad \text{for every } k, \tag{129}$$

and nondecreasing convex function g.

Now suppose the original system is a $GI/G/c$ queue with $\rho < 1$, and that the modified system uses *cyclic assignment*. Each of the c single-server queues created by cyclic assignment is a $GI/G/1$ queue with service distribution G and interarrival distribution that is the *c-fold* convolution of A. Let D_c and D_{1i} be the corresponding stationary random variables, where i refers to any of the single server queues. Let D' be the corresponding stationary quantity generated by the D'_n in (125). Now from (64) and (129),

$$D' \le_c D_{1i},$$

and from (126),

$$D_c \le_{st} D'.$$

Combining, we have

$$D_c \le_c D_{1i}. \tag{130}$$

From (130), or $L = \lambda w$ applied to (116),

$$d_c \le d_{1i}, \tag{131}$$

and, applying Kingman's upper bound (15), we get

$$d_c \le (\lambda/c)(c\sigma_a^2 + \sigma_g^2)/2(1 - \rho), \tag{132}$$

where (131) and (132) were proposed but not proven by Kingman [1970]. Of course, we really should use one of Daley's bounds, (19) or (21), to improve on (132).

Remarks. It is tempting to conjecture that (130) also holds as a stochastic ordering, but Whitt [1981] has shown by counterexample that this is not true in general. The place to look is $P(D_c = 0)$. Several authors have conjectured that (130) holds as a stochastic ordering at least for $M/G/c$. It is easily shown that for $GI/D/c$, (130) holds as an *equality*, i.e., $D_c =_{st} D_{1c}$. We didn't mention it at the time, but $D_F \le_c D_L$, and also for the delay under other alternatives to FIFO in Section 5-14.

Using bounds in this section, we now extend conditions for finite $GI/G/1$ delay moments to $GI/G/c$ queues with $\rho < 1$; all systems start empty.

Let $V_n^{(f)}$ be the work in system found by C_n in the fast single server queue generated by $\{T_n, S_n/c\}$. Other quantities for this queue will be superscripted in the same way. We now denote (92) by $V_n^{(c)}$. Inequality (101) holds everywhere, including arrival epochs, but should we have $A(0) > 0$, we include only the work associated with C_j, $j < n$. In any event, $V_n^{(f)} \leq V_n^{(c)}$ holds on sample paths, but we need only $\leq_{\text{st}}$.

For an upper bound on work, we now assign customers to the servers *randomly*, rather than cyclically, which has the advantage of creating a collection of c *symmetric* single server queues. From the above and (123),

$$V_n^{(f)} \leq_{\text{st}} V_n^{(c)} \leq_{\text{st}} \sum_{i=1}^{c} V_{ni}^{(r)}, \qquad n = 1, 2, \ldots, \tag{133}$$

where $V_{ni}^{(r)}$ is the work found at server i by C_n, whether or not C_n is assigned there, under random assignment.

Now $V_n^{(f)} = D_n^{(f)}$, FIFO delay at the fast single server. For random assignment, it is convenient to pretend that C_n is served by every server, but that the service time is zero at all servers except the one where C_n is actually assigned. Looking at each server separately, this means that C_n has service distribution

$$(c - 1)U_0/c + G/c, \tag{134}$$

there, where U_0 is the unit step at the origin, and $V_{ni}^{(r)} \sim D_n^{(r)}$, where $D_n^{(r)}$ is the delay of the nth arrival at a $GI/G/1$ queue with inter-arrival distribution A and service distribution (134).

Because these systems start empty, $\{V_n^{(f)}\}$, $\{V_n^{(c)}\}$, and $\{V_{ni}^{(r)}\}$ are stochastically increasing sequences, and for each sequence, their kth moment "converges" (finite or not) to the kth moment of their limiting distribution. Comparing the fast single server with each of the random assignment queues, they have the same arrival rate (positive or zero), and either finite or infinite kth service moment. From this and Theorem 11 in Chapter 9 (and a technical detail about moments of the *sum* in (133), see Wolff [1984]), we have

Theorem 2. For a $GI/G/c$ queue with $\rho < 1$, and any $k > 0$, the kth moment of the stationary distribution of (arrival-average) work in system is finite if $E(S^{k+1}) < \infty$, and when $\lambda > 0$, only if $E(S^{k+1}) < \infty$.

Because $D_n^{(c)}$ is the smallest component of work *vector* $\mathbf{V}_n^{(c)}$, $D_n^{(c)} \leq V_n^{(c)}/c$, and we have

Theorem 3. For a $GI/G/c$ queue with $\rho < 1$, and any $k > 0$, the kth moment of the stationary delay distribution is finite if $E(S^{k+1}) < \infty$.

Important Remarks. By the same argument used to prove Theorem 2, and from Theorem 13 in Chapter 9, we can show that when $0 < \rho < 1$, the stationary distribution of *time-average* work in system for the $GI/G/c$ queue has finite kth moment if and only if $E(S^{k+1}) < \infty$. However, we no longer have monotone convergence, and instead rely on the regenerative nature of all three systems, in the more general sense discussed at the end of Section 11-6. Sigman [1988a] shows that the random assignment queue is regenerative.

More important is the restriction of these and other results to renewal arrivals. In the context of tandem and network queues, arrival processes at the stations are not renewal when they are not Poisson. The bounding systems we have developed can handle general arrival processes, but we don't know the appropriate conditions for finite moments in the single channel case. For some limited results in this direction, see Wolfson [1984].

11-8 APPROXIMATIONS FOR THE *M/G/c* QUEUE; QUEUE LIMITS AND RETRIALS

All the bounds in Section 11-7 apply to $M/G/c$, but because of the importance of this model in applications, approximations of a more ad hoc nature are also of interest. Assume $\rho < 1$.

We begin by recalling that if this model operates under a *symmetric* rule such as *processor sharing* (*PS*), certain performance measures are *insensitive* to the service distribution. To be specific, stationary $N_{PS} \sim N_{M/M/c}$, and given N_{PS}, the remaining service of the customers in system are i.i.d. $\sim G_e$. This means that stationary (total) work has the representation

$$V_{PS} = \sum_{j=1}^{N_{PS}} S_{ej}, \tag{135}$$

where the S_{ej} are i.i.d. $\sim G_e$.

Let V_F be the corresponding stationary work under FIFO. We get an approximation on first-moment performance measures by equating expected work under these two rules,

$$E(V_F) = Q_F E(S) + c\rho E(S_e) \approx E(V_{PS}) = L_{M/M/c} E(S_e), \tag{136}$$

where $L_{M/M/c} = Q_{M/M/c} + c\rho$ and $E(S_e) = E(S^2)/2E(S)$. We get

$$Q_F \approx Q_{M/M/c} E(S^2)/2E^2(S) = Q_{M/M/c}(1 + c_g^2)/2, \tag{137}$$

an appealingly simple approximation that has been "derived" several times in the literature by various arguments, this being the simplest. The easiest way to compute $Q_{M/M/c}$ is to use $Q = \lambda d$, where $d = P(D_{M/M/c} > 0)/(c\mu - \lambda)$, and this probability is computed recursively, as in Problem 5-12.

We stress that (137) is an approximation because even when work is conserved in the sense that it is neither created nor destroyed by the rules of operation, work in system is not sample-path invariant (see Section 10-1), and time-average work will be rule dependent.

The accuracy of this approximation has been investigated (see Section 11-12 for some references), and it is easily shown to work well in heavy traffic (it is asymptotically equivalent to bounds that work well there, see Section 11-10). In this range, there is an even easier alternative: The fast single server in Section

11-6 is an $M/G/1$ queue, and lower bound (103) can be computed exactly. (This is not recommended for large c; see Section 11-10.) The approximation (137) seems to work very well for Erlang service where, empirically, it is a lower bound, and the percent error increases with the number of phases and $1/\rho$. It is easily shown to be exact for service distribution

$$(1 - \alpha)U_0 + \exp(\alpha\mu),$$

so it may work well for irregular service also.

Approximation (137) can be poor in light traffic, but there are other more complicated approximations that work well in this range, e.g., Boxma et al., at least for Erlang service. See Section 11-12 for references that review this work, and the references therein.

At present, general conditions under which the approximation in (137) is either an upper or lower bound are not known.

The same idea can be used on the $M/G/c$ queue with *queue limit,* where arrivals finding $\ell \geq c$ customers in system are lost. We make the *second assumption* that the arrival rate of those served,

$$\lambda_s, \textit{ is the same under FIFO and PS.} \tag{138}$$

With the additional subscript ℓ to denote queue limit.

$$E(V_{F\ell}) = Q_{F\ell}E(S) + \lambda_s E(S^2)/2 \approx E(V_{\mathrm{PS}\ell}) = L_{M/M/c/\ell}E(S_e), \tag{139}$$

$L_{M/M/c/\ell} = Q_{M/M/c/\ell} + \lambda_s E(S)$, where under PS, λ_s is the same for general and exponential service. We have

$$Q_{F\ell} \approx Q_{M/M/c/\ell}(1 + c_g^2)/2. \tag{140}$$

For later reference, call what we did here the *PS method.*

Assumption (138) is a dicey proposition because the work in queue associated with a customer under FIFO may be quite different from work under PS. With constant service, for example, FIFO work per customer in queue is twice expected PS work. If they have the same average total work, and we subtract $\lambda E(S)$ customers from both, there will on average be half as many FIFO customers as PS customers. Thus we expect that FIFO will hit the queue limit less often, and hence have a larger λ_s. This shows that it is incompatible to equate both time-average work and λ_s for these two models. For empirical evidence that the approximations by the PS method can be poor, and a better but more complicated approximation, see Miyazawa [1986].

Abstracting what we have just done, suppose that for an $M/G/c/\ell$ queue, we can approximate the fraction of customers lost $f_\ell(\lambda) = 1 - \lambda_s/\lambda$ as a function of λ, e.g., by the PS method, or in some other way. For the PS method, f_ℓ is increasing and concave, with $f_\ell(0) = 0$, and $f_\ell(\lambda) \to 1$ as $\lambda \to \infty$.

Suppose we actually have a *retrial* version of the same model (Section 7-9), where customers finding ℓ customers in system are not necessarily lost; they may instead go *into orbit* and come back to try again later. Customers returning from orbit are called *retrials*. Suppose arrivals who find ℓ customers in system enter orbit (and come back later as retrials) with probability $\alpha_1 > 0$, and that retrials who find ℓ customers in system reenter orbit with probability α_2, independent of the number of prior retrials. Customers who find the system full and choose not to enter orbit are lost.

As in Chapter 7, we make the assumption (approximation) that *retrials see time averages* (*RTA*). (As explained in Chapter 7, there is reason to expect this to be a good approximation if retrials don't return quickly. No automatic redialing!) Under RTA, we find that by equating the rate λ_R that customers enter and leave orbit (return to the system), λ_R is a solution to the equation

$$(\alpha_1\lambda + \alpha_2\lambda_R)f_\ell(\lambda + \lambda_R) = \lambda_R, \qquad \lambda_R > 0. \tag{141}$$

Because λ_s for the PS model is the same as for $M/M/c/\ell$, the function f_ℓ under the PS method is also the same, and (141) reduces to a special case of equation (70) in Section 7-9. As stated there, it can be shown (see Greenberg [1986] for a proof) that (141) has a unique solution in this case.

Remarks. It may disturb some readers that we have said nothing about how customers behave while in orbit. In Section 7-9, it is assumed that the times customers spend in orbit are i.i.d. $\sim$ exp (γ), but the RTA approximation there is independent of γ. The approximation is poor when γ is large relative to other parameters (customers return quickly), but it improves as γ decreases, becoming exact as $\gamma \to 0$. (It has not been formally shown that the improvement is monotone in γ.) It should be clear that we don't really need *exponential* orbit times either. For definiteness, suppose orbit times are i.i.d. with some arbitrary distribution. Equating rates into and out of orbit requires that the number-of-customers-in-orbit process be stable. A sufficient condition for this is $\alpha_2 < 1$. It would be interesting to try out this approach on a better approximation, e.g., that by Miyazawa.

11-9 TANDEM QUEUES—SOME GENERAL AND LIGHT-TRAFFIC RESULTS

We briefly explore the effect on performance measures of

(i) interchanging the order of service operations (stations),
(ii) cxchanging distributions that are ordered by $\leq_c$, and
(iii) permitting the service times of the same customer at the different stations to be dependent random variables.

Throughout this section, we place no queue limits at the stations. We begin by recalling the product-form results in Chapter 6 under Poisson arrivals and exponential service:

Suppose we have two stable single-channel FIFO queues in tandem at service rates μ_i at station i, $i = 1, 2$. Think of a production line, where we might have a choice of which service operation to perform first. If we were to switch these rates to μ_2 followed by μ_1, what difference would it make? The joint stationary distribution of (N_1, N_2) would still be of product form, with the ρ_i switched. The distributions of $N = N_1 + N_2$ and $W = W_1 + W_2$ would be unchanged. This turns out to be an exception. The order of service operations usually does affect performance measures, and we now consider some cases where we can say that one ordering is preferred to another:

Suppose we have two single-channel FIFO queues in tandem, with arrival epochs $0 \leq t_1 \leq t_2 \cdots$. There are two service tasks A and B that can be performed in either order, and we have two possible arrangements of tandem queues. A first ($A \rightarrow B$), which we call *order 1,* and *order 2,* which is B first ($B \rightarrow A$). Whatever order we choose, it is the same for all customers. Let

S_n^a be the service time of C_n on task A, and

S_n^b be the service time of C_n on task B, $\quad n \geq 1$.

Assume both arrangements start empty.

For $n \geq 1$ and $\ell, j \in \{1, 2\}$, let $R_{jn}^{(\ell)}$ be the departure epoch of C_n from station j under ordering ℓ. In the expressions that follow, quantities with subscript $n = 0$ equal zero.

Suppress notation for a moment, and consider one single-channel FIFO queue. An easy analog of the relation between successive delays $\{D_n\}$ is

$$R_n = \max\,(t_n, R_{n-1}) + S_n, \qquad n \geq 1, \tag{142}$$

and it follows by induction on n that

$$R_n = \max_{1 \leq i \leq n} \left[t_i + \sum_{j=i}^{n} S_j \right], \qquad n \geq 1. \tag{143}$$

We now apply (143) twice for each ordering, where the departure epochs from station 1 are the arrival epochs at station 2: For $n \geq 1$,

$$R_{2n}^{(1)} = \max_{1 \leq j \leq n} \left[\max_{1 \leq i \leq j} \left(t_i + \sum_{k=i}^{j} S_k^a \right) + \sum_{k=j}^{n} S_k^b \right], \tag{144}$$

$$= \max_{1 \leq i \leq j \leq n} \left[t_i + \sum_{k=i}^{j} S_k^a + \sum_{k=j}^{n} S_k^b \right], \tag{145}$$

$$= \max_{1 \leq i \leq n} \left[t_i + \max_{i \leq j \leq n} \left(\sum_{k=i}^{j} S_k^a + \sum_{k=j}^{n} S_k^b \right) \right], \tag{146}$$

and for ordering 2, we simply interchange a and b,

$$R_{2n}^{(2)} = \max_{1 \le i \le n} \left[t_i + \max_{i \le j \le n} \left(\sum_{k=i}^{j} S_k^b + \sum_{k=j}^{n} S_k^a \right) \right]. \tag{147}$$

Basic Assumptions for (i) and Later for (ii). So far, we haven't made any stochastic assumptions. Now suppose the arrival process is arbitrary, $\{S_n^a\}$ and $\{S_n^b\}$ are i.i.d. sequences, independent of each other and the arrival process, and that these quantities are the same for both arrangements.

As we vary j for fixed i in (146) and (147), we are exchanging random variables of generic type S^a and S^b on a one-for-one basis in the overall sum. Now suppose service times of different tasks are *nonoverlapping*. That is, assume $P(S^a \ge S^b) = 1$, which means that task A takes at least as long as task B. When this is the case, the inner maximum is achieved when we have included as many S^a-type variables as possible. This occurs when $j = n$ in (146), and $j = i$ in (147), and these expressions become

$$R_{2n}^{(1)} = \max_{1 \le i \le n} \left[t_i + \sum_{k=i}^{n} S_k^a + S_n^b \right], \text{ and} \tag{148}$$

$$R_{2n}^{(2)} = \max_{1 \le i \le n} \left[t_i + \sum_{k=i}^{n} S_k^a + S_i^b \right]. \tag{149}$$

First observe that if the S_n^b are constant, (148) and (149) are equivalent, and have the same distribution. Now suppose the S_n^b are random, and let $\{i = u\}$ be the event that (148) is maximized when $i = u$ (not necessarily unique). We make the important (and obvious) observation that $\{i = u\}$ is independent of S_n^b. *Given* $\{i = u\}$, the conditional distribution of (148) is the same as that of the term $i = u$ in (149), but that term does not necessarily maximize (149). This means that for every u,

$$(R_{2n}^{(1)} \mid \{i = u\}) \le_{\text{st}} (R_{2n}^{(2)} \mid \{i = u\}).$$

Uncondition, and we have this result: When $P(S^a \ge S^b) = 1$,

$$R_{2n}^{(1)} \le_{\text{st}} R_{2n}^{(2)}, \qquad n = 1, 2, \ldots, \tag{150}$$

i.e., *the departure epoch of every customer is stochastically smaller when the longer task is performed at station 1,* and when shorter service S^b is constant, *the departure epochs do not depend on the ordering.*

Remarks. We ordinarily expect (150) to be strict when the S^b are random, and it is easy to show that this is true when S^a is unbounded. While not by itself the explanation of (150), notice that when the longer task is done first, there never is a queue at the second station.

We will obtain a similar result under different assumptions. Toward that end, we state without proof a result for *exchangeable* random variables (see Section 9-13; i.i.d. is a special case) from Tembe and Wolff [1974].

Nesting Lemma. For exchangeable random variables $X_1, \ldots, X_n$, every real number a_k, and every subset A_k of k integers from $\{1, \ldots, n\}$, $k = 1, \ldots, n$,

$$P\left(\sum_{i\in A_k} X_i \le a_k, k = 1, \ldots, n\right) \le P\left(\sum_{i=1}^{k} X_i \le a_k, k = 1, \ldots, n\right). \tag{151}$$

Remark. The sums on the right are *nested*; as we increase k the same random variables remain in the sum. Intuitively, if X_1 is "small," we expect $X_1 + X_2$ also to be "small," but not necessarily $X_2 + X_3$.

From (151), it is immediate that

$$\max{}_{1\le k\le n}\left[\sum_{i=1}^{k} X_i - a_k\right] \le_{\text{st}} \max{}_{1\le k\le n}\left[\sum_{i\in A_k} X_i - a_k\right]. \tag{152}$$

Returning to tandem queues, suppose that $S^a = s$, a constant, and let S^b be arbitrary, possibly overlapping S^a. Condition on $\{t_i\}$ in (144), and for fixed j, suppose that the inner maximum occurs at $i = u(j)$ (not necessarily unique). Performing this inner maximization, (144) becomes a maximization over nested sums,

$$(R_{2n}^{(1)} \mid \{t_i\}) = \max{}_{1\le j\le n}\left\{t_{u(j)} + [j - u(j) + 1]s + \sum_{k=j}^{n} S_k^b\right\}. \tag{153}$$

Under these assumptions, (147) becomes

$$R_{2n}^{(2)} = \max{}_{1\le i\le j\le n}\left[t_i + (n - j + 1)s + \sum_{k=i}^{j} S_k^b\right]. \tag{154}$$

The maximum over a selected subset of these terms cannot exceed (154). For the subset of terms in (154) containing a fixed number $n - j + 1$ of the S^b variables, we select the one corresponding to $i = u(j)$ in (153):

$$(R_{2n}^{(2)} \mid \{t_i\}) \ge \max{}_{1\le j\le n}\left\{t_{u(j)} + [j - u(j) + 1]s + \sum_{k=u(j)}^{n-j+u(j)} S_k^b\right\}. \tag{155}$$

Applying (152) to (154) and (155),

$$(R_{2n}^{(1)} \mid \{t_i\}) \le_{\text{st}} (R_{2n}^{(2)} \mid \{t_i\}), \tag{156}$$

and unconditioning,

$$R_{2n}^{(1)} \le_{\text{st}} R_{2n}^{(2)}, \qquad n = 1, 2, \ldots, \tag{157}$$

i.e., when one service task is constant, *the departure epoch of every customer is stochastically smaller if that task is performed at the first station.*

Remark. It might seem that (150) and (157) are contradictory results, but if the constant service task is nonoverlapping and shorter, the departure epochs are independent of the order.

We state without proof two generalizations of these results from Tembe and Wolff, again for an arbitrary arrival process:

(a) *For customers who have r tasks with nonoverlapping service times, to be performed by a collection of r single-channel queues in tandem, the departure epoch of every customer is stochastically the smallest when tasks are performed in the order: longest to shortest. If all the tasks (except possibly the longest) have constant service times, the departure epochs are the same for all r! arrangements of the tandem queues.*

(b) *(157) holds when the constant service task is performed in a multichannel station.*

Important Remarks. Weber [1979] showed that for $r \geq 2$ single-channel queues in tandem and exponential tasks, the departure process of all arrangements are (stochastically) the same. From (a), this is also true for constant service, but from (157), *this result will not hold when constant and exponential service distributions are mixed.* Note that the stochastic orderings in (a) and (b) hold *one customer at a time*; it is not necessarily true that the entire collection (or even two consecutive) departures are stochastically ordered. Weber's result is remarkable, because like (a) and (b), but unlike product-form results in Chapter 6, it holds for an arbitrary arrival process. Unfortunately, his proof gives little insight into why his result is true. A better understanding of his result may lead to generalizations of (a) and (b).

Recall that our results were derived by conditioning on the arrival process, and hold conditionally. (For (150), we conditioned only on $\{i = u\}$, but that was for notational convenience.) The total waiting time of C_n in a two-station tandem queue is

$$W_n = R_{2n} - t_n, \tag{158}$$

and a stochastic ordering on $(R_{2n}^{(\ell)} \mid \{t_i\})$ implies a corresponding ordering on waiting times,

$$(W_n^{(1)} \mid \{t_i\}) \leq_{\text{st}} (W_n^{(2)} \mid \{t_i\}), \text{ and}$$
$$W_n^{(1)} \leq_{\text{st}} W_n^{(2)}, \qquad n = 1, 2, \ldots. \tag{159}$$

The same is true for the (total) number-of-customers-in-system process, $\{N(t)\}$.

We summarize: Under the conditions in (a) and (b), *the waiting time of every customer and the number of customers in system at every epoch are each stochastically minimized for the stated ordering.*

For this reason, we will call these orderings *optimal.*

Note that we have not shown how to order performance measures between nonoptimal orderings, but we can apply these results to orderings where (say) the task assigned to the first station is fixed.

When the queues are stable and limiting distributions exist (in a time-average sense), the corresponding limiting distributions will be stochastically ordered.

Remark. As for multichannel queues, easy examples will show that stable tandem queues with renewal arrivals and i.i.d. service times may never empty, even in the single-channel case. Nevertheless, they are regenerative in the same sense as multichannel queues in Section 11-6. Nummelin [1981] was the first to show this for the single-channel case.

Unfortunately, constant service and nonoverlapping service are rather restrictive assumptions. They should be viewed as limiting versions of more general situations. Taken together, (a) and (b) suggest *design criteria* (when there is a choice): Tasks with service times that are long (large mean) and regular (small coefficient of variation) should be done before tasks with service times that are short and irregular. Keep in mind that mean and variance are crude measures in light and moderate traffic, and of course (a) and (b) don't suggest what to do when these criteria are in conflict. Intuitively, constant service in particular seems to "regularize" the departure process and improve performance downstream.

An alternative use of these results is to bound performance measures for one arrangement by another arrangement that is easier to analyze.

Remarks on Design Criteria

Pinedo [1982] and Whitt [1985] have proposed different design criteria from those preceding (and from each other); we briefly discuss Whitt only. His approach, which is developed for and applied to queueing networks in his [1983a,b] papers, is to approximate the departure process from each station as a renewal process. First, for the $GI/G/1$ queue itself, he proposes the approximation

$$d \approx \rho(c_a^2 + c_g^2)/2\mu(1 - \rho). \tag{160}$$

In [1983a], he uses a refinement of this expression proposed by Kramer and Langenbach-Belz [1976] when $c_a^2 < 1$. In this range, they approximate d by multiplying the approximation (160) by the factor

$$g = \exp\{-[2(1 - \rho)/3\rho][(1 - c_a^2)^2/(c_a^2 + c_g^2)]\}.$$

Notice that (160) is exact for the $M/G/1$ queue; $g < 1$ for $c_a^2 < 1$, and $g = 1$ when $c_a^2 = 1$.

From (7), (160) is equivalent to an approximation for $V(Y_\infty)$, which, when substituted for $V(Y_\infty)$ in the line above (12), gives Whitt's proposed approximation for

$$c_\tau^2 \approx \rho^2 c_g^2 + (1 - \rho^2) c_a^2, \tag{161}$$

where τ is a *stationary* inter-departure time. We remark that from (12), bounds on d earlier in this chapter immediately give bounds on c_τ^2.

At the end of Section 11-7, we remarked that in queueing networks, (composite) arrival processes at each station are not renewal when they are not Poisson. The fundamental reason for this is that except for *M*/*M*/1 and some trivial cases, the departure process of a *GI*/*G*/1 queue is not a renewal process. (See Daley [1976] and the references therein.) It is also true that the superposition of renewal processes is not a renewal process.

Nevertheless, Whitt makes the additional approximation that the inter-departure times are i.i.d. with c_τ^2 given by (161).

Starting with renewal input, i.i.d. service times at every station, and specified c^2 for every distribution, the expected total delay can be computed (approximated) for any of the $r!$ possible arrangements of r single-channel queues in tandem.

With these methods, Whitt arrives at what he calls these three *heuristic design principles* for minimizing total expected delay, where we number the tasks and subscript c^2 accordingly:

(P1) If all tasks have *the same mean service time,* order by increasing c_{gi}^2, i.e., most to least regular.

(P2) If $c_a^2 \le c_{g1}^2 = \cdots = c_{gr}^2$, order by *decreasing mean service time.*

(P3) If $c_a^2 \ge c_{g1}^2 = \cdots = c_{gr}^2$, order by *increasing mean service time.*

Our results are independent of the arrival process, and c_a^2 plays no role. Nevertheless, from (a) and (b), we expect (P1), have no objection to (P2), but find (P3) quite surprising. Putting (P2) and (P3) together, his approximations also predict that when all distributions have the same c^2, order will not matter. This is consistent with (a) and Weber's result. Using Whitt's approximations, an intuitive explanation of (P3) is this: When the arrival process is the primary source of variation, and ρ is large for some task, the $(1 - \rho)$ in (160) makes d for that task the dominant term, and putting it last minimizes the numerator.

Whitt presents evidence in support of these criteria for "smooth" distributions, but he also is careful to point out their heuristic nature, and gives examples, e.g., two point distributions, where the approximation is poor, and the criteria may lead to incorrect conclusions. He is also well aware of (a), but points out what he calls the *pipelining* effect of constant service, and that (a) may be misleading for smooth distributions. In spite of this, and the preceding intuitive explanation we remained skeptical of (P3), and decided to investigate:

Greenberg and Wolff [1988] investigated this question for two queues in tandem with Poisson arrivals and i.i.d. service times for each task, under *light traffic*. This means that the results are valid for fixed service distributions as λ (and the ρ_i) $\rightarrow 0$. We found the following:

For 2-Erlang service, the longer service should be first (contrary to (P3)), and for hyper-exponential service with mass at zero, the shorter service should be first (contrary to (P2)!). For hyper-exponential service distribution $H = (1 - \alpha)/\mu_1 + \alpha/\mu_2$, with *balanced means* (which means that $(1 - \alpha)/\mu_1 = \alpha/\mu_2$), our method was not accurate enough to determine which task should be first.

These results, which are found by a straightforward application of (170) and (173), are also contrary to Pinedo's criteria. Of course, they hold only in light traffic, where criteria based only on two moments may work poorly. On the other hand, E_2 is certainly smooth by almost any definition. Note that Whitt does not propose his criteria for heavy traffic. More work needs to be done to sort out the conflicting evidence we have presented here.

Comparison Methods for Tandem Queues

Stochastic ordering results for $GI/G/1$ and $GI/G/c$ queues, e.g., (61), are easily extended to tandem queues. Of greater interest is ordering $\leq_c$. We will need an easy extension by Bessler and Veinott [1966] of (64) to vectors that we state without proof:

For independent random variables $X_1, \ldots, X_n$ and $Y_1, \ldots, Y_n$,

$$X_i \leq_c Y_i,\ i = 1, \ldots, n \Leftrightarrow g(X_1, \ldots, X_n) \leq_c g(Y_1, \ldots, Y_n), \tag{162}$$

for all nondecreasing convex functions g.

We now compare two single-channel tandem queues (systems), each with r stations, i.i.d. service times at each station, and the same *arbitrary* arrival process that is independent of all else. Let $G_{\ell j}$ be the service distribution at station j of system ℓ, $j = 1, \ldots, r$, and $\ell = 1, 2$, where

$$G_{1j} \leq_c G_{2j}, \qquad j = 1, \ldots, r, \tag{163}$$

and let $R_{jn}^{(\ell)}$ be the departure epoch of C_n for station j of system ℓ. Assume both systems start empty and $E(t_n) < \infty$ for every n.

From (142) by induction or by writing out expressions such as (145), it is easy to see that the $R_{jn}^{(\ell)}$ are *nondecreasing convex functions of the arrival epochs and service times*. From (163), it follows immediately that for every j and n,

$$\{R_{jn}^{(1)} \mid \{t_n\}\} \leq_c \{R_{jn}^{(2)} \mid \{t_n\}\}, \tag{164}$$

and first applying (64) conditionally, we have

$$\{R_{jn}^{(1)}\} \leq_c \{R_{jn}^{(2)}\}. \tag{165}$$

Now let $W_n^{(\ell)}$ be the *total* waiting time of C_n in system ℓ, where

$$W_n^{(\ell)} = R_{rn}^{(\ell)} - t_n. \tag{166}$$

From (164) and shift,

$$\{W_n^{(1)} \mid \{t_n\}\} \leq_c \{W_n^{(2)} \mid \{t_n\}\}, \text{ and we have} \tag{167}$$

$$G_{1j} \leq_c G_{2j}, \quad j = 1, \ldots, r \Rightarrow W_n^{(1)} \leq_c W_n^{(2)}, \quad n = 1, 2, \ldots. \tag{168}$$

Important Remarks. We get (168) for an arbitrary arrival process, provided only that it be *the same* for both systems. This of course holds for $r = 1$, and deserves to be remembered in that context as well. When we have renewal arrivals for both systems, *with the same arrival rate,* where

$$A_1 \leq_c A_2,$$

we get (165) in one step by observing that the departure epochs are nondecreasing convex functions of service and inter-arrival times. We also have

$$-t_n^{(1)} \leq_c -t_n^{(2)},$$

but the terms on the right in (166) are not independent, and we cannot use convolution. Taking expectations, however, we have

$$E\{W_n^{(1)}\} \leq E\{W_n^{(2)}\}, \qquad n \geq 1, \tag{169}$$

which (apparently) is the best we can do. For $r = 1$ station, this is a weaker result than we were able to derive in Section 11-5.

Dependent Service Times in Light Traffic

We now treat the third topic listed at the beginning of this section. Suppose we have two single-channel queues in tandem, Poisson arrivals at rate λ, and i.i.d. service times at each station. To avoid needless subscripting, let C_n have service time $X_n \sim G$ at station 1, and service time $Y_n \sim H$ at station 2, with finite variances. Otherwise, we let X_n and Y_n have an arbitrary *joint* distribution (the same for all n). Thus, the service times of the same customer at different stations are permitted to be dependent random variables. Define ρ_i in the usual way.

Remarks. One motivation for this formulation is message transmission in a network, where transmission time is proportional to message length, and a message may have to be retransmitted several times to reach its destination. This suggests the special case of *equal* service times $X_n = Y_n$, the same random variable. Boxma [1979a,b] and Pinedo and Wolff [1982] are papers devoted exclusively to this special case. Their results are not limited to light traffic. It is easy to find other practical examples of dependence, if not this strong. In fact, some dependence ought to be expected in "real life"; the usual independence assumption is made for mathematical convenience.

Our objective is to analyze the effect of dependence on performance, and to compare it with independence. We obtain light traffic results from Wolff [1982b] by a method devised there that we now describe. As illustrated, the method has other applications.

When $\rho \equiv \rho_1 + \rho_2 < 1$, it is easy to show that total work in the tandem queue is bounded above by work in a corresponding *bounding queue*, which is a standard $M/G/1$ queue with arrival rate λ, where C_n has service time $X_n + Y_n$. This bound may seem crude, but it provides a convenient way to show (we skip the somewhat tedious details) that time-average work in the tandem queue has this representation:

$$E(V) = \rho_1 E(X_e + Y_x) + \rho_2 E(Y_e) + O(\rho^2) \qquad \text{as } \rho \to 0. \tag{170}$$

To understand (170), think of observing the tandem queue at a random moment. By $\rho \to 0$, we mean $\lambda \to 0$ for fixed service distributions. The $O(\rho^2)$ denotes a term with the property $O(\rho^2)/\rho^2 \to$ a finite limit, as $\rho \to 0$. Now ρ_i is simply the probability that station i is busy, and $X_e \sim G_e$ and $Y_e \sim H_e$ are the corresponding remaining service times, each with the familiar equilibrium distribution. The most interesting part of this expression is Y_x, which is the service time of a customer at the *second* station who is found in service at the first. The form of (170) allows us to ignore the effects of two or more customers in system at a random moment, when traffic is light.

With what by now should be familiar methods, it is straightforward to determine the distribution of Y_x:

$$P(Y_x > y) = \int_0^\infty P(X > x, Y > y)\, dx/E(X), \tag{171}$$

where $(X, Y) \sim (X_n, Y_n)$. From (171),

$$E(Y_x) = E(XY)/E(X). \tag{172}$$

If X and Y are independent, (171) reduces to $Y_x \sim H$, but otherwise, the length-biasing effect of finding a customer in service induces (171) for the distribution of that customer's service time at the second station.

Any dependence present will not affect the first station, a standard $M/G/1$ queue, and we direct attention to D_2, the stationary delay at the second station. From PASTA, time average (170) is also an arrival average. We "tag" a customer on arrival at the first station. By a somewhat involved argument to ensure that the error term that follows is $O(\rho^2)$, it can be shown that D_2 has expectation

$$d_2 = \rho_1 E\{(Y_x - X_a)_+\} + \rho_2 E\{(Y_e - X_a)_+\} + O(\rho^2) \qquad \text{as } \rho \to 0, \tag{173}$$

where $X_a \sim G$, the service time of the arriving customer, is independent of everything else.

Equation (173) is valid for arbitrary joint distributions, but to make comparisons, we need to consider the nature of any dependence that may be present. In expressions that follow, d_2 will denote some dependent case, and $d_2(I)$ will denote the corresponding independent case. We make the following

Assumption A$^+$

$$P(X > x, Y > y) \geq P(X > x)P(Y > y) \qquad \text{for all } x, y \geq 0, \tag{174}$$

which is the definition of random variables said to be *positively quadrant dependent* (Barlow and Proschan [1981], p. 142). When the moments exist, this is a stronger property than being positively correlated, and in fact, (174) is equivalent to

$$(Y \mid X > x) \geq_{st} Y \qquad \text{for all } x.$$

From (171), it is easily shown that

$$A^+ \Rightarrow Y_x \geq_{st} Y, \tag{175}$$

i.e., the length biasing at the first station induces a stochastic ordering at the second! Similarly, letting A^- be assumption (174) when the inequality is reversed, we have

$$A^- \Rightarrow Y_x \leq_{st} Y. \tag{176}$$

By comparing only the first term in (173), we get

$$A^+(A^-) \Rightarrow E\{(Y_x - X_a)_+\} \geq (\leq)\ E\{(Y - X_a)_+\}, \tag{177}$$

and whenever the inequality in (177) is strict, we get

$$d_2 > (<)\ d_2(I) \qquad \text{for sufficiently small } \rho. \tag{178}$$

We need a strict inequality because the $O(\rho^2)$ terms are different.

When $X = Y \sim \exp(\mu)$, it is easily shown that

$$d_2 = 7\rho/4\mu + O(\rho^2) \text{ and } d_2(I) = \rho/\mu + O(\rho^2) \qquad \text{as } \rho \to 0, \tag{179}$$

in agreement with the Pinedo and Wolff paper, and in the independent case, with product-form results.

Important Remarks. Our method works for an arbitrary number of stations, but gets involved. For the equal exponential service case worked out in Wolff [1982b], the expected delay at each successive station gets larger. The intuitive explanation for this is that because of length biasing, a customer found in service will progress more slowly through the stations than a "random draw." An arrival is more likely to catch up to, and be delayed by, this customer as the number of stations increases. In fact, for equal service times with an arbitrary distribution, Calo [1979] has shown that not counting the first station, the *actual* delay at station r is an increasing function of r. Calo's result is not confined to light traffic.

For two stations, the behavior of this model in heavy traffic is quite different. Some computations are possible in the equal service time case (Boxma), but we will discuss some simulation results. Mitchell et al. [1977] simulated a tandem queue with Poisson arrivals and a version of a bivariate exponential distribution for (X_n, Y_n); we discuss only the positively correlated case, which has property

A^+. For $\rho_1 = \rho_2 \geq .6$, they consistently get

$$d_2/d_2(I) < 1, \tag{180}$$

where the ratio decreases as the ρ_i and correlation each increase; ratios less than 1/2 were achieved. For limited simulations with more than two stations, they get gradual improvement after the second station, but still worse than the first, which is consistent with Calo. Pinedo and Wolff simulate the equal exponential case, obtaining (178) in light traffic, and a smooth curve for $d_2(\rho_i)$ when combined with Mitchell et al. The curves for the dependent and independent cases cross, with crossover estimated to be at .58, just below the range of the Mitchell et al. experiments! Choo and Conolly [1980] claim to show (180) analytically for the Mitchell et al. model, for $0 < \rho_i < 1$. However, their derivation is incorrect, and their numerical results for the dependent case are consistently below those from simulation.

Our understanding of light traffic is now pretty good, but what is the explanation for heavy-traffic behavior? The effect of Y_x, which determines light-traffic comparisons, is apparently overwhelmed by something else. Here is a try: When a customer with service time X_a arrives, let V_{ij} be the amount of work associated with customers at station i that will be performed at station j ($V_{21} = 0$). In heavy traffic, we expect

$$D_2 \approx (V_{22} + V_{12} - V_{11} - X_a)_+ .$$

Under equal service times, V_{12} and V_{11} are nearly equal (the effects under light traffic still apply to the customer found in service), and (almost) cancel out. In the independent case, their difference will be more disperse, and this will tend to make the positive part larger.

These results also show that we will not get the product-form results of Chapter 6 for dependent exponential service times. Inequality (178) implies time- and arrival-average work at the second station are *not* equal; e.g., see (159) and (160) in Chapter 5, even though the arrival process to that station is Poisson. (The first station is *M/M/*1.) Can you explain this?

In the context of Chapter 6, Kelly ([1979], p. 80) observes that dependent service times are permitted in quasi-reversible queues, because the service time of a customer at a station may depend on that customer's class. He is quite right, of course, provided that the stations operate as *symmetric* queues. FIFO is not a symmetric rule, and queueing networks under FIFO will not be quasi-reversible when service times are dependent (in the way discussed here), even when they are exponential.

11-10 *GI/G/*1 AND *GI/G/c* QUEUES IN HEAVY TRAFFIC

We have referred several times to approximations and bounds that are "good in heavy traffic." In this section, we discuss more precisely what we mean by that. Because of their technical nature, we reference formal proofs.

Suppose we have a sequence of stable *GI/G/*1 queues, where the *k*th queue in this sequence has (respectively) inter-arrival and service distributions A_k and G_k, and as $k \to \infty$, $A_k \xrightarrow{D} A$ and $G_k \xrightarrow{D} G$, where $\xrightarrow{D}$ denotes convergence in distribution. In this section only, let $T_k \sim A_k$, $S_k \sim G_k$, $X_k = S_k - T_k$, and D_k be the corresponding *stationary* delay for queue *k*. Similarly, we define λ_k, μ_k, ρ_k, and $\sigma_k^2 = V(X_k)$. Quantities without subscript *k* are for *A* and *G*. We now state

Theorem 4. If as $k \to \infty$, $\rho_k \to \rho = 1$, $E(S_k^2) \to E(S^2)$, and $E(T_k^2) \to E(T^2)$, where $\sigma^2 > 0$,

$$2(1 - \rho_k)D_k/\lambda\sigma_k^2 \xrightarrow{D} \exp(1) \qquad \text{as } k \to \infty, \tag{181}$$

an exponential distribution with mean 1.

See Asmussen [1987], page 197, for a proof based on characteristic function (14), and for other references. Kingman initiated this approach, e.g., see [1962a]. For the *M/G/*1 queue, where we fix *G* and let $\rho \to 1$, this is an elementary exercise; see Problem 8-14. Asmussen also shows that when $\rho_k \to \rho < 1$,

$$D_k \xrightarrow{D} D,$$

which is a justification for using phase-type distributions as approximations.

Notice that the coefficient on D_k in (181) is the reciprocal of Kingman's upper bound (15), which is the reason for earlier statements that (15) (and Daley's better bounds) are good in heavy traffic.

An intuitive explanation for (181) is this: For a stable *GI/G/*1 queue, *D* has a representation as a geometric sum of ascending ladder heights. As $\rho \to 1$, the mean of the geometric $\to \infty$. The contribution of each ladder height is small, and we expect only their mean to figure in the limit. Thus, we expect the limit to be exponential, as we earlier found for *M/G/*1, where the ladder heights that occur are $\sim G_e$, independent of ρ. We can guess the parameter of the exponential by showing that under the conditions of the theorem, $E(Y_\infty^2) \to 0$ in equation (7).

Under conditions somewhat stronger than for (181), Köllerström [1974] showed that for the *GI/G/c* queue, where $\rho_k \to \rho = 1$,

$$2(1 - \rho_k)D_k/\lambda(\sigma_{ak}^2 + \sigma_{gk}^2/c^2) \xrightarrow{D} \exp(1), \tag{182}$$

i.e., in *sufficiently heavy traffic,* multichannel delay has approximately the same distribution as single-channel delay, for the corresponding *fast single-server* queue used to obtain lower bound (103). In fact, his proof uses the fast single server as an approximation.

We emphasized *sufficiently heavy* because (182) needs to be used with care, as Example 11-8 shows.

Example 11-8

We compare *Q* and the corresponding heavy-traffic estimate Q_{HT}, for the *M/M/*1 and *M/M/*100 queues, with $\rho = .9$. We get

$$Q_{100} = 1.95, \quad Q_{1f} = 8.1, \quad \text{and} \quad Q_{HT} = 9.05.$$

When we compare Q_{1f} with Q_{HT}, the estimate is off by about 10 percent because $P(D_{1f} > 0) = .9$. But $P(D_{100} > 0) = .22$, and as an estimate of Q_{100}, Q_{HT} is unacceptable.

For large systems (large c), we have to be careful about what we mean by heavy traffic. As we discovered in Chapter 5, these systems can be very busy (as far as the servers are concerned) and yet give good customer service. This is why pooling is such a good idea when it is convenient to implement.

Let N_∞ be the stationary number-of-customers-in-system for the *infinite* channel queue corresponding to the 100-channel queue in Example 11-8, where N_∞ is Poisson with mean *and variance* 90. Now $P(N_\infty > 99) = .16$ (the normal approximation would give a number very close to this) is a lower bound on $P(D_{100} > 0)$, but it is a pretty good one.

More generally, let $a = c\rho = \lambda E(S)$ be the *offered load*. When both a and c are large, the magnitude of d_c is very sensitive to the *relative magnitude* of these quantities. (For other performance measures, this is also true in loss systems; see Section 7-8.) For $M/G/c$ (the usual case in applications), N_∞ is also Poisson, and $P(N_\infty \geq c)$ is easy to calculate or approximate. When $c - a \geq \sqrt{a}$, as in Example 11-8, (182) will be poor. See Newell [1973] for an analysis of the questions raised here.

An ad hoc approach to c-channel queues with (moderately) heavy traffic is to approximate the distribution of $(D_c \mid D_c > 0)$ by (182), and separately estimate $P(D_c > 0)$. If this probability is small, almost anything we do, e.g., use the exact value for $M/M/c$, will be better than (182) alone.

An alternative approach to heavy traffic is to use what are called *diffusion approximations*; see Newell [1982] for a fairly readable account of this theory, or Harrison [1985] for a more mathematical treatment and extensive references. Newell's [1973] and [1984] monographs also use this approach.

Tandem and network queues in heavy traffic also have been analyzed with diffusion methods; see Newell [1979] and Harrison. These models have complicated boundary conditions (queues cannot be negative), and even the tandem case is very difficult. In terms of formal solutions, Harrison [1973] shows that for tandem queues, the limiting joint delay distribution is a function f, say, only of means and variances, as in (181). Except in a few special cases, however, he is unable to find f. There are published "approximate" solutions for the network case, but we choose not to give them out of fear that as a result, someone might use them!

11-11 BOUNDS ON DELAY DISTRIBUTIONS FOR *GI/G/*1 QUEUES

These results are stuck here because, as in Section 11-10, we do not prove the main results.

Let D be the stationary delay in queue for a stable $GI/G/1$ queue, and S, T, and $X = S - T \sim F$ be the usual generic variables. Suppose $E(e^{\eta X}) = 1$ for some real $\eta \neq 0$. From Jensen's inequality (65) and $E(X) < 0$, this implies $\eta > 0$, and that $E(e^{\eta S}) < \infty$, so we are imposing conditions on S; e.g., S has finite moments of all orders.

When $\eta > 0$ exists, Kingman [1964, 1970] obtains the following exponential bounds on the tail distribution of D:

$$\alpha e^{-\eta t} \leq P(D > t) \leq e^{-\eta t}, \qquad t > 0, \tag{183}$$

where

$$1/\alpha = \sup_{t>0} E\{e^{\eta(X-t)} \mid X > t\}. \tag{184}$$

For the lower bound to be useful, we need $\alpha > 0$, which is easily seen to be true for the distributions commonly used in practice. [In the bounded case, where b is the smallest number such that $F(b) = 1$, the sup is over the interval $(0, b)$.]

It is easily shown (condition on T) that if α satisfies (184), then

$$1/\alpha \leq \sup_{t>0} E\{e^{\eta(S-t)} \mid S > t\} \equiv 1/\alpha', \tag{185}$$

and we can use α' in place of α in (183).

Ross [1974] obtained an upper bound slightly stronger than the one in (183):

$$P(D > t) \leq \beta e^{-\eta t}, \qquad t > 0, \tag{186}$$

where

$$1/\beta = \inf_{t>0} E\{e^{\eta(X-t)} \mid X > t\}.$$

As above

$$1/\beta \geq \inf_{t>0} E\{e^{\eta(S-t)} \mid S > t\} \equiv 1/\beta', \tag{187}$$

and we can use β' in place of β in (186).

In some cases, the α' and β' are easy to determine; see problems at the end of this chapter.

11-12 FURTHER DISCUSSION AND LITERATURE NOTES

As always, our main objective is to increase our understanding of queueing phenomena. The literature related to this chapter is not only enormous, but the incredible variety of methods and detailed results are difficult to summarize in a coherent way.

Fortunately, there exist specialized references for more detail and other references. For *GI/G/1*, *GI/G/c*, and *M/G/c* queues, Stoyan [1983] summarizes much of the theory, and Tijms [1986] has computational results about the performance of various heuristic, ad hoc, (or worse) approximations. Daley and Trengrove [1977] is also good.

Kingman has been very influential, as should be clear from references in earlier sections, and was the first to derive (15) and (22). Marshall [1968a] was the first to obtain (10), (12), and the bounds in Section 11-4, and in [1968b], provides the easy derivation of (14). Marshall assumed IFR (DFR) for (51), whereas Daley and Trengrove (and references therein) show that DMRL (IMRL) suffices.

Basic references for Section 11-3 and the partial orderings of distributions in Section 11-5 are Barlow and Proschan [1965, 1981] as well as Stoyan. Stoyan and coauthors (see references in Stoyan) are primarily responsible for the queueing applications in Section 11-5.

In addition to their fundamental contributions discussed in Section 11-6, Kiefer and Wolfowitz [1956] derived the finite moment results in Theorems 2 and 3 in Section 11-7. Our elementary derivation of these results is from Wolff [1984]. It has taken 30 years to find easy ways to do what they did in their classic papers!

Inequalities (101) and (102), and lower bound (103) are from Brumelle [1971a], but the concept of the fast single server goes back to Kiefer and Wolfowitz [1955]. The derivation of upper bounds in (113) through (116) are from Wolff [1977], and the derivation of (123) is from Wolff [1987]. Finding elementary upper bounds on various quantities for the $GI/G/c$ has been a stumbling block. There are numerous incorrect statements and proofs in the literature; see Stoyan [1976] for mention of some old ones, and Wolff [1987] for some more recent ones.

The upper bounds in Section 11-7 show that the standard $GI/G/c$ queue fed by a single FIFO queue is optimal for various performance measures with respect to the alternatives considered. For other senses in which this queue is optimal, see Foss [1980] and Daley [1987].

Approximation (137) for the $M/G/c$ queue was first proposed (apparently) by Lee and Longton [1957], and independently several times since, including Nozaki and Ross [1978], who also proposed queue-limit version (140). The elementary argument we give for (137) has been proposed independently by several authors. Because of the importance of this model in applications, many other approximations have been proposed.

For tandem queues, (150), (157), and their generalizations to (a) and (b) on page 510 are from Tembe and Wolff [1974]; these results for constant service, where order of service doesn't matter, were found earlier by Friedman [1965]. The derivations of (168) and (169) are from Niu [1981], and there are also interesting results in Niu [1980]. The light traffic theory and results for dependent service times are from Wolff [1982b].

Mathematically, tandem queues are the simplest examples of queueing networks, but by channeling all customers onto a single route, they also magnify the effects of deviations from such idealized assumptions as independent and exponential service times. Nevertheless, it is important to understand the nature of these effects.

Approximations that either ignore these effects or attempt to take them into account in ad hoc ways would be expected to work better for networks with diverse routes, provided that a small number of routes do not carry most of the traffic.

What is available to approximate queueing networks? We have the product-form results of Chapter 6, which hold exactly for Poisson arrivals, exponential service, and the usual independence assumptions, with general service permitted at stations that operate as symmetric queues. For Poisson arrivals, general service, and FIFO queues, we might approximate each station as an $M/G/1$ (or $M/G/c$) queue, as suggested by Kleinrock [1976]. This is worth trying when we have diverse routes, and *traffic is not heavy*. More ambitious is the method developed by Whitt [1983a,b], who approximates the composite arrival process at each station by a renewal process.

Approximating composite arrival processes by either renewal or Poisson processes is a tricky business. There is a well-known theorem (roughly) stating that the superposition of n renewal processes becomes Poisson, when properly rescaled, as $n \to \infty$. The quality of approximations based on this theorem depends not only on n (and of course the inter-arrival distribution), but also the application. If the composite process is the input into a queue, is traffic heavy or light? For light traffic, we need only the approximation to be good "locally", i.e., for the next few arrivals. For heavy traffic, the variability of the arrival process over a long period of time (many arrivals) is important, and a Poisson approximation can be very poor. This is the reason for the emphasis on *traffic is not heavy* in the preceding paragraph. See Newell, [1984b].

A similar issue arose in Section 11-10 with regard to heavy-traffic approximations for multichannel queues. Letting $\rho \to 1$, while holding the number of channels c fixed, fails to capture important aspects of systems with large c.

PROBLEMS

11-1. Find $E(I)$ in terms of busy cycles, and use it to find (6).

11-2. Find (14), and "naively" differentiate it to find (10).

11-3. Go through the algebra to get Kingman's upper bound (15).

11-4. By canceling third moments, find an expression for $E(D^2)$ for the $GI/G/1$ queue.

11-5. Let $R \sim F$ be nonnegative and discrete with successive point masses at $a < b$. For $x \in [a, b]$ we can write

$$(R - x)_+ = (R - b)_+ + (b - x)I_a,$$

where I_a is the indicator of the event $\{R > a\}$. From the preceding, show that

$$E\{(R - x)_+^2\} = E\{(R - b)_+^2 \mid R > a\}F^c(a) + g(x), \text{ and}$$

$$E^2\{(R - x)_+\} = E^2\{(R - b)_+ \mid R > a\}F^2(a) + g(x)F^c(a),$$

where

$$g(x) = 2(b - x)E\{(X - b)_+\} + (b - x)^2 F^c(a),$$

a decreasing function of x. From these expressions, show that

$$E\{(R - x)_+^2\}/E^2\{(R - x)_+\}$$

is nondecreasing in $x \in [a, b]$. Proceeding from point mass to point mass, we have (17) for discrete random variables when $\alpha = 2$. Why is it immediate that (17) also holds for continuous random variables when $\alpha = 2$?

11-6. Verify (23) and (24).

11-7. Show that mixtures of DFR distributions are DFR. For hyper-exponential distribution H with failure rate function $r(t) = h(t)/H^c(t)$, show that

$$\lim_{t \to \infty} r(t)$$

exists, and is equal to the *slower* of the two exponential rates.

11-8. Give an *easy* proof of this: Convolutions of NBU distributions are NBU.

11-9. Verify (40) analytically.

11-10. Apply results in Section 11-4 to bound c_τ^2 in (14).

11-11. Show that T has IFR (DFR) $\Rightarrow I_e \leq_{\text{st}} (\geq_{\text{st}})\ T_e$.

11-12. Show that for arbitrary distribution F and $U \sim$ uniform on (0, 1),

$$X \equiv F^{-1}(U) \sim F.$$

11-13. *Sketch of proof of (64).* It is obvious (why?) that the right-hand inequality implies the left. For the converse, let X and Y be nonnegative, where (63) implies $E(Y) < \infty$. Let g_n be a *piecewise linear*, convex, and nondecreasing function. That is, for

$$0 = a_1 < \cdots < a_n < a_{n+1} = \infty \quad \text{and} \quad 0 = b_{-1} \leq b_0 < \cdots < b_n < \infty,$$

g_n is of form

$$g_n(x) = g_n(a_j) + b_j(x - a_j) \qquad \text{for } x \in [a_j, a_{j+1}) \text{ and } j = 1, \ldots, n,$$

which can be written

$$g_n(x) = \sum_{j=0}^{n} (b_j - b_{j-1})(x - a_j)_+ .$$

From this, why does (64) hold for all piecewise linear functions? How would you complete the proof?

11-14. Derive (70).

11-15. In Example 11-7, show that the stationary number of customers in system for these queues are also stochastically ordered.

11-16. Kingman [1970] derives the following lower bound for the expected delay in a $GI/G/c$ queue generated by $\{T_n, S_n\}$:

$$d_c \geq d_{1K}/c\rho - [c\sigma_a^2 + \sigma_g^2 + (1 - c^{-1})E^2(S)]/2E(S),$$

where d_{1K} is the expected delay in a $GI/G/1$ queue generated by $\{cT_n, S_n\}$.

(a) How is this single-channel queue related to the fast one used to get lower bound (103)?
(b) Use (a) to write the preceding inequality in terms of $d_{1f} \equiv d_1$ in (103).
(c) Show that *if* d_{1f} were equal to upper bound (15) for this queue, Kingman's lower bound would be equal to the bound in (103).
(d) Of course d_{1f} is lower than upper bound (15). Use this fact to show that (103) is better (greater) than Kingman's lower bound.

11-17. If we use random assignment rather than cyclic to get an upper bound on d_c, compare the corresponding bound with (132).

11-18. Suppose we have two stable single-channel queues in tandem, with Poisson arrivals, *constant* service $1/\mu_1$ at station 1, and exponential service at rate μ_2 at station 2. Let w_2 be the average delay at the second station. Use results in Section 11-9 to show that

$$w_2 \leq 1/(\mu_2 - \lambda),$$

where the right-hand quantity is what the expected delay would be *if* arrivals at the second station were Poisson. Constant service at the first station seems to *regularize* arrivals at the second station. (Niu [1980] conjectures that w_2 decreases as the constant service time increases.)

11-19. When approximation (160) is substituted into (12), show that we get (161).

11-20. By any acceptable method, derive (171).

11-21. Verify (179).

11-22. At the end of Section 11-9, it is remarked that for two queues in tandem with Poisson arrivals, exponential service, and *dependent* service times (of the same customer), arrivals at the second station do not see time averages, even though that arrival process is Poisson. How can this be?

11-23. If you think Q_{HT} is a poor approximation of Q_{100} in Example 11-8, calculate (or estimate) the corresponding cyclic assignment upper bound. It is much worse. (The random assignment bound is about twice the cyclic bound.)

11-24. Find an expression for (185) when
(a) S is NBU, and
(b) S is DFR.
(c) Explicitly evaluate the result in (b) when S in hyper-exponential.

11-25. Find an expression for (187) when
(a) S is NWU, and
(b) S is IFR.
(c) Explicitly evaluate the result in (b) when S is k-Erlang.

11-26. *Time and arrival average queue lengths.* (Marshall and Wolff [1971]) For a stable FIFO $GI/G/c$ queue, let N_q and N_{qa} be the stationary number of customers in system at a random point in time, and as found by an arrival, respectively, with means Q and Q_a. Under FIFO, N_{qa} is the number of arrivals during a stationary delay D. Let $m_a(t)$ be the ordinary renewal function generated by the inter-arrival distribution A with positive arrival rate λ.
(a) Use a well-known bound on renewal functions (somewhere in Chapter 2), and

the preceding observations, to show that

$$Q_a \geq Q - 1.$$

(b) When A is BMRL-$\overline{1/\lambda}$, use another bound on renewal functions to show that

$$Q_a \leq Q.$$

Remark. When $c = 1$, results in (a) and (b) hold for N and N_a.

(c) When $m_a(t) \leq \lambda t$ for all t, the inequality is preserved under convolution, $m_a^{(r)}(t) \leq \lambda^r t^r / r!$ Use this fact and additional properties of renewal functions to show that when A is BMRL-$\overline{1/\lambda}$,

$$E\{N_{qa}(N_{qa} - 1) \cdots (N_{qa} - r + 1)\} \leq \lambda^r E(D^r), \qquad r \geq 2.$$

(d) For Poisson arrivals ($M/G/c$), show that

$$E\{N_q(N_q - 1) \cdots (N_q - r + 1)\} = \lambda^r E(D^r).$$

APPENDIX

We collect here certain concepts and results from mathematical analysis that are needed from time to time in this book. Our style will be to explain a mathematical "issue," and then to state without formal proof conditions that ensure the validity of whatever we want to do. The most frequent issue will be whether we can interchange the order of mathematical operations.

We deal only with real numbers and real valued functions, but many of the results carry over to vectors and more generally to what are called complete metric spaces.

A-1 SEQUENCES AND SERIES; LIMITS AND CONVERGENCE

Let $s_1, s_2, \ldots$ be a sequence of numbers. We say $\{s_n\}$ has *limit* ℓ, denoted by

$$\lim_{n\to\infty} s_n = \ell, \tag{1}$$

if for every $\varepsilon > 0$, there is an integer N such that for all $n > N$, $|s_n - \ell| < \varepsilon$, where ℓ is some finite number. We write (1) even when $\ell = \infty$ (or $-\infty$), where we mean that for every number M, there is an integer N such that for all $n > N$, $s_n > M$. In either case, we say the limit *exists*. For finite ℓ, we say $\{s_n\}$ *converges* to ℓ. For brevity, sometimes we will write limits such as (1) as $\lim s_n$.

Some examples are: For $s_n = 1/n$, $\ell = 0$; for $s_n = n$, $\ell = \infty$; and for $s_n = (-1)^n$, $\lim s_n$ *does not exist*.

Let $\{s_n\}$ and $\{t_n\}$ have finite limits ℓ and υ, respectively, and c be any number. Some standard properties of limits are: $\lim (s_n + t_n) = \ell + v$, $\lim cs_n = c\ell$, and $\lim s_n t_n = \ell v$.

A sequence that is monotone (say nondecreasing) will always have a (possibly infinite) limit. In the nondecreasing case, the limit is finite if and only if the sequence has a finite upper bound.

More generally, we say $\{s_n\}$ has *limit superior* (or *upper limit*) U, denoted by

$$U = \limsup_{n \to \infty} s_n,$$

if (i) for every $\varepsilon > 0$, there exists an integer N such that for all $n > N$, $s_n < U + \varepsilon$, and (ii) for every $\varepsilon > 0$ and integer M, $s_n > U - \varepsilon$ for some $n > M$. These conditions either determine a finite U uniquely, or no finite U exists, and we set $U = \infty$. Analogously, we define *limit inferior* (*lower limit*)

$$L = \liminf_{n \to \infty} s_n,$$

or by $L = -\limsup (-s_n)$.

Clearly, $U \geq L$, and (1) is equivalent to $U = L = \ell$. In applications, e.g., see Section 2-14, we establish (1) by showing that for some ℓ, $U \leq \ell$ and $L \geq \ell$. Note that there we had a function $s(t)$ say, for $t \in [0, \infty)$, but the concepts and definitions carry over in an obvious way as $t \to \infty$.

We call $\{s_n\}$ a *Cauchy sequence* if for every $\varepsilon > 0$, there is an integer N such that for all $n > N$ and $m > N$, $|s_n - s_m| < \varepsilon$. Another way of showing (1) is the *Cauchy criterion*: $\{s_n\}$ converges to a finite limit if and only if $\{s_n\}$ is a Cauchy sequence. Note that this method neither requires knowledge of nor explicitly determines the limit ℓ.

Now suppose $\{s_n\}$ is a sequence of partial sums of sequence $\{a_n\}$, i.e.,

$$s_n = \sum_{i=1}^{n} a_i, \qquad n = 1, 2, \ldots .$$

We say the *series* $\sum_{n=1}^{\infty} a_n$ *converges* to ℓ, and write

$$\sum_{n=1}^{\infty} a_n = \ell, \tag{2}$$

if (1) holds for finite ℓ. Otherwise, we say the series *diverges*. Thus, divergence includes two cases: $\{s_n\}$ has a well-defined but infinite limit, and $\{s_n\}$ has no limit. When $\{s_n\}$ has a well-defined but infinite limit, we will continue to write (2), and whether ℓ is finite or not, we say that the series has limit ℓ when sequence $\{s_n\}$ does. Sometimes the sum in (2) will begin with $n = 0$ (or we can add a term $a_0 \equiv 0$), and for brevity, we denote series (2) by $\sum a_n$.

The Cauchy criterion applied to a series is that $\sum a_n$ converges if and only if for every $\varepsilon > 0$, there is an integer N such that for all $m \geq n > N$

$$\left| \sum_{i=n}^{m} a_i \right| < \varepsilon. \tag{3}$$

Setting $m = n$, (3) implies that $\sum a_n$ converges *only if*

$$\lim_{n\to\infty} a_n = 0. \tag{4}$$

However, (4) is not sufficient for convergence, as illustrated by the well-known result

$$\sum_{n=1}^{\infty} 1/n = \infty. \tag{5}$$

In fact, we state without proof a useful generalization of this result:

$$\sum_{n=1}^{\infty} 1/(n)^p \text{ converges if } p > 1, \text{ and diverges if } p \leq 1. \tag{6}$$

Does ℓ in (2) depend on the order in which the terms in $\{a_n\}$ are summed? A standard example that illustrates what can go wrong is

$$a_n = (-1)^n/n, \qquad n = 1, 2, \ldots. \tag{7}$$

The Cauchy criterion (3) implies that $\sum a_n$ in (7) converges. However, it is easy to show that

$$-1 - 1/3 + 1/2 - 1/5 - 1/7 + 1/4 - 1/9 - 1/11 + \cdots,$$

i.e., the rearranged sum, where two negative terms are followed by one positive term (from their original order), cannot have the same limit as $\sum a_n$, although it does converge.

We have cancellation of positive and negative terms, where the cumulative effect of this cancellation depends on the order of the terms in the sum. Not only does (5) diverge, the sum of only the positive terms and of only the negative terms in (7) are each infinite.

The following important property ensures that separate sums of the positive and negative terms are each finite.

Definition. Series $\sum a_n$ *converges absolutely* if $\sum |a_n|$ converges.

The Cauchy criterion may be used to show that all summed rearrangements of $\{a_n\}$ *converge to the same finite limit* if and only if $\sum a_n$ converges absolutely. In effect, we can sum the positive and negative terms separately, and then subtract their magnitudes.

More generally, if $a_n \geq 0$, all summed rearrangements of $\{a_n\}$ *have the same (possibly infinite) limit,* because no cancellation occurs. In fact, this can be weakened further: For each n, let $a_n^+ = \max(a_n, 0)$ and $a_n^- = -\min(a_n, 0) = (-a_n)^+$, where these quantities are called the *positive part* and the *negative part* of a_n, respectively, and $a_n = a_n^+ - a_n^-$. Now suppose $\sum a_n^+$ and $\sum a_n^-$ have limits ℓ^+ and ℓ^-, respectively. All summed rearrangements of $\{a_n\}$ have well-defined but possibly infinite limit

$$\ell = \ell^+ - \ell^-, \tag{8}$$

if and only if either $\ell^+ < \infty$, or $\ell^- < \infty$. Thus we can get different "answers" only when both ℓ^+ and ℓ^- are infinite, and (8) becomes $\infty - \infty$.

As an example, let X be a discrete random variable with $P(X = x_n) = p_n$, $n = 1, 2, \ldots$. The expected value of X is

$$E(X) = \sum x_n p_n,$$

which is of the preceding form with $a_n = x_n p_n$. Because the p_n are nonnegative, absolute convergence is written $\sum |x_n| p_n = E(|X|) < \infty$, which ensures that $E(X)$ is finite. More generally, $E(X)$ is well defined but possibly infinite if either $E(X^+) < \infty$ or $E(X^-) < \infty$.

Series $\sum c_n$ is called the *convolution* (or *Cauchy product*) of series $\sum a_n$ and $\sum b_n$ if

$$c_n = \sum_{i=0}^{n} a_i b_{n-i}, \qquad n = 0, 1, \ldots. \tag{9}$$

If $\sum a_n$ converges to A, and $\sum b_n$ converges to B, where at least one of these series converges absolutely, it can be shown that $\sum c_n$ converges to C, where

$$C = AB. \tag{10}$$

Now suppose we want to interchange the order of a double (or multiple) sum. An interchange is a rearrangement of the original order of summation, and a sufficient condition for

$$\sum_{i=1}^{\infty} \sum_{j=1}^{\infty} a_{ij} = \sum_{j=1}^{\infty} \sum_{i=1}^{\infty} a_{ij} \tag{11}$$

is absolute convergence of either $\sum_i \sum_j |a_{ij}|$ or $\sum_j \sum_i |a_{ij}|$. If the a_{ij} are nonnegative, the interchange is always valid, even when the sum is infinite, and the interchange is valid if we have absolute convergence of either the positive or negative parts.

An example of (11) is (117) in Chapter 1.

A-2 INTEGRATION; *dF* NOTATION

Integrals arise as limits defined in terms of weighted sums.

In elementary calculus, what is called the *Riemann* integral has the physical interpretation of an area under some function $g(t)$ on some finite interval $[a, b]$,

$$\text{area} = \int_a^b g(t)\,dt,$$

as in Figure A-1, where $g(t)$ is bounded on $[a, b]$. If $g(t)$ can be negative, the integral is the difference between "positive" and "negative" areas.

The formal definition of the Riemann integral is in terms of upper and lower sums of rectangular areas: Partition $[a, b]$ into n subintervals with end points

$$a = t_0 \leq t_1 \leq \cdots \leq t_{n-1} \leq t_n = b,$$

let M_i and m_i be the respective least upper bound and greatest lower bound of $g(t)$ on subinterval i, and let $\Delta t_i = t_i - t_{i-1}$, $i = 1, \ldots, n$. Now write

$$\sum_{i=1}^{n} m_i\,\Delta t_i \leq \text{would-be area} \leq \sum_{i=1}^{n} M_i\,\Delta t_i. \tag{12}$$

Now let $n \to \infty$ and the $\Delta t_i \to 0$ (the formal limit process is a bit more involved). If the upper and lower sums converge *to the same limit* ℓ, we define the Riemann integral

$$\int_a^b g(t)\,dt = \ell\,. \tag{13}$$

More generally, let $\alpha(t)$ be a monotone nondecreasing function of t on the interval $[a, b]$, and $\Delta\alpha_i = \alpha(t_i) - \alpha(t_{i-1})$, $i = 1, \ldots, n$. In place of (12), write

$$\sum_{i=1}^{n} m_i\,\Delta\alpha_i \leq \text{would-be integral} \leq \sum_{i=1}^{n} M_i\,\Delta\alpha_i. \tag{14}$$

Now let $n \to \infty$ and the $\Delta t_i \to 0$ in (14). If the upper and lower sums converge *to the same limit* $\ell(\alpha)$, we define the *Riemann-Stieltjes* (*RS* or just *Stieltjes*) integral

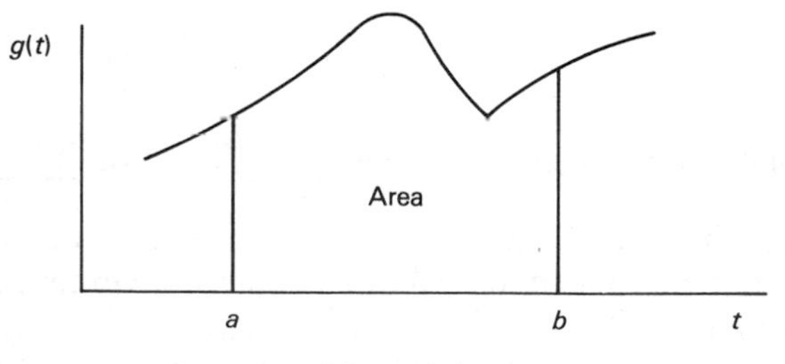

Figure A-1

of g with respect to α over $[a, b]$ to be

$$\int_a^b g(t)\, d\alpha(t) = \ell(\alpha). \tag{15}$$

Thus the Riemann integral is the special case we get by choosing $\alpha(t) = t$.

If $\alpha(t)$ is differentiable, with derivative $\alpha'(t)$, (15) becomes

$$\int_a^b g(t)\, d\alpha(t) = \int_a^b g(t)\alpha'(t)\, dt,$$

where the right-hand expression is a Riemann integral of the function $g(t)\alpha'(t)$.

We assume that $\alpha(t)$ is right-continuous, meaning that at any point $t = \tau$,

$$\lim_{t \to \tau^+} \alpha(t) = \alpha(\tau),$$

where the notation $t \to \tau^+$ denotes a right-hand limit, i.e., t approaches τ from the right. Owing to monotonicity, left-hand limits exist,

$$\lim_{t \to \tau^-} \alpha(t) \equiv \alpha(\tau^-).$$

If $\alpha(t)$ is discontinuous at τ, the size of the jump there is $\alpha(\tau) - \alpha(\tau^-)$. (A function is *continuous* at a point if the right- and left-hand limits at that point exist and are equal.)

Recall that by convention, distribution functions $F(t)$ are right-continuous, and consequently, so are renewal functions $m(t)$. If $X \sim F$ is a discrete random variable, $P(X = \tau) = F(\tau) - F(\tau^-)$.

An RS integral over an infinite interval (a so-called *improper* integral) is defined as a limit, e.g.,

$$\int_0^\infty g(t)\, d\alpha(t) = \lim_{b \to \infty} \int_0^b g(t)\, d\alpha(t), \tag{16}$$

whenever the limit exists.

Think of (15) as the limit of a "weighted sum" of g, where the weights are *changes* in the function α. In fact, if α is a step function with jumps $\Delta(\alpha_i) = \alpha(t_i) - \alpha(t_{i-1})$, (15) becomes

$$\int_a^b g(t)\, d\alpha(t) = \sum_{i: t_i \in (a,b]} g(t_i)\, \Delta\alpha_i. \tag{17}$$

On the other hand, if $\alpha(t) = F(t)$, a probability distribution function with density function $f(t)$, (15) becomes the Riemann integral

$$\int_a^b g(t)\, dF(t) = \int_a^b g(t) f(t)\, dt, \tag{18}$$

which can be thought of as "differentially" weighting $g(t)$ by $f(t)\, dt$.

Thus, the Stieltjes integral has been popular in applied probability because it combines the continuous and discrete cases. However, there are some technical problems with the definition of an RS integral.

A minor problem is that because of right continuity, the contribution of a jump in $\alpha(t)$ at the left-hand end point a is excluded from (15). In this book, this comes up when we are integrating with respect to the distribution of a nonnegative random variable, e.g., a convolution

$$\int_0^t G(t - u)\, dF(u). \tag{19}$$

Clearly, if F has probability mass at 0, i.e., $F(0) - F(0^-) = F(0) > 0$, we want to include $G(t)F(0)$ in (19). The way out is the 0^- (or sometimes a^-) convention, meaning that we interpret the lower limit of integration to be to the left of 0, so that this point *is* included.

More serious is the question of whether the limit that defines the RS integral (15) will exist. It always will when g is continuous, or when g is monotone and α is continuous. However, if g and α have a discontinuity at the *same* point τ, the LS integral will not exist because the upper and lower sums in (14) will differ by at least (jump in g) $\cdot$ (jump in α) at that point. If α is a distribution function, and the intent is to evaluate g *at* a point of positive probability, the correct contribution of this point to the integral is

$$g(\tau)[\alpha(\tau) - \alpha(\tau^-)].$$

Thus any "dF" and "dm" integral in this book may be thought of as an RS integral with these modifications.

As an example of our notation, in Chapter 1, we introduced the Laplace-Stieltjes transform of a nonnegative random variable $X \sim F$,

$$E(e^{-sX}) \equiv \tilde{F}(s) = \int_0^\infty e^{-st}\, dF(t),$$

which we now see is just a particular form of an RS integral. Our notation calls attention to the fact that it is a transform of F, which has the advantage of being well defined whether or not F has a density function f. In the density function case, we have

$$\int_0^\infty e^{-st}\, dF(t) = \int_0^\infty e^{-st} f(t)\, dt,$$

where the right-hand integral is the *Laplace* transform of the function f, i.e., it is a different kind of integral transformation, but also of a *different* function.

The theory of RS integration can be extended to functions α that are of *bounded variation,* which, on a finite interval, means that $\alpha(t)$ can be represented as the difference of two monotone functions. When α and g are both of bounded variation and at least one of the integrals in (20) exists, then so does the other, and we state the formula for *integration by parts*,

$$\int_a^b g(t)\, d\alpha(t) + \int_a^b \alpha(t)\, dg(t) = g(b)\alpha(b) - g(a)\alpha(a). \tag{20}$$

As an application of (20) and our 0^- convention, we derive the relation between the Laplace-Stieltjes transform of the distribution F of a nonnegative random variable, and the *Laplace* transform of F (not f):

$$\int_0^b e^{-st}\, dF(t) - s \int_0^b F(t)e^{-st}\, dt = e^{-sb}F(b) - e^{-s0}F(0^-) = e^{-sb}F(b).$$

Now let $b \to \infty$, and we have

$$\int_0^\infty e^{-st}F(t)\, dt = \int_0^\infty e^{-st}\, dF(t)/s, \tag{21}$$

a result needed to obtain equation (48) in Chapter 2.

As noted, (15) may fail to exist, and this is one reason the theory of RS integration has gone out of fashion in abstract treatments of probability theory. It has been replaced by a measure-theoretic approach that we briefly describe.

We now discuss integration with respect to a measure. In probability theory, a random variable $X(\omega)$ is function defined on points in a sample space Ω. A probability *measure* P is a function that assigns nonnegative weights (probabilities) to *sets* of points called *events*, where we require that $P(\Omega) = 1$. The expected value of X is the integral of the function $X(\omega)$ with respect to probability measure, denoted by

$$E(X) = \int_\Omega X(\omega)\, dP(\omega). \tag{22}$$

The set of events is closed under countable set operations (union, intersection). We call X a random variable only when it is *measurable*, which means that certain sets of points are events, e.g., it is sufficient to require that for all real x, $\{\omega\colon X(\omega) > x\}$ are events. This condition is needed in order to define (22).

We call random variable X_s a *simple function* if it has only a finite number of distinct values, x_i, $i = 1, \ldots, n$, where for each i, we let $p_i = P\{\omega\colon X_s(\omega) = x_i\}$. The expected value of a simple function is defined as

$$E(X_s) = x_1 p_1 + \cdots + x_n p_n, \tag{23}$$

i.e., each possible value of X_s is weighted by the probability that it occurs. The notation in (22) is intended to convey this idea. If X_s is nonnegative and the x_i are ordered, $0 \le x_1 < \cdots < x_n$, (23) can be written

$$E(X_s) = x_1 + (x_2 - x_1)P(X_s > x_1) + \cdots + (x_n - x_{n-1})P(X_s > x_{n-1})$$

$$= \int_0^\infty P(X_s > t)\, dt. \tag{24}$$

the integral of the tail of the distribution of X_s!

Now suppose that X is a nonnegative but otherwise arbitrary random variable. It is easy to find nonnegative simple functions with the property that for all points ω,

$$0 \leq X_s(\omega) \leq X(\omega). \tag{25}$$

In fact, for any nonnegative X, it can be shown that there exists a nondecreasing sequence of simple functions $\{X_{sn}\}$ that satisfy (25) such that for every ω,

$$\lim_{n\to\infty} X_{sn}(\omega) = X(\omega).$$

In terms of this sequence, we *define*

$$E(X) \equiv \lim_{n\to\infty} E(X_{sn}) = \lim_{n\to\infty} \int_0^\infty P(X_{sn} > t)\,dt$$

$$= \int_0^\infty \lim_{n\to\infty} P(X_{sn} > t)\,dt = \int_0^\infty P(X > t)\,dt, \tag{26}$$

where this limit exists but may be infinite. This is the way we found $E(X)$ in Section 1-14.

If X is not sign constrained, we define

$$E(X) = E(X^+) - E(X^-), \tag{27}$$

provided at least one of the terms on the right is finite. If both of these terms are infinite, we say $E(X)$ *does not exist*.

There are some gaps to fill in to make the above rigorous. In particular, in Section A-3, we justify the interchange of limit and integral in (26). Note that from the right-hand integral in (26), $E(X)$ depends only on properties of X, and not on any particular sequence of simple functions.

Integration with respect to other measures can be defined in a similar manner. In one dimension, *Lebesgue* measure assigns measure $\mathcal{L}\{(a, b)\} = b - a$ to any finite open interval, where $\Omega = (-\infty, \infty)$. *Lebesgue-Stieltjes* measure assigns measure $\alpha(b) - \alpha(a)$ to any interval of form $(a, b]$, where α is nondecreasing and right-continuous. These measures can be used to generalize the theory of Riemann and RS integration.

Lebesgue measure is infinite ($\mathcal{L}(\Omega) = \infty$), but it assigns finite measure to finite intervals. Probability measure is finite; in fact, $P(\Omega) = 1$ is needed for the rearranged sum in (24) and for the representation of $E(X)$ as a tail integral in (26).

Integration, including expectation, is a differential form of summation. We can think of the sums in Section A-1 as integrals with respect to a measure that assigns a weight of 1 to every integer.

As we found for sums at the end of Section A-1, *absolute convergence* is a sufficient condition for interchanging the order of multiple integrals, which includes any of the symbols $\int$, $\sum$, and E. The order also may be interchanged when we are integrating nonnegative functions. In what are called product spaces, this result is known as *Fubini's Theorem*.

A-3 INTERCHANGING LIMIT AND INTEGRAL

For random variables $X \sim F$ and $X_n \sim F_n$, $n = 1, 2, \ldots$, suppose sequence $\{X_n\}$ converges to X. Under what conditions is it true that $\lim_{n\to\infty} E(X_n) = E(X)$? The mode of convergence (with probability 1, in probability measure, or in distribution) is irrelevant, because $E(X_n)$ depends only on F_n. For simplicity, suppose the random variables are nonnegative. In this case, we want to find conditions for the validity of the interchange of lim and $\int$:

$$\lim_{n\to\infty} E(X_n) = \lim_{n\to\infty} \int_0^\infty F_n^c(t)\,dt = \int_0^\infty \lim_{n\to\infty} F_n^c(t)\,dt = \int_0^\infty F^c(t)\,dt = E(X). \quad (28)$$

(*Technical remark:* The limit in the middle expression in (28) may not exist at points t where F has a jump, but this has no effect on the right-hand integral. In any event, right-continuity determines F at all t.)

If the integrals on the left in (28) were over a finite interval, the interchange would be valid (the F_n^c are bounded functions). However, these integrals are improper, and in this case, *uniform convergence* is needed to ensure that the interchange is valid, which will also imply that the right-hand integral is finite. The Cauchy criterion version of uniform convergence of integrals over $[0, \infty)$ of functions $g_n(t)$ is this: For every $\varepsilon > 0$, there is a finite T such that for all $v > u > T$,

$$\int_u^v g_n(t)\,dt < \varepsilon \qquad \text{for } all\ n.$$

In particular, (28) is valid if for some T, $X_n < T$ for all n. (In general, sequence $\{X_n\}$ is said to be *uniformly bounded* if $|X_n| < T < \infty$ for all n.)

If the X_n are stochastically increasing, i.e., $F_n(t)$ is nonincreasing in n for each t, (28) is valid but may be infinite. This condition holds for sequence $\{X_{sn}\}$ in (26), which ensures the validity of the interchange there.

If the X_n are not sign constrained, we can deal separately with the positive and negative parts.

We now mention a related result that is often stated in ways that may obscure the relationship:

The *Dominated Convergence Theorem* states that if $\{X_n\}$ converges to X and $|X_n| < Y$ w.p.1 for all n, where $E(Y) < \infty$, then

$$\lim_{n\to\infty} E(X_n) = E(X).$$

If Y is a constant, this result is called the *Bounded Convergence Theorem*. Notice that the integrability of Y ensures that the integrals corresponding to $\{E(X_n)\}$ are uniformly convergent.

The expected-value versions of the Elementary Renewal Theorem and of the Renewal-Reward Theorem (Sections 2-2 and 2-3) are not direct consequences of their random-variable versions, and require separate proofs, because the quantities $M(t)/t$ and $C(t)/t$ are not bounded. In some applications, e.g., finding the limiting distribution of excess, $C(t)/t$ is a fraction of time, which is bounded by 1 for all t. In this case, the Bounded Convergence Theorem applies, and the expected-value version of the Renewal-Reward Theorem is an immediate consequence of the random-variable version.

When we don't have uniform convergence, what happens? *Fatou's Lemma* states that if a sequence of nonnegative random variables $\{X_n\}$ converges to X,

$$\liminf_{n\to\infty} E(X_n) \geq E(X). \tag{29}$$

Thus the contribution of rare "tail" values of X_n to $E(X_n)$ is lost in the limit.

Interchanging limit and an infinite sum is essentially the same operation as interchanging limit and an improper integral; it requires that the limit be uniform with respect to the summation variable. For example, in the derivation of the backward equations in Section 4-6, we write

$$\lim_{h\to 0} \sum_{k\neq i} P_{ik}(h)P_{kj}(t)/h = \sum_{k\neq i} \lim_{h\to 0} P_{ik}(h)P_{kj}(t)/h, \tag{30}$$

where we know that

$$\sum_{k\neq i} P_{ik}(h) = 1 - P_{ii}(h) = a_i h + o(h). \tag{31}$$

The limit is *uniform* with respect to k if for every $\varepsilon > 0$, there is a positive δ such that for all $h < \delta$,

$$|\, a_{ik} - P_{ik}(h)/h \,|\, P_{kj}(t) < \varepsilon \qquad \text{for } all\ k. \tag{32}$$

From (31), left-hand sum in (30) on $k > N$ can be made arbitrarily small by letting N get large, which implies (32) and that the interchange is valid.

For the corresponding forward equations, we want to find

$$\lim_{h\to 0} \sum_{k\neq j} P_{ik}(t)P_{kj}(h)/h, \tag{33}$$

but there is no analogue of (31) that ensures uniform convergence. What is sometimes done is to *assume* uniform convergence, i.e., that $\lim_{h\to 0} P_{kj}(h)/h$ converges uniformly in k; in this case, the interchange can be made in (33), and the deviation of the forward equations is valid. However, this assumption is unacceptably restrictive, e.g., it does not hold for the $M/M/\infty$ queue. This is why we sidestepped this question in Section 4-6 and dealt with it in a different way in Section 4-9.

A-4 DIFFERENTIATING INTEGRALS AND SUMS

For us, the issue first arises in Section 1-12, where we differentiate expressions such as $G(z) = E\{z^X\}$ and $\tilde{F}(s) = E\{e^{-sX}\}$ with respect to the transform variable.

The expectation is represented as either a sum or an integral, and we wish to interchange derivative and expectation.

If X is integer-valued with distribution $\{p_n\}$, we can write

$$G(z) = \sum_{n=0}^{\infty} p_n z^n,$$

which is a power series. Interchanging derivative and expectation here means differentiating the series term by term. Power series can be differentiated term by term *inside* the interval of convergence, which here means at least for $|z| < 1$. Hence we have

$$G'(z) = \sum_{n=1}^{\infty} np_n z^{(n-1)} \qquad \text{for } |z| < 1. \tag{34}$$

In order to find $E(X)$, we would like to set $z = 1$ in (34), which is on the boundary of the interval of convergence. We can do this when it is clear that $G'(z)$ is continuous at $z = 1$ (approached from the left); otherwise, we appeal to

Abel's Limit Theorem: If $A(z) = \sum_{n=0}^{\infty} a_n z^n$ converges for $|z| < r$, and also at $z = r$, then

$$\lim_{z \to r^-} A(z) = \sum_{n=0}^{\infty} a_n r^n. \tag{35}$$

The converse, i.e.,

$$\left| \lim_{z \to r^-} A(z) \right| < \infty \Rightarrow (35), \tag{36}$$

can be false. A sufficient condition for (36), which will cover our needs, is $r = 1$ and $a_n \geq 0$ for all n.

By applying (35) and (36), we have

$$E(X) = \lim_{z \to 1^-} G'(z), \tag{37}$$

whether or not this limit is finite, and higher moments may be found in a similar manner from higher derivatives.

The theory for transforms of nonnegative random variables,

$$\tilde{F}(s) = \int_0^{\infty} e^{-st}\, dF(t), \tag{38}$$

is essentially the same, where we may differentiate (38) inside the integral for $s > 0$, and we find moments by taking limits as $s \to 0^+$, e.g.,

$$E(X) = - \lim_{s \to 0^+} \tilde{F}'(s), \tag{39}$$

whether or not this limit is finite.

Suppose we want to find a limit, as in (37) and (39), of some function $h(x) = f(x)/g(x)$ as $x \to a$, say, where $f(x) \to 0$ and $g(x) \to 0$ as $x \to a$. We have (loosely) $\lim_{x \to a} h(x) = 0/0$, which is sometimes called an indeterminate expression. What to do?

Fortunately, we have *L'Hospital's Rule*: Let $f(x)$ and $g(x)$ be differentiable in some interval (a, b), and $g'(x) \neq 0$ for all $x \in (a, b)$, where $-\infty \leq a < b \leq \infty$. If

$$f'(x)/g'(x) \to \ell \qquad \text{as } x \to a, \tag{40}$$

where ℓ is well defined but possibly infinite, and either (1) $f(x) \to 0$ and $g(x) \to 0$ as $x \to a$, or (2) $g(x) \to (+ \text{ or } -)\ \infty$ as $x \to a$, then

$$h(x) \to \ell \qquad \text{as } x \to a. \tag{41}$$

Note: This is a one-sided limit, $x \to a^+$, and the result also holds for $x \to b^-$. If the limit in (40) is also indeterminate, the process can be repeated if f' and g' satisfy the conditions of the theorem.

Now suppose we want to differentiate an integral with respect to a variable that not only appears inside the integral, but also in the limits of integration. The most general case for a Riemann integral is of the form

$$G(x) = \int_{a(x)}^{b(x)} g(x, t)\, dt, \tag{42}$$

where both a and b are functions of x.

Fix a rectangular set of points $R = \{(x, t): c \leq x \leq d,\ u \leq t \leq v\}$ so that $[u, v]$ is large enough to include the range of integration in (42). Let D_x denote the partial derivative of a function with respect to x, and assume that both g and $D_x g$ are continuous on R. Also assume that $a(x)$ and $b(x)$ have finite derivatives on $[c, d]$. Under these assumptions, the derivative of $G(x)$ exists on $[c, d]$, and is given by

$$G'(x) = \int_{a(x)}^{b(x)} D_x g(x, t)\, dt + g[x, b(x)]b'(x) - g[x, a(x)]a'(x). \tag{43}$$

When neither a nor b varies with x, the derivative in (43) has a single term, and (43) is valid for RS as well as Riemann integrals. If either a or b is infinite, as in (38), (43) is valid if, in addition, the integral in (43) is uniformly convergent. On the other hand, an even more elementary example of (43) is the relation between a distribution function and the corresponding density function, e.g., in the nonnegative case,

$$F(x) = \int_0^x f(t)\, dt,$$

where $F'(x) - f(x)$. Thus integration and differentiation are in this sense inverse operations, which is the fundamental theorem of calculus.

REFERENCES

ALEKSANDROV, A. M. [1974]. A Queueing System with Repeated Orders. *Eng. Cyb.* **12,** No. 3, 1–3.

ASMUSSEN, S. [1987]. *Applied Probability and Queueing*. John Wiley, New York.

AVI-ITZHAK, B., AND P. NAOR [1963]. Some Queuing Problems with Service Station Subject to Breakdown. *Oper. Res.*, **11,** 303–320.

AVI-ITZHAK, B., W. L. MAXWELL, AND L. W. MILLER [1965]. Queuing with Alternating Priorities. *Oper. Res.*, **13,** 306–318.

BARBOUR, A. D. [1976]. Networks of Queues and the Method of Stages. *Adv. Appl. Prob.*, **8,** 584–591.

BARLOW, R. E., AND F. PROSCHAN [1965]. *Mathematical Theory of Reliability*. John Wiley, New York.

BARLOW, R. E., AND F. PROSCHAN [1981]. (Reprint of original 1975 edition, with corrections.) *Statistical Theory of Reliability and Life Testing*. TO BEGIN WITH, Silver Spring, MD.

BENEŠ, V. E. [1963]. *General Stochastic Processes in the Theory of Queues*. Addison-Wesley, Reading, MA.

BESSLER, S. A., AND A. F. VEINOTT, JR. [1966]. Optimal Policy for a Multi-Echelon Inventory Model. *Naval Res. Logist. Quart.*, **13,** 355–389.

BILLINGSLEY, P. [1965]. *Ergodic Theory and Information*. John Wiley, New York.

BOROVKOV, A. A. [1967]. On Limit Laws for Service Processes in Multi-Channel Systems. *Siberian Math. J.*, **8,** 746–763 (English translation).

BOXMA, O. J. [1979a]. On a Tandem Queueing Model with Identical Service Times at Both Counters, I. *Adv. Appl. Prob.*, **11,** 616–643.

BOXMA, O. J. [1979b]. On a Tandem Queueing Model with Identical Service Times at Both Counters, II. *Adv. Appl. Prob.*, **11**, 644–659.

BOXMA, O. J., J. W. COHEN, AND N. HUFFELS [1979]. Approximations of the Mean Waiting Time in an *M/G/s* Queueing System. *Oper. Res.*, **27**, 1115–1127.

BROCKMEYER, E. [1954]. The Simple Overflow Problem in the Theory of Telephone Traffic. *Teleteknik*, **5**, 361–374.

BROWN, M., AND S. M. ROSS [1972]. Asymptotic Properties of Cumulative Processes. *SIAM J. Appl. Math.*, **22**, 93–105.

BRUMELLE, S. L. [1971a]. Some Inequalities for Parallel-Server Queues. *Oper. Res.*, **19**, 402–413.

BRUMELLE, S. L. [1971b]. On the Relation between Customer and Time Averages in Queues. *J. Appl. Prob.*, **8**, 508–520.

BRUMELLE, S. L. [1973]. Bounds on the Wait in a *GI/G/*1 Queue. *Mgmt. Sci.*, **20**, 773–777.

BRUMELLE, S. L. [1978]. A Generalization of Erlang's Loss System to State Dependent Arrival and Service Rates. *Math. Oper. Res.*, **3**, 10–16.

BURKE, P. J. [1956]. The Output of Queuing Systems. *Oper. Res.*, **4**, 699–704.

BURKE, P. J. [1969]. The Dependence of Sojourn Times in Tandem *M/M/s* Queues. *Oper. Res.*, **17**, 754–755.

BURKE, P. J. [1972]. Output Processes and Tandem Queues. *Proceedings of the Symposium on Computer-Communication Networks and Teletraffic*, 419–428. J. Fox, ed. Polytechnic Press, Brooklyn. Distributed by Wiley, New York.

CALO, S. B. [1979]. Delay Properties of Message Channels. *Proceedings of the 1979 International Conference on Communications*, Boston, 43.5.1–43.5.4.

CHANDY, K. M., U. HERZOG, AND L. WOO [1975]. Parametric Analysis of Queuing Networks. *IBM J. Res. Dev.*, **19**, 36–42.

CHARLOT, F., M. GHIDOUCHE, AND M. HAMAMI [1978]. Irréductibilité et récurrence au sens de Harris des 'Temps d'attente' files *GI/G/q*. *Z. Wahrsch.*, **43**, 187–203.

CHOO, Q. H., AND B. CONOLLY [1980]. Waiting Time Analysis for a Tandem Queue with Correlated Service. *Eur. J. Oper. Res.*, **4**, 337–345.

CHOW, Y. S., AND T. L. LAI [1979]. Moments of Ladder Variables for Driftless Random Walks. *Z. Wahrsch.*, **48**, 253–257.

CHUNG, K. L. [1967]. *Markov Chains with Stationary Transition Probabilities*, 2nd ed. Springer-Verlag, Berlin.

CHUNG, K. L. [1974]. *A Course in Probability Theory*, 2nd ed. Academic Press, New York.

ÇINLAR, E. [1975]. *Introduction to Stochastic Processes*. Prentice-Hall, Englewood Cliffs, NJ.

COBHAM, A. [1954]. Priority Assignment in Waiting Line Problems. *Oper. Res.*, **2**, 70–76.

COCHRAN, W. G. [1977]. *Sampling Techniques*, 3rd ed. John Wiley, New York.

COHEN, J. W. [1957a]. Basic Problems of Telephone Traffic Theory and the Influence of Repeated Calls. *Philips Telecomm. Rev.*, **18**, No. 2, 49–100.

COHEN, J. W. [1975b]. The Full Availability Group of Trunks with an Arbitrary Distribution of the Inter-Arrival Times and a Negative Exponential Holding Time Distribution. *Simon Stevin*, **31**, 169–181.

COHEN, J. W. [1969]. *The Single Server Queue*. North-Holland, Amsterdam.

COX, D. R. [1955]. The Analysis of Non-Markovian Stochastic Processes by the Inclusion of Supplementary Variables. *Proc. Camb. Philos. Soc.*, **51,** 433–441.

COX, D. R., AND V. ISHAM [1980]. *Point Processes*. Chapman & Hall, London.

COX, D. R., AND W. L. SMITH [1961]. *Queues*. John Wiley, New York.

CRANE, M. A., AND A. J. LEMOINE [1977]. *An Introduction to the Regenerative Method for Simulation Analysis*. Springer-Verlag, New York.

DALEY, D. J. [1976]. Queueing Output Processes. *Adv. Appl. Prob.*, **8,** 395–415.

DALEY, D. J. [1977]. Inequalities for Moments of Tails of Random Variables, with Queueing Applications. *Z. Wahrsch.*, **41,** 139–143.

DALEY, D. J. [1987]. Certain Optimality Properties of the First-Come First-Served Discipline for *G/G/s* Queues. *Stoc. Proc. Applic.*, **25,** 301–308.

DALEY, D. J., AND T. ROLSKI [1984]. Some Comparability Results for Waiting Times in Single- and Many-Server Queues. *J. Appl. Prob.*, **21,** 887–900.

DALEY, D. J., AND C. D. TRENGROVE [1977]. Bounds for Mean Waiting Times in Single-Server Queues: A Survey. Statistics Dept. (IAS), Australian National University.

DONEY, R. A. [1980]. Moments of Ladder Heights in Random Walks. *J. Appl. Prob.*, **17,** 248–252.

FAKINOS, D. [1981]. The *G/G/*1 Queueing System with a Particular Queue Discipline. *J. Roy. Statist. Soc.*, **B43,** 190–196.

FELLER, W. [1971]. *An Introduction to Probability Theory and Its Applications*, Vol. 2, 2nd ed. John Wiley, New York.

FOSS, S. G. [1980]. Approximation of Multichannel Queueing Systems. *Siberian Math. J.*, **21,** 851–857.

FRANKEN, P., D. KÖNIG, U. ARNDT, AND V. SCHMIDT [1982]. *Queues and Point Processes*. John Wiley, New York.

FREDERICKS, A. A. [1980]. Congestion in Blocking Systems—A Simple Approximation Technique. *Bell Sys. Tech. J.*, **59,** 805–827.

FREDERICKS, A. A. [1983]. Approximating Parcel Blocking via State Dependent Birth Rates. *Proceedings of the Tenth International Teletraffic Congress*, Montreal, paper 5.3.2.

FREDERICKS, A. A., AND G. A. REISNER [1979]. Approximations to Stochastic Service Systems with an Application to a Retrial Model. *Bell Sys. Tech. J.*, **58,** 557–576.

FRIEDMAN, H. D. [1965]. Reduction Methods in Tandem Queueing Systems. *Oper. Res.*, **13,** 121–131.

FUHRMANN, S. W. [1984]. A Note on the *M/G/*1 Queue with Server Vacations. *Oper. Res.*, **32,** 1368–1373.

FUHRMANN, S. W., AND R. B. COOPER [1985]. Stochastic Decompositions in the *M/G/*1 Queue with General Vacations. *Oper. Res.*, **33,** 1117–1129.

GAVER, D. P. [1962]. A Waiting Line with Interrupted Service, Including Priorities. *J. Roy. Statist. Soc.*, **B25,** 73–90.

GREENBERG, B. S. [1986]. *Queueing Systems with Returning Customers and the Optimal Order of Tandem Queues*. Doctoral dissertation, Dept. of IEOR, University of California, Berkeley.

GREENBERG, B. S., AND R. W. WOLFF [1987]. An Upper Bound on the Performance of Queues with Returning Customers. *J. Appl. Prob.*, **24**, 466–475.

GREENBERG, B. S., AND R. W. WOLFF [1988]. Optimal Order of Servers for Tandem Queues in Light Traffic. *Mgmt. Sci.*, **34**, 500–508.

GREENBERG, B. S., R. C. LEACHMAN, AND R. W. WOLFF [1988]. Predicting Dispatching Delays on a Low Speed Single Track Railroad. *Trans. Sci.*, **22**, 31–38.

HARRISON, J. M. [1973]. The Heavy Traffic Approximation for Single Server Queues in Series. *J. Appl. Prob.*, **10**, 613–629.

HARRISON, J. M. [1985]. *Brownian Motion and Stochastic Flow Systems*. John Wiley, New York.

HARRISON, J. M., AND A. J. LEMOINE [1976]. On the Virtual and Actual Waiting Time Distributions of a *GI*/*G*/1 Queue. *J. Appl. Prob.*, **13**, 833–836.

HEYMAN, D. P., AND M. J. SOBEL [1982]. *Stochastic Models in Operations Research*, Vol. 1. McGraw-Hill, New York.

HEYMAN, D. P., AND S. STIDHAM, JR. [1980]. The Relation between Customer and Time Averages in Queues. *Oper. Res.*, **28**, 983–994.

HOOK, J. A. [1969]. Some Limit Theorems for Priority Queues. Technical Report No. 91, Department of Operations Research, Cornell University, Ithaca, NY.

JACKSON, J. R. [1963]. Jobshop-like Queueing Systems. *Mgmt. Sci.*, **10**, 131–142.

JACKSON, R. R. P. [1954]. Queueing Systems with Phase-type Service. *Oper. Res. Quart.*, **5**, 109–120.

JAISWAL, N. K. [1968]. *Priority Queues*. Academic Press, New York.

JONIN, J. G. [1984]. The Systems with Repeated Calls: Models, Measurements, Results. *Proceedings of the Third International Seminar on Teletraffic Theory*, 197–208.

KAMAE, T., U. KRENGEL, AND G. L. O'BRIAN [1977]. Stochastic Inequalities on Partially Ordered Spaces. *Ann. Prob.*, **5**, 899–912.

KARLIN, S. [1968]. *Total Positivity*. Stanford University Press, Stanford, CA.

KATZ, S. [1967]. Statistical Performance Analysis of a Switched Communications Network. *Proceedings of the Fifth International Teletraffic Congress*, New York, 566–575.

KEILSON, J., J. COZZOLINO, AND H. YOUNG [1968]. A Service System with Unfilled Requests Repeated. *Oper. Res.*, **16**, 1126–1137.

KELLEY, F. P. [1979]. *Reversibility and Stochastic Networks*. John Wiley, New York.

KELLY, F. P. [1982]. Networks of Quasi-Reversible Queues. *Applied Probability—Computer Science: The Interface*, Vol. 1, R. L. Disney and T. J. Ott, eds., Birkhauser, Boston.

KELLY, F. P. [1983]. Invariant Measures and the *Q*-Matrix. *Probability, Statistics, and Analysis*, J. F. C. Kingman and G. E. H. Reuter, eds., Cambridge University Press, Cambridge.

KIEFER, J., AND J. WOLFOWITZ [1955]. On the Theory of Queues with Many Servers. *Trans. Amer. Math. Soc.*, **78**, 1–18.

KIEFER, J., AND J. WOLFOWITZ [1956]. On the Characteristics of the General Queueing Process with Applications to Random Walk. *Ann. Math. Statist.*, **27**, 147–161.

KINGMAN, J. F. C. [1961]. Two Similar Queues in Parallel. *Ann. Math. Statist.,* **32,** 1314–1323.

KINGMAN, J. F. C. [1962a]. On Queues in Heavy Traffic. *J. Roy. Statist. Soc.,* **B24,** 383–392.

KINGMAN, J. F. C. [1962b]. Some Inequalities for the *GI/G/*1 Queue. *Biometrika,* **49,** 315–324.

KINGMAN, J. F. C. [1962c]. The Effect of Queue Discipline on Waiting Time Variance. *Proc. Camb. Philos. Soc.,* **58,** 163–164.

KINGMAN, J. F. C. [1964]. A Martingale Inequality in the Theory of Queues. *Proc. Camb. Philos. Soc.,* **59,** 359–361.

KINGMAN, J. F. C. [1970]. Inequalities in the Theory of Queues. *J. Roy. Statist. Soc.,* **B32,** 102–110.

KINGMAN, J. F. C., AND S. J. TAYLOR [1966]. *Introduction to Measure and Probability.* Cambridge University Press, London.

KLEINROCK, L. [1976]. *Queueing Systems,* Vol. II: *Computer Applications.* John Wiley, New York.

KÖLLERSTRÖM, J. [1974]. Heavy Traffic Theory for Queues with Several Servers. I. *J. Appl. Prob.,* **11,** 544–552.

KOSTEN, L. [1937]. Über Sperrungswahrscheinlichkeiten bei Staffelschaltungen. *E.N.T.,* **14,** 5–12.

KRAMER, W., AND M. LANGENBACH-BELZ [1976]. Approximate Formulae for the Delay in the Queueing System *GI/G/*1. *Eight International Teletraffic Congress,* Melbourne, 235/1–8.

LAVENBERG, S. S., AND M. REISER [1980]. Stationary State Probabilities at Arrival Instants for Closed Queueing Networks with Multiple Types of Customers. *J. Appl. Prob.,* **17,** 1048–1061.

LAW, A. M., AND W. D. KELTON [1982]. *Simulation Modeling and Analysis.* McGraw-Hill, New York.

LEE, A. M., AND P. A. LONGTON [1957]. Queueing Processes Associated with Airline Passenger Check-In. *Oper. Res. Quart.,* **10,** 56–71.

LEMOINE, A. J. [1974]. On Two Stationary Distributions for the Stable *GI/G/*1 Queue. *J. Appl. Prob.,* **11,** 849–852.

LEMOINE, A. J. [1976]. On Random Walks and Stable *GI/G/*1 Queues. *Math. Oper. Res.,* **1,** 159–164.

LINDLEY, D. V. [1952]. On the Theory of Queues with a Single Server. *Proc. Camb. Philos. Soc.,* **48,** 277–289.

LINDVALL, T. [1977]. A Probabilistic Proof of Blackwell's Renewal Theorem. *Ann. Prob.,* **5,** 482–485.

LITTLE, J. D. C. [1961]. A Proof for the Queuing Formula: $L = \lambda W$. *Oper. Res.,* **9,** 383–387.

LOYNES, R. M. [1962]. The Stability of a Queue with Non-independent Inter-arrival and Service Times. *Proc. Camb. Philos. Soc.,* **58,** 497–520.

LUBACZ, J., AND J. ROBERTS [1984]. A New Approach to the Single Server Repeat Attempt

System with Balking. *Proceedings of the Third International Seminar on Teletraffic Theory*, 290–293.

MARSHALL, K. T. [1968a]. Some Inequalities in Queuing. *Oper. Res.*, **16**, 651–665.

MARSHALL, K. T. [1968b]. Some Relationships between the Distributions of Waiting Time, Idle Time, and Interoutput Time in the *GI/G/*1 Queue. *SIAM J. Appl. Math.*, **16**, 324–327.

MARSHALL, K. T. [1979]. The Length-Bias Paradox: First Words in Printed Lines Are Longer Than Average. *Math. Sci.*, **4**, 63–68.

MARSHALL, K. T., AND R. W. WOLFF [1971]. Customer Average and Time Average Queue Lengths and Waiting Times. *J. Appl. Prob.*, **8**, 535–542.

MILLER, D. R. [1972]. Existence of Limits in Regenerative Processes. *Ann. Math. Statist.*, **43**, 1275–1282.

MILLER, R. G., JR. [1963]. Stationary Equations in Continuous Time Markov Chains. *Trans. Amer. Math. Soc.*, **109**, 35–44.

MITCHELL, C. R., A. S. PAULSON, AND C. A. BESWICK [1977]. The Effect of Correlated Service Times on Single Server Tandem Queues. *Naval Res. Logist. Quart.*, **24**, 95–112.

MIYAZAWA, M. [1979]. A Formal Approach to Queueing Processes in the Steady State and Their Applications. *J. Appl. Prob.*, **16**, 332–346.

MIYAZAWA, M. [1986]. Approximation of the Queue-Length Distribution of an *M/GI/s* Queue by the Basic Equations. *J. Appl. Prob.*, **23**, 443–458.

MORI, M. [1975]. Some Bounds for Queues. *J. Oper. Res. Soc. Japan*, **18**, 151–181.

NEUTS, M. F. [1981]. *Matrix-Geometric Solutions in Stochastic Models*. Johns Hopkins Univerisity Press, Baltimore, MD.

NEWELL, G. F. [1973]. *Approximate Stochastic Behavior of n-Server Service Systems with Large n*. Springer-Verlag, New York.

NEWELL, G. F. [1979]. *Approximate Behavior of Tandem Queues*. Springer-Verlag, New York.

NEWELL, G. F. [1982]. *Applications of Queueing Theory,* 2nd ed. Chapman & Hall, New York.

NEWELL, G. F. [1984a]. *The* $M/M/\infty$ *Service System with Ranked Servers in Heavy Traffic*. Springer-Verlag, New York.

NEWELL, G. F. [1984b]. Approximations for Superposition Arrival Processes in Queues. *Mgmt. Sci.*, **30**, 623–632.

NIU, S.-C. [1980]. Bounds for the Expected Delays in Some Tandem Queues. *J. Appl. Prob.*, **17**, 831–838.

NIU, S.-C. [1981]. On the Comparison of Waiting Times in Tandem Queues. *J. Appl. Prob.*, **18**, 707–714.

NOZAKI, S. A., AND S. M. ROSS [1978]. Approximations in Finite-Capacity Multi-Server Queues with Poisson Arrivals. *J. Appl. Prob.*, **15**, 826–834.

NUMMELIN, E. [1981]. Regeneration in Tandem Queues. *Adv. Appl. Prob.*, **13**, 221–230.

NUMMELIN, E. [1984]. *General Irreducible Markov Chains and Non-negative Operators*. Cambridge University Press, London.

O'Donovan, T. M. [1974]. Distribution of Attained Service and Residual Service in General Queuing Systems. *Oper. Res.*, **22,** 570–575.

Oliver, R. M., and A. H. Samuel [1962]. Reducing Letter Delays in Post Offices. *Oper. Res.*, **10,** 839–892.

Pinedo, M. [1982]. On the Optimal Order of Stations in Tandem Queues. *Applied Probability—Computer Science: The Interface,* Vol. 2, 307–325, R. L. Disney and T. J. Ott, eds. Birkhauser, Boston.

Pinedo, M., and R. W. Wolff [1982]. A Comparison between Tandem Queues with Dependent and Independent Service Times. *Oper. Res.*, **30,** 464–479.

Pollett, P. K. [1986]. Connecting Reversible Markov Processes. *Adv. Appl. Prob.*, **18,** 880–900.

Prabhu, N. U. [1980]. *Stochastic Storage Processes. Queues, Insurance Risk, and Dams.* Springer-Verlag, New York.

Rapp, Y. [1964]. Planning a Junction Network in a Multiexchange Area. *Ericsson Tech.*, **1,** 77–130.

Reich, E. [1957]. Waiting Times When Queues Are in Tandem. *Ann. Math. Statist.*, **28,** 768–773.

Riordan, J. [1962]. *Stochastic Service Systems.* John Wiley, New York.

Ross, S. M. [1974]. Bounds for the Delay Distribution in *GI/G/*1 Queues. *J. Appl. Prob.*, **11,** 417–421.

Rubinstein, R. Y. [1981]. *Simulation and the Monte Carlo Method.* John Wiley, New York.

Rudin, W. [1964]. *Principles of Mathematical Analysis*, 2nd ed. McGraw-Hill, New York.

Sakata, M. S., S. Noguchi, and J. Oizumi [1969]. Analysis of a Processor-Sharing Queueing Model for Time-Sharing Systems. *Proceedings of the Second Hawaii International Conference on Systems Science*, 625–628.

Sauer, C. H., and K. M. Chandy [1981]. *Computer Systems Performance Modeling*, Prentice-Hall, Englewood Cliffs, NJ.

Schrage, L. E. [1968]. A Proof of the Optimality of the Shortest Remaining Processing Time Discipline. *Oper. Res.*, **16,** 687–690.

Schrage, L. E., and L. W. Miller [1966]. The Queue *M/G/*1 with the Shortest Remaining Processing Time Discipline. *Oper. Res.*, **14,** 670–684.

Sevast'yanov, B. A. [1957]. An Ergodic Theorem for Markov Processes and its Application to Telephone Systems with Refusals. *Teor. Verojatnost. i Primenen,* **2,** 106–116 (Russian with English summary).

Shanthikumar, G. J. [1987]. On Stochastic Comparison of Random Vectors. *J. Appl. Prob.*, **24,** 123–136.

Sigman, K. [1988a]. Regeneration in Tandem Queues with Multiserver Stations. *J. Appl. Prob.*, **25,** 391–403.

Sigman, K. [1988b]. Queues as Harris Recurrent Markov Chains. *Queueing Syst.*, **3,** 179–198.

Sigman, K. [1989]. One-Dependent Regenerative Processes and Queues in Continuous Time. Submitted to *Math. Oper. Res.*

SMITH, D. R. [1978]. A New Proof of the Optimality of the Shortest Remaining Processing Time Discipline. *Oper. Res.*, **26**, 197–199.

SMITH, W. L. [1955]. Regenerative Stochastic Processes. *Proc. Roy. Soc.*, **A232**, 6–31.

SMITH, W. L. [1958]. Renewal Theory and Its Ramifications. *J. Roy. Statist. Soc.*, **B20**, 243–302.

STIDHAM, S., JR. [1974]. A Last Word on $L = \lambda W$. *Oper. Res.*, **22**, 417–421.

STIDHAM, S., JR. [1982]. Sample-Path Analysis of Queues. *Applied Probability—Computer Science: The Interface,* Vol. 2, 41–70, R. L. Disney and T. J. Ott, eds. Birkhauser, Boston.

STOYAN, D. [1976]. A Critical Remark on a System Approximation in Queueing Theory. *Math. Operationsforch. Statist.*, **7**, 953–956.

STOYAN, D. [1983]. *Comparison Methods for Queues and Other Stochastic Models,* English ed., edited and revised by D. J. Daley. John Wiley, New York.

SUZUKI, T., AND Y. YOSHIDA [1970]. Inequalities for Many-Server Queue and Other Queues. *J. Oper. Res. Soc. Japan,* **13**, 59–77.

SYSKI, R. [1960]. *Introduction to Congestion Theory in Telephone Systems.* Oliver & Boyd, London.

TAKÁCS, L. [1962]. *Introduction to the Theory of Queues.* Oxford University Press, New York.

TAKÁCS, L. [1963]. The Limiting Distribution of the Virtual Waiting Time and the Queue Size for a Single-Server Queue with Recurrent Input and General Service Times. *Sankhyā*, **A25**, 91–100.

TEMBE, S. V., AND R. W. WOLFF [1974]. The Optimal Order of Service in Tandem Queues. *Oper. Res.*, **24**, 824–832.

TIJMS, H. C. [1986]. *Stochastic Modeling and Analysis; A Computational Approach.* John Wiley, New York.

TSOUCAS, P., AND J. WALRAND [1983]. A Note on the Processor-Sharing Queue in a Quasi-reversible Network. *Adv. Appl. Prob.*, **15**, 468–469.

WALRAND, J. [1983]. A Note on Norton's Theorem for Queuing Networks. *J. Appl. Prob.*, **20**, 442–444.

WALRAND, J., AND P. VARAIYA [1980]. Sojourn Times and the Overtaking Condition in Jacksonian Networks. *Adv. Appl. Prob.*, **12**, 1000–1018.

WATANABE, S. [1964]. On Discontinuous Additive Functionals and Levy Measures of a Markov Process. *Japan J. Math.*, **34**, 53–70.

WEBER, R. R. [1979]. The Interchangeability of $\cdot/M/1$ Queues in Series. *J. Appl. Prob.*, **16**, 690–695.

WELCH, P. D. [1964]. On a Generalized $M/G/1$ Queuing Process in Which the First Customer of Each Busy Period Receives Exceptional Service. *Oper. Res.*, **12**, 736–752.

WHITT, W. [1972]. Embedded Renewal Processes in the $GI/GI/s$ Queue. *J. Appl. Prob.*, **9**, 650–658.

WHITT, W. [1980]. The Effect of Variability in the $GI/G/s$ Queue. *J. Appl. Prob.*, **17**, 1062–1071.

WHITT, W. [1981]. On Stochastic Bounds for the Delay Distribution in the $GI/G/s$ Queue. *Oper. Res.*, **29**, 604–608.

Whitt, W. [1983a]. The Queueing Network Analyzer. *Bell Sys. Tech. J.,* **62,** 2779–2815.

Whitt, W. [1983b]. The Performance of the Queueing Network Analyzer. *Bell Sys. Tech. J.,* **62,** 2817–2843.

Whitt, W. [1984a]. Heavy-Traffic Approximations for Service Systems with Blocking. *AT&T BLTJ,* **63,** 689–708.

Whitt, W. [1984b]. Open and Closed Models for Networks of Queues. *AT&T BLTJ,* **63,** 1911–1979.

Whitt, W. [1984c]. Minimizing Delays in the *GI/G/*1 Queue. *Oper. Res.,* **32,** 41–51.

Whitt, W. [1985]. The Best Order for Queues in Series. *Mgmt. Sci.,* **31,** 475–487.

Wilkinson, R. I. [1956]. Theories for Toll Traffic Engineering in the U.S.A. *Bell Sys. Tech. J.,* **35,** 421–514.

Wolff, R. W. [1970a]. Work-Conserving Priorities. *J. Appl. Prob.,* **7,** 327–337.

Wolff, R. W. [1970b]. Time Sharing with Priorities. *SIAM J. Appl. Math.,* **19,** 566–574.

Wolff, R. W. [1977a]. An Upper Bound for Multi-Channel Queues. *J. Appl. Prob.,* **14,** 884–888.

Wolff, R. W. [1977b]. The Effect of Service Time Regularity on System Performance. *Computer Performance,* 297–304, K. M. Chandy and M. Reiser, eds. North-Holland, New York.

Wolff R. W. [1982a]. Poisson Arrivals See Time Averages. *Oper. Res.,* **30,** 223–231.

Wolff, R. W. [1982b]. Tandem Queues with Dependent Service Times in Light Traffic. *Oper. Res.,* **30,** 619–635.

Wolff, R. W. [1984]. Conditions for Finite Ladder Height and Delay Moments. *Oper. Res.,* **32,** 909–916.

Wolff, R. W. [1987]. Upper Bounds on Work in System for Multi-channel Queues. *J. Appl. Prob.,* **24,** 547–551.

Wolff, R. W. [1988]. Sample-Path Derivations of the Excess, Age, and Spread Distributions. *J. Appl. Prob.,* **25,** 432–436.

Wolff, R. W., and C. W. Wrightson. [1976]. An Extension of Erlang's Loss Formula. *J. Appl. Prob.,* **13,** 628–632.

Wolfson, B. [1984]. Some Moment Results for Certain Tandem and Multiple-Server Queues. *J. Appl. Prob.,* **21,** 901–910.

Yamazaki, G. [1984]. Invariance Relations of *GI/G/*1 Queueing Systems with Preemptive-Resume Last-Come-First-Serve Queue Discipline. *J. Oper. Res. Japan,* **27,** 338–346.

INDEX

A

B

C

D

E

F

G

M

N

O

P

Q

R

S